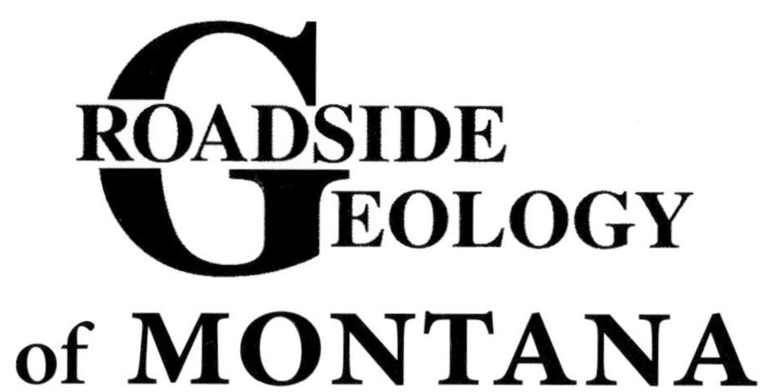

# of MONTANA

Second Edition

DONALD W. HYNDMAN
AND ROBERT C. THOMAS

Illustrated by Chelsea M. Feeney

2025

Boulder, Colorado

First printing, February 2020
Fourth Printing, January 2025

Geologic maps constructed by Chelsea M. Feeney using geologic data from the Montana Bureau of Mines and Geology, mbmg.mtech.edu. Do not reproduce or modify without the permission of the MBMG.

Photos by authors unless otherwise credited.

Cover photo: The Mission Range in northwest Montana.
Back cover photo: Medicine Rocks State Park in far southeastern Montana.

Library of Congress Cataloging-in-Publication Data

Names: Hyndman, Donald W., author. | Thomas, Robert C. (Robert Curtiss), 1962- author. | Alt, David D. Roadside geology of Montana.
Title: Roadside geology of Montana / Donald W. Hyndman and Robert C. Thomas ; illustrated by Chelsea M. Feeney.
Description: Second. | Missoula, Montana : Mountain Press Publishing Company, 2020. | Series: Roadside geology series | Revised edition of: Roadside geology of Montana / David Alt and Donald W. Hyndman. c1986. | Includes bibliographical references and index. | Summary: "Nearly 50 years after the first Roadside Geology of Montana was published, comes a newly revised full-color edition. Few other places have such amazingly diverse and well-exposed rocks. Hundreds of students and visitors from around the country swarm Montana's outcrops every summer to take it all in." —Provided by publisher.
Identifiers: LCCN 2019054034 | ISBN 9780878426966 (paperback)
Subjects: LCSH: Geology—Montana—Guidebooks.
Classification: LCC QE133 .A674 2020 | DDC 557.86—dc23
LC record available at https://lccn.loc.gov/2019054034

PRINTED IN THE UNITED STATES BY VERSA PRESS

3300 Penrose Place • P.O. Box 9140 • Boulder, CO 80301-9140
303-357-1000, option 3 • 800-472-1988
gsaservice@geosociety.org • www.geosociety.org

We dedicate this new, second edition of

Roadside Geology of Montana

to Dr. David Alt,

a perceptive and imaginative thinker,

a stimulating speaker,

and an inspiring teacher.

Dave not only had a big role in the first edition but also in several other books in the Roadside Geology series. Dave Alt and Don Hyndman came up with the idea of roadside geology books more than thirty years ago as a means of conveying their passion for geology to their students and to the public in general. What better way to learn about the formation of the landscapes around us and the history of the Earth recorded in the rocks under our feet than to get out there and see it firsthand?

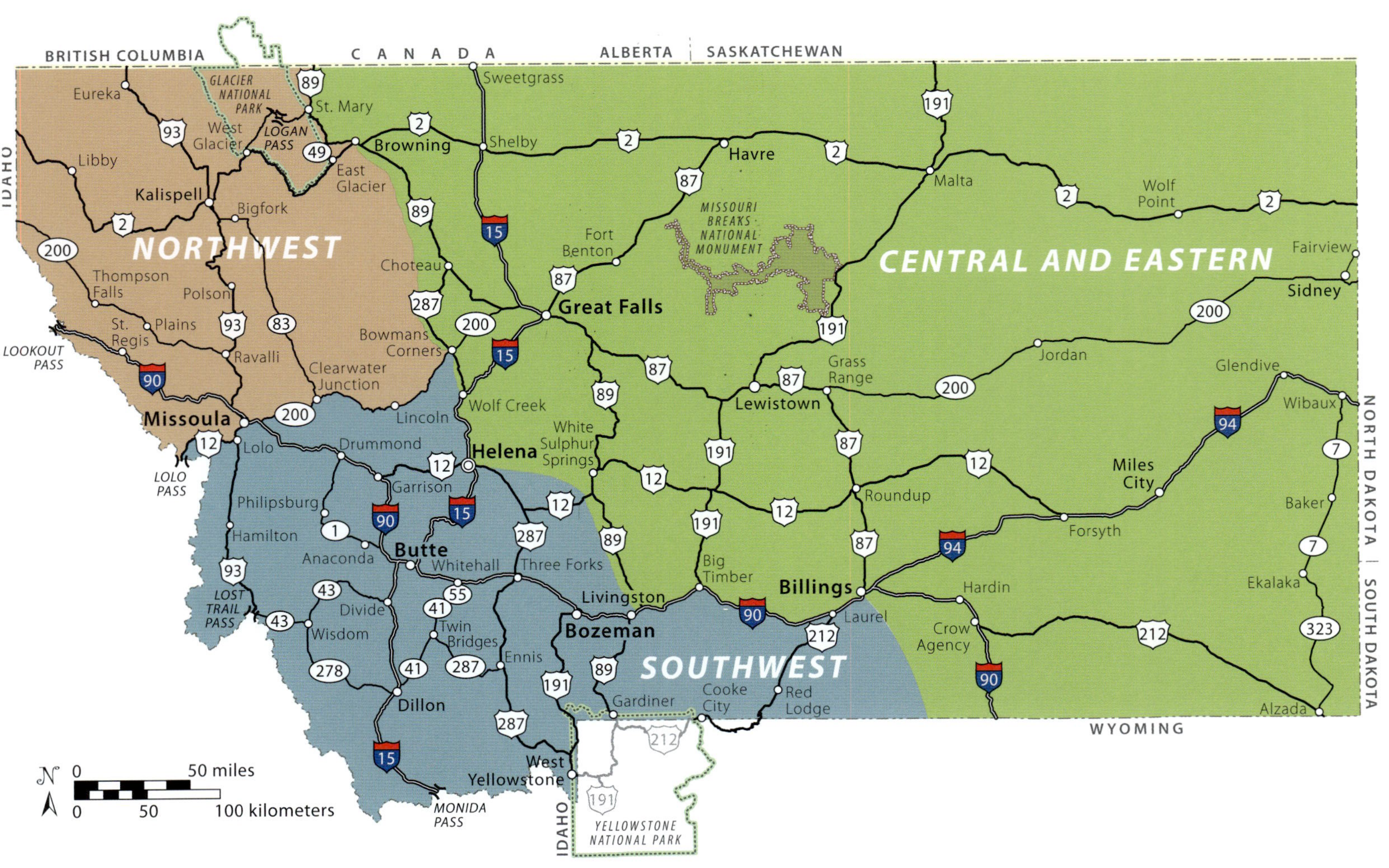

BRITISH COLUMBIA
C A N A D A
ALBERTA
SASKATCHEWAN
IDAHO
NORTH DAKOTA
SOUTH DAKOTA
WYOMING
NORTHWEST
CENTRAL AND EASTERN
SOUTHWEST
GLACIER NATIONAL PARK
LOGAN PASS
MISSOURI BREAKS NATIONAL MONUMENT
YELLOWSTONE NATIONAL PARK
LOOKOUT PASS
LOLO PASS
LOST TRAIL PASS
MONIDA PASS
Eureka
Libby
Kalispell
West Glacier
St. Mary
Browning
East Glacier
Bigfork
Thompson Falls
Polson
St. Regis
Plains
Ravalli
Missoula
Lolo
Clearwater Junction
Bowmans Corners
Choteau
Sweetgrass
Shelby
Havre
Fort Benton
Great Falls
Malta
Wolf Point
Fairview
Sidney
Jordan
Glendive
Wibaux
Baker
Ekalaka
Alzada
Miles City
Forsyth
Roundup
Grass Range
Lewistown
Lincoln
Wolf Creek
Helena
White Sulphur Springs
Drummond
Garrison
Philipsburg
Hamilton
Anaconda
Butte
Whitehall
Three Forks
Divide
Wisdom
Twin Bridges
Dillon
Ennis
West Yellowstone
Bozeman
Livingston
Big Timber
Billings
Laurel
Hardin
Crow Agency
Gardiner
Cooke City
Red Lodge
0 50 miles
0 50 100 kilometers
N

CONTENTS

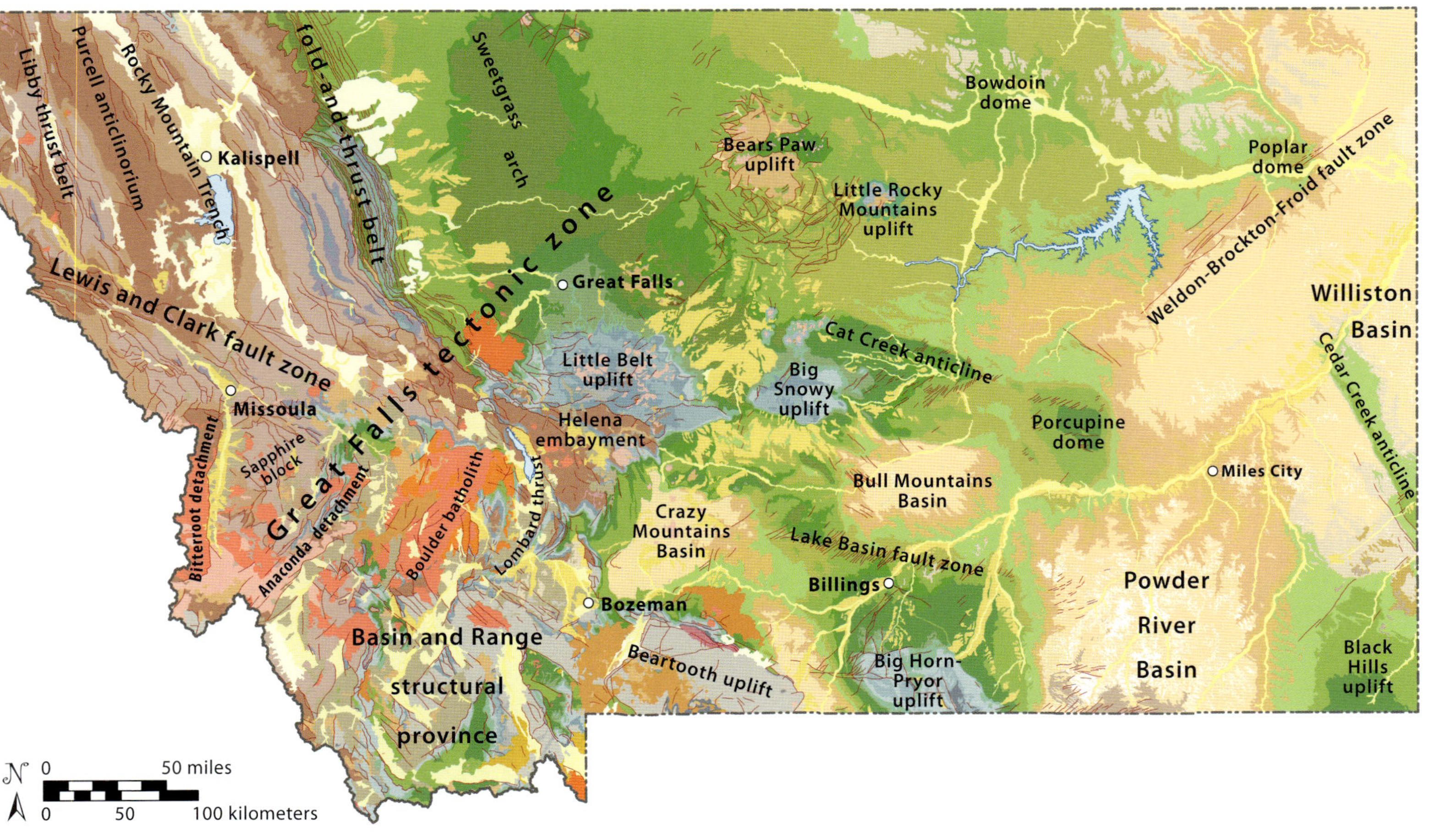

Geologic map of Montana and major tectonic features.

Geologic Units on Maps in *Roadside Geology of Montana*

CENOZOIC

QUATERNARY

- Qal alluvium
- Qs sediment
- Qg glacial deposits
- Qgr sediment and gravel deposits; higher benches and Yellowstone River terrace gravel

QUATERNARY-TERTIARY

- QTgr gravels of High Plains Surface
- QTcl clinker deposits

TERTIARY

- Tgr pediment gravel; Flaxville Gravel
- Ts Sixmile Creek and Renova Formations
- Tk Kishenehn Formation
- Tw Wasatch and Arikaree Formations
- Fort Union Formation
  - Tfsb Sentinel Butte Member
  - Tfli Linley Member
  - Tftr Tongue River Member
  - Tfe Ekalaka Member
  - Tfld Ludlow Member
  - Tfle Lebo Member
  - Tft Tullock Member

MESOZOIC

- TKb Tertiary-Cretaceous Beaverhead Group

CRETACEOUS — LATE

- Klks St. Mary River, Horsethief, and Willow Creek Formations
- Khf Lance, Hell Creek, and Fox Hills Formations
- Kb Bearpaw and Pierre Shales
- Kjc Two Medicine, Judith River, and Gammon Formations, and Claggett Shale
- Ket Eagle and Telegraph Creek Formations
- Kvet Virgelle; Eagle through Telegraph Creek Formations
- Kmr Marias River and Niobrara Shales of Colorado Group

CRETACEOUS — EARLY

- Kbl Blackleaf; Cody, through Thermopolis Formations
- Kk Kootenai Formation

- KJŦs Cretaceous-Jurassic-Triassic sedimentary rocks; includes Morrison Formation and Ellis Group

PALEOZOIC

- PIPMs Permian-Pennsylvanian-Mississippian sedimentary rocks
- Mbs Mississippian Big Snowy Group
- Mississippian Madison Group
  - Mmc Mission Canyon Limestone
  - Mml Lodgepole Limestone
- MD€s Mississippian-Devonian-Cambrian sedimentary rocks
- DO€s Devonian-Ordovician-Cambrian sedimentary rocks
- €s Cambrian sedimentary rocks

MIDDLE PROTEROZOIC

BELT SUPERGROUP

- Ymi Missoula Group
- Ypg Piegan Group
- Yr Ravalli Group
- Ybl Lower Belt Group

EARLY PROTEROZOIC–ARCHEAN

- Xm Neihart Quartzite, basement mylonite, and basement metamorphic rocks
- XAm basement metamorphic rocks

IGNEOUS AND VOLCANIC ROCKS

QUATERNARY-TERTIARY

- Qhr Huckleberry Ridge Tuff
- Qt Lava Creek tuff
- QTv volcanic rocks
- Qb basalt

TERTIARY

- Tsct Timber Hill Basalt Member of Sixmile Creek Formation
- Tv volcanic rocks; includes Dillon Volcanics
- Tlc Lowland Creek Volcanics
- Tav Absaroka Volcanics
- Ti granitic intrusions
- Tsh granitic intrusions; includes shonkinite

TERTIARY-CRETACEOUS

- TKv volcanic rocks
- TKi granitic intrusions
- TKi Skalkaho intrusion

CRETACEOUS

- Kv volcanic rocks
- Klsr Sliderock Mountain Formation of the Livingston Group
- Kamv Adel Mountains Volcanics
- Kem Elkhorn Mountains Volcanics
- Ksh alkalic intrusions; shonkinite
- Ki granitic intrusions

CAMBRIAN

- €i granitic intrusions

LATE PROTEROZOIC

- Zi diabase intrusions

MIDDLE PROTEROZOIC

- Yi anorthosite intrusions

ARCHEAN

- Asw gabbroic intrusion of the Stillwater Complex

- fault
- strike-slip fault
- normal fault; ball on down-dropped side
- detachment fault; hachures on down-dropped side
- thrust fault; teeth on upthrown side

GENERALIZED STRATIGRAPHIC COLUMN FOR MONTANA

ERA	PERIOD		FORMATION		DESCRIPTION OF UNITS	DEPOSITIONAL ENVIRONMENT
CENOZOIC	QUATERNARY		sediment and alluvium		gravel, sand, silt, clay	streams, lakes
			glacial deposits		till, outwash, lake deposits	glacial
	TERTIARY	NEOGENE	Sixmile Creek Formation		conglomerate, tuffaceous sandstone	rivers, alluvial fan
		PALEOGENE	Renova / Kishenehn Fm.		tuffaceous siltstones, ash beds	rivers, lakes
			Fort Union Formation		sandstone, shale, coal, clinker	rivers, swamps
MESOZOIC	CRETACEOUS		see detailed section for the Cretaceous and Tertiary Periods on page 254			in and near the Western Interior Seaway
	JURASSIC		Morrison Formation		mudstone, sandstone, limestone	rivers, floodplain
			Ellis Group		platy, fossiliferous, sandy limestone	shallow marine
	TRIASSIC		Thaynes Formation		limestone with interbedded siltstone, shale, and sandstone	marine
			Chugwater Formation		tan siltstone, limestone, shale; some gypsum and local limestone on top	shallow marine, nonmarine
			Dinwoody Formation		tan siltstone, limestone; grades eastward toward red shale siltstone	marine shelf
PALEOZOIC	PERMIAN		Phosphoria Formation		sandstone, shale, limestone, chert phosphate rock	shallow sea deposits
	PENNSYLVANIAN		Quadrant Formation		sandstone	beach; dunes
			Tensleep Formation		interbedded carbonate, shale, anhydrite	marine
			Amsden Formation		carbonates; dissolved into caverns	marine
			Tyler Formation		sandstone	river
	MISSISSIPPIAN		Snowcrest Range Group		siltstone, sandstone, minor limestone	marine
			Big Snowy Group		shale, limestone, gypsum, coal	tidal, coastal plain, marine
			Tendoy Group		chert-bearing limestone	marine
			Madison Group	Mission Canyon Limestone	pale-blue-gray, massive limestone	shallow marine
				Lodgepole Limestone	cherty limestone and dolomite	shallow marine
	DEVONIAN		Three Forks Formation		dolomitic siltstone, shale; fossiliferous limestone	marine with evaporite basins
			Jefferson Formation		fetid brown dolomite	marine
			Maywood Formation		thin-bedded dolomitic limestone, shale	shallow marine
	SILURIAN				no rocks	
	ORDOVICIAN		Bighorn Dolomite		massive cliff-forming light-gray sandy dolomite	offshore shelf
	CAMBRIAN		Snowy Range Formation		limestone and shale	shallow marine
			Pilgrim Limestone		limestone	shallow marine
			Park Shale		grayish-green micaceous shale	offshore shelf
			Meagher Limestone		gray-blue limestone; conglomerate	shallow marine
			Wolsey Shale/Gordon Sh.		dark-green micaceous shale; glauconitic	offshore shelf
			Flathead Sandstone		pinkish-gray,glauconitic, pebbly	marine shoreface
PROTEROZOIC	MIDDLE PROTEROZOIC		Belt Supergroup	see detailed section for Belt Supergroup on page 29		
				Missoula Group	red to purple mudstone, some green, gray	
				Piegan Group	carbonate rocks with some mudstone, quartzite	
				Ravalli Group	white to dark-gray quartzite, siliceous mudstone	
				Lower Belt Group	dark-gray mudstone and muddy quartzite	
ARCHEAN			basement rock (granite, gneiss, and schist) of Wyoming and Medicine Hat blocks			

General stratigraphic column for Montana. See page 29 for Belt Supergroup stratigraphy and page 254 for more detailed stratigraphy of Cretaceous and Tertiary rocks.

The Chinese Wall in the Bob Marshall Wilderness is an upturned slab of Cambrian limestone and some shale in the Overthrust Belt. —Courtesy of Rick and Susie Graetz

PREFACE

If we look under the greenery, the spectacular scenery of Montana is supported by rocks that have an amazing history that began more than 3 billion years ago. That mountain, the river, the distinctive rock formation—how did they get there and when? Many geological processes operate so slowly that we never see them happen, others suddenly and unexpectedly. Why? Unravelling the story has been a daunting but incredibly fascinating task for geologists, but their efforts have uncovered the underlying framework and processes, exposing them in ways that we can all understand. The events are now clear and easy for us to understand if we know what to look for.

Looking back, we couldn't have imagined how much would be learned about the geology of Montana in the last thirty-odd years. Yes, the rocks are still there, at least most of them. Some have been removed by construction, and some new ones have been exposed. We have learned so much that the once-scattered pieces of the story now nicely stitch together to make a coherent and straightforward historical account.

Most of what has been learned is thanks to thousands of geologists of the Montana Bureau of Mines and Geology and US Geological Survey, along with university faculty and students not only in Montana but all over the country. Petroleum and mining geologists have enhanced our understanding as they seek out petroleum and valuable minerals, analyzing the geological environment to understand the context and processes involved in their formation.

When Dave Alt and Don Hyndman wrote the previous edition of this book, the understanding of plate tectonics, the ongoing movement of Earth's handful of lithospheric plates, was still in its infancy. Those plates pull apart, collide, and shift laterally, in different places and at different times. These processes lead to the formation of mountains and valleys, rise of plateaus, and depression of broad areas below sea level where they collect new sediments. We now have a good idea of what has caused such big changes and when. In this second edition we analyze and simplify this vast trove of information for people who are not geologists while also trying to keep the information scientifically correct and up-to-date for geologists seeking a basic background in Montana geology. We hope our readers find that this book increases their interest and enjoyment in exploring Montana.

This new second edition maintains the same general approach as the first: an introductory chapter to outline the broad view of geology of the region, followed by three separate large chapters that describe the rocks and geology along every major highway in the state; however, the text is almost entirely rewritten. Although most of the photos are our own, many others were graciously contributed by colleagues. Entirely new, up-to-date color geological maps are simplified from new detailed maps from the Montana Bureau of Mines and Geology and the US Geological Survey, expertly enhanced by Chelsea Feeney. We are indebted to geologists everywhere who contributed to our understanding of Montana geology; though we can't thank them all individually, we are grateful for their work.

Discussions with Ginette Abdo, Dick Berg, Colleen Elliott, Dave Foster, Chris Gammons, Debbie Hanneman, Jack Horner, Billie Jo Juneau, Tom Kalakay, Jeff Kuhn, Rebekah Levine, Paul Link, Jeff Lonn, Tom Lowe, Stacia Martineau, Marli Miller, Gregory Retallack, Sheila Roberts, Spruce Schoenemann, Jim Sears, Mike Stickney, David Varricchio, Susan Vuke, Don Winston, and many others were invaluable. They are not, however, responsible for any errors.

Many people and organizations provided photographs or artwork that materially improved this book: Jeff Albrecht, Rod Benson, Dick Berg, Christopher Boyer, Norm Dwyer, Bryan Gartland, Richard Gibson, Rick and Susie Graetz, Dale Greenwalt, Debbie Hanneman, Jack Horner, Hal James, Dave Mogk, Marli Miller, Chris Pace, Ken Pierce, Sheila Roberts, Parker Webb, Craig Zaspel, the Montana Bureau of Mines and Geology, Mansfield Library, and the Montana Historical Society. We thank them all. Producing the book would not have been possible without the knowledge and ability of our editor, Jennifer Carey, and detailed editing by James Lainsbury. Finally, to our spouses, Shirley and Anneliese, we are grateful for their patience, support, and help on long field trips throughout Montana.

We thank the thousands of readers of our first edition for their kind words over the years and hope that this new book provides even more enjoyment, fascination, and understanding of our spectacular state.

—Missoula, Montana, January 2020

The Block Mountain basalt (in the distance) flowed over sedimentary rocks (in the foreground) that were folded and thrust faulted during the Sevier orogeny. Note the large, open fold in Triassic limestone in the foreground.

GEOLOGIC TIME SCALE

ERA or EON	PERIOD		EPOCH	AGE (mya)	IMPORTANT GEOLOGIC EVENTS IN MONTANA	
CENOZOIC	QUATERNARY		HOLOCENE	0.01		Yellowstone hot spot, 16.5 mya to present; crustal extension, 50 mya to present
			PLEISTOCENE		ice sheets advance south across Montana; glaciers form in mountains modern streams begin to flow	
	TERTIARY	NEOGENE	PLIOCENE	**2.6**	deposition of Sixmile Creek, Flaxville Gravel, pediments	pediments form
			MIOCENE	5.3	wet tropical climate	
		PALEOGENE	OLIGOCENE	23	camels, rhinos, horses, and tapirs roamed semiarid grasslands	Renova Formation, 48-20 mya
			EOCENE	33.9	mountains form in central Montana, 50 mya; Central Montana Alkalic Province Lowland Creek Volcanics, Absaroka Volcanics Supergroup, Dillon Volcanics stable period forming laterite soil	
			PALEOCENE	56	Fort Union Formation; coal	Sevier and Laramide deformation, 90–50 mya
				66	extinction of dinosaurs	
MESOZOIC	CRETACEOUS				conglomerate, sandstone, shale deposited in Western Interior Seaway, 80–65 mya Elkhorn Mountains Volcanics, Boulder batholith, Idaho batholith; 80–70 mya	
	JURASSIC			145	Sundance Seaway shale, limestone, and sandstone	Atlantic Ocean begins to open
	TRIASSIC			201		
PALEOZOIC	PERMIAN			**252**	major extinction event at end of Permian	Pangaea forms
	PENNSYLVANIAN			299	sandstone, limestone, chert, and phosphorite	shallow ocean
	MISSISSIPPIAN			323	sandstone, limestone, and shale	
	DEVONIAN			359	Madison Group limestones	
	SILURIAN			419	limestone and shale	
	ORDOVICIAN			444	no rocks	
	CAMBRIAN			485	sandy dolomite sandstone and limestone first abundant animal fossils	
PROTEROZOIC				**541**	continental rifting, 750 mya Belt sedimentary formations, 1,470–1,400 mya Big Sky orogeny: collision between Medicine Hat block and Wyoming craton, 1,850 mya	
ARCHEAN				**2,500**	Stillwater layered intrusion, 2,700 mya most basement rocks in Montana are older than 2,700 mya oldest basement rock in Montana, 3,500 mya	

*mya = millions of years ago

Geologic timescale and important events in Montana's geologic history.

MONTANA

THE FIRST FOUR BILLION YEARS

The story in the rocks is there for all of us to read. The rock record reads like a book, only from back to front. It's not hard to read if you know some basic rules for determining the relative ages of rocks and structures: The oldest rock is on the bottom of a stack of sedimentary layers unless disrupted by tectonic forces. Most sedimentary rocks were originally horizontal, so if they're tilted or folded, that deformation occurred after the layers formed. Any rock cut by a fault must be older than the fault, and rock intruded by or surrounded by magma must be older than the intrusion. Maps that show the distribution of past environments of similar ages are constructed using fossils, layers of volcanic ash of a known age, or radiometric dates from rocks.

The age of many igneous and metamorphic rocks can be measured by using the decomposition rates of some radioactive atoms ("parent isotope" atoms emit atomic particles to form "daughter isotopes"). The half-life of such an isotope—that is, the time for half the parent isotope to decay to the daughter isotope—is now well-known for most radioactive elements, so geoscientists can determine the age of some rocks. The clock begins ticking for a parent isotope when the rock cools to a certain level: the point at which a magma crystallizes or a metamorphic rock recrystallizes.

The Earth is 4.5 billion years old, so old that even Montana's oldest rocks, in the Beartooth Plateau, weren't around at the beginning. The really old Belt Supergroup rocks that cover much of northwestern Montana are a third of the age of the Earth. The vast majority of Montana's rocks, still hundreds of millions of years old, are less

The layered metamorphic rock enclosed within the granitic rock must be older.

This folded rock north of Dillon contains snail and clam fossils that show it was once sediment on the bottom of a lake or estuary. Deep burial and compression during mountain building folded the rock. Uplift and erosion made it possible for us to see it at the surface today.

than one-twentieth the age of Earth. The first ice advance to leave substantial deposits in this part of the world appeared only 200,000 years ago, and *Homo sapiens* arrived in Montana as recently as 13,500 years ago.

The big picture of Montana geology depends heavily on understanding the theory of plate tectonics. This well-established theory is backed by vast amounts of data, experiments, and calculations. Subjected to every conceivable critique, it has stood the test of time. Even predictions of what should be true using the theory have turned out to be correct. It really should be called the *law* of plate tectonics.

Let's start with the basics. Like a golf ball, Earth has a small central core, a thick mantle that surrounds the core, and a thin outer crust. The thick mantle makes up most of the Earth. It consists of heavy black rocks called peridotite and has two different zones: the lithosphere and the asthenosphere. Although solid, most of the mantle (the asthenosphere) is hot enough that its rocks can slowly spread and flow like warm tar. The cooler, more rigid outer part of the Earth, the lithosphere, consists of the uppermost mantle and the crust, the thinner outer shell of the planet that we see at the surface. Under the oceans the crust is mostly black basalt that's only 4 or so miles thick; the crust of the continents is 25 or so miles thick and is composed of granitic and metamorphic rocks of similar composition, along with a much thinner skin of sedimentary and volcanic rocks.

The more rigid lithosphere, comprising Earth's crust and uppermost mantle, slowly drifts over the underlying, relatively soft asthenosphere. It has broken into a dozen or so large plates and several smaller ones. Oceanic lithosphere is 35 to 40 miles thick and denser than continental lithosphere, which averages 120 miles thick. The plates drift around at 1 to 2 inches per year, much too slowly for us to watch, but in 1 million years that adds up to about 16 miles! This entire process is called plate tectonics.

Most of the geologic action is at the plate boundaries, where the plates collide, are pulled apart, or slide past one another. Where plates collide, denser, heavier oceanic lithosphere generally dives under continental lithosphere or lighter oceanic lithosphere, a process called subduction. The descending slab gradually heats and releases

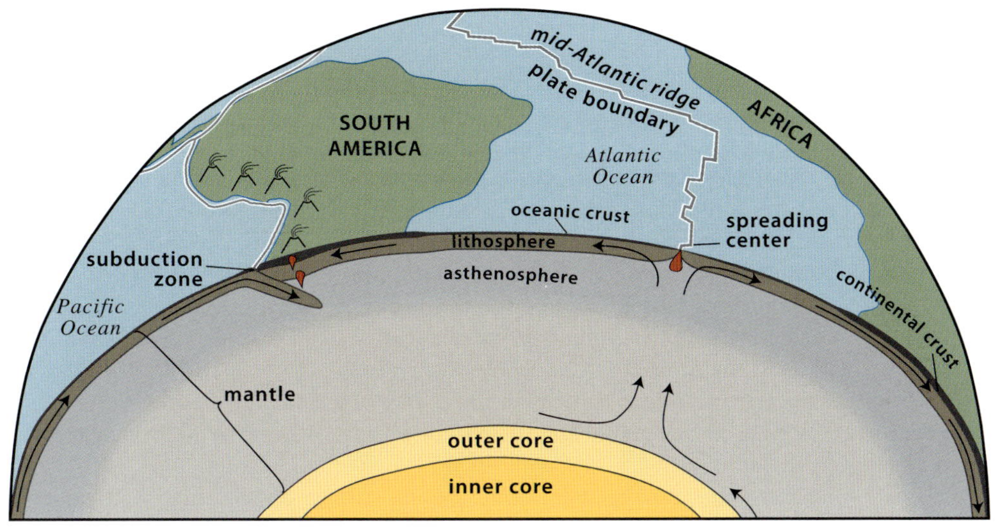

The Earth as it would look if it were sawed in half. —Modified from US Geological Survey

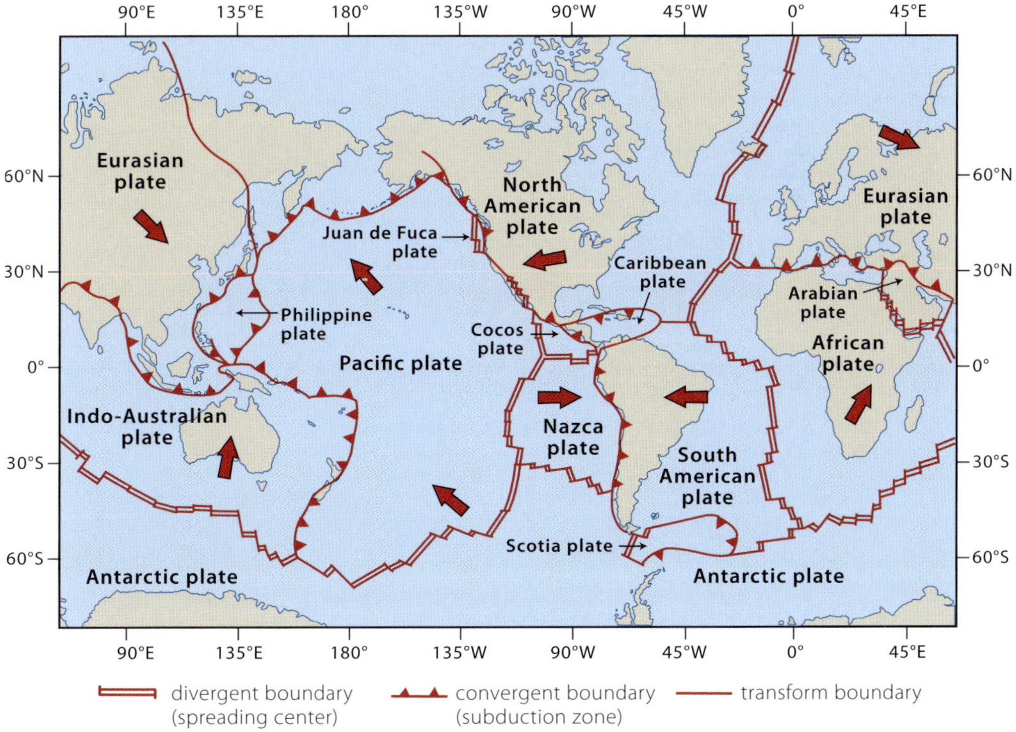

The major tectonic plates of the Earth. The arrows show the relative movement of the plates.
—Modified from US Geological Survey

water that causes the overlying mantle to melt at lower temperatures, working much like salt does to melt ice. The melted rock rises and ponds beneath crustal rocks, melting them to create silica-rich magmas that form chains of volcanoes, such as those in the Cascades and in Japan.

When continents collide, they crumple and thicken to create mountain ranges. These ranges have deep roots composed of rocks that are metamorphosed by intense heat and pressure. Some rocks get so hot that they melt and intrude overlying metamorphic rocks. When plates are pulled apart, new basins can form. Upwelling mantle can cause continental lithosphere to extend and break into parallel basins and ranges. If the process continues, a new ocean basin will form. As the lithosphere thins, the reduction in pressure below it causes the underlying mantle to partially melt, forming basaltic magma. Little magma is generated where plates slide past one another, but earthquakes at these boundaries are common, as they are at all plate boundaries. Hot spots originate deep in the mantle and seem to be mostly stationary and unrelated to tectonic plate motion.

Plate tectonics—the movement of plates at the surface of the Earth and their mountain-building collisions and basin-forming rifts—has created what we see today in Montana. The geological map of Montana shows rock patterns that suggest there are three regions of rocks that have different ages and different histories. Northwest Montana is dominated by 1.47-to-1.4-billion-year-old folded and thrust-faulted sedimentary rocks that were sliced into long northwest-trending ranges and valleys in Tertiary time. Southwest Montana has isolated ranges, spacious valleys, ancient to very young rocks, and complex deformational structures. Central and eastern Montana, spreading out eastward from the high Rocky Mountains as the Great Plains, is dominated by sedimentary rocks that are about 150 million years old or younger, except where odd igneous rocks and older sedimentary rocks punched up from below

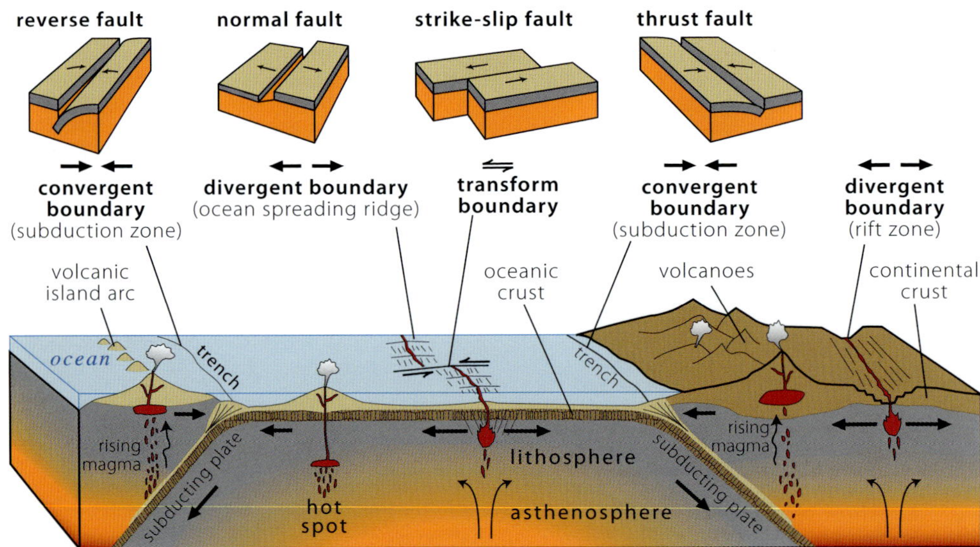

The types of boundaries between tectonic plates at the surface of the Earth and the types of faults that are common at each boundary. —Modified from US Geological Survey

to form isolated mountain ranges. We discuss the rocks and landscapes of these three regions in more detail in the following chapters.

ARCHEAN TIME

4,600 TO 2,500 MILLION YEARS AGO

Montana's oldest rocks are mostly buried deep in the continental crust but are exposed in some mountain ranges in southwest and central Montana. The deep crustal rock began as ancient sedimentary and igneous rocks that were heated and stretched under high pressure and high temperature to form mostly schist and gneiss, streaky metamorphic rocks that may preserve some semblance of their original layers. Where these metamorphic rocks got hot enough to melt, the magma later cooled and crystallized to form light-colored granitic rocks. The chemical compositions of much of the Archean granite, gneiss, and schist in Montana are similar to those of ordinary sand and mud—the stuff from which they formed.

This streaky, layered rock in Bear Trap Canyon probably started as deeply buried shale or sandstone that was then heated, softened, smeared out, and folded.

Geologists often call large areas of Archean and Proterozoic rock of the continental crust "basement rock" because it seems to continue down to great depth. If you drill deep enough almost anywhere on the continent, the hole will eventually penetrate basement rock. It lies beneath virtually all of Montana, but it's exposed at the surface primarily in southwest Montana, mostly south of I-90 and in small areas in the Little Rocky Mountains east of Hays and the Little Belt Mountains northeast of Helena.

Basement rocks in Montana are not uniform. They record past environments and plate tectonic events. Field mapping, deep-sounding geophysics, a few drill holes, and many radiometric age dates have distinguished terranes of quite different ages, compositions, and origins. For example, the huge Wyoming craton, exposed in the Beartooth Mountains and parts of southwest Montana, consists of three distinct crustal blocks with a wide variety of rock types. Some are as old as 3.5 billion years and contain minerals that are 4 billion years old. The Medicine Hat block, which appears to be exposed only in the Little Rocky Mountains in north-central Montana, consists mostly of 3.3-to-2.6-billion-year-old rocks.

PROTEROZOIC TIME

2,500 TO 541 MILLION YEARS AGO

The Archean Medicine Hat block was sutured to the Wyoming craton during the closure of an ancient ocean roughly 1.9 to 1.7 billion years ago—a mountain-building episode geologists call the Big Sky orogeny. The merging of these blocks in early Proterozoic time formed a small part of the ancient geological core of the North American continent called Laurentia. The collision formed a major northeast-trending suture zone called the Great Falls tectonic zone, which is made up mostly of rocks metamorphosed during the event. The closure of the ocean basin between the Medicine Hat block and Wyoming craton resulted in subduction of ocean floor and the generation of magma in early Proterozoic time. These igneous rocks are exposed in the Little Belt Mountains in central Montana. Eventually, the heavier, older, and colder Wyoming craton subducted under the edge of the Medicine Hat block, suturing them together.

Around 1.5 billion years ago, the ancient, deeply eroded continental crust of Montana was pulled apart, creating a basin within the continent—called the Belt Basin. It filled with a shallow inland sea (the Belt Sea) fed by numerous rivers draining a supercontinent called Nuna. Thick deposits of sandy, muddy, and limy sediments accumulated in the basin for about 70 million years. These early sediments, known as the Belt Supergroup, were so named because they were first studied in the Big Belt Mountains of central Montana. Quartzites, dolomitic limestones, and slightly metamorphosed mudstones and siltstones of the Belt Supergroup are well exposed in northwest Montana, including in Glacier National Park.

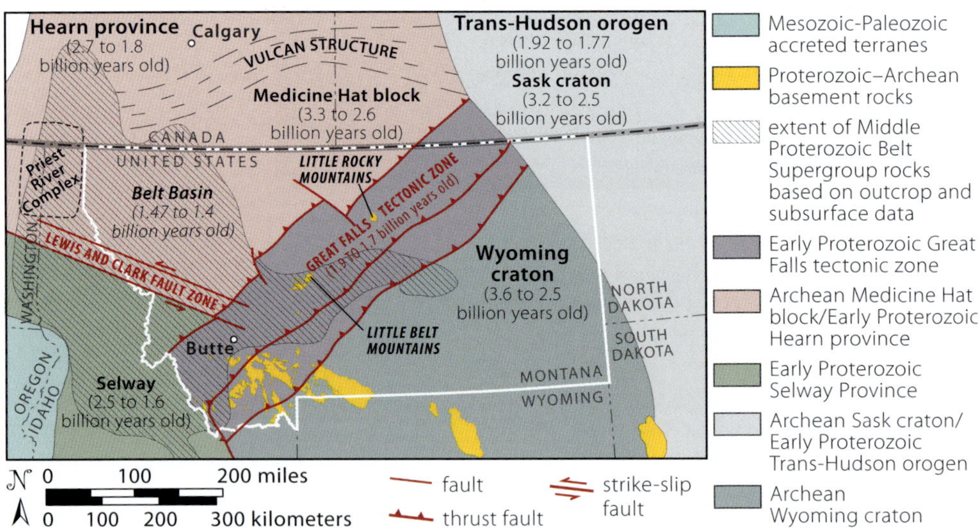

The early Proterozoic suture of the Medicine Hat block to the Wyoming craton, called the Great Falls tectonic zone, contains rocks metamorphosed during the collision around 1.9 to 1.7 billion years ago (Big Sky orogeny). Early Proterozoic igneous rocks in the Little Belt Mountains were generated during the subduction of ocean floor as an ocean basin closed between the continents. The yellow areas are places where Archean and early Proterozoic rocks are exposed. —Modified from Foster and others, 2006

The Belt formations of Montana contain abundant fossils of primitive plants that formed structures called stromatolites, but otherwise they contain no real trace of animal life. Without animals to churn up the sediments, structures such as mud cracks, ripple marks, and raindrop impressions are well preserved in the Belt rocks.

After hundreds of millions of years, a supercontinent called Rodinia rifted apart, splitting the Belt Basin about 750 million years ago and generating basalt magma that cooled as diabase sills and dikes, including the prominent Purcell Sill in Glacier National Park. The new western coast of the continent ran through northeastern Washington and western Idaho. Rocks of the Belt Basin continued to erode for another 300 million years. No land plants existed at that time, so the landscape was barren and probably eroded more quickly than it does today.

PALEOZOIC TIME

541 TO 252 MILLION YEARS AGO

At least 1 billion years of erosion left most of Montana a featureless plain by the start of the Paleozoic Era, about 541 million years ago. About 520 million years ago, an ocean spread over the land from west to east, depositing first beach sand, then continental shelf mud, and eventually offshore limestone as the water became deeper. Montana was located near the equator, and the marine sediments of this age are loaded with the shells of trilobites, brachiopods, and other tropical marine life, which experienced an explosion in diversity at the beginning of Paleozoic time. The contact between the 520-million-year-old oceanic sedimentary rocks and the underlying Archean and Proterozoic rocks is called the Great Unconformity because it marks a tremendous gap in the rock record. An unconformity represents a period of time during which either no sediments accumulated, or sediments were eroded. The Great Unconformity is well exposed in southwest Montana.

Sea level rose and fell during the remainder of Paleozoic time, resulting in times of marine deposition and times of exposure, erosion, and no deposition. During a major rise in sea level in Mississippian time, the fossil-rich Madison Group limestone was deposited across much of Montana. Fossils of a variety of tropical marine animals, including corals, can now be found in these limestones from the mountains to the plains. Because limestone resists erosion in arid climates, it forms prominent landmarks, such as Gates of the Mountains near Helena and Beaverhead Rock near Twin Bridges. During wet periods, the Madison Group limestone dissolved, forming caverns, such as Lewis and Clark Caverns in the Jefferson River canyon.

During Pennsylvanian time, regional uplift caused the ocean to retreat. Coastal dunes left in its wake are preserved today as sandstone of the Quadrant Formation. The rock consists of very large cross beds that formed when well-rounded and well-sorted sand collected on the downwind face of migrating dunes. The sea rose again in Permian time, and sediment eroded from uplifts to the south and volcanic islands to the west was deposited on the coastal shelf to form chert and black shale. Phosphate formed where deeper, nutrient-rich water was pulled toward the surface by winds and upwelling. The bones and feces of animals living in this upwelling zone were turned into phosphate-rich rocks of the Phosphoria Formation. It is mined today for use as fertilizer.

The Paleozoic Era ended with the extermination of more than 96 percent of all ocean species and 70 percent of land vertebrate species. Known as the Great Dying,

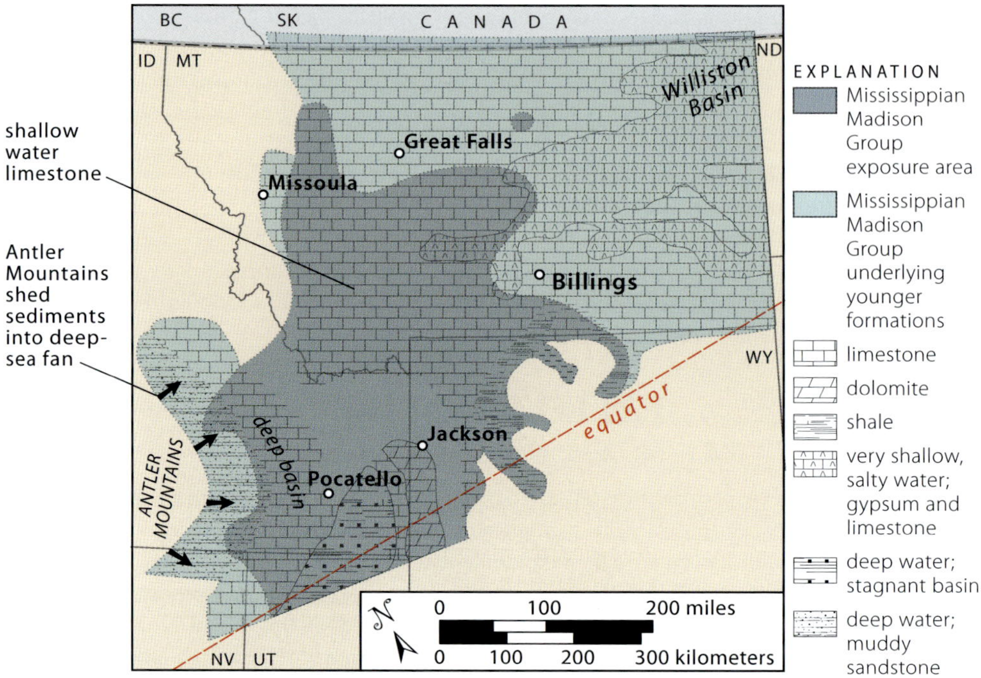

BC SK CANADA
ID MT ND
Williston Basin
Great Falls
Missoula
Billings
shallow water limestone
Antler Mountains shed sediments into deep-sea fan
WY
equator
deep basin
Jackson
ANTLER MOUNTAINS
Pocatello
NV UT
0 100 200 miles
0 100 200 300 kilometers

EXPLANATION
Mississippian Madison Group exposure area
Mississippian Madison Group underlying younger formations
limestone
dolomite
shale
very shallow, salty water; gypsum and limestone
deep water; stagnant basin
deep water; muddy sandstone

Montana during deposition of the Mississippian Madison Group limestone. —Modified from Alt and Hyndman, 1995

Tubular-shaped colonial coral fossils in the Madison Group limestone in Gallatin Canyon.

the extinction likely involved many environmental changes occurring all at once, including massive basalt eruptions in Siberia, an asteroid impact, global warming, changes in ocean chemistry, and perhaps the release of vast amounts of carbon dioxide from shallow-water methane ice. The extinction boundary is not preserved in Montana, but brachiopods, snails, and coiled squids survived and lived in shallow seas in Montana about 15 million years after the extinction; their fossils are preserved in marine shale and limestone of the Triassic Dinwoody Formation in southwest Montana. Complex systems such as reefs took millions of years to recover, and trees were so decimated that no coal formed anywhere in the world for about 10 million years after the mass extinction.

MESOZOIC TIME

252 TO 66 MILLION YEARS AGO

Shortly before the beginning of Mesozoic time, the shifting lithospheric plates joined to assemble most of the Earth's continental crust into a single supercontinent called Pangaea. It was short-lived. A few tens of millions years after it formed, a continental rift split the supercontinent along a line that survives today as the trace of the mid-Atlantic ridge, a 10,000-mile-long mostly submarine ridge that runs generally north to south beneath the Atlantic Ocean. Rocks along the east coasts of North and South America and the west coasts of Europe and Africa contain diabase sills and dikes that cooled from basalt magma that rose through faults along the rift. The ages of these diabase intrusions indicate that the Atlantic Ocean began to open around 180 million years ago.

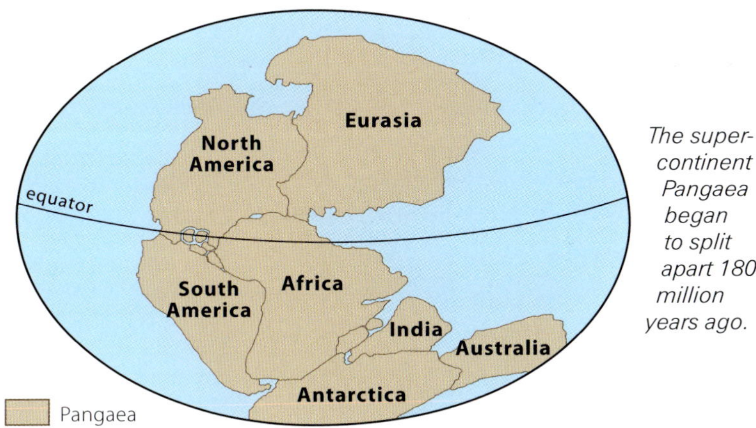

The supercontinent Pangaea began to split apart 180 million years ago.

Unlikely as it may seem, those remote events happening thousands of miles away started the long series of crustal movements in western North America that first raised the Rocky Mountains. The western margin of North America was a very thick, solid buttress left behind by the rifting of the Belt Basin near the western edge of Idaho 750 million years ago. In western Montana, that buttress consisted of tens of miles of basement gneisses of the ancient core of the continent capped by a 10-mile-thick pile of Belt sedimentary rocks. As the plate bearing the North American continent began to move west, away from the newly opening Atlantic Ocean, this hard buttress collided head-on with the solid floor of the Pacific Ocean, an oceanic plate known as the Farallon plate. Something had to give.

The denser Farallon plate bent downward as it subducted to the east into the mantle beneath the coast of western North America, which at that time ran through eastern Washington and Oregon. This collision between an immovable continent and a dense ocean floor was slow and ongoing. The descending slab was heated and dehydrated, generating magma that fed an overlying volcanic arc, while the western edge of the continent was compressed like an accordion to create a broad highland. Geologists refer to such a zone of crumpled and fractured rocks as a fold-and-thrust belt.

The weight of all that thick crust in western Montana depressed the crust to the east to form a north-south trough known as a foreland basin. In Jurassic time, the basin was filled with water, creating the Sundance Seaway. Marine sediments of the Ellis Group, some with star-shaped crinoid fossils, accumulated in the seaway. By the end of Jurassic time, the Sundance Seaway had retreated, and sediments eroded from the rising western mountains were spread eastward to become the Morrison Formation. North of Dillon, Morrison stream deposits have yielded fossil bones of juvenile sauropods that likely failed to cross a swollen river about 150 million years ago.

By Cretaceous time, ocean water filled the foreland basin to create the Western Interior Seaway, a shallow inland sea that extended from the Arctic Ocean to the Gulf of Mexico. Vegetation grew in coastal swamps adjacent to the seaway, where it was gradually buried and compressed to become coal. Dinosaurs roamed the coastal plain, and swimming reptiles and coiled squids ruled its waters. It was a new and strange time in Montana.

Star-shaped crinoid fossils occur in marine rocks of the Jurassic Ellis Group in southwest Montana. —Courtesy of Marli Miller

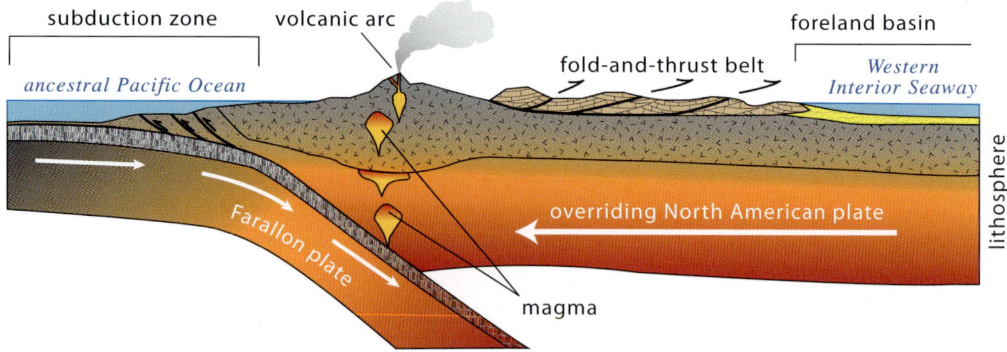

Cross section showing the development of the Mesozoic subduction zone, volcanic arc, fold-and-thrust belt, and interior seaway. —Modified from Hendrix, 2011

The conglomerate (bottom) of the Cretaceous Kootenai Formation was deposited in a foreland basin by streams flowing east off western highlands. Most of it was eroded from the older Quadrant Formation (light-colored pebbles) and Phosphoria Formation (dark-colored pebbles), showing that they were exposed in the fold-and-thrust belt at the time. The transition from gravel to sand (top) records the lateral migration of the river across its floodplain. This outcrop is north of Dillon.

Sediments eroded from the mountains rising to the west were deposited in coastal plain and oceanic environments to the east. Coarse sediments were deposited close to the mountains, while streams carried fine-grained silt and clay into the seaway. The ocean waxed and waned in Cretaceous time, resulting in a thick stack of interbedded marine and nonmarine sediments as the foreland basin subsided. Major marine incursions are marked by the Mowry, Claggett, and Bearpaw Shales, while sand deposits of the Frontier, Eagle, Judith River, and Fox Hills Formations record times of relative sea-level drop.

Rocks throughout most of Montana were complexly deformed during the Mesozoic plate collision. They were fractured, faulted, folded, crumpled, raised, dropped, and torn in a variety of directions. Some of the deformation involved the old Archean and Proterozoic basement, while some only the sedimentary cover. There are two broad patterns to the deformation, which geologists call "thin skin" and "thick skin."

Sevier Thin-Skin Deformation

In western North America from about 140 to 50 million years ago, horizontal compression driven by subduction from the west took advantage of weak bedding planes in sedimentary rocks above the igneous and metamorphic basement rocks. Packages of sedimentary rocks were shoved to the east along low-angle faults. Where these thrust faults could not continue forward, they broke upward toward the surface, forming folds over the tips of the faults and placing older rocks over younger rocks. A broad wedge of rocks piled up, thickest in western Montana and thinning along the east front of the Rocky Mountains, where the pile is called the Overthrust Belt. Geologists refer to this thin-skin deformation style as Sevier deformation, and this mountain-building event as the Sevier orogeny. In western Montana, some of the big thrust faults of this event include the Libby, Lewis, Lombard, Medicine Lodge, and Grasshopper faults. East of the Overthrust Belt, deformation was more subdued in the softer, more easily eroded Cretaceous sedimentary rocks; that area is often referred to as the Disturbed Belt. In some areas of southwest Montana, granitic magma intruded folds forming over the tops of rising thrust faults.

Thrust faults grew eastward with time, so the youngest folds and faults tend to be found to the east. Sediments shed off the front of the growing Sevier highlands were, in turn, folded and faulted. In places, deposits eroded from the front of the growing

View northwest in the heart of the Overthrust Belt, looking over Pentagon Mountain in the Bob Marshall Wilderness west of Choteau. Thick beds of limestone (cliffs) were thrust over younger shales (hidden under talus slopes). —Courtesy of Rick and Susie Graetz

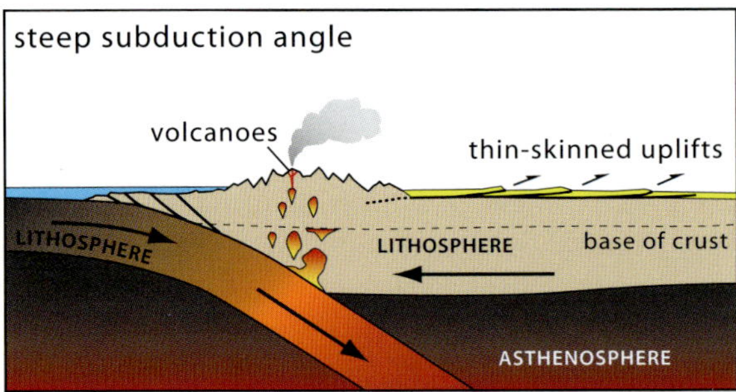

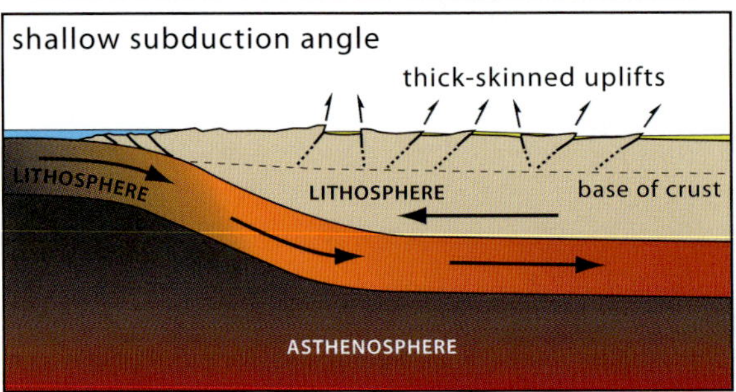

The relationship between subduction angle and deformational style—either thin-skinned Sevier or thick-skinned Laramide—in the Mesozoic subduction zone. —Modified from Hendrix, 2011

fold-and-thrust belt were overrun by the thick slabs of older rock. For example, cobbles of the Cretaceous-Tertiary Beaverhead Conglomerate of southwest Montana were crushed by overriding rock.

Laramide Thick-Skin Deformation

As the Mesozoic Era came to an end, the angle of subduction of the Farallon plate decreased, causing steep reverse faults to deform rocks deep in the continental plate, raising basement rocks to the surface. This thick-skin style of deformation was part of the mountain-building event known as the Laramide orogeny. Paleozoic and Mesozoic sedimentary rocks were draped in large folds over the basement uplifts. The sedimentary rocks have mostly eroded off the tops of these blocks but are locally preserved along the flanks as steeply dipping beds, often called palisades because they resemble the palisade walls built to defend forts. The Beartooth Mountains near Red Lodge were initially raised along faults at this time. Most of the Laramide deformation is a bit younger than that of the Sevier and is concentrated east of the Sevier deformation, but overlap occurred in both time and location.

Sediments eroded from the Laramide uplifts were deposited in an adjacent foreland basin to the east. Eventually, the tectonic compression pushed the basement rocks over these basin sediments along faults, folding and faulting them as well. Fault movement continued into early Tertiary time, affecting the widespread Fort Union Formation of central and eastern Montana.

Some of the Laramide structures occur within parts of Montana pulled apart by later basin and range extension, which started around 17 million years ago. As a result, many of these extensional ranges now expose basement rocks that were first exposed during the Laramide orogeny. Since broken rock provides weaknesses that can be used by later tectonic movements, we see evidence that some of the Laramide reverse faults were reactivated as normal faults when the crust was pulled apart in Tertiary time.

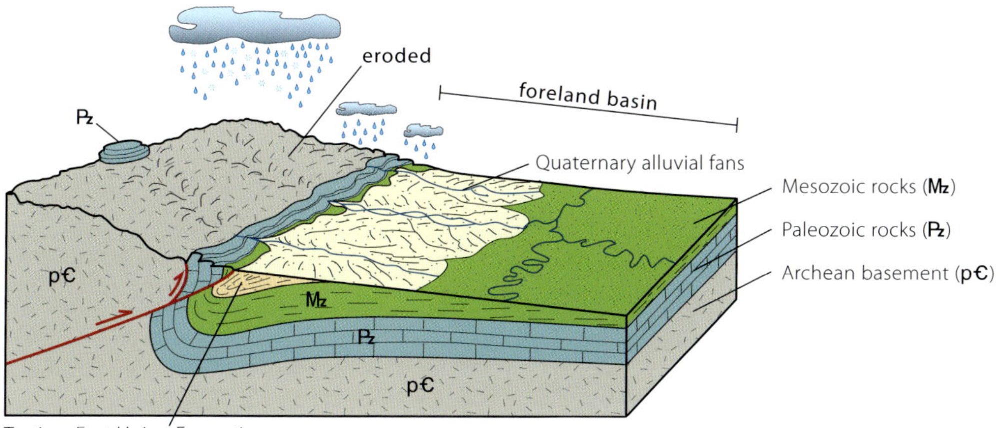

Schematic diagram of Laramide deformation. Steep reverse faults brought basement rocks to the surface. Paleozoic and Mesozoic sedimentary rocks were folded into palisades along the flanks of the basement and flatten out under the sediments in the foreland basin. Sediments eroded from the raised blocks were deposited in the foreland basin and eventually overridden by the basement rocks. —Modified from Wise, 2000

Cretaceous Intrusions

While sedimentary rocks were being shuffled eastward and stacked at the surface, subduction-generated magma rose to the west in the subsurface. Molten basalt and flaming hot steam rising from the oceanic crust that was sinking through the subduction zone torched the lower part of the continent. The intense heat melted Archean basement rocks to form granite magma. Between 90 and 70 million years ago, enormous volumes of this magma rose in the continental crust of the mountain ranges east of the old western coastline, from southern California north through Idaho and western Montana, and on through British Columbia. That magma crystallized as large and small plutons that together form major batholiths, including the Idaho batholith in central Idaho and the Bitterroot Mountains of western Montana. Smaller satellite masses of that granite are scattered eastward in southwest Montana.

The Boulder batholith, between Butte and Helena, is the same age, about 80 to 70 million years old, but is 100 miles east of the long trend of granite masses that parallels the old continental margin. In addition, instead of crystallizing deep in the crust, some of the Boulder batholith magma erupted to form the enormous Elkhorn Mountains volcanic pile. (The mysterious origin of this batholith is discussed in the Southwest Montana chapter.)

Much of the gold, copper, silver, zinc, lead, and other minerals of the Treasure State was deposited by processes related to these granitic intrusions. Where the hot magma came into contact with sedimentary rocks, watery fluids reacted to form mineral-rich deposits, often in quartz veins. Sometimes the granite reacted with limestone to form garnet, epidote, and other minerals in skarn. Elsewhere, as in Butte, the granite itself was altered to clay and mica in large, rich mineral deposits.

In southwest Montana, granite of the Cretaceous Pioneer batholith (speckled rock on the bottom) intruded Paleozoic sedimentary rocks (gray-green rock above), baking and enriching them with economically valuable minerals through the exchange of fluids between magma and rock.

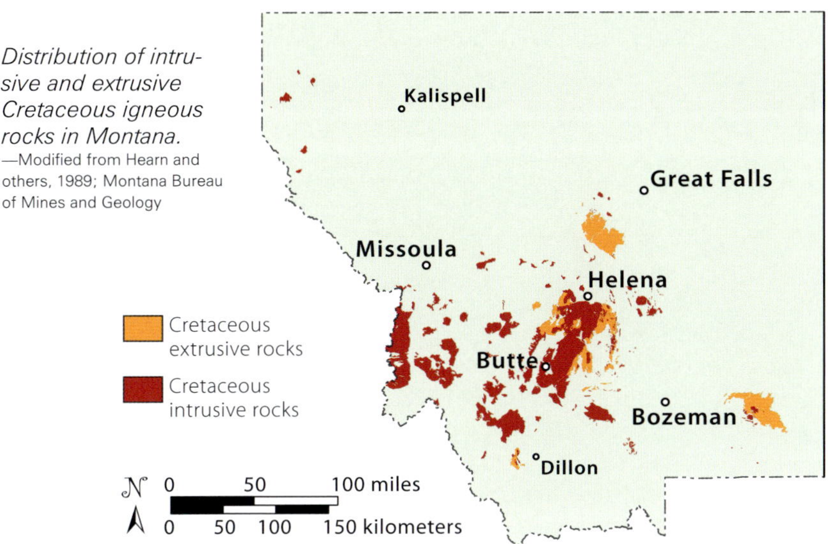

Distribution of intrusive and extrusive Cretaceous igneous rocks in Montana. —Modified from Hearn and others, 1989; Montana Bureau of Mines and Geology

CENOZOIC TIME

66 MILLION YEARS AGO TO PRESENT

After the 20 million or so years of intense folding, faulting, granite intrusion, and volcanic eruption of Late Cretaceous time, Montana settled into about 20 million years of geologic quiet. The Western Interior Seaway retreated from North America. Then, 66 million years ago, the dinosaurs that roamed its coastal margins and the creatures that swam in its waters were wiped out by the global environmental devastation caused by a large asteroid that struck the Yucatán Peninsula of eastern Mexico. It's estimated that up to 85 percent of all species were exterminated. The extinction boundary is well exposed in eastern Montana at places such as Iridium Hill, north of Jordan; Bug Creek, south of Glasgow; and Makoshika State Park, south of Glendive.

From about 100 to 50 million years ago, slices of continental crust carried on the Farallon plate or transported along the plate margin were added to the western margin of North America to form much of today's Washington State. Those additions forced the subduction zone to jump farther west to form a new coastline. The subduction zone of Cenozoic time was so far west of Montana that it's difficult to imagine any connection between it and events here. Nevertheless, the Montana part of the continent continued its mountain-forming dance.

About 50 million years ago in Eocene time, a brief but widespread period of tectonic activity and faulting was accompanied by the intrusion of granites and the eruption of explosive volcanoes. The resulting Absaroka Volcanics Supergroup covers large parts of the Absaroka and Gallatin Ranges and regions south into Wyoming, and the Dillon Volcanics cover areas around Dillon. At about the same time, intense Lowland Creek volcanic activity occurred in the area of the Boulder batholith, and a broad area of scattered alkali-rich volcanoes erupted in central Montana. Some geologists attribute the volcanism to nearly flat subduction of the Farallon plate, so that magmas formed inland as far as central Montana. Others find this difficult to

The Cretaceous-Tertiary boundary at Makoshika State Park. The geologist points to the base of a coal bed that marks the time horizon when dinosaurs and other animals and plants were exterminated, about 66 million years ago. The rocks below the coal belong to the Cretaceous Hell Creek Formation, famous for its dinosaur fossils. The overlying Paleocene (Tertiary) Fort Union Formation does not contain dinosaur fossils but does contain broadleaf plant fossils. —Courtesy of Sheila Roberts

imagine. Typically, volcanoes form parallel to the trench that develops where a plate is subducted beneath another, and directly above the subducted slab, but the Eocene volcanic centers are widely spread out, and most of their rocks are much richer in alkali minerals than those of typical subduction zone volcanoes.

Other researchers relate the volcanism to crustal extension. They point to an enormous swarm of large dikes, vertical fractures filled with magma, that trends northeast from south-central Idaho through southwestern Montana. Because those dikes fill fractures, we can be sure that the Earth's crust at that time was stretching in a direction perpendicular to their trend. Since crustal stretching can reduce pressure on hot rocks at depth, it's possible that at least some of the Eocene volcanic centers may have been triggered by decompression melting of the still-hot Cretaceous granitic rocks when the thick stack of Sevier thrust sheets collapsed. Perhaps the crust of the Sevier thrust sheets relaxed as the subduction zone jumped west when new continental masses were added to the west coast of North America. As would happen when a giant bulldozer backs away from a pile of dirt, the crust that had been thickened by the Sevier deformation would have collapsed as the compression relaxed. Whatever their origin, big, explosive stratovolcanoes spewed out thick piles of volcanic debris, especially in central Idaho, southwest Montana, and northwest Wyoming. The climate was tropical in Montana at the time, and the lush vegetation growing near the volcanoes was rapidly buried by volcanic debris, permeated with silica-rich water, and turned into petrified wood.

Volcanism continued in parts of western Montana into Oligocene and Miocene time, but the magma was now basaltic. Some geologists interpret the chemical change as evidence that the subducting plate was progressively sagging into the mantle, stretching

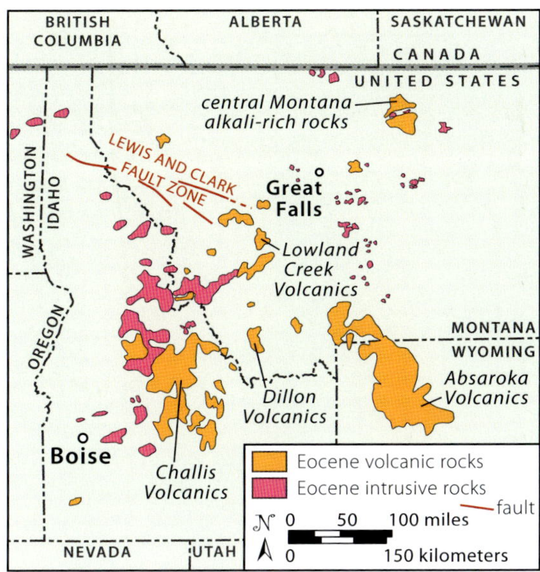

Intrusive and volcanic Eocene-age rocks.

and causing extension in the overlying crust, therby allowing basaltic magma to rise directly from the mantle. Those volcanoes are now largely eroded away, but lava, ash, and debris flows deposited in stream valleys are preserved. Later tectonic extension left many of the old lava-filled stream valleys behind as topographic highs because the tough volcanic lava flows resisted erosion better than surrounding sediments. What were once topographic lows became high ridges on the modern landscape.

Extension at this time formed some of the valleys in westernmost Montana, including the Big Hole, Bitterroot, Mission, Flathead, and Kishenehn, which progressively filled with thick stream, alluvial fan, and lake sediments of the Renova and Kishenehn Formations. To the east, Renova sediments also accumulated in less tectonically active basins.

Beginning about 17 million years ago, extension produced north-to-northeast-trending basins and ranges in parts of western Montana when the Pacific plate began moving northwest, away from the North American plate. This northward shearing of the continental margin formed the infamous San Andreas fault of California. Farther inland, the movement stretched the continent, especially in Nevada and Utah, thinning the continental plate and permitting the rise of the hot asthenosphere. The heat caused the brittle lithosphere to expand and thin, forming high-standing ranges separated by broad basins—the Basin and Range Province. The western edge of that spreading is the Sierra Nevada of California; the eastern edge is the Wasatch Front at Salt Lake City. In Montana, a narrower zone of spreading stretches from the western edge of Yellowstone National Park through southwest Montana to Helena, and on up to the Flathead Valley.

Around 17 million years ago, the Yellowstone hot spot emerged in northern Nevada, causing the crust to thermally expand, stretch, and crack over a broad region of the western United States, including in Montana. The ancestral Missouri River developed and flowed north off this crustal bulge in Miocene time. It followed several

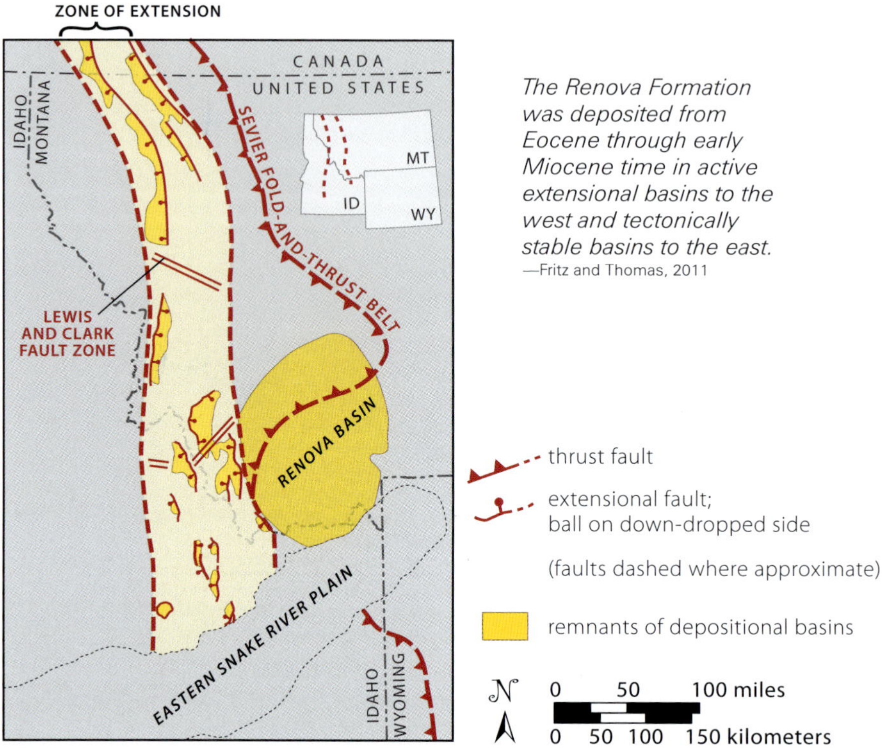

The Renova Formation was deposited from Eocene through early Miocene time in active extensional basins to the west and tectonically stable basins to the east.
—Fritz and Thomas, 2011

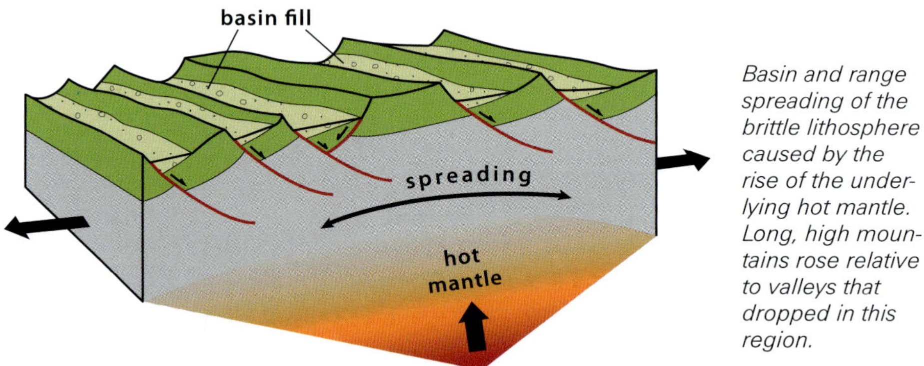

Basin and range spreading of the brittle lithosphere caused by the rise of the underlying hot mantle. Long, high mountains rose relative to valleys that dropped in this region.

northeast-trending extensional basins into southwest Montana, filling them with gravel, hot-spot volcanic ash, and basalt of the Sixmile Creek Formation.

As the North American plate moved slowly (roughly 2 centimeters per year) to the southwest, the hot spot and its thermal bulge moved northeastward, closer to southwest Montana. Around 4.4 million years ago, changes in crustal stresses from the encroaching hot spot caused the crust to break into northwest-trending basins and ranges. New streams developed and mountains were raised in the path of the old drainages, creating the Missouri River drainage we see today. Gravel-covered benches that slope gently toward the valley bottoms are Pliocene pediment surfaces (old valley

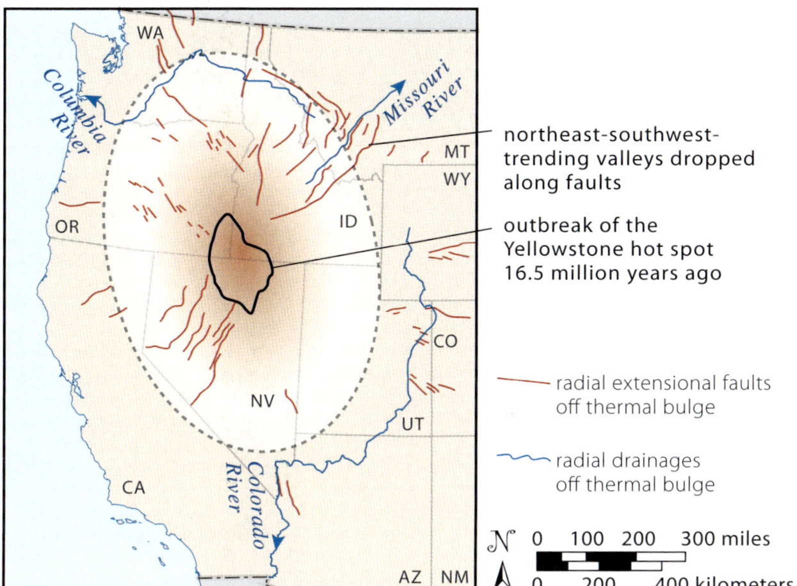

Path of the ancestral Missouri River starting around 16.5 million years ago.
—Modified from Sears and others, 2009

floor surfaces) eroded by sheetlike floods during intense desert storms. The benches were eroded as the land continued to rise and the climate became wetter during Pleistocene time.

Crustal extension is still happening in Montana, and earthquakes pose a real threat to the many older communities in the state, with their buildings constructed of inflexible rock and brick. Montana's many earthquakes make it the fourth-most seismically active state in the union. The fourteenth-largest earthquake in the contiguous United States occurred in 1959 at Hebgen Lake. The 7.5 magnitude temblor caused a massive rockslide that buried a campground and blocked the Madison River, forming Earthquake Lake.

Pleistocene Glaciers and Landscapes

During the Pleistocene Epoch, enormous glaciers formed in high mountains everywhere, and ice sheets formed on flat land in northern Europe and central Canada. They repeatedly advanced and retreated. Over the last 11,700 years, the Earth has been in an interglacial period—the Holocene Epoch.

Glaciers covered many of the mountains in Montana and a large portion of the northern plains, but the big valleys outside of northwest Montana remained free of ice except for the Paradise Valley, south of Livingston, which hosted a tongue of ice fed from a 3,000-foot-thick ice cap over the Yellowstone Plateau. Advancing glaciers dammed the Yellowstone, Missouri, and Clark Fork Rivers, impounding huge glacial lakes. Glacial meltwater formed enormous alluvial fans, such as the geologically famous Cedar Creek alluvial fan, about 10 miles east of Ennis. Roaming the valleys and plains of Montana were Columbian mammoths, dire wolves, saber-toothed cats, horses, camels, and bison. Humans arrived as early as 13,000 years ago. The valleys

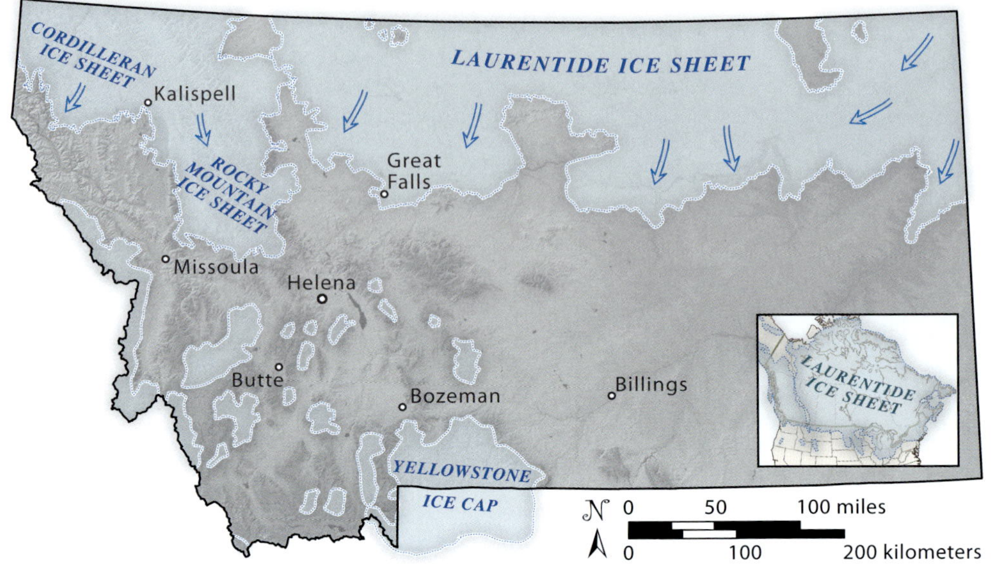

Maximum ice coverage in Montana and North America (inset) during Pleistocene time.
—Modified from Fullerton and others, 2004; 2012; Vuke, 2015

were much wetter than today, with lush grasses in the uplands and cottonwoods and birches along the rivers.

People often wonder how wet or how cold it was during these times. Most estimates place the drop in mean annual temperature during the last ice age in the range of a few degrees—noticeable, but nothing that our wardrobes couldn't handle. All of us are, after all, direct descendants of people who camped out all their lives, generation after generation, during the last ice age.

Much of our modern landscape shows, in one way or another, the mark of the ice. Our high mountain peaks owe their jagged form to glacial sculpting, and the lower valleys in those mountains, as well as large expanses of nearby low areas, contain deep deposits of glacial till and outwash gravel. Continental ice creeping out of central Canada covered most of central and eastern Montana north of the Missouri River and left moraines, deposits of sediment that mark the ice margin's exact outline.

Like any carving tool, glaciers make a distinctive cut. Glaciers straighten mountain valleys and deeply gouge them into something approximating a U-shaped profile. The heads of those valleys end in deep scoops, called cirques, that, from a distance, look as though some mythical giant wielding an ice cream scoop gouged a helping of rock out of the top of the peak. Lakes nestled in cirques are called tarns.

Where several glaciers gouge cirques into the top of a mountain, they reduce the peak to a gnarled pinnacle of rock called a horn—as in Switzerland's Matterhorn. Thick, heavy ice converted the higher ranges of northwest, southwest, and central Montana into serrated rows of craggy horn peaks, each one dropping off into cirques that lead downward into deeply gouged valleys. Many of those glaciated valleys are so straight that you can stand in their lower reaches and look directly at the distant peaks that spawned the glacier. Unglaciated river valleys never provide such long views.

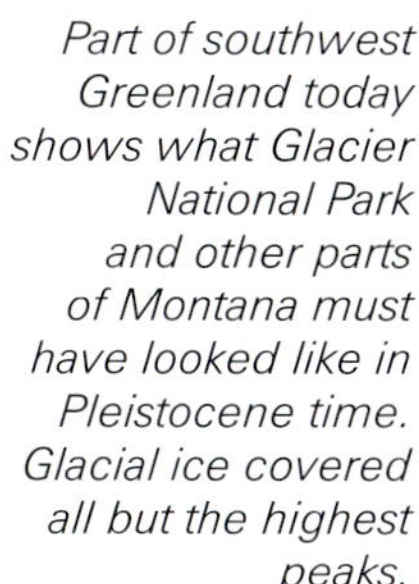

Part of southwest Greenland today shows what Glacier National Park and other parts of Montana must have looked like in Pleistocene time. Glacial ice covered all but the highest peaks.

Iceberg Lake in Glacier National Park is a tarn in a glacial cirque. —Courtesy of Rod Benson, Bigskywalker.com

As mountain glaciers descend their valleys, and as continental glaciers move farther south, they advance into warmer climates. Eventually, they reach an area where the climate is warm enough to melt the ice as fast as it moves forward. That balance establishes the end of the glacier. Ice continues to move down its valley, carrying its load of rocks and sand. Once the conveyor belt of ice reaches the downstream end, where melting keeps up with forward movement, it deposits its load of sediment in a lumpy deposit called till. This process is continuous. Till, a chaotic mass of rocks and dirt mixed together, looks like something a bulldozer scraped up and dumped. Till is often dotted with large boulders that are not from the local bedrock. Such far-traveled rocks are called glacial erratics, which is derived from the Latin *errare*, which means "to wander."

Any deposit of glacial till is a moraine. Mountain glaciers deposit a bench-like lateral moraine along the valley wall near their lower sides and a hummocky terminal

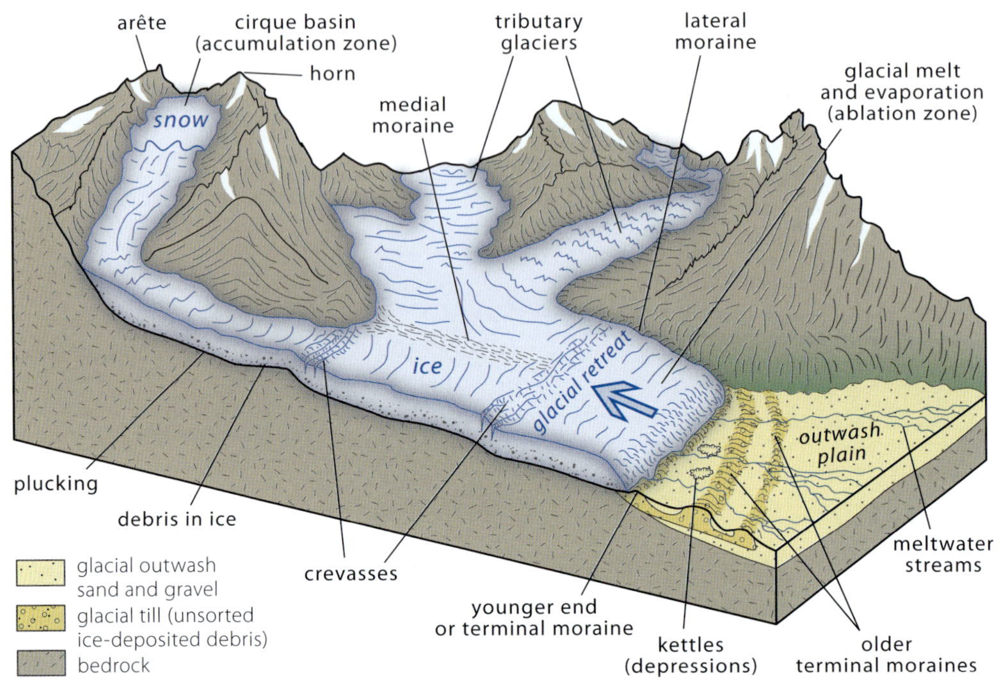

The typical depositional features of a glacier.

moraine across their front. As the climate warms, less snow collects on the upper slopes and the glacier thins enough that it can no longer flow downslope. It stagnates and gradually melts, dropping whatever dirt, sand, and rocks it carried; that deposit is called ground moraine. Continental ice sheets like the one that covered much of the northern part of central and eastern Montana are not confined within valley walls, so they leave only terminal moraines that trace the old ice front for hundreds of miles, along with thin ground moraine.

As long days full of sunshine mellowed the ice-age summers, vast torrents of water poured off the melting glaciers, carrying enormous loads of sediment melted out of the ice. Meltwater rivers crossed the terminal moraines and dumped their loads of sediment downstream as deposits of well-sorted sand and gravel called outwash. Outwash floors the lower parts of Montana's glaciated mountain valleys, and there are ribbons of gravel on the plains.

For many years, most geologists agreed that during Pleistocene time there were four great ice advances separated by interglacial episodes, in which most of the land ice melted. Evidence recently obtained from deep-sea cores shows that there were many ice advances, perhaps as many as twenty. In Montana, we see clear evidence of just two major episodes of glaciation, both late in Pleistocene time. In the mountainous regions of Montana, the older and slightly larger alpine ice advance is called the Bull Lake glaciation, the younger the Pinedale glaciation. On the plains, where the large Laurentide continental ice sheet flowed south from Canada, the older event is called the Illinoian, and the younger the Late Wisconsin.

Age data from the glacial deposits show that the Bull Lake glaciers lasted from about 160,000 to about 130,000 years ago, with the ice reaching its maximum extent around 136,000 years ago. The Illinoian is roughly the same age, ranging from about 190,000 to 130,000 years and reaching its maximum around 140,000 years ago. Glaciers of the Pinedale glaciation lasted from about 30,000 to 12,000 years ago, reaching their maximum extent between 19,000 and 15,000 years ago. On the plains, the Late Wisconsin ice lasted from around 35,000 to 11,700 years ago, reaching its maximum around 23,000 years ago.

Near the end of Pleistocene time, a milder climate caused glaciers to retreat around the globe. Between 10,000 and 6,000 years ago, a period of time known as the Holocene Climate Optimum, temperatures warmed significantly, and most of the glaciers in Montana melted completely. Vegetation returned to landscapes once covered by ice, and pollen studies show that vegetation in mountainous terrain transitioned from tundra to spruce forests and eventually pine as the climate became warm and dry. In the valleys, alluvial fans and pediment surfaces supported grass and even sage in drier parts of the state. Birch, willow, and cottonwoods persisted along the river corridors.

Climatic fluctuations continued throughout the Holocene, with several warming and cooling periods. Following the Medieval Warm Period, a warming between roughly AD 950 and 1250, the climate cooled a few degrees from AD 1300 to 1870, a period referred to as the Little Ice Age. Surviving glaciers in places such as Glacier National Park grew in size, and small glaciers redeveloped elsewhere in Montana. All of the glaciers started melting again during the last century, leaving moraines, rock glaciers (glacial masses choked with rocks), and some small glaciers tucked high

Grinnell Glacier is an example of Glacier National Park's shrinking glaciers.
—Courtesy of T. J. Hileman (1938) and Lingsey Bengtson (2009), US Geological Survey

up in drainages. Glacial ice retreat has more recently sped up as a result of human-induced climate change. Grinnell Glacier, the most-visited glacier in Glacier National Park, has lost nearly half its footprint in the last fifty years. Another ice age may return to Montana, but not for thousands or millions of years—definitely not in our lifetimes. Go see the glaciers in Glacier National Park while you still can.

Anthropocene

In recent years, some scientists have suggested that, due to the impact humans have had on the planet, we've entered a new geologic time that should be called the Anthropocene. Though Montana is still relatively wild and untouched relative to some other parts of the world, its landscapes, too, show signs of human modification.

By the 1860s, Montana had been discovered by those seeking gold, copper, silver, and other economic minerals. Most were pulled from deposits that formed during Cretaceous and early Tertiary time. The people who came built communities and provided needed resources for growing markets, but they also tore up the landscape and left a mixed legacy. From historic mine waste to streams turned inside out, the Treasure State has a long history of environmental impacts caused by humans.

Not all the rocks and landforms in Montana are natural. In mining areas, in particular, be on the lookout for ways humans have modified the landscape. Some changes are extremely obvious, such as tailings piles and mining pits. Other changes are more subtle—water pollution from acid mine drainage, saline seeps in agricultural fields, and straightened river channels. Despite these changes, much of Montana looks like it did 1,000 years ago. On your travels, imagine what it might have looked like 10,000 years ago, 100,000 years ago, or even 1 billion years ago.

Acid mine drainage from the lower Elkhorn Mine adit (tunnel), just west of Coolidge in the Pioneer Mountains of Southwest Montana.

NORTHWEST MONTANA

GLACIATED VALLEYS IN FAULTED BELT ROCK

Belt Supergroup sedimentary formations dominate northwest Montana. These sediments were deposited in a basin (Belt Basin) on the Archean and Proterozoic basement of the North American continent between 1.47 and 1.4 billion years ago. At the time, most of the world's continents were assembled into one supercontinent called Nuna. The origin of the Belt Basin is not well understood, but it formed through crustal stretching, or extension, and was located well within the supercontinent. The basin was huge, and some sedimentary deposits covered large areas of it. Today the Belt rocks cover about 77,220 square miles in western Montana, northern Idaho, and eastern Washington, with even more in southern British Columbia and Alberta, where rocks of the Belt Basin are called the Purcell Supergroup.

As the basin slowly subsided, sand, silt, clay, and limestone accumulated along the basin's margins and within the shallow inland sea that filled it, reaching a total thickness of 11 miles (59,000 feet). The deeply buried sediment compressed and warmed enough to be slightly metamorphosed. Sandstone, siltstone, and claystone were

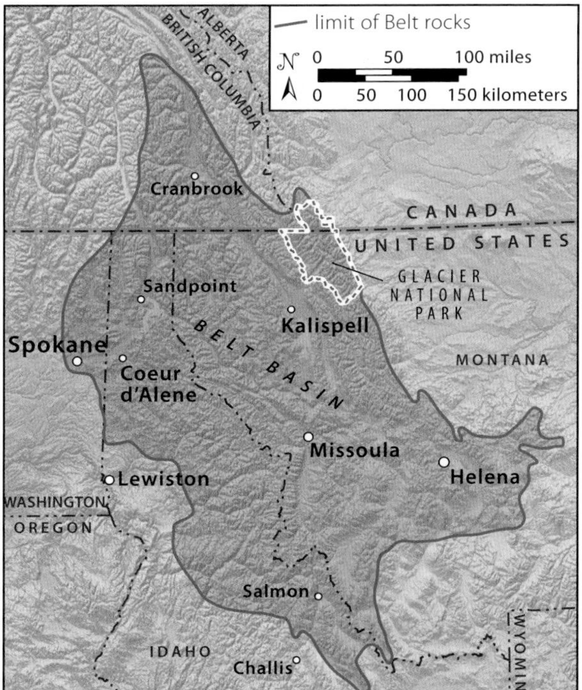

Approximate outline of the Belt Basin based on the distribution of Belt Supergroup rocks. —Modified from Winston, 1989

transformed into quartzite, siltite, and argillite. An eastward arm of the basin called the Helena Embayment reached well into central Montana, while coarse conglomerate shed off highlands to the south (the Dillon block) mark the basin's southeastern margin. Sandstone lenses derived from eroding highlands to the southwest show that there was land surrounding most, perhaps all, of the Belt Basin.

Belt sediments were deposited during a time very unlike the one we know today. Imagine a hot, hostile world with an unbreathable carbon dioxide–rich atmosphere and a barren landscape without animals, plants, or soil. The absence of vegetation and soil made it easy for wind and water to transport sediment for long distances. Giant aprons of alluvium spreading beyond the highlands were episodically flooded by water that deposited sediment over vast distances. Waning flow formed asymmetrical ripples, and sediment in shallow standing water shifted back and forth to form symmetrical ripples. Where the water receded, clay-rich sediments dried to form mud cracks, and raindrops from passing storms occasionally pockmarked the mud. Subsequent floods or storms ripped up clay chips, some of which were rolled into mudballs during transport. When water evaporated, it sometimes left behind salt crystals, now preserved in mudstone as square impressions about one-eighth inch

Symmetrical ripples from wind-generated currents in the Snowslip Formation (Missoula Group) in Glacier National Park.

Mud cracks in the Snowslip Formation of Glacier National Park formed when clay-rich sediments dried, contracted, and cracked.

Cube-shaped salt casts in the Grinnell Formation (Ravalli Group) in Glacier National Park formed when water evaporated on a mudflat.
—Courtesy of Marli Miller

across. In deeper parts of the basin, silt and clay slowly settled from the water, depositing dark, organic-rich laminated sediments.

These sedimentary structures are exquisitely preserved in the Belt rocks because there were no animals around to churn the sediment while searching for food. But the Belt Basin was anything but lifeless. The water teemed with cyanobacteria, a primitive form of single-celled plant life. The microbes, which still exist today in environments where little else can survive, tend to grow in mats that trap and cement particles of sediment together, making a layered, sometimes dome-shaped structure called a stromatolite. Limestone Belt rocks are full of them. The cyanobacteria helped make the limestone by consuming carbon dioxide, thus reducing the acidity of the water around them, and then precipitating calcium carbonate.

Cross-sectional view of a dome-shaped stromatolite in the upper part of the Wallace Formation (Piegan Group) in Glacier National Park.

Based on the distinctive characteristics of Belt Supergroup rocks, geologists have divided the supergroup into four groups of formations (from oldest to youngest): Lower Belt Group, Ravalli Group, Piegan Group, and Missoula Group. Each group is made up of several formations, which change in character across the basin, resulting in different formation names in each group depending on the rock's location in the basin.

Distinguishing formations in the Belt Supergroup can be difficult due to the lack of fossils to help geologists identify and correlate the rocks from outcrop to outcrop. The rock colors of the Belt are striking, but coloring is often caused by chemical changes that occur after sediment is deposited. However, color can be imparted during deposition, and it can help us better understand the original environment the sediment was deposited in. Regardless of origin, color is a distinctive characteristic of Belt rocks, and it can be used to differentiate formations.

Black and dark-gray colors, such as those found in some Lower Belt Group rocks, can be caused by organic carbon and iron sulfide minerals, especially in shale deposited in low-oxygen environments. Many of the other colors come from tiny amounts of iron oxides. Ferric iron minerals, such as hematite, color rock red, brown, or purple, whereas ferrous iron colors rock gray or green. Hydrous ferrous oxide minerals, such as limonite, color rock yellow or brown. These are common colors in many Ravalli and Missoula Group rocks. Dolomitic limestone, like that in the lower Piegan Group, can be tan due to weathering and the replacement of magnesium with iron.

The earliest Belt sediments in northwest Montana, the Prichard Formation (Lower Belt), laid down on Archean basement rocks, are mostly monotonous dark-colored mudstones and siltstones that were deposited in relatively deep water. Diabase sills (basalt) that were injected into the Prichard are related to the continental extension that created the Belt Basin itself, so they are found throughout these old sedimentary rocks. Throughout northwest Montana, the Belt formations contain numerous dikes and sills composed of diabase, a black igneous rock identical to basalt except that its mineral grains are large enough to see without a microscope. Wherever geologists find evidence of continental extension, swarms of diabase dikes and sills are part of the picture. Some of the basalt magma that rose from the Earth's mantle in response to the extension squirted into the continental crust as it broke apart, and some even erupted onto the surface as basalt lava flows.

The dikes and sills in the upper parts of the Prichard Formation were obviously injected after the sediments they intrude were deposited; in fact, some of the magma boiled its host sediments before solidifying. Most of the Belt sills formed early in the deposition of the Belt, a bit more than 1.45 billion years ago, though many are younger. For example, the well-known Purcell Sill, the prominent dark band visible in high cliffs of the Piegan Group in Glacier National Park, was injected around 780 million years ago, well after the deposition of the Belt rocks. Some magmas generated

Red, green, yellow, and even white sediments in the Belt Supergroup —Courtesy of Marli Miller

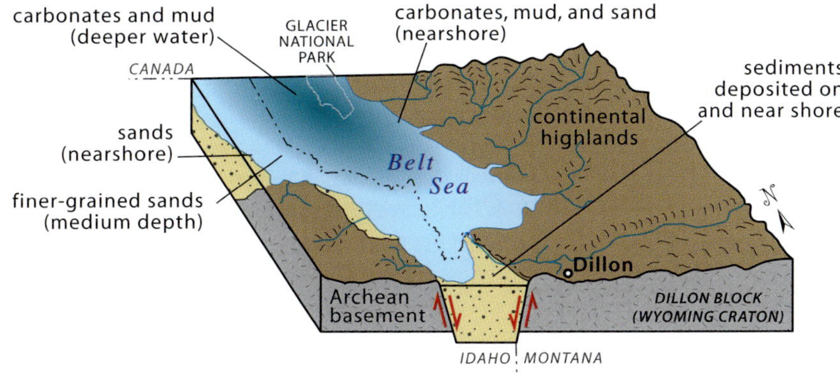

Generalized reconstruction of the Belt Basin during deposition of the Piegan Group. Sediment deposition varied laterally at any given time, depending on water depth and distance from shore. —Modified from Winston, 1989

STRATIGRAPHIC COLUMN FOR PROTEROZOIC BELT SUPERGROUP

ERA	GROUP	NORTHWEST MONTANA		GLACIER NATIONAL PARK	SOUTHWEST MONTANA		CENTRAL MONTANA	
MIDDLE PROTEROZOIC	Missoula	Libby Formation (6,000-7,550 ft)	Garnet Range Formation (0-8,200 ft)		Pilcher Formation	(Missoula Group 1,000 to 10,700 ft)		
			McNamara Formation (100-5,415 ft)	McNamara Formation	Garnet Range Formation			
		Bonner Formation (500-1,900 ft)		Bonner Formation			no rocks	
		Mount Shields Formation (1,000-6,560 ft)		Mount Shields Formation				
		Shepard Formation (600-3,600 ft)		Shepard Formation				
		Snowslip Formation (0-5,450 ft)		Snowslip Formation				
	Piegan	Wallace Formation		Wallace Formation (1,000-7,000 ft)	Wallace Formation (0-3,000 ft)			
		Helena Formation		Helena Formation (330-3,000 ft)	Helena Formation (3,000 ft)		Helena Formation (4,000 ft)	
	Ravalli	St. Regis Formation (1,000-3,000 ft)		Empire Formation (0-2,000 ft)	Empire Formation (1,150 ft)		Empire Formation (790 ft)	
		Revett Formation (500-2,500 ft)		Grinnell Formation (2,600-3,800 ft)	Spokane Formation (230-460 ft)		Spokane Formation (5,000 ft)	
		Burke Formation (2,500-7,500 ft)						
	Lower Belt	Prichard Formation (16,400 ft)		Appekunny Formation (1,000-3,000 ft)	LaHood Formation (3,000-7,850 ft)	Greyson Formation (0-4,265 ft)	LaHood Formation (7,000-10,000 ft)	Greyson Formation (5,000-7,900 ft)
				Altyn Formation (2,300-2,600 ft)		Newland Formation (0-6,500 ft)		Newland Formation (8,000 ft)
								Chamberlain Formation
								Neihart Quartzite

Groups and formations of the Belt Supergroup in Montana, arranged by region.

during the deposition of the Belt escaped onto the submerged sea bottom, cooling into pillow-lava flows, such as those in the Missoula Group (Snowslip Formation), which are well exposed above Granite Park Chalet in Glacier National Park.

Like the Purcell Sill, younger diabase sills intrude Belt formations along the Rocky Mountain front down to Rogers Pass, and in the Garnet Range east of Missoula. These sills are 780 million years old and were intruded during rifting of a supercontinent called Rodinia near the end of Proterozoic time, long after the Belt Basin had filled and its rocks had been deeply buried.

The considerable extra load of hundreds of feet of thick diabase sills, which were 20 to 25 percent heavier than the basin's sediments, pushed the Belt Basin down, deepening the water in the Belt Sea. After the Prichard was deposited and its sills had intruded, the basin stabilized and additional sediments quickly filled the basin so that the younger Belt sediments were deposited in much shallower water.

Belt sediments are thick along the western margin of the basin and thin progressively eastward. The basin's rocks abruptly end to the west because continental rifting of the supercontinent Rodinia split the basin around 750 million years ago. The

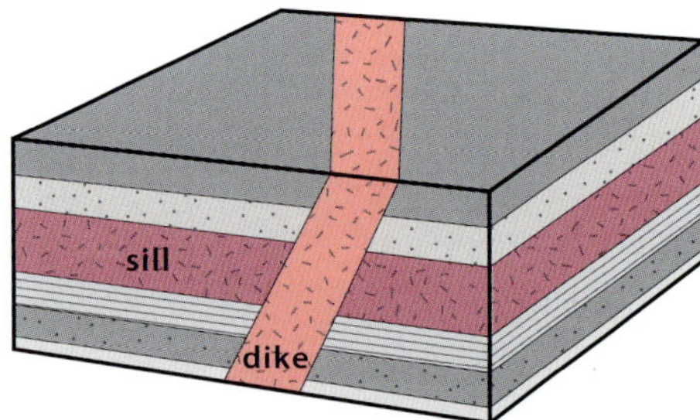

Dikes are intrusions of igneous rocks that cut across older rock layers, and sills are intrusions that parallel older rock layers.

The Purcell Sill is dark diabase that intruded parallel to lighter-colored limestone layers of the Wallace Formation (Piegan Group) in Glacier National Park about 780 million years ago. White bands of marble on either side of the diabase sill were limestones baked and bleached by magma of the sill. —Courtesy of Marli Miller

resulting new edge of the continent at that time trended from northeastern Washington south through western Idaho and eastern Oregon. That line was to remain the west coast of North America until almost 200 million years ago.

Where is the missing western piece of the Belt Supergroup? Continental crust is too light to sink into the mantle, so the piece that split off North America about 750 million years ago must still exist somewhere. Continental basement rocks in part of northeastern Asia closely resemble those in the northern Rocky Mountains, and they are the same age. The Proterozoic sedimentary rocks of central Siberia also look like the Belt rocks of the northern Rocky Mountains, complete with diabase dikes and sills. Many geologists strongly suspect that the detached piece of North America now forms a large region of Asia, north of China. Other proposals suggest that the missing piece is now in Australia. Stay tuned.

Initial Rise of the Rockies

Long after the Belt rocks had been deposited, most of Earth's continents came together once again to form a new supercontinent, known as Pangaea. It began to break apart not long after it formed, and the North Atlantic Ocean started developing in the rift zone about 175 million years ago. The South Atlantic Ocean opened around 140 million years ago as Africa separated from South America. The relative westward drift of North America led it to collide with the Pacific seafloor of the Farallon plate. This caused the leading edge of the less dense continent to crumple, buckle, and rise as the denser ocean floor sank beneath it in a subduction zone. Those events led, ultimately, to the initial rise of the Rocky Mountains, parallel to the deforming continental margin, from about 90 to 55 million years ago, in Late Cretaceous and earliest Tertiary (or Cenozoic) time.

Just west of Montana, rocks of the Spokane dome in northern Idaho and northeasternmost Washington were intensely sheared, forming a rock called mylonite, and pushed eastward along a major shear zone at great depth and at high temperatures. Big slabs of mostly Proterozoic and Paleozoic sedimentary rocks in Montana also moved east along thrust faults, so it seems likely that the deep shear zone continues under these thrust faults. Radiometric age dates suggest this was happening about 75 million years ago.

The slabs that moved at this time formed the Montana Overthrust Belt. This belt of thin-skin Sevier folds and thrusts at the eastern edge of northwest Montana makes

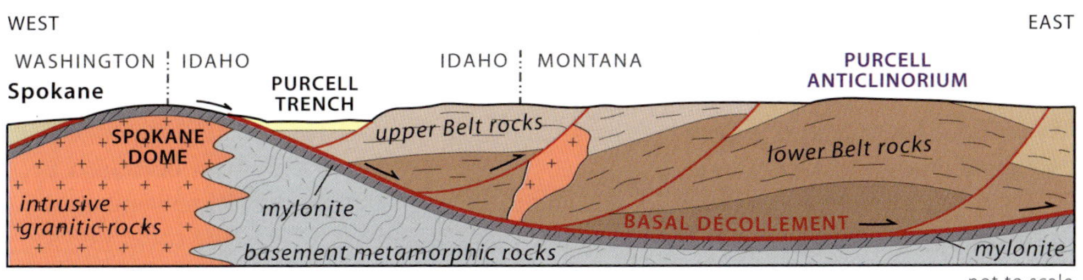

Cross section through Northwest Montana along a line approximately from Spokane to Eureka. Thick slabs of rock moved from west to east, likely along a deep shear zone that separates underlying higher-grade metamorphic rocks from lower-grade thrust slabs above.

up the spectacular high ranges of the Rockies from Glacier National Park south to the Helena area. *Overthrust* refers to the older Proterozoic and Paleozoic sedimentary rocks that were thrust eastward up and over younger Cretaceous sedimentary rocks. Individual fault slices of rock were stacked slab on slab like shingles on a roof. Southeast of the Overthrust Belt the same faults are more subdued and form lower and less rugged mountains.

In northwest Montana east of the Spokane dome, that eastward movement of much of the sedimentary pile involved old, hard Belt rocks that were deeply buried at the bottom of the pile. The drag on rocks beneath them caused big sections of Belt rock to shear upward from a basal fault, or décollement, to the Earth's surface on major thrust faults. The younger Paleozoic and Mesozoic sedimentary rocks that once covered them moved farther east and were then cut by thrust faults carrying Belt rocks that were shoved up and over their trailing ends. For the most part, the thrust faults progressed from west to east over time, and individual faults moved rocks as much as 100 miles to the east of their original positions.

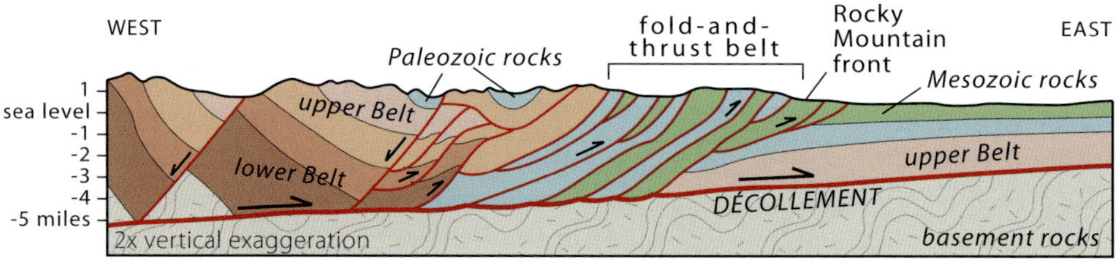

West-east cross section across northwest Montana.

Lewis and Clark Fault Zone

The northwest trend of the rugged mountains and narrow valleys between Missoula and the Idaho border, including a broad area around I-90, MT 200, and the Clark Fork River, forms a discordant transition zone between the long north-trending ranges and valleys to the north and the more broken-up, isolated ranges south of I-90. A broad view of western Montana looks like what you might expect if a gigantic hand had shoved the southwestern quarter of Montana some tens of miles to the east along horizontally moving strike-slip faults (faults similar to the San Andreas fault in California). In essence, the south end of the northern mountains was dragged to the east, pushing those to the south still farther east. So, what gives?

The area of displacement consists of a broad zone of faults, up to 50 miles wide, that trends from northeastern Washington southeast to beyond Helena. This is the long-lived and enigmatic Lewis and Clark fault zone. Faults within this zone are predominantly strike-slip, in which rocks move horizontally against each other along the fault, and dip-slip, in which they move vertically. Some faults can be traced a whopping 150 miles along their length, so this is a major area of lateral movement of rocks in Montana. As might be expected across a major fault zone, the rocks north and south of it are distinctly different. There are very few intrusions of granite north of the fault zone, but there are numerous Cretaceous to early Tertiary granite intrusions, some small to extremely large, within the complex south of the fault zone. A greater variety of rocks and structures also occurs south of the fault zone.

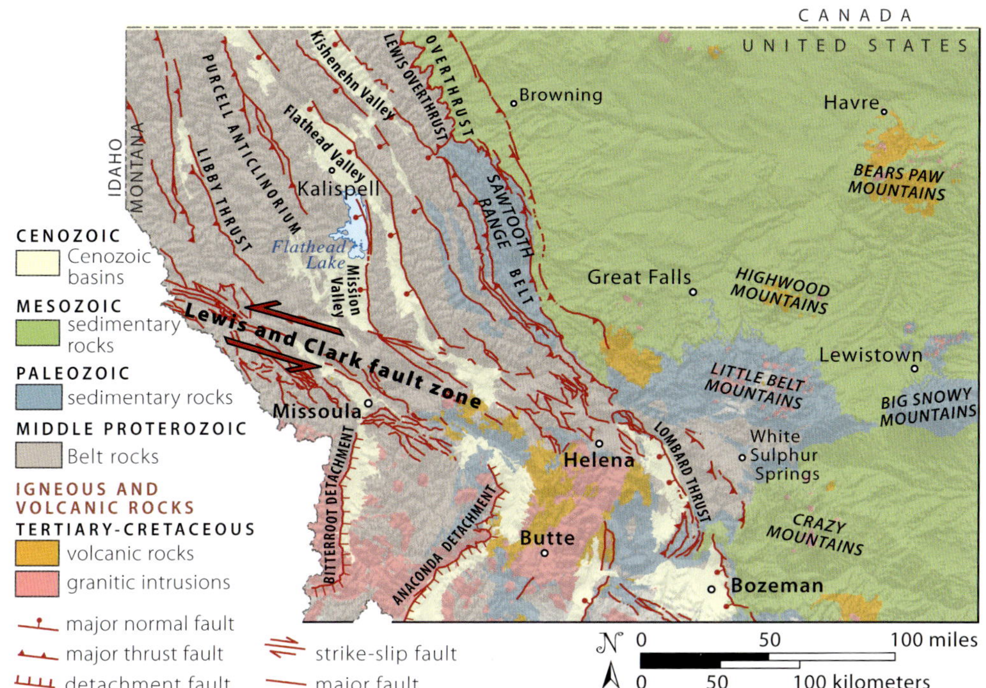

Major faults in western Montana and Cenozoic basins that formed along normal faults.
—Modified from Montana Bureau of Mines and Geology

No one knows the total displacement across all the faults in the entire Lewis and Clark fault zone, but it must be large, probably tens of miles at the very least. Geologists disagreed for years about when and in what direction the zone's faults moved. Imagine looking across one of the faults and watching it move. Does the rock across the fault move to the left (left-lateral fault) or to the right (right-lateral fault)? Most now accept that the overall movement on the faults has been to the left, making it a left-lateral fault zone. That is also the direction the fault zone seems to displace the old continental margin.

The faults break folds that appear to have formed while the region's big masses of granite were invading the crust, so they probably moved 70 million years ago. However, some faults in the zone are much older—early Proterozoic in age. But there is more to the story.

In the Coeur d'Alene mining district of northern Idaho, some of the faults of the Lewis and Clark fault zone cut igneous rocks that are only 50 million years old and moved fault blocks on the south side of them to the west, so it seems that those faults reversed their direction of movement. One of the faults in the Missoula Valley breaks rocks that are probably no more than about 25 million years old, so it was active until at least then. Faults to the north and to the east of Missoula, however, are still moving slightly, accompanied by small-to-medium-sized earthquakes.

Although this fault zone has been around for a long time, it remains active. On July 6, 2017, a magnitude 5.8 earthquake struck southeast of Lincoln, probably the strongest earthquake in Montana since 1964. The quake was on a strike-slip fault only 3 miles underground. Shaking knocked items off shelves and caused relatively minor damage to masonry. One couple was nearly impaled by a falling antler from a rack hung over their bed!

Cenozoic Extensional Basins

Sometime after the events of 50 million years ago, a new generation of faults began to pull apart the tortured rocks of the Northern Rockies. This extension cracked and thinned the crust so that some fault blocks moved up to form mountain ranges, whereas others moved down to form broad basins and valleys. Thinning of the crust caused the underlying hot asthenosphere to rise, along with the whole region. Sections of crust that drop and are bounded by faults on at least two sides are called grabens, which form valleys; if the block drops along a fault on only one side of its valley, it forms a half graben. Most of these faults trend northwest, a few northeast. Many are still active because they occasionally generate earthquakes. This change in the tectonic picture happened as the compressional forces generated in the subduction zone along the west coast decreased. New large basins formed in northwest Montana during this time, including the Mission, Flathead, and Kishenehn Valleys, and filled with Tertiary stream and lake sediments. (See map on page 33.)

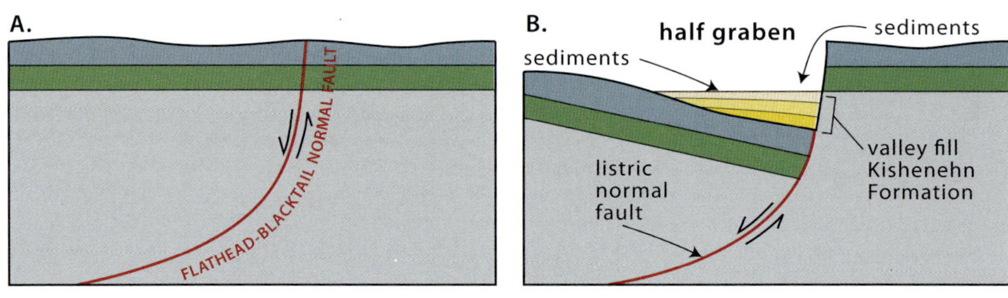

Formation of the Kishenehn Basin during Eocene time along the western margin of Glacier National Park. —Modified from National Park Service

The western front of the Mission Range south of Flathead Lake is a big listric normal fault. The valley first dropped during Cenozoic time, extensional movement that appears to be ongoing today.

The Tertiary sedimentary rocks consist of conglomerate, sandstone, siltstone, limestone, and even coal and oil shale. The limestones likely formed in intermittent ponds and restricted lakes and might record a relatively dry climate. However, abundant petrified wood shows that the climate was wet enough to support plant life. Fossils of freshwater invertebrates, fish, and mammals, including primates, have been found. Volcanic ash in these sedimentary layers was likely derived from explosive volcanoes in the Western Cascades of Oregon and Washington.

Exposures of the Tertiary Kishenehn Formation along the Middle Fork of the Flathead River have yielded some of the most exceptionally preserved insect fossils in the world. Around 46 million years ago, a north-south-trending lake formed in an extensional basin along what is now the western margin of Glacier National Park. A pile of sediment more than 1-mile-thick accumulated in the lake, including carbon-rich oil shale that preserved the insects. Researchers from the Smithsonian Institution collected more than 16,000 fossil insects, including many groups of flies, dragonflies, butterflies, termites, and even those annoying mosquitoes!

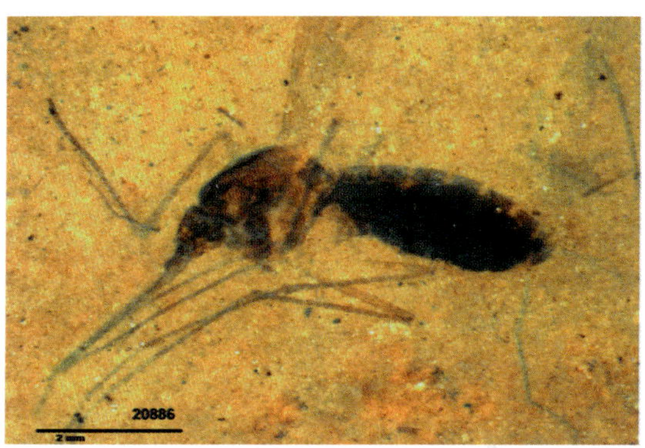

A fossilized female mosquito in a paper-thin piece of shale in the Kishenehn Formation. The 46-million-year-old insect drew blood for its last meal, was blown into a lake in what is now northwest Montana, and sank, belly still full. —D. Greenwalt Photo, Courtesy of Smithsonian Institution

By 20 million years ago, the oceanic Farallon plate beneath the Pacific Ocean began moving northwest relative to the North American plate. This northwestward shearing has stretched the continent, forming basin and range topography, which extends all the way from Nevada into northwest Montana (see the book's first chapter for more on this topic). The Basin and Range Province comprises linear mountain ranges and valleys separated by normal faults; the mountains are rising relative to the dropping valleys. This spreading is still active as far north as Kalispell, evidenced by frequent small earthquakes in the area. We know that large earthquakes remain possible. The fact that the steep western fronts of the Mission and Swan Ranges come right to the valley floor indicates that the faults have been active relatively recently. The Mission fault north of St. Ignatius shows a 20-foot offset cutting 10,000-year-old Glacial Lake Missoula sediments. That size offset would have formed during a magnitude 7.5 earthquake.

Glacial Lake Missoula and Its Floods

During the Pleistocene ice age, the Laurentide ice sheet spread outward from the area of Hudson Bay, and the smaller Cordilleran ice sheet covered British Columbia. The two giant ice sheets were separated by the high peaks of the Canadian Rockies. The Cordilleran ice sheet flowed east toward the plains, so it may have butted up against the Laurentide ice sheet at times. The southern fringes of the Cordilleran ice sheet spread long fingers down the big north-trending valleys. One lobe poured down the Purcell Valley of northern Idaho as far as Sandpoint and Lake Pend Oreille. In doing so, the 20-mile-wide glacier crossed the path of the Clark Fork River, blocking it with ice some 2,000 feet thick.

That ice dam backed up water all the way upstream into the Missoula Valley, forming massive Glacial Lake Missoula, and sent fingers of water into all of the surrounding valleys. As the water rose higher against the ice blockage, it ultimately crept under the ice, finally floating the glacier and flushing downstream in a catastrophic flood over southeastern Washington. Glacial Lake Missoula and its colossal floods are by now well-known to almost all Montanans, but one hundred years ago that was not the case.

J Harlen Bretz, a geology professor at the University of Washington and later the University of Chicago, spent several years in the 1920s studying highly unusual erosional features all across southeastern Washington. In spite of the dry climate and the lack of modern streams, he quickly concluded that the big channels, potholes, and other features of the region had been eroded by some kind of gigantic flood. His ideas were broadly rejected, especially amongst those in elite Ivy League circles who were focused on the belief that present-day geologic processes demonstrated how past features formed. Catastrophic floods didn't fit their narrative. In addition, Bretz could not provide a source for his flood.

At about that time, US Geological Survey geologist Joseph T. Pardee, working in western Montana, was studying evidence for a huge multi-tentacled lake that had

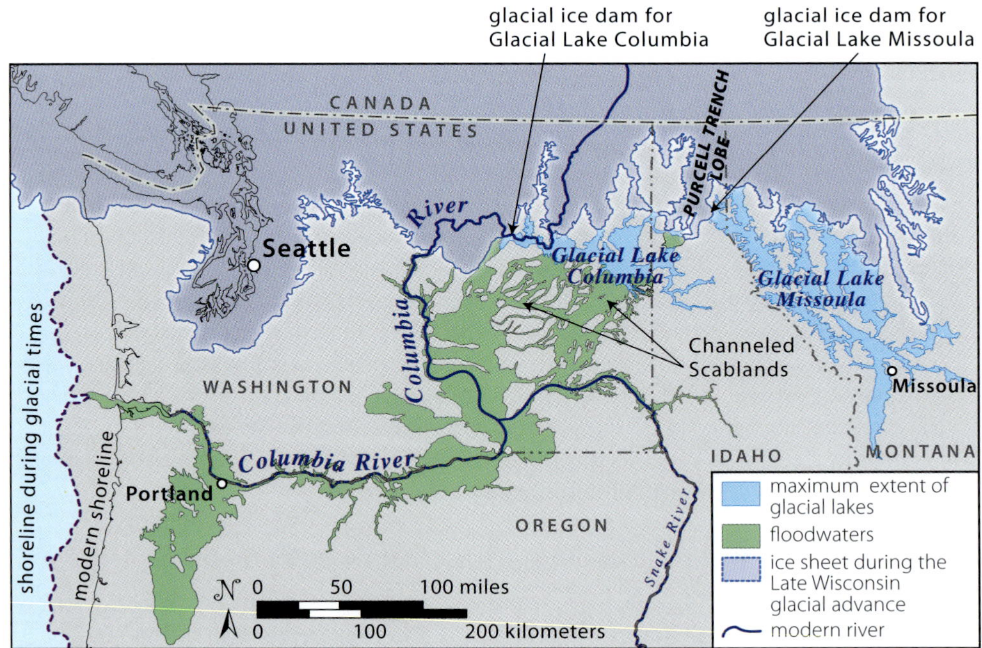

Glacial Lake Missoula (blue) and the Channeled Scablands of the Pleistocene ice age.
—Modified from the Ice Age Floods Institute

been centered on Missoula. Pardee recognized the many strandlines marking high stands of the lake, and gigantic ripple marks and gravel flood bars that formed when the lake drained. He showed that the lake drained in catastrophic floods through the area of Lake Pend Oreille and Spokane, and on downstream to the Columbia River. Suddenly Bretz had the source of his flood and vindication for his catastrophic flood hypothesis.

The highest shoreline at Missoula is at about 4,200 feet. Downstream at the ice dam the water would have been about 2,130 feet deep. Because ice is about 10 percent less dense than water, the ice filling the valley had to have been more than 2,360 feet thick or it would have floated in this depth of water. As the lake continued to fill, at some point it was high enough to float the ice and drain—catastrophically! It seems likely that the ice dam would have floated rather than simply breaking up because the ice would have been tens of miles wide—relatively easy to float but difficult to push away or break up without first being floated.

When the ice dam in northern Idaho floated, Glacial Lake Missoula emptied in a matter of days, releasing the greatest flood known in the geologic record. The volume of the lake was about 500 cubic miles—comparable to that of modern-day Lake Ontario. The latest calculations suggest that in narrow gorges the water may have reached 80 miles per hour, with flow rates of about 2.4 cubic miles per hour.

After the ice dam washed out and the lake had drained, the glacier continued to flow south until it established a new ice dam. Then the impounded Clark Fork River again flooded the mountain valleys of western Montana to form a new Glacial Lake Missoula that deepened year after year until it, too, floated its ice dam and drained. The ice also impounded the Kootenai River farther north to form another ice-dammed glacial lake about 15,000 years ago, during the Pinedale glaciation. The same damming of drainages must have happened during the earlier Bull Lake glaciation, possibly on an even larger scale.

Glacial lakes stored their archives in thin layers of light and dark sediment called varves. Glacial meltwater is typically milky with finely ground rock flour that forms as rocks embedded in the moving glacier grind against underlying bedrock. Rock flour

Glacial Lake Missoula shorelines in Missoula are accentuated by melting snow.

accumulates on the floors of glacial lakes during the summer when large volumes of ice melt. Meanwhile, algae and microscopic animals flourish in the sunlit surface waters of the lake. The coming of winter ends the melting, thus cutting off the supply of rock flour, and the algae and animals that thrived during the long summer days die with the freeze. Their remains settle to the lake floor during the winter to become a layer of dark sediment. Each pair of light/summer and dark/winter layers records the seasons of one year of the glacial lake.

The varves in Glacial Lake Missoula sediments in western Montana show that the earliest and deepest filling of the lake that was preserved lasted 58 years. Each successive filling lasted a shorter period of time, with the final one lasting only 9 years. In all of its 36 fillings during the Pinedale glaciation, the lake existed for fewer than 1,000 years.

The lake likely grew deeper and larger with every year the ice dam lasted. Because each successive filling lasted fewer years, each shoreline is likely younger than the one above it and older than the one below. Therefore, the highest shoreline should correspond to the earliest filling, which lasted for the longest time. The flood deposits in the valleys of eastern Washington reflect the same pattern. As the climate began to warm, the glacier in the Purcell Valley began to thin. As it thinned, it floated in shallower water, so each successive filling of the lake drained at a lower level and after fewer years of existence. Of course, we have no record of any floods whose shorelines were obliterated by later floods.

Glacier National Park

GOING-TO-THE-SUN ROAD

St. Mary—West Glacier

50 miles

Glacier National Park is as geologically spectacular as it is scenic, and there is no better cross section of the park's geology than that along Going-to-the-Sun Road. Building the famous road proved to be a challenge, to put it mildly. Surveying and early construction involved getting to the site and around the cliffs of the Garden Wall on foot. Horses and mules brought in supplies as the trail widened. Eventually Model T dump trucks, steam shovels, and tractors took over. Dynamite was heavily used on the hard, unforgiving Belt rock. Crews started at Avalanche Creek, near the head of Lake McDonald, in 1924, and didn't reach the top of Logan Pass until 1929. The east side, including a 400-foot tunnel through the cliff on the flank of Piegan Mountain, wasn't completed until 1933.

The road transects an enormous thrust slab, today some 2 miles thick, of Belt rocks that were pushed up from the west and moved east at least 80 miles along a nearly horizontal fault surface called the Lewis overthrust. The Belt rocks, 1.47 to 1.4 billion years old, now lie on top of Cretaceous sedimentary rocks. This amazing, almost horizontal fault structure carried almost all of Glacier National Park eastward over the top of the High Plains and over older faults of the Overthrust Belt. The Belt rocks are bent downward into a broad fold called the Akamina syncline, the axis of which is along the Continental Divide. From St. Mary to Logan Pass, the road climbs up-section through the southwest-dipping limb of the fold, which exposes, in succession, the

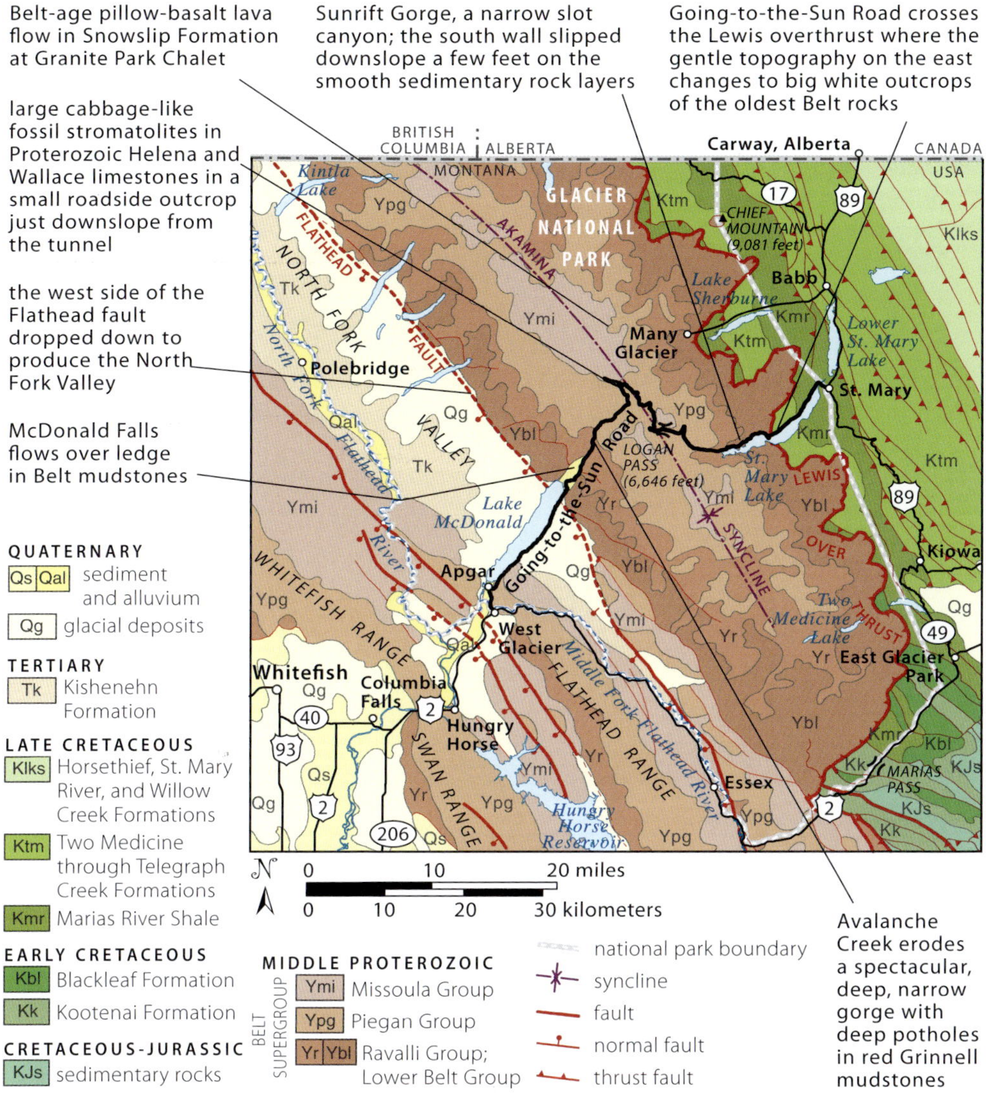

Belt-age pillow-basalt lava flow in Snowslip Formation at Granite Park Chalet
Sunrift Gorge, a narrow slot canyon; the south wall slipped downslope a few feet on the smooth sedimentary rock layers
Going-to-the-Sun Road crosses the Lewis overthrust where the gentle topography on the east changes to big white outcrops of the oldest Belt rocks
large cabbage-like fossil stromatolites in Proterozoic Helena and Wallace limestones in a small roadside outcrop just downslope from the tunnel
the west side of the Flathead fault dropped down to produce the North Fork Valley
McDonald Falls flows over ledge in Belt mudstones
Avalanche Creek erodes a spectacular, deep, narrow gorge with deep potholes in red Grinnell mudstones
BRITISH COLUMBIA
ALBERTA
MONTANA
CANADA
USA
Carway, Alberta
GLACIER NATIONAL PARK
Kintla Lake
FLATHEAD FAULT
AKAMINA
NORTH FORK VALLEY
North Fork Flathead River
Polebridge
CHIEF MOUNTAIN (9,081 feet)
Babb
Lake Sherburne
Lower St. Mary Lake
Many Glacier
St. Mary
LOGAN PASS (6,646 feet)
Going-to-the-Sun Road
St. Mary Lake
LEWIS OVERTHRUST
SYNCLINE
Lake McDonald
Apgar
West Glacier
WHITEFISH RANGE
Two Medicine Lake
Kiowa
East Glacier Park
Whitefish
Columbia Falls
Hungry Horse
Middle Fork Flathead River
FLATHEAD RANGE
SWAN RANGE
Hungry Horse Reservoir
Essex
MARIAS PASS
0 10 20 miles
0 10 20 30 kilometers
QUATERNARY
Qs Qal sediment and alluvium
Qg glacial deposits
TERTIARY
Tk Kishenehn Formation
LATE CRETACEOUS
Klks Horsethief, St. Mary River, and Willow Creek Formations
Ktm Two Medicine through Telegraph Creek Formations
Kmr Marias River Shale
EARLY CRETACEOUS
Kbl Blackleaf Formation
Kk Kootenai Formation
CRETACEOUS-JURASSIC
KJs sedimentary rocks
MIDDLE PROTEROZOIC
BELT SUPERGROUP
Ymi Missoula Group
Ypg Piegan Group
Yr Ybl Ravalli Group; Lower Belt Group
national park boundary
syncline
fault
normal fault
thrust fault

Geology along Going-to-the-Sun Road in Glacier National Park and US 89 east of the park.

WEST-SOUTHWEST
EAST-NORTHEAST
WHITEFISH RANGE
APGAR MOUNTAINS
LIVINGSTON RANGE
Lake McDonald
LOGAN PASS
EAST FLATTOP MOUNTAIN
North Fork Flathead River
Apgar
St. Mary
LEWIS OVERTHRUST
FLATHEAD FAULT
Archean basement
0 5 10 miles
0 5 10 15 kilometers
former thrust fault was reactivated as a normal fault in Tertiary time
thrust fault
normal fault
TERTIARY
Kishenehn Formation
CRETACEOUS
sedimentary rocks
PALEOZOIC
sedimentary rocks
MIDDLE PROTEROZOIC
upper Belt rocks
lower Belt rocks

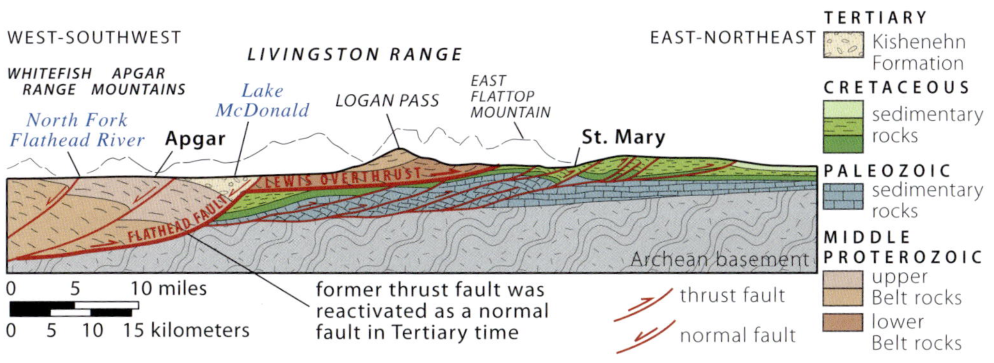

Cross section along the line of Going-to-the-Sun Road.

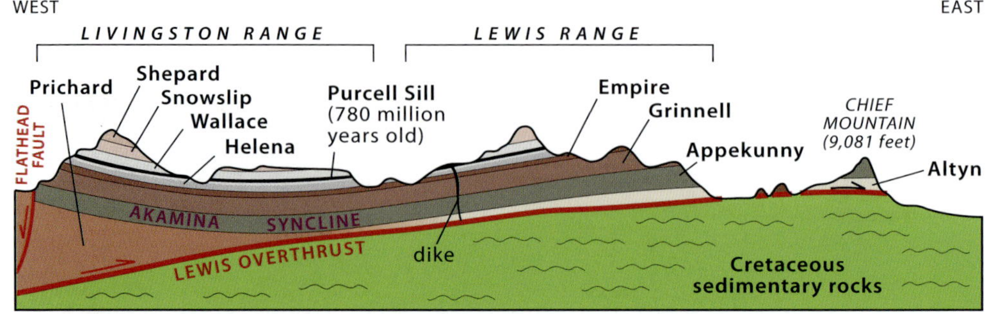

Generalized cross section through Glacier National Park.

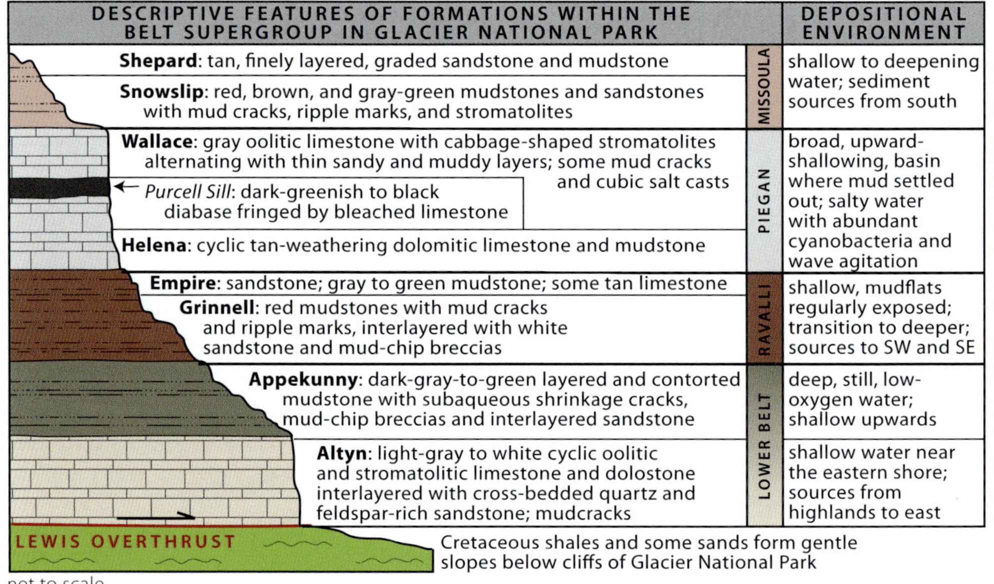

Stratigraphic column for Glacier National Park.

Altyn, Appekunny, Grinnell, Empire, Helena, Wallace (with the Purcell Sill), Snowslip, and Shepard Formations.

The Lewis overthrust occurs near the east end of Going-to-the-Sun Road, at the base of the first big outcrop of Proterozoic Belt rocks a few miles west of the entrance station, but the soft Cretaceous Two Medicine Formation sediments below are not well exposed. The oldest Belt rocks belong to the Altyn Formation. This dolomitic limestone, exposed in a big roadcut just west of the Rising Sun Campground, is tan in weathered outcrops, white in fresh roadcuts, and everywhere full of sand grains. The quartz sand was derived from the continent Laurentia (the eastern two-thirds of Canada and United States), which was to the east, some of it possibly recycled from the Neihart Quartzite, which covered a large part of the continent prior to the formation of the Belt Basin. Some of those sand-sized grains are made of concentrically laminated balls of limestone called *ooids*, an ancient Greek word for "egg"; shallow,

Cross-bedded oolites in sandy limestone of the Altyn Formation. The downslope direction of the cross-bedding layers point in the direction of flow. Different flow directions, such as those seen here, were caused by shifting currents, possibly related to waves or tides. —Courtesy of Marli Miller

nearshore waves formed and shaped them. The Altyn also contains stromatolites, big cabbage-like sedimentary structures made by photosynthetic bacteria that lived in the shallows along the edge of a water body.

Big roadcuts near the western end of St. Mary Lake expose the Appekunny Formation, some 3,000 feet of green mudstone that lies above the Altyn Formation. It consists largely of laminated sediments that settled in standing water, and of layers of contorted sediments that were deposited in clouds of coarser sediment that were rapidly deposited on the water-saturated mud, deforming it. The fine-grained, laminated deposits suggest that the basin deepened following the deposition of the nearshore Altyn Formation. The equivalent mudstones on the west side of the park are almost black and look like another deepwater rock unit called the Prichard Formation everywhere else in Northwest Montana, even though it may be the Appekunny. The Appekunny contains peculiar imprints of fossils (*Horodyskia*) that look like strings of beads; they are considered some of the oldest known fossil eukaryotes, or cells with a nucleus. The green Appekunny grades into the overlying red Grinnell Formation where the rocks alternate between red and green mudstone.

The Grinnell Formation, some 2,500 feet thick, consists of red mudstones with mud cracks and ripple marks, showing that the water they were deposited in became shallow. In some places waves and stream currents broke up the mud cracks, mixing them with sand to form a distinctive rock with red mud chips and mud balls in a white quartz-sand matrix. The clay-rich muds were deposited from suspension in shallow water on vast mudflats that were regularly exposed to the air, which caused them to dry and crack. The coarse quartz sand was derived from highlands to the southeast, and the finer-grained sand was deposited by streams emanating from highlands to the southwest.

The overlying Empire Formation is poorly exposed, but a grayish-green outcrop occurs on the north side of a curve near where Going-to-the-Sun Road begins to climb uphill. The Empire in Glacier National Park is only a few hundred feet thick and is transitional between the underlying mudflats of the Grinnell and the overlying deeper-water limestones of the Helena Formation. As a result, it contains rocks similar to both these formations.

Mud chips and mud balls in white quartzite of the Grinnell Formation next to Going-to-the-Sun Road near the St. Mary Falls Trailhead.
—Courtesy of Marli Miller

Concentric patterns of the eroded tops of the domal stromatolites at Logan Pass. See photo on page 27 for a side view of the stromatolite's cabbage-like interior.
—Courtesy of Marli Miller

From here to Logan Pass and down the west side of the pass almost to The Loop, the road is in the Helena and overlying Wallace Formations. The Helena is dolomitic limestone composed of alternating thin laminae of carbonate and noncarbonate mud that can be traced for great distances across the Belt Basin. The water of the basin must have risen again, moving the source of land-derived sediment farther away and allowing limestone to precipitate through evaporation and with the help of photosynthetic bacteria. The distinctive gray-colored stromatolitic limestone mounds of the Wallace Formation look a bit like groups of petrified cabbages. Like the Altyn Formation, the Wallace contains concentrically laminated ooids that look like fish eggs less than 1 millimeter across. Wave and current agitation formed them, and their presence at the top of this formation, where the section transitions to the overlying Snowslip Formation, indicates the basin's water shallowed.

A thick black basalt sill, the Purcell Sill, intruded limestones of the Wallace Formation around 780 million years ago, a time when the supercontinent Rodinia was beginning to break apart. It occurs throughout the higher elevations in the park, including in the cliffs above Grinnell Glacier and along the Highline Trail close to Logan Pass. The Purcell Sill is a slab of black diabase, a grainy variety of basalt, as much as 100 feet thick. The magma squirted between rock layers in the upper part of the Wallace Formation, baking the dark organic matter out of the limestone, both above and below the sill, bleaching it to white and turning it to marble. Look for this dark layer fringed above and below with thin ribbons of white in the high cliffs of the park. This amazing sheet of magma stretches for more than 50 miles, from near Two Medicine Lake to north of the Canadian border! In a few areas it split into a second sill that cuts layers and ramps up into younger units elsewhere in the park.

At Granite Park, on the Highline Trail north of Logan Pass, there is neither granite nor any rock that looks like granite. Instead, Belt-age basalt magma erupted there during the deposition of the Snowslip Formation to become a basalt lava flow. It appears as pillows, rounded structures that form when lava erupts either underwater or into wet sediments, so we know that basaltic magma was still being generated during the later stages of sediment accumulation in the Belt Basin.

Basalt pillow lavas in the Snowslip Formation 20 feet below Granite Park Chalet in Glacier National Park.

The easiest place to see formations of the Missoula Group is along the trail from the visitor center at Logan Pass to the Hidden Lake overlook. The Belt Basin shallowed with the deposition of the Snowslip Formation, and limestone of the Wallace Formation was replaced by land-derived sediment. These sediments, deposited by vigorous streams from the south onto vast mudflats that were intermittently exposed and submerged by water, include colorful mudstones and sandstones in various shades of red, green, beige, and yellow caused by iron-bearing minerals in the rock. Ripple marks and mud cracks cover many of the bedding surfaces, telling tales of sand shifting under gentle waves or currents, and of mud drying and cracking in the sun. Storms sometimes swept over these surfaces, ripping up and redepositing mud chips formed by the mud cracks. Occasionally you may see delicate little prints left by sharp cubes of crystallized salt that grew in soft mud; apparently at least some of the water in this depositional environment was salty. As with many of the Belt formations, these rocks contain beautiful stromatolites but no animal life, which had not yet evolved and likely would have destroyed the exquisite sedimentary structures had they been around at the time.

Mud cracks in the Snowslip Formation on the top of Piegan Mountain in Glacier National Park. —Courtesy of Marli Miller

The lower portion of the Shepard Formation caps the tops of several high peaks, such as Clements and Reynolds Mountains, both of which are visible from the Hidden Lake Trail. Finely laminated sand and mud suggest a deepening of the basin and flooding of the Snowslip mudflats. The sediment was brought in by streams and settled in the water, forming graded couplets that were sometimes exposed and then cracked. Farther offshore, limestone precipitated directly from the water, and stromatolites formed where light could penetrate to the bottom.

The Hidden Lake Trail is an excellent place to take in the glacial features sculpted by ice that covered the park during Pleistocene glaciation. Streams that drain east and west from the Continental Divide at Logan Pass were filled with massive glaciers that plucked and carried rocks from the high peaks, transporting them to the valleys and plains below. When the ice retreated at the end of Pleistocene time, the landscape had been thoroughly transformed. Previously rounded mountains had been eroded into rugged pointed peaks (horns), knife-shaped ridges (arêtes), and bowl-shaped depressions (cirques), many of which filled with lakes, including Hidden Lake. Ice descending stream-eroded V-shaped valleys left them U shaped, with much-steepened sidewalls and widened bottoms. Many smaller side valleys were left hanging at higher elevations because their smaller glaciers could not keep pace with the deeper erosion of the big valley glaciers. Today, these hanging valleys are easily recognized by their spectacular waterfalls, such as Bird Woman Falls across from Going-to-the-Sun Road west of Logan Pass. Where the ice descending the big valleys moved over bumps in the topography, they scoured out large depressions that later filled with water, creating chains of long, linear lakes.

At the down-valley end of each glacier, rock debris was deposited as glacial till and outwash. The glacial till forms mounds of rock material of various sizes that was dumped by—and once outlined—a glacier. An example is the small, more recent (Holocene) moraine at the base of Clements Mountain on the Hidden Lake Trail. The outwash was deposited by meltwater streams and formed nearly flat surfaces, such as

View of the hanging valley of Bird Woman Falls from Going-to-the-Sun Road.

A relatively small glacial moraine along the base of Clements Mountain at Logan Pass is next to the Hidden Lake Trail.

that below upper St. Mary Lake. Very fine sediment particles, or rock flour, impart the milky greenish color to the park's water bodies. The paler the water's color, the more rock flour it contains. Be on the lookout for all of these features as you cross the park on Going-to-the-Sun Road.

These days many wonder if the park's glaciers are doomed. The short answer is yes. Of the estimated 150 glaciers in Glacier National Park in 1850, most remained in 1910, but only a shrunken 25 remain today. The US Geological Survey estimates that most could disappear by 2030. The culprit is Earth's warming climate. Snowfields come and go with the season and vary year by year, but the long-term ice tells the real story. Grinnell, Sperry, and Jackson Glaciers, among the better known, have been monitored for more than a century, and all are decreasing in size. Note that a glacier

Potholes drilled in the bedrock floor of McDonald Creek, near a large pull-off just up the valley from the head of Lake McDonald.

is defined as a long-term mass of ice that flows downslope under its own weight. Ice less than about 160 feet thick remains brittle and won't flow, so thinner or stagnant ice and permanent snowfields are not glaciers.

West of Logan Pass, the Belt rocks dip to the northeast in the west limb of the Akamina syncline. As you pass down the west side back through all the formations that were exposed on the east, some look a bit different because they were deposited in a different part of the Belt Basin.

The hard Belt sedimentary rocks are famous for breaking along smooth, flat bedding surfaces. Steep stretches of the park's energetic streams scour out any loose rocks, gravel, or sand, so the water skitters over flat bare rock. Where vigorous eddies or whirlpools form, rocks caught in the current can scour the bottom in one spot, slowly drilling a circular hole in the solid bedrock. Wonderful examples of such potholes are beautifully displayed at low water in McDonald Creek, not far upstream from the northeast end of Lake McDonald.

Lake McDonald occupies a basin scoured by glacial ice mostly in Tertiary sedimentary rocks of the North Fork Valley. Near Lake McDonald Lodge, the road crosses the Flathead fault, an active northwest-trending extensional fault that started forming a half graben in Eocene time, creating the Kishenehn Basin on the west side of the fault. This basin filled with stream, lake, and swamp deposits that preserved a variety of fossils, including early primates and spectacularly preserved insects. The listric normal fault curves at depth, tilting all the rocks down to the northeast. At depth, it appears to merge with the Lewis overthrust.

MT 49 AND US 89
EAST GLACIER PARK—ALBERTA BORDER
49 miles

All of MT 49 and US 89 between East Glacier and the Canadian line lies just east of Glacier National Park. Bedrock near the road is soft Cretaceous sedimentary rocks. North from East Glacier to Kiowa, winding MT 49 requires one's full attention on the road and is closed in winter. The road passes over the folded and thrust-faulted

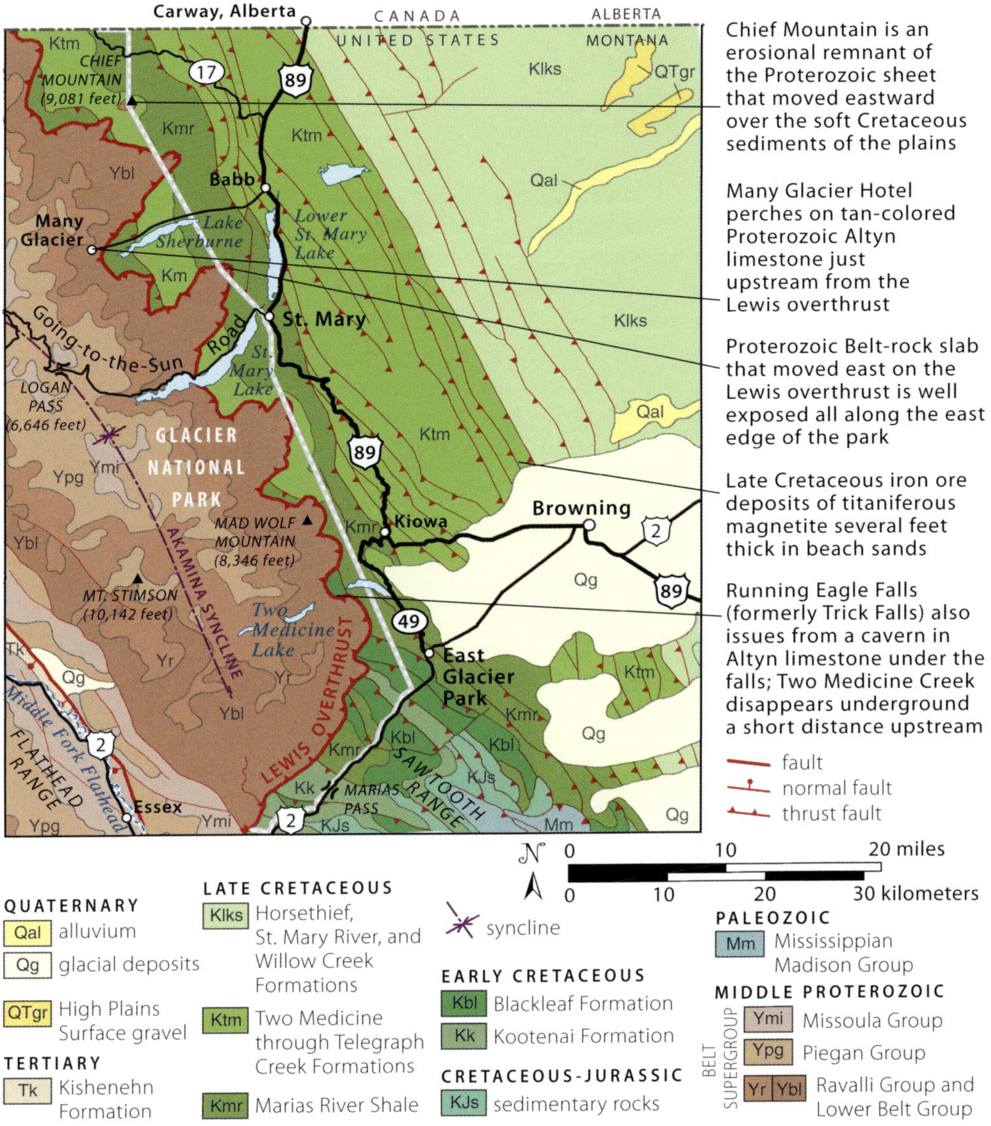

Carway, Alberta
CANADA
UNITED STATES
ALBERTA
MONTANA
CHIEF MOUNTAIN (9,081 feet)
Chief Mountain is an erosional remnant of the Proterozoic sheet that moved eastward over the soft Cretaceous sediments of the plains
Many Glacier Hotel perches on tan-colored Proterozoic Altyn limestone just upstream from the Lewis overthrust
Proterozoic Belt-rock slab that moved east on the Lewis overthrust is well exposed all along the east edge of the park
Late Cretaceous iron ore deposits of titaniferous magnetite several feet thick in beach sands
Running Eagle Falls (formerly Trick Falls) also issues from a cavern in Altyn limestone under the falls; Two Medicine Creek disappears underground a short distance upstream
Babb
Many Glacier
Lake Sherburne
Lower St. Mary Lake
St. Mary
Going-to-the-Sun Road
St. Mary Lake
LOGAN PASS (6,646 feet)
GLACIER NATIONAL PARK
AKAMINA SYNCLINE
MAD WOLF MOUNTAIN (8,346 feet)
MT. STIMSON (10,142 feet)
Two Medicine Lake
LEWIS OVERTHRUST
Kiowa
Browning
East Glacier Park
SAWTOOTH RANGE
MARIAS PASS
Middle Fork Flathead
FLATHEAD RANGE
Essex
17
89
2
49
fault
normal fault
thrust fault
0 10 20 miles
0 10 20 30 kilometers
QUATERNARY
Qal alluvium
Qg glacial deposits
QTgr High Plains Surface gravel
TERTIARY
Tk Kishenehn Formation
LATE CRETACEOUS
Klks Horsethief, St. Mary River, and Willow Creek Formations
Ktm Two Medicine through Telegraph Creek Formations
Kmr Marias River Shale
syncline
EARLY CRETACEOUS
Kbl Blackleaf Formation
Kk Kootenai Formation
CRETACEOUS-JURASSIC
KJs sedimentary rocks
PALEOZOIC
Mm Mississippian Madison Group
MIDDLE PROTEROZOIC
BELT SUPERGROUP
Ymi Missoula Group
Ypg Piegan Group
Yr Ybl Ravalli Group and Lower Belt Group

Geology along US 89 between Browning and the Alberta border.

WEST
EAST
MT. STIMSON (10,142 feet)
MAD WOLF MOUNTAIN (8,346 feet)
Proterozoic Belt rocks
LEWIS OVERTHRUST FAULT
Kiowa
Browning
Cretaceous sedimentary rocks
Paleozoic sedimentary rocks
Archean basement
0 5 10 miles
0 5 10 15 kilometers

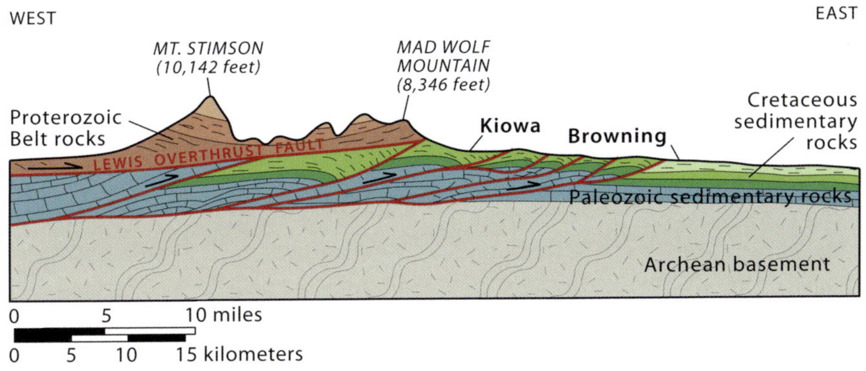

Cross section through Glacier National Park.

Cretaceous Marias River Shale and is prone to landslides. There are some great outcrops but few good places to stop. About 2 miles from East Glacier, the road begins following the meandering Two Medicine River, with its stunted forest of aspens and gnarled pines. From the east end of Two Medicine Lake, it climbs over Looking Glass Hill, cutting through remnants of the old High Plains Surface that existed before it was deeply dissected by glaciers and streams during the last 3 million years. Looking Glass Hill is at the north end of Two Medicine Ridge, which is held up by slightly more resistant sandstone of the Cretaceous Virgelle and Two Medicine Formations.

Glacial till covers most of the landscape, but greenish-brown Cretaceous sandstone and black mudstone crop out in places. The distinct reddish tint of some of the till is a result of longer weathering; it's till of an earlier glaciation. The till of the most recent ice advance, the Pinedale glaciation, is pale brownish-gray. Glaciers of the

Layers of glacial outwash and glacial lake sediments standing on end where they were deformed by a later advance of another glacier. The outcrop is high on a ridge just north of East Glacier on MT 49.

A roadcut on MT 49, 2 miles north of East Glacier, north of the Two Medicine valley. Beautiful folds and small faults are exposed in layers of brown Cretaceous sandstone. These layers may have been rumpled beneath the Lewis overthrust as the Proterozoic Belt rocks moved eastward over them. Those overlying Belt rocks have since eroded away.

recent advance removed most exposures of the older till. In one roadcut on the south side of the ridge, beds of glacial outwash sand and gravel stand on end, presumably pushed up by advancing ice of the Pinedale glaciation. That shouldn't be possible, of course, because loose sand and gravel should just fall apart. It must have been solidly frozen when the ice bulldozed it.

At the top of the pass above Kiowa are sweeping views west up the Two Medicine valley, north along the east front of the park, and east over the High Plains. The view is truly spectacular and worth a stop if you can find a safe place to do so.

TWO MEDICINE ROAD

About 4 miles north of East Glacier Park is the turnoff to Two Medicine Lake. Just past the upper end of Lower Two Medicine Lake, about 8 miles west of the turnoff, is the parking lot for Running Eagle Falls. A nice trail through the woods leads about 0.5 mile to the falls in the Altyn limestone of the Belt Supergroup, just above the Lewis overthrust. During wet weather of late spring or early summer, lots of water pours over the lip of the falls, sometimes obscuring the lower falls that pours out through a cavern one-third of the way up the cliff. As the water level in the stream above drops, the upper falls dwindles, leaving the lower falls dominant. In late season the upper falls dries up.

The road to Two Medicine Lake follows a remarkably deep trough that ice-age glaciers gouged out. The road crosses the Lewis overthrust and climbs into the hard Proterozoic rocks at the Two Medicine River. The bedrock of this area is comprised of Cretaceous formations to the east and Proterozoic Belt rocks to the west.

Two Medicine Lake lies where three tributary glaciers joined to form a single, much larger glacier that gouged a deep basin. After the Pinedale glacial advance ended, the thick mass of ice that filled this deeper area of valley floor lingered after thinner parts of the glacier had melted, preventing the basin from filling with outwash sediments.

Cliffs above Two Medicine Lake are an especially good place to see the former depth of the ice. Look high for the line that separates glacially smoothed rock below from much rougher, craggy outcrops above the former level of the ice. The rough outcrops owe their ragged form to cold temperatures that freeze water in cracks, which then expands, prying blocks of rock loose.

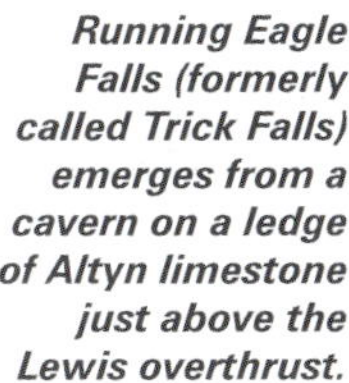

Running Eagle Falls (formerly called Trick Falls) emerges from a cavern on a ledge of Altyn limestone just above the Lewis overthrust.

North of Kiowa, US 89 crosses ridges capped by glacial till left by melting glaciers. The northern ridge, about 6 miles south of St. Mary, marks the Hudson Bay Divide, from which all water drains to the north into Hudson Bay rather than to the Missouri-Mississippi River basin. Near the mountain front the glacial moraines and outwash were derived from mountain glaciers draining the big mountain valleys in Glacier National Park. Farther east, the largest of those glaciers coalesced to form a broad piedmont glacier, a glacier that spreads ground moraine (till) across flatter ground that rests below and downstream of a mountain glacier. Boulders (erratics) of rocks deposited by both types of glaciers litter the landscape.

The views of the Rocky Mountain front between Kiowa and St. Mary are spectacular, especially from the crests of the ridges north of Kiowa. The steep front consists of Proterozoic Belt rocks that were thrust over Cretaceous sedimentary rocks along the nearly flat Lewis overthrust from about 75 to 59 million years ago. Thrust faults with far less displacement occur in the Cretaceous rocks on the plains; they are much harder to see due to poor exposure. To get a good view of the Lewis overthrust, stop at the crest of one of the ridges and look at White Calf and Divide Mountains to the west, where Lower Belt rocks of the Altyn Formation rest on Cretaceous shale.

Lower St. Mary Lake floods an area where a thick mass of stagnant glacial ice lay unmelted for some time after other parts of the valley were clear. Meltwater could not deposit outwash where the ice lingered, so when it did finally melt, that part of the valley floor was a deep basin that today holds water. For miles along US 89, from north of St. Mary, along Lower St. Mary Lake, to north of Babb, roadcuts consist of beige glacial till loaded with rocks and boulders of all sizes. Some are angular and some, which may have been rolled around in a glacial outwash stream, are distinctly rounded. Watch for boulders with distinct straight scratches, or striations, left where the rocks were scraped over bedrock. Perhaps you can identify Belt sedimentary rock carried eastward from Glacier National Park. If you find some made of granite, or coarse-grained pale metamorphic rocks farther east on the plains, those are not from the park but were carried southward from far to the northeast on the Canadian Shield by the continental ice sheet.

The Lewis overthrust seen to the west from US 89 between Kiowa and St. Mary. —Courtesy of Marli Miller

MANY GLACIER ROAD

From US 89 to Swiftcurrent Lake, Many Glacier Road follows the valley of Swiftcurrent Creek, crossing the Lewis overthrust about 1 mile east of Many Glacier. Except in the Many Glacier area, Cretaceous sedimentary rocks form the bedrock beneath the valley floor, with Proterozoic Belt sedimentary rocks making up the steep valley walls. All of the Proterozoic rocks exposed near the road to Many Glacier are tan, magnesium-rich (dolomitic) limestones of the Altyn Formation. Green and red mudstones of the Appekunny and Grinnell Formations make broad color bands on the higher slopes. Distant views of the high cliffs of the Helena and Wallace limestones to the west show the black ribbon of the Purcell Sill, a 780-million-year-old diabase intrusion, with its borders of limestone that were baked to white marble by the intruding magma.

The Lewis overthrust is exposed in the valley walls about halfway up along the length of Lake Sherburne, where hard Proterozoic sedimentary rocks rest on soft Cretaceous sandstone and shale. The best exposure is on Wynn Mountain to the southwest. The deeply buried Proterozoic rocks were compressed and shoved up along a high-angle fault through the entire stack of sedimentary rocks, flattening out and sliding eastward as much as 80 miles on a horizontal fault in clay-rich and possibly wet Cretaceous sediments.

Before Glacier National Park was established in 1910, it was prospected for mineral and other geologic resources, just like other parts of Montana. In 1892, prospectors found copper minerals in and near diabase dikes in the Altyn limestone around Swiftcurrent Lake. The area was then under the jurisdiction of the Blackfeet Tribe, so developers pushed to change the tribal land boundary so they could start mining the ore in 1898. The town of Altyn sprang up on the flats about three-quarters of a mile down-valley from Many Glacier Hotel. Many of the diabase dikes and sills in the Proterozoic sedimentary formations throughout western Montana deposited small amounts of copper, but not enough to support significant mining operations. The mining around Swiftcurrent Lake was no different, so Altyn never grew beyond a population of one hundred people and was deserted by 1910.

In 1902, a miner prospecting for copper near Swiftcurrent Lake found oil seeping into his workings. Drilling began in 1903, and a well sunk to a depth of 550 feet in 1905 actually produced small amounts of oil, the first ever found in Montana! Another well finished in 1909 produced natural gas, which was used until 1914 to heat one house, the only customer the field ever had. Other efforts to establish production were even

View of Swiftcurrent Lake from the Many Glacier Hotel. The two drainages on either side of Grinnell Point (center) were deeply glaciated during Pleistocene time, leaving a knife-shaped ridge called an arête. The mountains are made of the Belt Supergroup. On the left, Mt. Gould has a dark layer, the Purcell Sill.

less successful. The boom had already fizzled by the time the park was established in 1910, and Lake Sherburne flooded the site in 1919. The water from the reservoir is the principal water storage for the US Bureau of Reclamation's Milk River Project, which provides irrigation water to north-central Montana farms.

The Many Glacier Hotel complex was built in 1915. Near the hotel, exposures of Altyn limestone consist of egg-shaped sand grains of calcite called ooids, which formed in the waves on a beach more than 1.45 billion years ago. The area is very popular with hikers, providing access to Grinnell Glacier, Iceberg Lake, and the Ptarmigan Wall, among other sites.

The hike to Grinnell Glacier is about 6 miles from the trailhead at Swiftcurrent Lake; it crosses sedimentary rocks of the Appekunny, Grinnell, Empire, Helena, and Wallace Formations of the Belt Supergroup. Grinnell Glacier, a fraction of the size it was one hundred years ago, is nestled in a glacial cirque carved into limestone of the Wallace Formation. The Purcell Sill forms a prominent dark layer with white baked zones above and below.

CHIEF MOUNTAIN

No one who is fortunate enough to see Chief Mountain ever forgets this great stump west of MT 17 between Babb and the Canadian border. Rising more than 4,000 feet above the plains, Chief Mountain demands attention and has long played a prominent role in the cultures of indigenous people. The Blackfeet word for Chief Mountain is *ninnāastŭkoō*. The peak typically appeared on early maps of the area, and in 1901 Bailey Willis of the US Geological Survey recognized that Chief Mountain and the other Belt rocks had moved east many tens of miles along the Lewis overthrust. Erosion left the block isolated from the other rocks above the thrust. It's what geologists call a klippe.

Given that the fault carried the 300-mile-long, probably well over 2-mile-thick Proterozoic slab of Belt rock as much as 80 miles east over the Cretaceous sediments on a virtually horizontal surface, that surface must have been slippery! Once the thrust slab stopped moving around 56 million years ago, the overlying rock started eroding away. That huge pile of Belt rock above the Lewis overthrust had to be removed to expose the Altyn and Appekunny Formations seen on Chief Mountain today. The great thrust slab may have extended even farther east, but any evidence has been eroded away. The large scar on the north face of Chief Mountain formed when weakened rock collapsed during a 4.1 magnitude earthquake on July 2, 1992. The landscape is a work in progress, always changing.

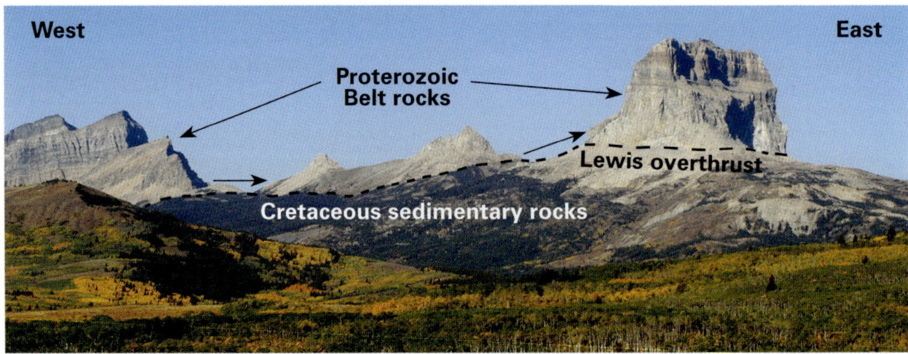

Chief Mountain is a raft of Proterozoic Belt rock (Altyn and Appekunny Formations) that was shoved eastward over Cretaceous shale. —Courtesy of National Park Service

US 2
KALISPELL—BROWNING
101 miles

James Hill's Great Northern Railway, built in 1893 along present-day US 2, skirted the southern boundary of today's Glacier National Park en route to Seattle. Plans for logging along Lake McDonald and interest in mining elsewhere were thwarted when pioneering naturalist George Bird Grinnell joined with James Hill to promote a new national park. Hill saw opportunity to use the pristine and spectacular region to lure tourists west on his trains. President William H. Taft signed off on Glacier National Park in May 1910. In cooperation with the government, Hill built chalets, big new hotels, roads, and trails in the park. Initially, access to the park's interior was by horseback or on foot. With better roads, the completion of a highway across Marias Pass (US 2) in 1930 and of Going-to-the-Sun Road in 1932, and the introduction of the Ford Model A in 1927, access improved and visitation to the park boomed.

US 2 crosses the flat floor of the Flathead River valley from Kalispell northeast to Columbia Falls, partly on flat sands and gravels deposited by the river, and partly on low, irregularly surfaced, rocky sediments left by the giant Cordilleran ice sheet that filled the Rocky Mountain Trench more than 12,000 years ago. That immense, enigmatic valley trends north-south for more than 1,000 miles from northern British Columbia to the Flathead Valley—and perhaps even the Mission Valley. Looming in the distance just east of Columbia Falls are the imposing walls of the Whitefish Range to the north and the Swan Range to the south, both composed of Proterozoic Belt rocks.

The unusually steep mountain front suggests active faulting. In fact, crustal spreading has been causing the valley to drop to the west relative to the ranges for millions of years, and the faults along the base of the ranges are still active. Small earthquakes occur regularly, and larger ones happened in the recent past. In the 1970s, the Creston area was hit by a magnitude 5.0 quake, and in 2014 an earthquake measuring 4.5 struck about 5 miles beneath Whitefish. This is the northern end of the Intermountain Seismic Belt, a region of active crustal spreading in the Intermountain West that regularly produces earthquakes. Even larger earthquakes are possible.

East of Columbia Falls to Hungry Horse, US 2 crosses hard Belt rocks of the Piegan Group in Bad Rock Canyon. Since the Flathead River cut the canyon directly across the trend of the mountains, the river was probably here first and got stuck in the hard Belt rocks as the mountains rose, cutting through them as they grew. Between Hungry Horse and West Glacier, the highway crosses low forested hills of glacial till left by ice-age glaciers.

During the Pleistocene ice advances, the glacier that filled the Rocky Mountain Trench pushed south through the Flathead Valley and into the Mission Valley. When it melted, it left a deep fill of glacial till and outwash on the floor of the Flathead Valley. The valley of the Middle Fork of the Flathead River filled with ice coming down from the mountains on either side of it. That glacier ground its way west through Bad Rock Canyon to join the huge glacier that filled the Rocky Mountain Trench. Glacial deposits, glacial striations on rocks, and hanging valleys on the mountainsides show that the ice that filled the Flathead Valley was about 6,000 feet thick at the Canadian border. The ice thinned so rapidly southward that the glacier ended not far south of

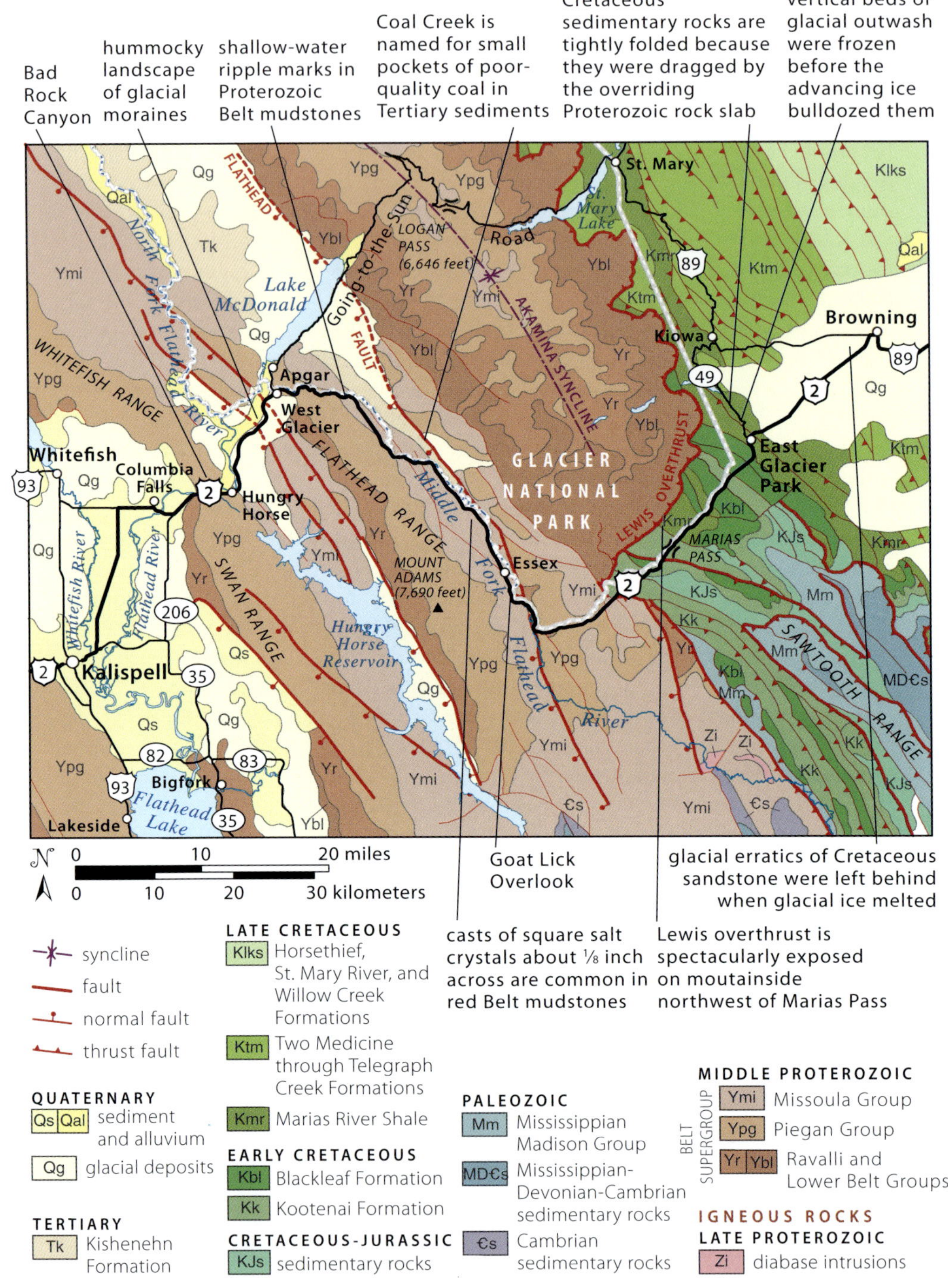

Geology along US 2 from Kalispell to Browning.

Flathead Lake. Evidently the ice-age summers were warm enough to melt enormous volumes of ice that must have released torrents of muddy meltwater.

Early settlers could recognize bear skins from the North Fork of the Flathead River valley because they smelled like kerosene. They solved that mystery of the odor during the 1890s after finding several oil seeps, in which the bears wallowed. A well drilled at the head of Kintla Lake, in the northwest corner of Glacier National Park, in 1901, penetrated only Belt rock and found no trace of oil. Several wells drilled in the valley since then did find shows of oil and gas, but there was no actual production. Might oil-bearing Paleozoic sediments lie below the Belt rock? Drilling much deeper in very hard rocks such as the Belt would be very expensive. And how much deeper? Geologists have found at least several hundred feet of oil shale in the Tertiary Kishenehn Formation exposed along the North and Middle Forks of the Flathead River west of Glacier National Park, but development of oil resources in such sensitive environments as those around Glacier National Park would certainly prove to be controversial.

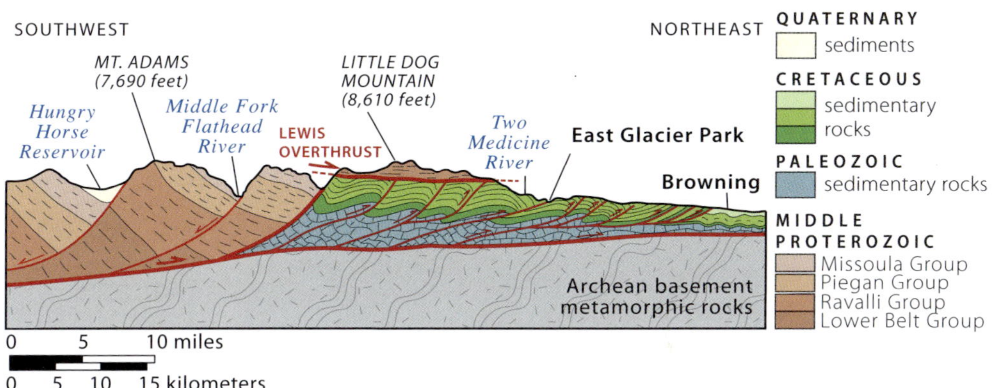

Cross section through the southern edge of Glacier National Park, just north of US 2.

The 564-foot-high Hungry Horse Dam, 3 to 4 miles upstream of the community of Hungry Horse on the South Fork of the Flathead River, was built between 1948 and 1953, primarily for flood control and electric power generation. Cheap power from the dam prompted the construction of the Columbia Falls aluminum plant and smelter in 1955. Bauxite aluminum ore comes from the tropical weathering of rocks in the Caribbean, so the raw materials had to be shipped a great distance to the plant. After several stages of expansion, the plant was producing 1 million pounds of aluminum per day. By 1969, researchers from the US Forest Service and the University of Montana, under contract from the National Park Service, documented damage in Glacier National Park from the plant's fluoride emissions. Only after a local orthodontist filed a class-action suit did the plant work to reduce emissions. The plant permanently closed in 2015 and was declared a Superfund site in September 2016. Contaminants at the site include fluoride, cyanide, arsenic, chromium, lead, and selenium, and there is concern that they could leach into the groundwater and surface waters of the Flathead River, which flows south into Flathead Lake.

From West Glacier, US 2 follows the Middle Fork of the Flathead River along the southwest boundary of Glacier National Park. The river flows northwest, following a structural depression formed by extensional faulting that probably continues today. For the first several miles from West Glacier the route passes through the Piegan Group, then red and green sedimentary rocks of the Missoula Group to just east of Essex. A few miles southeast of West Glacier, between mileposts 157 and 158, giant slabs of dark-red sandstone hang ominously over the highway on the south. One of these days, freezing water may loosen one of those big slabs to come crashing down onto the highway.

About halfway between West Glacier and the Isaak Walton Inn at Essex, there's a spectacularly red, strikingly layered roadcut in Missoula Group sandstone and mudstone. The layers, originally laid down as horizontal muds and sands, were tilted during the deformation of the Overthrust Belt, the front range of the Rockies. As with many exposures of Missoula Group rocks, you may see mud cracks, ripple marks, and perhaps even cubic impressions of one-eighth-inch casts of salt crystals.

Prominent layers of red mudstone and sandstone of the Missoula Group, about 14 miles east of West Glacier near milepost 168, dip about 45 degrees to the east.

At Essex, consider stopping to see the Izaak Walton Inn, a Tudor revival structure built by the Great Northern Railway in 1939 to house railway workers. Approximately 3 miles south of Essex is a parking area west of the highway for the Goat Lick Overlook, a place where geology and animal behavior are distinctly related. The goat lick is located in gray-colored clay exposed in a steep cliff wall on a meander loop of the Middle Fork of the Flathead River. The clay formed along the north-trending Roosevelt fault, an active normal fault along which the rocks west of the road dropped, forming the basin the river now follows. The fault movement shattered the Piegan Group Belt rocks, forming wet seeps that concentrate sulfate minerals such as gypsum, along with potassium, sodium, and phosphorous, which are essential for mountain goats.

Craggy peaks in Glacier National Park can be seen from US 2 just east of Essex and the Izaak Walton Inn. The prominent spire of Mt. Saint Nicholas (9,376 feet) stood above the height of the ice cap during the Pinedale glacial advance, but its precipitous flanks were scoured by those glaciers.

US 2 leaves the Middle Fork of the Flathead River and turns sharply northeast for several miles as it climbs toward Marias Pass. Having built the railway across northern Montana to Havre, Lewis Hill, James Hill's son and successor to the Northern Pacific

Railway, sent a surveyor out to find a crossing of the Rockies. After several attempts to locate a suitable railroad pass, John F. Stevens finally succeeded, crossing what is now Marias Pass on December 10, 1889, in spite of below-zero temperatures and deep snow.

About 4 or 5 miles west of Marias Pass, you'll see the last of the hard, Belt rocks and begin seeing gray roadcuts of dark sandstone separated by similarly dark-gray layers of soft platy shale. Those shales and sandstones contain fossils that indicate they were laid down between about 110 and 80 million years ago. Somehow, we skipped over 1.4 billion years of geologic time to reach Cretaceous sedimentary rocks, which here lie stratigraphically under the Proterozoic Belt rocks. As any beginning geology student knows, sedimentary rocks are laid down with younger rocks on top of the older rocks. What's going on?

Stop at the big parking area at Marias Pass, where the answer to the "Proterozoic over Cretaceous" rock dilemma becomes very clear. Look north to the treeless mountain range. The top half of the mountain, above a thin, indistinct, cream-colored horizontal layer, is made of hard cliff-forming rocks. These are the Proterozoic Belt sedimentary rocks. The bottom half is soft Cretaceous shales and sandstones that don't form cliffs. The cream-colored layer is actually the Altyn limestone, the bottom of the Belt sedimentary pile and the oldest rock in the park. The very sharp, horizontal base of this formation is the world-famous Lewis overthrust.

Because the soft Cretaceous sediments must have been there when the Proterozoic Belt rocks slid over them a distance of about 80 miles, then the movement must have happened after the Cretaceous sediments had been deposited, in this case sometime after about 75 million years ago but before 59 million years ago. Volcanic rocks that are 76 million years old are cut by the fault, and 59-million-year-old volcanic rocks are not cut by the fault, bracketing the age of fault movement. How could a thick slab of Belt rocks move that far on a horizontal surface? Most geologists conclude that the friction on the fault surface must have been very low. If the Cretaceous clays were

Cretaceous Kootenai Formation shale and sandstones in beds about 8 feet thick, halfway between mileposts 196 and 197 a couple of miles west of Marias Pass.

View to the northwest of the Lewis overthrust from 5 miles east of Marias Pass. The cream-colored Altyn limestone, gray Appekunny mudstone, and paler-gray Grinnell mudstone lie above the fault. Below are soft Cretaceous shales and sands.

young enough to be somewhat wet, loading them with heavy rocks on top would squeeze out water to lubricate movement, at least near the leading edge of the moving slab. In addition, if those clays contained some expandable clay (smectite), friction would drop to almost nothing near the leading edge of the slab and sliding would be almost inevitable, even with only a slight slope. In fact, we know that such swelling clay is widespread in the Cretaceous sedimentary formations of central and eastern Montana. Ranchers and others who drive unpaved dirt roads during wet weather are more than aware of how slippery that stuff is!

From where we first see Cretaceous sedimentary rocks, a few miles west of the pass, to East Glacier Park, all of the mountains south of US 2 are part of the Overthrust Belt. There, Paleozoic and Mesozoic sediments are stacked as repeated sections along thrust faults, like groups of cards dipping steeply west. (We discuss them in the Central and Eastern Montana chapter.)

The same dark-gray to brown Cretaceous shales and sands appear in scattered exposures almost to East Glacier Park. Some of the black shales contain small fossil ammonites that at first glance look like big snails, but they were actually more like an octopus that lived in a shell. They were close relatives of the modern pearly nautilus. Ammonites flourished in great numbers in the shallow seas that flooded much of Montana during Cretaceous time, then died in the same calamity that annihilated the dinosaurs.

Hummocky moraines littered with scattered boulders and kettle lakes between East Glacier Park and Browning record an enormous piedmont glacier. It formed as glaciers flowing east from the big canyons in the Rocky Mountains spread a large sheet of ice over the High Plains as far as several miles east of Browning. Piedmont glaciers spreading across such a large area on a gentle slope move so slowly that their ice is nearly stagnant. Their moraines are vast fields of irregular humps and hollows that show little of the shaping influence regularly seen with moving ice.

Just west of Browning the low rolling hills are pockmarked with big angular boulders dropped from the broad piedmont glacier that spread out from the valley glaciers in Glacier National Park.

I-90
Missoula—Lookout Pass
105 miles

Lewis and Clark came through the Missoula Valley on their way west in September 1805 and again on their return trip east in June and July 1806. The Missoula portion of the Mullan Road, built by Lieutenant John Mullan in 1860, was constructed along the north side of the Clark Fork River to open up the territory to settlers. The Northern Pacific Railway arrived in 1883. In 1908 the Milwaukee Road railroad followed roughly the same path as the Mullan Road, as did I-90 in the 1960s.

Between Missoula and Frenchtown, the highway follows the length of the Missoula Valley, a broad structural basin that formed during extensional and possibly strike-slip crustal movements that created many of the mountains and valleys seen today in western Montana. The high and glacially carved Rattlesnake Mountains north of the valley contain complexly folded Proterozoic Belt sedimentary formations broken along several large faults. The much lower and unglaciated Sapphire Mountains, which rise along the southeast end of the valley, are the trailing edge of the Sapphire detachment block. They contain less-deformed Belt formations. See MT 200: Missoula—Rogers Pass—Bowmans Corners (US 287 Junction) for more on the Sapphire block.

From Missoula to Frenchtown and Huson, the steep slopes on the north edge of I-90 mark the eroded edge of valley-fill sediments of the Renova Formation, deposited in collapsed basins that formed from Eocene to early Miocene time, 48 to 20 million years ago. These sands and gravels were initially deposited by rivers and in lakes in a warm, humid environment that became progressively drier and more desertlike in Miocene time. Capping the Renova sediments is the Sixmile Creek Formation, deposited as gravels 17 to 4 million years ago while the region was truly a desert. At the time, the ground sloped gradually south toward the center of the valley, much like big desert valleys in Nevada.

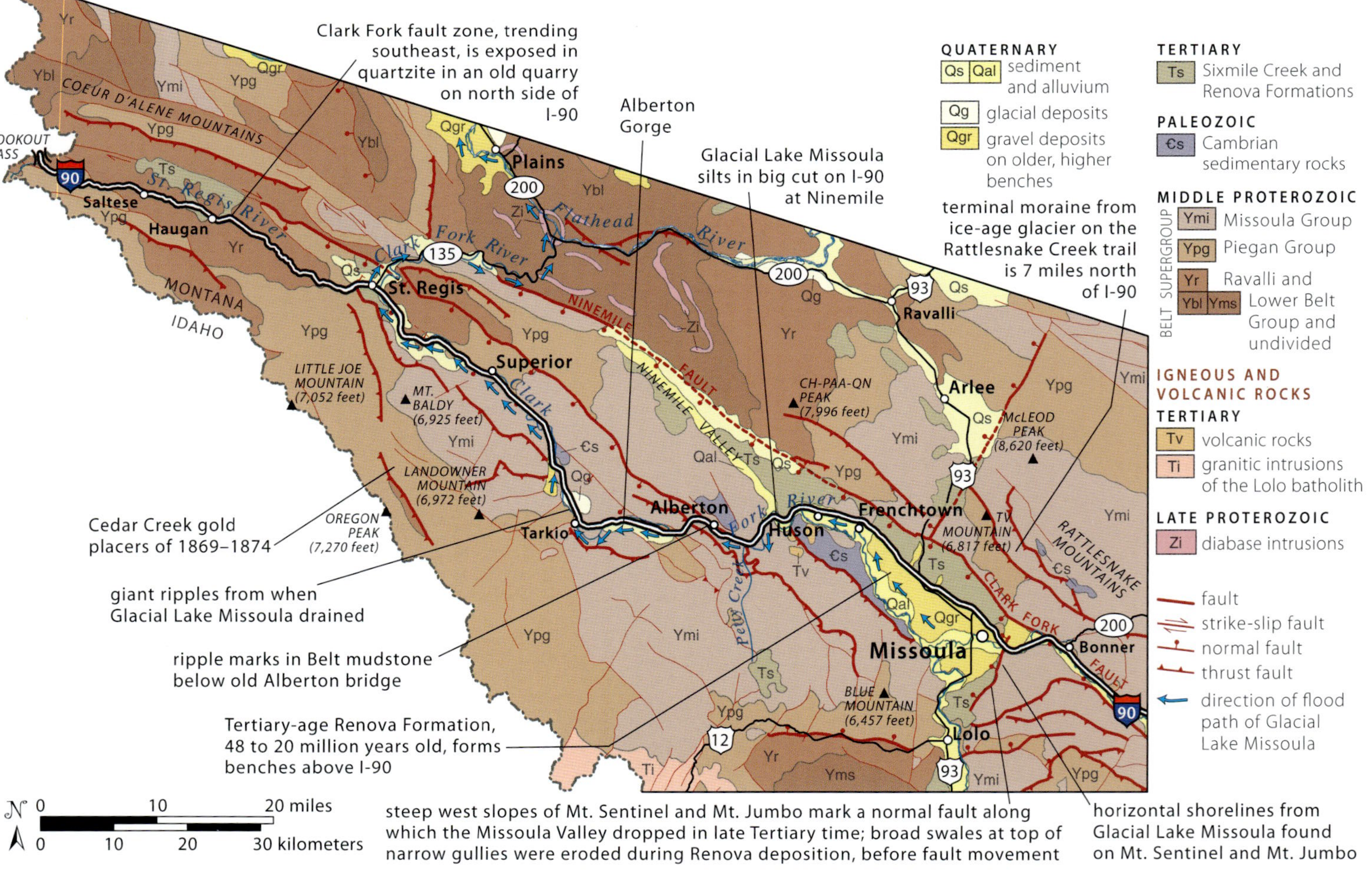

Geology along I-90 between Missoula and Lookout Pass.

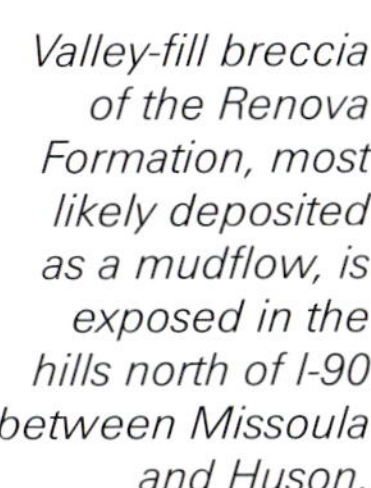

Valley-fill breccia of the Renova Formation, most likely deposited as a mudflow, is exposed in the hills north of I-90 between Missoula and Huson.

The striking and prominent horizontal lines on the steep faces of the local mountains, Sentinel and Jumbo, at the eastern end of the Missoula Valley are ancient shorelines of Glacial Lake Missoula, the largest glacial lake known to have existed anywhere. A glacier dammed the Clark Fork River near the Idaho border, and it backed up to create an ice-age lake centered on Missoula. As the lake filled and water at the ice dam deepened, it caused the lighter glacial ice to float and eventually break up, triggering floods of epic proportions. The great rushes of water that drained Glacial Lake Missoula poured down the Clark Fork Valley en route to the Pacific Ocean. In the narrower stretches of valley between Frenchtown and St. Regis, watch for ragged outcrops of bedrock along the valley floor and for lower valley walls with nearly all the soil scrubbed off.

Although the fault-bounded Missoula Valley is several miles wide, with a deep fill of Tertiary-age sediments, farther west the Clark Fork River valley floor is less than 1 mile wide. I-90 is perched on a nearly flat terrace tens to more than 100 feet higher than the river and hemmed in by high mountains with steep valley walls. The last flood of Glacial Lake Missoula deposited the flat terrace. In the 10,000 or so years since, the river has eroded a new channel down through it.

Where the river is visible from the highway, it is often incised into young river gravels; elsewhere it slices through solid Belt bedrock, as in the spectacular Alberton Gorge from just east of Alberton (exit 75) to just east of Tarkio (exit 61). Famous to whitewater boaters for its awesome river scenery, absolutely vertical canyon walls, and challenging rapids, the lower gorge is accessible from Cyr (exit 70) downstream to Tarkio. The bedrock channel of the Clark Fork River today is close to where it was while Glacial Lake Missoula drained from its highest lake levels.

Downcutting by the river exposed a thick section of Glacial Lake Missoula sediments at Ninemile. The big roadcuts in pinkish soft silts just east of the I-90 bridge across the Clark Fork at Ninemile contain a remarkably complete record of the declining floods. The section consists of sequences of varves (lake-bed deposits) alternating with river sediments. There are thirty-six sequences of lake beds, each recording a separate filling of Glacial Lake Missoula. The coarser-grained river

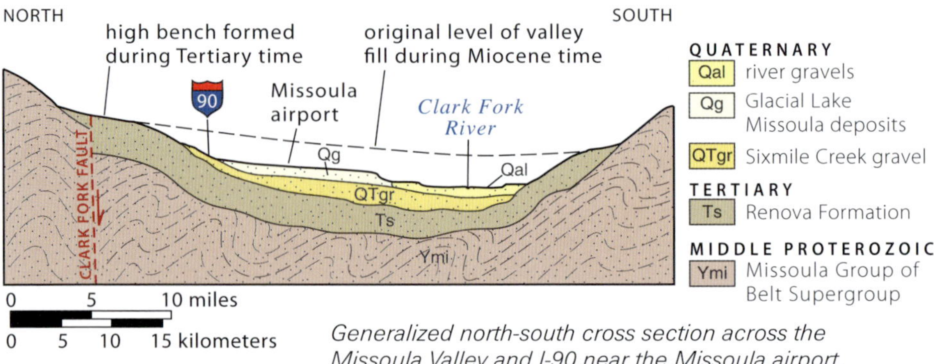

Generalized north-south cross section across the Missoula Valley and I-90 near the Missoula airport.

deposits that separate them accumulated after each drainage of the lake. Thick layers at the bottom of the section accumulated first, followed by each successively higher layer. The lowest, oldest layer was deposited over fifty-seven years; it represents the highest lake level, when the ice dam in Idaho was thickest. Each higher, later layer developed for fewer years, representing successively thinner ice dams. By the stage represented by the top layer, the lake only lasted for six years. This roadcut clearly tells us that the ice dam was thinning and receding. There must have been an earlier series of sediment layers that recorded the exact opposite sequence of events when the thin, frontal edge of the glacier was advancing. Those must have been swept clean out of the valley by the later higher-level floods.

The eastern exit to Alberton provides a short, easy detour down to an old steel bridge across the Clark Fork. The canyon is very narrow here and exposes Belt mudstone with ripples and mud cracks. Floodwater here would have been fast and energetic, carrying coarse rocks and scouring deeply, removing everything down to bare bedrock.

Most of the hard, cliff-forming Proterozoic rocks around Alberton and almost all the way to Superior are colorful green and red Belt rocks belonging to the Missoula Group, the youngest of the incredibly thick pile of Belt sedimentary rocks.

A huge roadcut in Glacial Lake Missoula lake beds (dark layers) and river deposits (pinkish-beige layers) near milepost 82 at Ninemile.

The mudstones contain beautifully preserved ripple marks, mud cracks, raindrop imprints, and even little square and triangular dents that the faces and corners of cubical salt crystals made as they imprinted the soft mud.

Cambrian formations on the ridge overlooking Alberton must be the source of the tumbled blocks of dolomite that stand north of the road at the Alberton exit. These Cambrian rocks lie at the southeast end of a long trend that extends north through Thompson Falls to near Libby. The Cambrian formations were deposited in shallow tropical oceans that slowly spread over the eroded Proterozoic rocks about 520 million years ago. They were later sandwiched between the older rocks along faults that probably moved about 82 million years ago.

Downstream, 5 miles west of Alberton and just east and west of exit 65, are enormous roadcuts in rounded Glacial Lake Missoula flood gravels. The contrast between these and the very fine silts at Ninemile reflects the high energy of the deeper and faster floodwaters that flowed through the more confined gorge. The flood dumped all of the eroded material, including large amounts of bedrock, as its flow slowed where the valley widened. Similar flood gravels continue as far west as Superior.

Just west of Tarkio, between mileposts 62 and 61, distinct giant ripples are preserved in grassy fields. Glacial Lake Missoula floodwaters left these as they swept through here.

Giant fold (dashed lines) in dark-gray Missoula Group rocks, south of I-90, just east of exit 77.

Raindrop imprints formed by a storm passing over an exposed mudflat about 1.4 billion years ago, now preserved in the Missoula Group near Alberton.

Glacial Lake Missoula flood gravels about 5 miles downstream and west of Alberton.

Northwest of Missoula, I-90 follows a series of big valleys eroded along nearly parallel faults of the Lewis and Clark fault zone, a major but poorly understood 20-to-40-mile-wide zone of faults. It extends not only the full length of I-90 from Missoula to Lookout Pass, but east to Helena and west through the Coeur d'Alene district of northern Idaho. The west-northwest zone of structural weakness may even extend across the Columbia River Plateau as far as the Okanagan Valley of Washington. One fault of the fault zone sliced some of the Renova Formation into Mount Jumbo, on the northeast side of Missoula, so this fault must have moved as recently as 25 million years ago, the age of the youngest Renova here. (See the introduction to this chapter for more information about the fault zone.)

The Superior area appears to be the easternmost extremity of the trend of the heavily mineralized Coeur d'Alene district that follows the Lewis and Clark fault zone

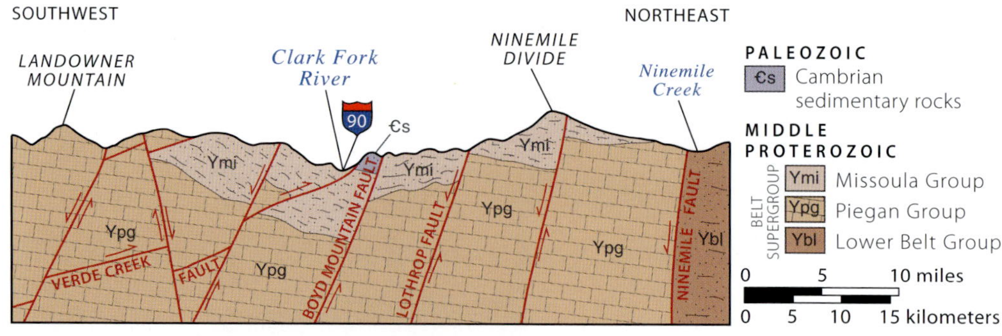

Cross section across the highway between Tarkio and Superior, 58 miles west of Missoula. Except for the small area of Paleozoic formations on the crest of the ridge north of the highway, all the rocks are Proterozoic Belt formations. The steep normal faults are younger and cut the older thrust faults. The Ninemile fault is part of the Lewis and Clark fault zone.

through northern Idaho. Numerous mines have produced lead and silver from the hills around Superior, along with small quantities of gold.

The Iron Mountain Mine, about 6 miles north of Superior on Flat Creek, was one of the largest in the district. Prospectors found outcrops of a vein about 40 feet wide there in 1888, and the mine, along with the little town of Pardee, went into full operation when the railroad came in 1891. Ore went down the mountain to a mill in Superior on an aerial tramway of huge buckets running on steel cables. The mine closed in 1897, then opened again a few years later to limp along off and on for years until about 1930.

Cedar Creek, south of Superior, was once the host of a regular gold rush. Prospectors found placer gold in the creek in late 1869, inspiring a midwinter stampede in which the entire gulch was staked in one day, and almost as quickly it acquired a population of some three thousand people who apparently had little else to do. That was the biggest gold rush ever in western Montana. Some estimates place the total production during those years at around $2 million, when an ounce of gold was worth about $20. Considering the number of years invested in digging gravel in Cedar Creek, that return could not have made many people rich. With literally thousands of would-be miners, most of those who became wealthy were merchants or liquor dealers.

At St. Regis (exit 33), the Clark Fork River leaves the interstate and turns northeast in a broad meander before flowing northwest along MT 200 into Idaho. This turn might be an inherited older path of the river that was established while the Renova Formation was being deposited, 48 to 20 million years ago, before the river cut down to the level where we see it now. From St. Regis to Lookout Pass, the interstate follows a narrow gorge the St. Regis River carved into Belt rocks of the Ravalli Group.

The Lewis and Clark fault zone is exposed just north of the highway near St. Regis. Look for a big rock cliff with chunks of solid rock in a matrix of crushed rock that's been bleached to pale colors by hot water circulating through it.

At Haugan (elevation 3,200 feet), 18 miles up the St. Regis River, fine pebble gravels are exposed in a large gravel pit. Whether these are from an early stage of the St. Regis River or are back-eddy deposits from Glacial Lake Missoula floods that

Folded Ravalli Group rock along I-90 at DeBorgia (exit 18).

backed up while making the abrupt clockwise turn at St. Regis (elevation 2,635 feet) is not clear, but the lack of distinct layers suggests they're back-eddy deposits from a big draining of Glacial Lake Missoula. The lake's high stand was at 4,200 feet, 1,000 feet above Haugan, so floodwaters raged here when the ice dam broke.

Well-layered dark-gray Ravalli Group rocks are beautifully exposed in roadcuts at DeBorgia, where they also show some big folds imposed by movement along the Lewis and Clark fault zone. Just east of Haugan (exit 16) are big rusty cuts in Ravalli Group sediments, badly broken and stained by water migrating along faults in the Clark Fork fault.

US 2
Kalispell—Idaho Border
121 miles

Except for one small section of Cambrian sediments preserved below a slice of thrust-faulted rock near Libby, all the bedrock between Kalispell and the Idaho border is Proterozoic Belt rock, sedimentary rocks deposited between 1.47 and 1.4 billion years ago in a vast basin within the continent. Along much of the route the bedrock lies beneath glacial debris left by melting glaciers at the end of Pleistocene glaciation. The Proterozoic formations are broken into a series of large slices, or thrust slabs, that moved east on big thrust faults more than 100 million years ago.

Thrust faults intrigue petroleum geologists because they tend to place older formations over younger ones, leaving them to wonder what they may conceal. Some geologists speculate that the Belt formations we see at the surface in northwest Montana may lie on top of much younger Paleozoic rocks. A slice of Cambrian limestone, best exposed in a big white roadcut about 15 miles south of Libby, is the uppermost and youngest part of a miles-thick west-dipping slab of Proterozoic Belt bedrock in the region. However, a wildcat well drilled on the Purcell anticlinorium (a giant arch in formations with many internal folds) near Island Lake in 1894 went down 17,774 feet in Proterozoic rocks without finding oil.

Kalispell lies in the Flathead River valley, near the south end of the Rocky Mountain Trench. The valley is filled with glacial moraines and outwash left 15,000 to

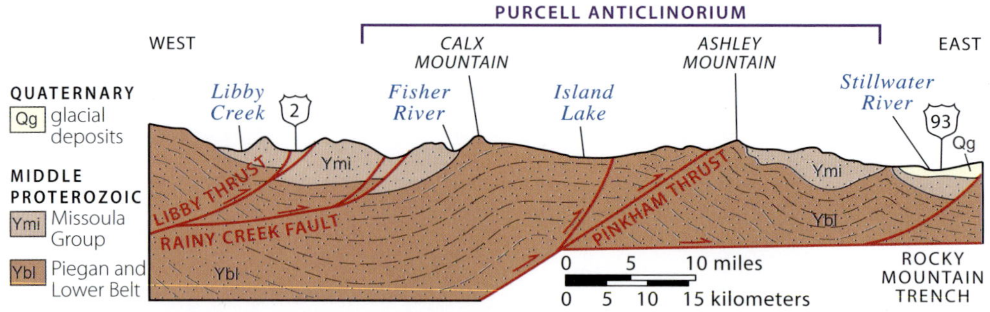

Cross section along a line north of US 2 from about 15 miles south of Libby to Kalispell (generally a few miles north of US 2). The Belt formations were warped into a broad fold as they were shoved east along deep thrust faults.

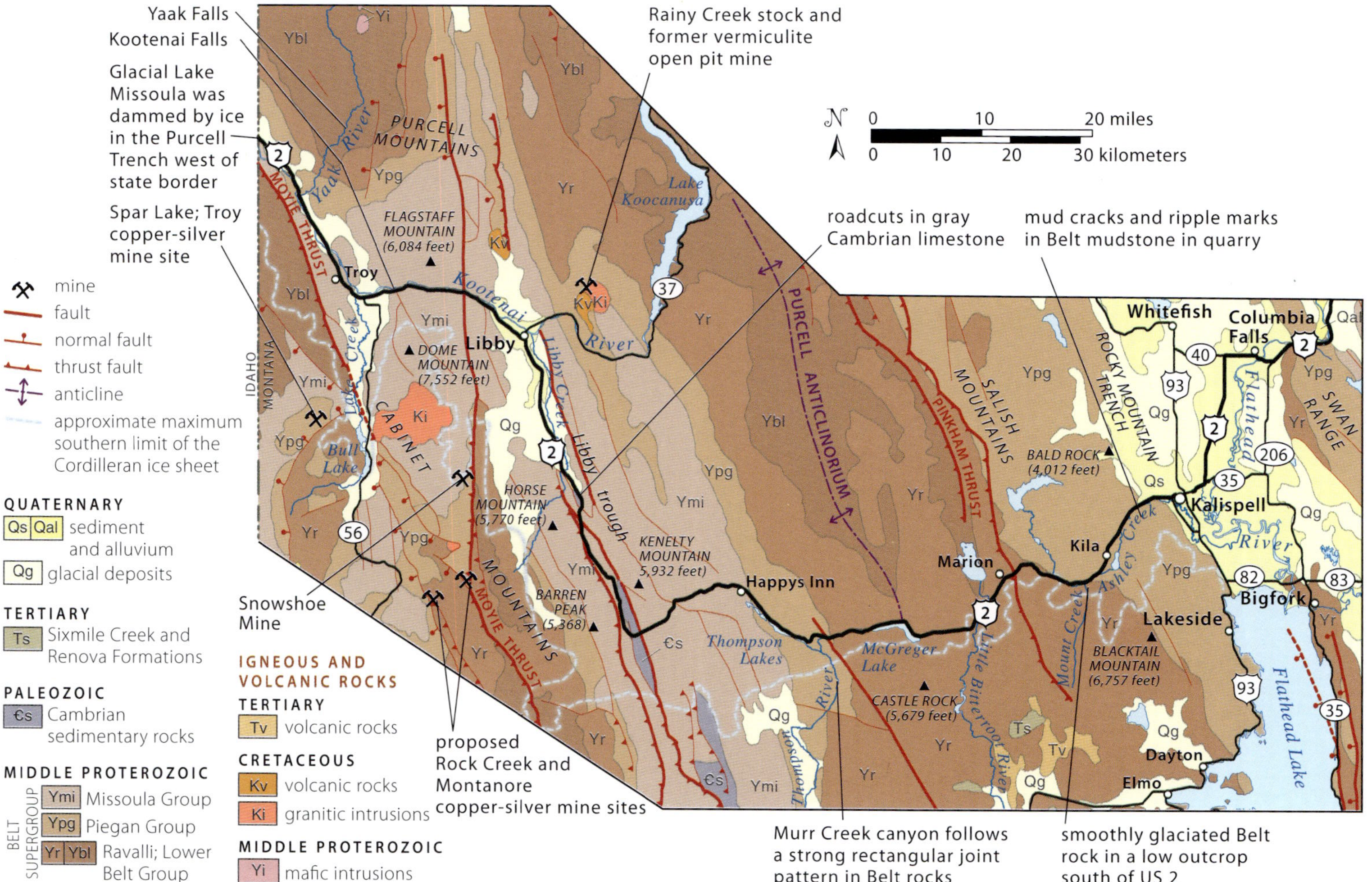

Geology along US 2 between Kalispell and the Idaho border.

12,000 years ago by a thick tongue of ice, the Flathead lobe of the Cordilleran ice sheet. Based on features on nearby mountainsides, it's estimated that the ice near Kalispell reached a maximum thickness of 2,500 feet. The Flathead lobe split into separate fingers at Kalispell. One headed south down the Swan Valley, one continued south through Flathead Lake and the Mission Valley, and another part, to the west, turned southwest and crept up Ashley Creek, passing through Kila and Marion and following the valley of the Little Bitterroot River. Meltwater streams from the western glacial finger flowed southwestward around its south edge against the high-standing Salish Mountains and into Glacial Lake Missoula.

US 2 west of Kalispell lies just north of the northern edge of the highest level of Glacial Lake Missoula, so we don't see deposits related to it. Instead, we see outcrops and the tops of roadcuts that were scoured flat and striated by rocks embedded in the base of overriding ice. Some of the abrading rocks still remain on the smoothly scoured surfaces, often with rocky unsorted deposits of glacial till on top of them. Where you see striations, they are parallel to the downhill—and sometimes the uphill—direction the ice was moving at that location. More than one direction of striations documents different ice-flow directions at different times, perhaps related to the thickness of the ice.

In some small valleys, such as that of the Little Bitterroot River near Marion, the ice locally moved upslope to elevations above the Flathead Valley floor at Kalispell. How can ice flow uphill? Ice-flow direction is controlled by the upper surface of the ice, in the same way that the water surface of a stream dictates the flow direction, even where the stream bottom may locally slope upstream. So, for example, starting with a 3,000-foot ice surface to the north, the downslope gradient to the south would push the ice down through valleys and up over adjacent ridges, allowing the ice to locally flow uphill.

West from Kalispell, US 2 follows the northern edge of a big glacial outwash flat on which gravel spread south from the melting ice sheet. About 12 to 13 miles southwest of Kalispell and west of Kila, big outcrops of pale-pink quartzite at the side of the road

A glacially sculpted outcrop of Belt rock (Helena or Wallace Formation) about 6 miles east of Marion (just west of milepost 107). The prominent, nearly horizontal lines visible in this photo are sedimentary layers; the glacial striations slope gently upward to the right across the layers on the right.

Rocks embedded in glacial ice scratched the underlying bedrock to form the glacial striations near milepost 80. At least two directions of ice movement are evident.

were sculpted by a glacier moving parallel to the road. If you stop to look closely at the glacial scratches, you'll notice that they were cut across—at an angle to—the more prominent sedimentary layers.

Watch for hummocky moraines littered with boulders, clear evidence that a glacier once lay there. About 3 miles west of Marion, US 2 crosses the Little Bitterroot River, which heads south for almost 50 miles to its mouth at the Flathead River, 15 miles south of Polson. Farther west, US 2 follows McGregor Lake and the Thompson Lakes, which occupy the valley of an ice-margin stream, an ideal low-gradient path for the highway. About 4 or 5 miles west of Marion are more outcrops that were rounded and sculpted by ice. Six miles southwest of Marion are big cuts in unlayered, unsorted glacial till with huge boulders that a glacier carried and then dropped when it melted.

Near milepost 80, 8 miles southeast of Happys Inn, is a prominent long roadcut with a top that was scoured by glacial ice. Glacial striations on this smooth surface are nearly parallel to the highway, indicating the direction of ice flow. Unfortunately, the sand and small pebbles on the smooth surface act like ball bearings, making it dangerously slippery, even when dry.

When the Pleistocene ice advances were at their maximum, glaciers pushing south out of British Columbia nearly buried the Purcell Mountains north of Libby so that only the highest peaks rose above the ice cap. Meanwhile, the Cabinet Mountains, just south of Libby, escaped such general coverage. There, mountain glaciers gouged out the valleys, leaving a dramatically carved alpine landscape of ragged peaks and ridges. Although the Proterozoic sedimentary formations north and south of the Kootenai River are nearly identical, the jagged Cabinet Mountains look quite different from their more ice-rounded counterparts north of the river.

Up near the Canadian border, northwest of Kalispell and west of Eureka, one long tongue of ice, probably a prong of the Flathead lobe when the last glacial advance was at its maximum, pushed south past Rexford down the valley of the Kootenai River that Lake Koocanusa, the reservoir behind Libby Dam, now floods. The southern

The pale-beige glacial lake silts on US 2 near milepost 5, 1.5 miles east of the junction with County Road 508, were deposited in Glacial Lake Kootenai. They are capped in the upper right by a thin deposit of old outwash gravels.

edge of the Cordilleran ice sheet reached south across the Kootenai River, extending about 14 miles up the valley of Libby Creek. Smaller glaciers flowed down the mountain valleys on either side of those lobes of ice and joined them as tributaries. Dense forests almost certainly cloaked the mountainsides above the ice-filled valleys, just as they do today in Alaska.

For several miles west of Libby, US 2 rides on a high, flat bench of glacial outwash gravels above the narrow rock-walled gorge of the Kootenai River; this bench was the smooth floor of the Kootenai River valley at the end of the Pinedale glacial advance. In the last 12,000 years or so the river has carved its canyon back down through those gravels to reexpose the Belt bedrock in the river channel. Giant cuts in gray to rusty Belt sedimentary rocks stand at steep angles along the south side of the highway.

The Purcell Valley of northern Idaho (10 or 20 miles west of the Idaho border) filled with an enormous glacier almost comparable to the one that filled the Rocky Mountain Trench in Montana. Based on erosional features on nearby mountains, the ice was about 5,500 feet thick in the Purcell Trench. That ice blocked the drainage of the Kootenai River, just as it did the Clark Fork River farther south, creating a lake that flooded the Kootenai River drainage and left deep deposits of glacial lake sediments. West of Troy near milepost 6, US 2 crosses the Yaak River to climb west out of the Kootenai River valley. Big cuts in beige lake silts near the top of the hill show that Glacial Lake Kootenai was higher than these silt deposits, meaning the ice that impounded it was even higher. The thin layer of gravel capping the lake silts is probably glacial outwash gravel that later spread out over the lake deposits after the lake had drained.

Mining near Libby and Troy

The Libby-Troy area has a long history of mining. Early prospectors found placer gold in Libby Creek about 20 miles south of Libby in 1886, and a town, Old Libby, sprang up in the next year. When the Great Northern Railway reached Libby in 1891, businesses around the placer mines on Libby Creek moved to the present location of

Libby to be near the railroad. The miners built a high flume from upstream to deliver water under pressure to hydraulic nozzles that could wash down the higher gravel banks and flush their contents through sluices to recover gold. The gravel was rich for a distance of several miles. Placer mining in Libby Creek continued off and on until fairly recently, but production since the early 1900s, probably less than $100,000, has been pitifully small considering the effort expended. The bedrock source of the gold was discovered in 1887 near the contact of a granite intrusion in the Belt rocks in the Cabinet Mountains, west of Libby Creek. Some of the mines, including the Snowshoe Mine, 12 miles southwest of Libby, worked for several decades and produced silver and lead, as well as gold.

In 1888, prospectors found deposits of copper, silver, lead, and gold at Sylvanite, on the Yaak River nearly 20 miles north of Troy. A narrow-gauge railroad hauled ore from the mines at Sylvanite to a concentrating mill in Troy. The mines produced a total of more than $4 million of metal before the mill burned in 1927 and the mines closed.

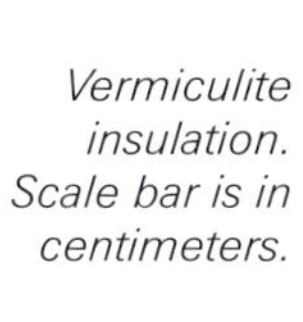

Vermiculite insulation. Scale bar is in centimeters.

The largest vermiculite mine in the country began producing in 1925 from a big open pit high on Vermiculite Mountain, about 7 miles northeast of Libby. Vermiculite is an altered variety of mica that expands like popcorn when heated to occupy about fifteen times its original volume. The yellowish-brown puffs have found many uses, mostly for fireproof insulation, acoustical plaster, and soil conditioner. By 1963, when W. R. Grace bought the company, the Libby mine was producing 80 percent of the world's supply of vermiculite pellets to be sold as insulation for people's homes.

Libby's claim to fame turned to infamy when not only miners but others began coming down with lung cancer. Vermiculite contains microscopic needles of the mineral tremolite that are released as dust when the rock is crushed. Tremolite, a rigid form of asbestos, lodges as an irritant in people's lungs to create asbestosis, a chronic, deadly disease. Miners carried it home on their clothes, and vermiculite was spread on playgrounds; it was everywhere. By 2015, 400 residents of Libby, a town of 2,600 people, had died of asbestosis; thousands more have symptoms, and many will die from it in the next few decades. Finally declared a federal Superfund site, much of the contamination has been cleaned up. Libby—surrounded by towering peaks, clean streams full of native trout, and pristine lakes—is now well-known for endless tourist opportunities.

The Rainy Creek stock, from which the vermiculite was mined, is an extraordinary igneous intrusion, most of it pyroxenite, an unusual black rock composed mainly of pyroxene. The core is a mass of rock that originally consisted almost entirely of black biotite mica, which was altered to yellow vermiculite. A cap of white syenite, mostly feldspar and the last part of the magma to solidify, covers part of the complex. The Skalkaho stock east of Hamilton is strikingly similar.

Magmas that crystallize as such rocks can come only from the mantle, deep beneath the continental crust. Several age dates suggest that the Rainy Creek stock formed about 90 million years ago, a time that doesn't fit well into the general geologic history of the region that geologists have constructed. Because such rocks are notoriously difficult to date, the age of intrusion may not be correct.

The rugged Cabinet Mountains south of Libby contain some of the largest deposits of copper and silver in the world. Since the 1980s two massive mines, the Rock Creek and Montanore, have been proposed as immense underground excavations under the Cabinet Mountains Wilderness. They are controversial, pitting economic profit and employment against long-term environmental damage and pollution. The large Troy mine at Spar Lake, about 15 miles south of Troy, mined silver and copper from a very large ore body in Proterozoic Belt sandstone (Revett Formation) of the Ravalli Group. Ore minerals are copper-iron sulfides with smaller amounts of lead and silver. It produced ore between 1981 and 1983, and 2005 and 2012, before it finally closed because of low metals prices and unstable rock underground. Most silver deposits are considerably younger than the rocks that contain them. These ore minerals are scattered along beds of sandstone, apparently deposited near the surface by metal-bearing brine that rose through carbon dioxide, hydrogen sulfide, and methane (natural gas) associated with old Belt-age oil. Similar deposits are now known in a zone south from here, through Trout Creek on I-90, to east of Wallace, Idaho. The Rock Creek and Montanore Mines north of Trout Creek have similar deposits.

Kootenai and Yaak Falls

About midway between Troy and Libby, the Kootenai River plunges a combined 200 feet over two beautiful waterfalls and down a spectacular series of steep rapids. The big parking area at Kootenai Falls, near milepost 21, is worth a stop. An easy trail, though not handicapped accessible, leads downslope, crossing the Burlington Northern Railroad on a series of steps, toward the river to where the trail branches. The branch upstream, to the right, leads less than a half mile to the end of the trail at the falls. Crystal-clear mountain water cascades and drops over horizontal layers of Mount Shields Formation (Missoula Group) mudstone and sandstone, deposited here about 1.4 billion years ago.

Look carefully at the rocks around the falls for ripple marks, salt casts, mud cracks, and stromatolites. If you backtrack to the trail junction and then continue a half mile downstream, there's a suspension footbridge across the Kootenai River. The bridge, hung by thin steel cables, looks and feels precarious, living up to its name, the Swinging Bridge. Under the south (near) end of the bridge are well-preserved fossil stromatolites, formed when mats of sticky cyanobacteria (bacteria that photosynthesize) trapped and precipitated sediment to form domal structures with crinkly internal layers; they look a bit like cabbage heads.

Picturesque Kootenai Falls spills over near-horizontal mudstone and sandstone of the Proterozoic Missoula Group.

Ripple marks on the bedding surface of this Missoula Group mudstone at Kootenai Falls formed when gentle winds blew across shallow water on a mudflat, moving the sediment back and forth into symmetrical ridges, or ripples.

To visit Yaak Falls, turn north onto County Road 508 from US 2 at milepost 3.7 and drive 6 or 7 miles to the labeled pull-off overlooking the falls. Across the highway are big outcrops of gray-brown middle Belt rocks dipping about 20 degrees south. The full flow of the Yaak River slides down this smooth bedding surface in a series of steps through a very narrow rock-bound slot. Though it looks like a huge waterslide, it ends in a dangerous churning cauldron.

US 12
Lolo—Lolo Pass (Idaho Border)
32 miles

US 12 follows the historic Lolo Trail, used for centuries by Native Americans as the northern route across the rugged Bitterroot Mountains. Lewis and Clark followed this route in mid-September 1805 with the help of Western Shoshone war chief Pikee Queenah, or Old Toby. The expedition noted that the hot springs near the pass came from the "intertices of a grey freestone rock," and that the "Indians had made a whole to bathe." The expedition made specific plans to camp at the hot springs on their return journey in 1806 just to enjoy a rare hot bath!

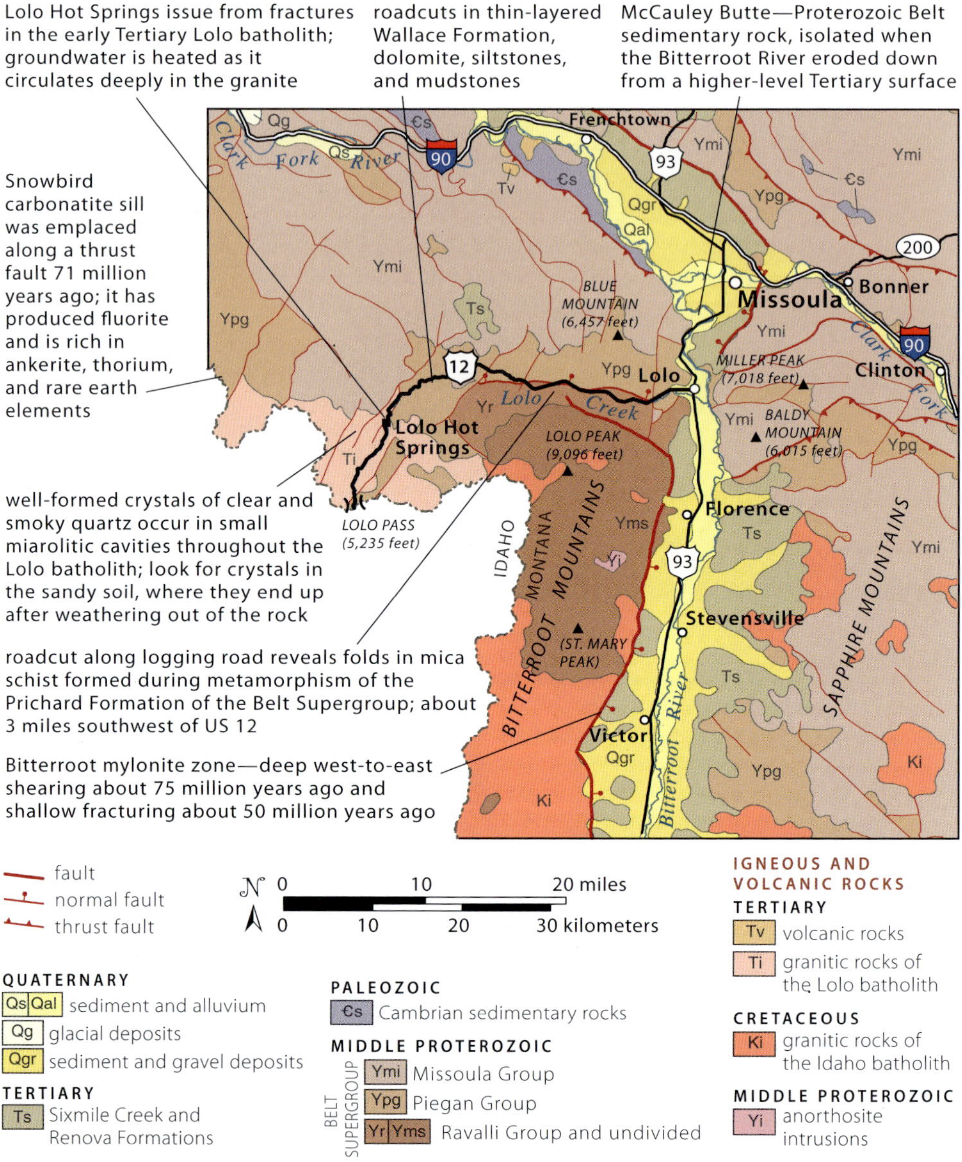

Lolo Hot Springs issue from fractures in the early Tertiary Lolo batholith; groundwater is heated as it circulates deeply in the granite
roadcuts in thin-layered Wallace Formation, dolomite, siltstones, and mudstones
McCauley Butte—Proterozoic Belt sedimentary rock, isolated when the Bitterroot River eroded down from a higher-level Tertiary surface
Snowbird carbonatite sill was emplaced along a thrust fault 71 million years ago; it has produced fluorite and is rich in ankerite, thorium, and rare earth elements
well-formed crystals of clear and smoky quartz occur in small miarolitic cavities throughout the Lolo batholith; look for crystals in the sandy soil, where they end up after weathering out of the rock
roadcut along logging road reveals folds in mica schist formed during metamorphism of the Prichard Formation of the Belt Supergroup; about 3 miles southwest of US 12
Bitterroot mylonite zone—deep west-to-east shearing about 75 million years ago and shallow fracturing about 50 million years ago
Clark Fork River
Frenchtown
90
93
200
12
Missoula
Bonner
Clinton
Clark Fork
Lolo
Lolo Creek
Lolo Hot Springs
LOLO PASS (5,235 feet)
BLUE MOUNTAIN (6,457 feet)
MILLER PEAK (7,018 feet)
BALDY MOUNTAIN (6,015 feet)
LOLO PEAK (9,096 feet)
(ST. MARY PEAK)
Florence
Stevensville
Victor
Bitterroot River
IDAHO
MONTANA
BITTERROOT MOUNTAINS
SAPPHIRE MOUNTAINS
Qg
Qs
Qgr
Qal
Ꞓs
Tv
Ts
Ti
Ki
Yi
Ymi
Ypg
Yr
Yms
fault
normal fault
thrust fault
N
0 10 20 miles
0 10 20 30 kilometers
QUATERNARY
Qs Qal sediment and alluvium
Qg glacial deposits
Qgr sediment and gravel deposits
TERTIARY
Ts Sixmile Creek and Renova Formations
PALEOZOIC
Ꞓs Cambrian sedimentary rocks
MIDDLE PROTEROZOIC
BELT SUPERGROUP
Ymi Missoula Group
Ypg Piegan Group
Yr Yms Ravalli Group and undivided
IGNEOUS AND VOLCANIC ROCKS
TERTIARY
Tv volcanic rocks
Ti granitic rocks of the Lolo batholith
CRETACEOUS
Ki granitic rocks of the Idaho batholith
MIDDLE PROTEROZOIC
Yi anorthosite intrusions

Geology along US 12 between Lolo and Lolo Pass.

SOUTH
NORTH
BITTERROOT MOUNTAINS
LOLO PEAK (9,096 feet)
Lolo Creek
12
near BLUE MOUNTAIN
Deep Creek
Frenchtown
90
Ki
Yb
Ypg
Ymi
Ꞓs
LOLO FAULT
BLUE MOUNTAIN THRUST
ALBERT CREEK THRUST
CLARK FORK FAULT
not to scale
CRETACEOUS
Ki granitic rocks of the Idaho batholith
PALEOZOIC
Ꞓs Cambrian sedimentary rocks
MIDDLE PROTEROZOIC
BELT
Ymi Missoula Group
Ypg Piegan Group
Yb Ravalli Group and undivided Belt

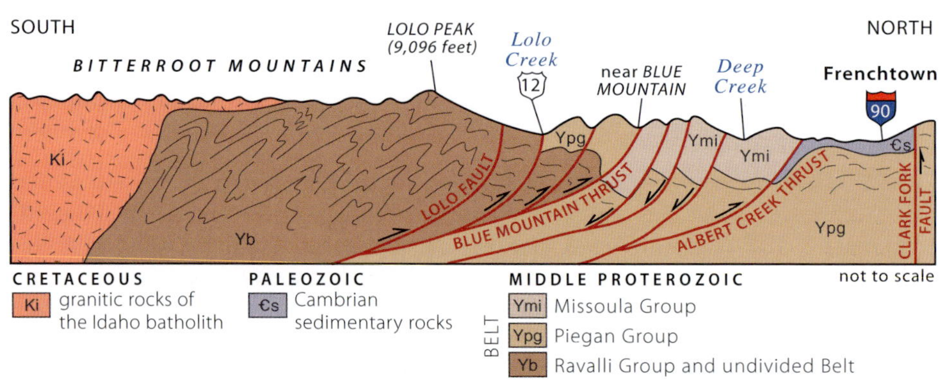

Cross section across US 12 several miles west of Lolo.

US 12 lies mostly in lightly metamorphosed Proterozoic Belt Supergroup sedimentary rocks, with older Belt rocks to the south and younger formations to the north. Although the rocks of this region have not been traced continuously from here into well-established Belt formations farther north, their rock types and stratigraphic sequence strongly resemble the sequence of Prichard Formation to Ravalli Group to Wallace Formation (Piegan Group) to Missoula Group seen elsewhere, so we use those terms here. Lolo Peak, elevation 9,096 feet, the highest point in the northern Bitterroot Mountains, lies 5 miles south of US 12. It is held up by a complex of Prichard Formation schist and gneiss heavily intruded by granite of the northernmost part of the Idaho batholith. These deepest-level Belt Supergroup schists and gneisses show moderate levels of metamorphism, probably related to their proximity to the heat of the Idaho batholith.

About 1 or 2 miles south of the highway, rocks on the south side of an east-west fault have risen relative to those on the north side, exposing white Ravalli Group quartzite with a few dark mica schists. US 12 follows Lolo Creek, which has eroded along and close to another fault, along which rocks on the south have risen against fine-grained, well-layered rocks of the Wallace Formation. Thus, we have older parts of the Belt Supergroup that metamorphosed at a higher temperature to the south, with younger parts of the Belt that were exposed to lower temperatures progressively farther north.

Thin-layered Wallace Formation (Piegan Group) is exposed in roadcuts from 17.5 to 19.5 miles west of Lolo. The whitish layers are quartz sandstone; dark layers are impure dolomite.

Between Lolo Hot Springs and Lolo Pass, the landscape changes. Scattered pine trees, often with little undergrowth, grow among big, rounded outcrops of massive, especially grainy granite. The soil consists of coarse sand derived from decomposed granite. The 50-million-year-old Lolo Hot Springs Granite weathers into giant rounded boulders.

The Lolo Hot Springs Granite is well-known to mineral collectors for the well-formed crystals of quartz that grew in miarolitic cavities throughout the granite, especially near dikes of aplite, a very pale fine-grained granite. These cavities formed as the magma crystallized and steam collected along newly formed fractures in the shrinking, solidifying magma. The magma crystallized at a shallow depth of perhaps a few thousand feet, which permitted steam bubbles to form. Excess silica precipitated at scattered points in the watery fluid to form feldspar and quartz crystals. The quartz crystals are typically clear and colorless, but some are very smoky to even black in color. Most of the cavities are tiny, too small to see without a magnifier, but

Lolo Hot Springs Granite, 50-million-year-old intrusive rocks exposed west of Lolo. Weathering along vertical and horizontal joint surfaces gradually leaves huge "boulders" right where they were formed, far from a stream. Inset: This miarolitic cavity (inset) in the Lolo Hot Springs Granite contains well-formed crystals of feldspar that grew in an open cavity in the rock. —Courtesy of Sheila Roberts

some are a few inches across. Some collectors have success finding quartz crystals by scouring dirt roads after a hard rain. Off the highway just north of Lolo Pass, rhyolite ash of similar age to the granite erupted at the Earth's surface, also indicating that the magma was near the surface.

Lolo Hot Springs emerges near the contact of Belt rock and the granite. Because the magma crystallized 50 million years ago, it has long since cooled below the necessary temperature to heat the water. But the Earth's crustal temperature increases by almost 2 degrees Fahrenheit for every 100 feet of depth. At a depth of 6,000 feet, its temperature would be around 120 degrees Fahrenheit, the temperature of the springs. Rainwater soaking into the ground to circulate as groundwater in fractures deep in the granite is heated to such a temperature before again rising to the surface. Almost all hot springs are near streams because groundwater intersects the surface at streams, which mark the top of the water table.

US 93
Missoula—Polson—Kalispell
121 miles

Three miles north of its junction with I-90, US 93 cuts through old high terraces of the Tertiary Renova and Sixmile Creek Formations and crosses the Ninemile fault—a major fault in the Lewis and Clark fault zone—onto hard, broken Belt bedrock. The fault is estimated to have had up to 7.5 miles of vertical offset, raising the Rattlesnake

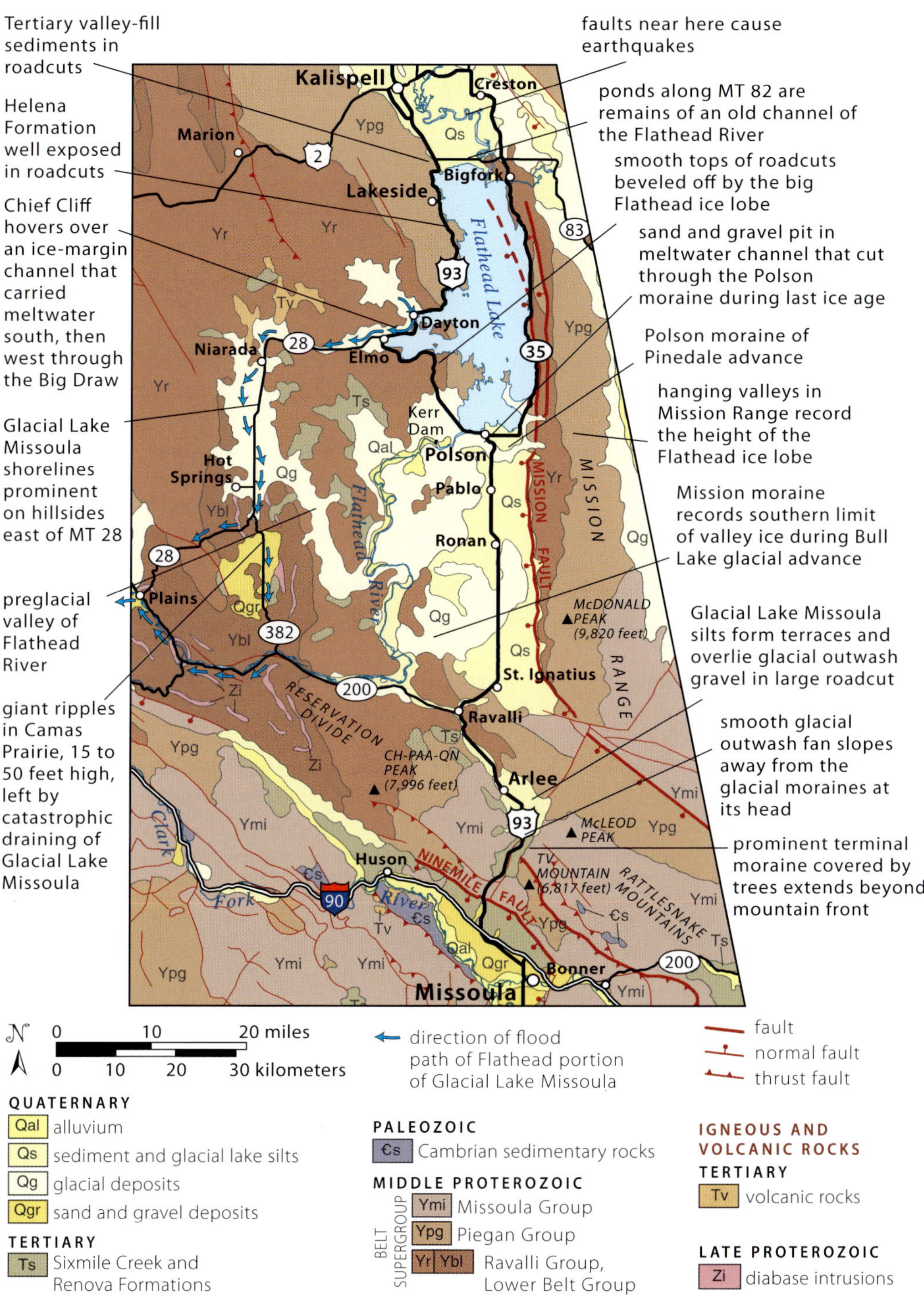

Geology along US 93 between Missoula and Kalispell.

Mountains north of Missoula in the process, but geoscientists think some of this movement was lateral sliding as well. Fortunately for Missoula, there is no evidence that the fault has moved for several million years. North of the fault, the highway climbs through Evaro canyon and the Rattlesnake Mountains to Arlee, in the Jocko Valley, one of the smaller extensional basins of the Northern Rockies. The low pass between Missoula and Evaro (7 or 8 miles north of I-90) is just barely low enough to have permitted the water of Glacial Lake Missoula's highest stand to connect with that in the Mission and Flathead Valleys to the north.

Large exposures of brownish gravel north of the divide between the Missoula and Jocko Valleys appear to be Pliocene Sixmile Creek deposits laid down during the last long period of dry climate prior to the Pleistocene ice advances. Big sheets of gravel merged the Missoula and Jocko Valleys in this area. Since the Sixmile Creek Formation was deposited, streams have hauled much of the gravel out of both valleys, leaving this high remnant.

Craggy peaks of the glaciated Rattlesnake Mountains make a bold wall along the southern margin of the Jocko Valley east of Arlee. The Jocko hills on the east and the Reservation Divide on the west, gently rounded hills that never contained glaciers, all consist of Proterozoic Belt rock. Big roadcuts in beige to pink rocks of the Missoula Group are exposed 2 miles north of Arlee, near the Jocko River Bridge.

Fine-grained beige silts deposited in Glacial Lake Missoula appear in big roadcuts between Arlee and Ravalli. The prominent terrace at eye level on the east side of the valley is a remnant of Glacial Lake Missoula silt deposits; it's a few tens of feet thick. The Jocko River, running through the valley, has eroded the edge of the terrace. The distinct pass between Ravalli and St. Ignatius was more than 1,000 feet below the highest stand of Glacial Lake Missoula.

The experience of seeing the steep wall of the Mission Range rise in the distance from the crest of the pass that descends eastward into St. Ignatius is one that is not easily forgotten. Towering more than 6,000 feet above the valley floor, the mountains are stunning. Between St. Ignatius and Polson, the deeply glaciated Mission Range

The low wooded hill in the middle ground is a terminal moraine of a glacier that reached the valley floor on the northwest side of the Rattlesnake Mountains, as seen from US 93 at the south end of the Jocko Valley, south of Arlee. It probably dates to the Bull Lake glaciation. It is most easily viewed by southbound travelers.

Along US 93 about 5 miles south of Ravalli are very thin silt layers deposted in Glacial Lake Missoula. Some layers are probably varves, alternating summer and winter layers.

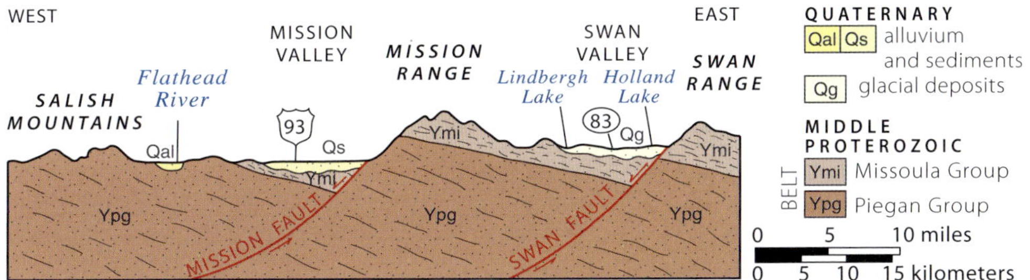

Cross section across the Mission and Swan Valleys north of St. Ignatius. The Mission and Swan Ranges basically consist of great slabs of Proterozoic sedimentary formations (Ravalli and Piegan Groups) tilted gently down to the east.

continues to loom to the east, its jagged skyline as dramatic as any in the region. The St. Mary fault, an eastern extension of the Lewis and Clark fault zone, abruptly chops off the southern end of the range near St. Ignatius just as it reaches its greatest elevation. The rocks are all Proterozoic Belt rocks, Ravalli Group quartzites in the lower exposures and Piegan Group rocks higher up.

The Mission fault, one of the big listric normal faults of the Northern Rockies, is responsible for the abrupt and steep range front. The limited amount of erosional debris at the base of the mountains has developed with time as sediment is trapped against the base of the range as the basin rapidly subsides. Where the Flathead lobe occupied the valley to the north, minimal sediment accumulated from the mountain drainages, in part because they were blocked by ice during glacial advances.

Like other big, basin-bounding faults in northwest Montana, the Mission fault dips steeply to the west, becoming flatter at greater depths. Such faults in northwest Montana have been active since Eocene time, but for a long time geologists were uncertain about how active they remain, and if they posed a significant seismic risk. That is until 1992, when the US Bureau of Reclamation studied the safety of old

The Mission Range rises high above the Mission Valley. Viewed from the rest area on US 93. Note Mission Falls, the curving white line midway up the mountain face, below the snow, on the right side of photo.

reservoir dams in the smaller valleys draining the Mission Range. To test the possible hazard from earthquakes, they excavated deep trenches across moraine and glacial lake sediments at the base of the range. What they found came as a shock: a 20-foot vertical offset in glacial lake silts, which suggests a major earthquake of perhaps magnitude 7.5 after Glacial Lake Missoula drained.

Dating suggests the event happened about 7,700 years ago. Geologists calculated the average recurrence interval—that is, how frequently such earthquakes occur—to be between 4,000 and 8,000 years. In other words, any time now! The southern segment of the fault extends from south of St. Ignatius to north of Ronan, the northern segment from north of Ronan to the north end of Flathead Lake. The Mission fault is clearly active and a seismic hazard for the region. A large-magnitude earthquake could result in ground rupture, shaking, collapse of water-saturated ground, landslides, and even significant disruption of the water in Flathead Lake, similar to what happened during the Hebgen Lake earthquake near West Yellowstone in 1959. The usual scenario for such an extensional normal fault is for the valley to drop three times as much as the range rises in a single earthquake event.

The landscape between St. Ignatius and Polson contains clear evidence of major glaciation during the Bull Lake ice advance around 140,000 years ago. A large glacier reached the Ninepipe Reservoir area, leaving the enormous Mission moraine there and an apron of glacial outwash on the floor of the Mission Valley downslope to the south. That moraine and outwash plain show their age in their more eroded and subdued appearance compared to the younger Polson moraine to the north. Glaciers advanced again during the Pinedale ice advance, which climaxed about 15,000 years ago, reaching only as far as the south end of Flathead Lake near Polson.

Low and hummocky hills that cross the mouths of the U-shaped valleys in the southern part of the Mission Range are glacial moraines deposited on the valley floor by alpine glaciers. The high peaks east of US 93 overlooking the Mission Valley stood above the huge glacier filling the valley during the Bull Lake glaciation, but smaller alpine glaciers eroded the peaks above the ice, leaving them ragged and sharp crested.

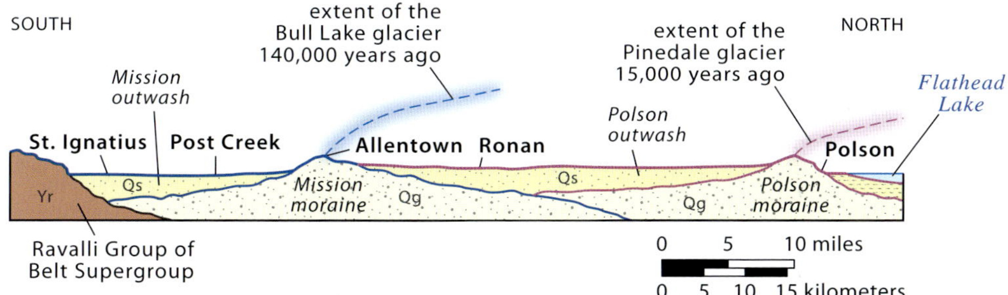

Profile of the Mission Valley between St. Ignatius and Polson showing moraines and outwash of the Bull Lake and Pinedale ice advances. The maximum advance of the Pinedale glacier, shown in pink, partially buried older deposits left by the Bull Lake glacier, shown in blue.

From the south end of Flathead Lake heading north, the peaks are lower and more rounded because the Flathead lobe of the Bull Lake glacial advance overrode them and ground them down. Bedrock outcrops on the tops of those rounded mountains are covered with deep scratches gouged by the passing ice and are littered with boulders that the ice left as it melted.

Innumerable distinctive potholes, or kettle lakes, surrounded by cattails mark the floor of the Mission Valley. They range in size and shape, from small round ponds 30 or 40 feet across to irregular, reedy ponds a half mile or so across. Clearly, they have something to do with the waning stages of the glacier that filled the valley. Most people interpret these to be places where blocks of stagnant ice were buried by outwash sand and gravel from upstream and then melted, leaving a depression that later filled with water.

More recent research suggests they may be pingos, which are common in the permafrost of the Arctic. Pingos form differently. Ice in the permafrost thaws in summer, then refreezes in winter. The pools of water underground freeze from the top down, bulging the overlying ground. Annual repetition of the process gradually bulges the mound higher, while sediment accumulates on the surface around the mound. When the ice cores within the mounds melt for good, they collapse and form depressions that can later fill with water and be hard to distinguish from kettle lakes. The perfectly round ponds in the Mission Valley are more likely to be pingos, because the ice in pingos typically swells as a round cone; the generally larger and irregularly shaped ponds are more likely from ice that was stranded near the toe of the glacier and later covered by outwash. With pingos, adjacent sediment layers curve up at the edge of the pond, but with kettles they curve down at the edge.

Soft hills on the west side of the valley preserve shorelines from Glacial Lake Missoula; its tentacles spread up all of the adjacent valleys in all directions. The repeated shorelines represent different fill levels between sudden drainings. Big boulders are scattered here and there, dropped by icebergs that broke off the leading edge of the Flathead lobe that was floating in Glacial Lake Missoula during the Pinedale glaciation. Little hills covered with pine trees in the Pablo area are old sand dunes that marched across the outwash plain during the Pinedale glacial advance, driven before the cold wind that constantly drained off the glacier.

A couple of miles south of the junction of US 93 with MT 35 at Polson is a scenic turnout on the east with a spectacular view north over Flathead Lake, the largest freshwater lake in the western states. Here we are standing on the crest of the Polson moraine, a series of prominent lumpy hills that spreads to the east. It marks the terminus of the Flathead lobe during the maximum advance of ice during the Pinedale glaciation around 15,000 years ago. It's estimated that the ice at the moraine was

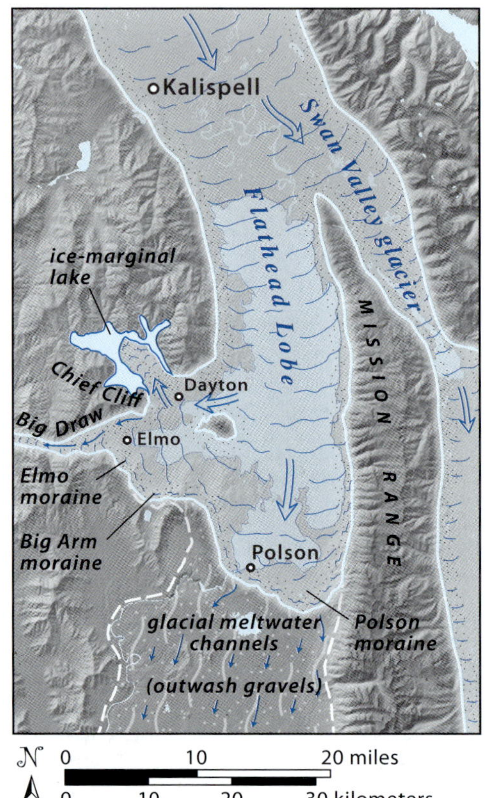

Ice flowing south through the Rocky Mountain Trench split at the north end of the Mission Range to send separate lobes down the Mission and Swan Valleys during both the Bull Lake and Pinedale glacial advances. This figure represents the Pinedale advance.

These circular ponds near the Ninepipe Reservoir may be glacial kettles or pingos.

850 to 1,000 feet above the level of the lake. The gravel in a pit to the west across US 93 was deposited in a glacial outwash channel that cut through the moraine. Note that both the moraine and the gravel outwash stand well above the level of the present lake.

Flathead Lake began to emerge at the end of the Pinedale glacial advance as the Flathead lobe retreated and water backed up behind the earth-fill dam created by the Polson moraine. The lake followed the southern margin of the retreating lobe, getting bigger and bigger as the ice retreated. The Flathead Lake basin is a 300-foot-deep hole; perhaps the great weight of the glacier itself created the depression, or it formed after stagnant ice left behind and buried in glacial deposits melted. As the ice melted and the lake filled to capacity, lake water overflowed the moraine and began to cut a channel through it. A bedrock hill beneath the moraine prevented the water from completely cutting through it, so the lake never completely drained.

North of Polson, US 93 detours to the west around the Big Arm of Flathead Lake. West of Elmo, MT 28 follows the Big Draw, a preglacial valley of the Flathead River. When the Flathead lobe reached south to Polson, it also flowed west at Elmo into the Big Draw, where it left the high Elmo moraine, the western finger of the Big Arm moraine. That moraine blocked the old valley of the Flathead River, forcing the modern river to find its present course west and then south from the southern end of Flathead Lake once the ice advances had ended. During the ice advances, streams north of Elmo drained east to the west side of the ice lobe and collected at the ice's margin; they then flowed south along it and then west into the Big Draw. In time, they gradually eroded an ice-margin valley (between bedrock and the giant glacier) well above the level of present-day Flathead Lake

When the ice receded north, the ice-margin valley remained perched high on the flank of the mountain, looking like an old irrigation ditch. The former ice-margin stream curved around the mountainside, eroding a rock-bound valley at the base of Chief Cliff, then flowed west across the Elmo moraine and spread gravelly

MT 35 (EAST SHORE OF FLATHEAD LAKE)

MT 35 hugs the limited space between the lakeshore and the steep side of the northern Mission Range. That steep slope is the fault scarp along which the Flathead Lake basin has dropped against the northern Mission Range. In a few places the road climbs over a terrace—a lateral moraine—before again dropping to the lake. All of the solid bedrock along this scenic drive between the south end of Flathead Lake and Bigfork is gray Proterozoic Belt rock. Many Belt outcrops are badly broken up as a result of movement on the Mission and Flathead faults.

Finley Point, about 3 miles north of the sharp turn of MT 35 to the north, is a long, hooked terminal moraine left by the Flathead lobe as it began to recede. Belt rock anchoring the moraine is exposed at the outer north end of the point. A scrap of moraine protrudes southward out into the lake at Yellow Bay State Park, halfway up the lake, as does another somewhat larger one at Woods Bay; these are the stubs of lateral moraines left by small tributary glaciers descending from the northeast.

Around the junction of MT 35 with MT 83, north of Bigfork, the lumpy low hills are also glacial moraine deposits. The smooth-topped gravel surfaces are glacial outwash left by the glacier as it receded to the north. North of Creston the abrupt base of the mountainside on the east is marked by very small alluvial fans built of material eroded from the mountains.

outwash down into the valley to the west toward Niarada. Even though the outwash deposits accumulated at least 12,000 years ago, the meltwater stream channels are still so clearly visible that they look as though they might have formed last year. The Montana Highway Department was so impressed that it built culverts where MT 28 crosses several of these channels, even though it's possible they haven't carried water since the Pinedale glaciation.

Lake Mary Ronan is west of US 93, 6 miles northwest of Dayton. A moraine that the Flathead lobe left on the valley wall blocks the drainage, impounding the lake. The ice-age ancestor of Lake Mary Ronan was a much larger lake trapped between the glacier and the hills to the west. Overflow from that lake, a substantial river, poured south along the edge of the ice, eroding a deep channel in the bedrock along the edge of the glacier. The ice is gone now, but long stretches of that abandoned channel were eroded into the slopes west of the highway, northwest of Elmo.

This big outcrop of glacial till along US 93 between Polson and Elmo is an extension of the moraine that blocked meltwater from flowing west through the Big Draw; it's the same moraine that MT 28 crosses west of Elmo.

Chief Cliff, a prominent bluff rising above a dry, abandoned stream channel, overlooks Big Arm Bay.

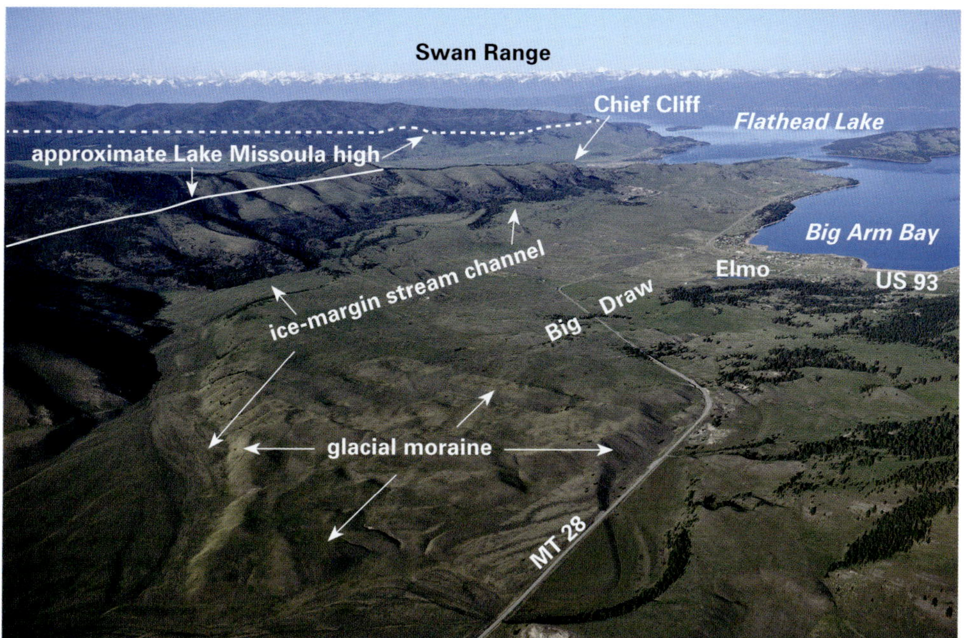

Aerial view, looking northeast, of glacial features near the west side of Flathead Lake. —Courtesy of Dave Bennett

Channels—some at road level, some below or above it—parallel US 93 through the rocky hills near mileposts 105 and 106. Ice-margin streams eroded them, and sometimes glacial tongues occupied them, scouring their sides to leave striations and grooves.

This big roadside outcrop north of Lakeside has glacial grooves and striations left by rocks embedded in glacial ice that passed by here.

GIANT RIPPLES OF CAMAS PRAIRIE

To visit the world's largest ripple marks, which formed during the catastrophic drainage of Glacial Lake Missoula, from Elmo head west on MT 28 and then south to about 3 miles south of the Hot Springs turnoff. Continue south on MT 382, 2 miles to Markle Pass, where a deep, funnel-shaped hole in bedrock on the east side of the road contains a little pond that dries up most summers. When Glacial Lake Missoula drained, water that filled the Little Bitterroot River valley north of here spilled south through Markle Pass and a pass at Wilks Gulch, 2 miles to the east, and onto Camas Prairie, then on down to the Flathead River. The torrent scoured holes into the bedrock in both passes.

The view south from Markle Pass onto Camas Prairie is nothing short of amazing. The whole valley downslope is covered with gigantic ripples. Joseph Pardee of the US Geological Survey first recognized and interpreted them as current ripples in 1942. That led to a complete understanding of the flood-scoured Channeled Scablands in Washington that J Harlen Bretz had studied decades before. MT 382 traverses a series of roadcuts through the ripples.

Although they become lower and harder to see as you head southward, the ripples remain clearly visible almost to the lower end of Camas Prairie. They appear most vividly when the sun is at a low angle, or in the spring when the grass is just beginning to grow. Like all the large valleys in the Northern Rockies, Camas Prairie acquired a deep accumulation of basin-fill sediments during the dry climate of middle Eocene through early Miocene time, and again in Pliocene time. But Camas Prairie is unique in not having contained a stream during the last several million years. It is a perfectly preserved fossil, a desert valley that still looks almost exactly as it must have looked at the end of Pliocene time, some 2 to 3 million years ago, except, of course, for the giant ripple marks.

Giant ripples in Camas Prairie, west of Flathead Lake. Note the farm building at bottom, for scale. **—Courtesy of Dave Bennett**

Cross section of a single ripple in a gravel roadcut on MT 382.

US 93
Kalispell—Eureka—British Columbia Border
74 miles

Between Kalispell and the Canadian border, US 93 follows the Rocky Mountain Trench, a gigantic valley that extends for 1,000 miles or more from Montana's Flathead Valley up the western side of British Columbia's Rockies all the way into Alaska. In spite of its size, its origin is not entirely clear, even after many decades of intensive study. The northern part appears to be controlled primarily by strike-slip faulting (lateral movement parallel to the trench), and the southern part by normal faulting (vertical slip). In a few places thrust faulting complicates the picture, but deep geophysical sensing seems to confirm this overall interpretation.

The zone of faults in the Rocky Mountain Trench follows a ramp, or step, in deep basement rocks. In Jurassic time, older Paleozoic and Mesozoic continental margin sediments were crushed against the continent and thrust up and over the stable continental basement to the east. Rivers eroded the trench along zones of broken-up rocks that marked north-to-northwest-trending faults, or along layers of softer rocks on steeply dipping limbs of folds. The alignment of the trench's valleys was fortuitous for the thick, southward-flowing Cordilleran ice sheet, and it amplified, straightened, and widened the trench to between 2 and 12 miles during the Pleistocene ice advances. The trench floor is now marked by low hills of till and outwash deposited by the ice sheet as it melted back to the north.

Much of the broad valley between Kalispell and Whitefish is floored by glacial outwash gravels; between Kalispell and the MT 40 junction, outwash rests between the Whitefish and Flathead Rivers, both east of US 93. The small rounded hills of outwash gravel are drumlins. Left by the huge Flathead lobe of the Pinedale glaciation as it stagnated and receded north, these mounds were shaped by the underside

Satellite view of the Rocky Mountain Trench (between the yellow arrows) and Front Range of the Rocky Mountains, from the Flathead Valley to northern British Columbia, looking west.
—Courtesy of National Aeronautics and Space Administration

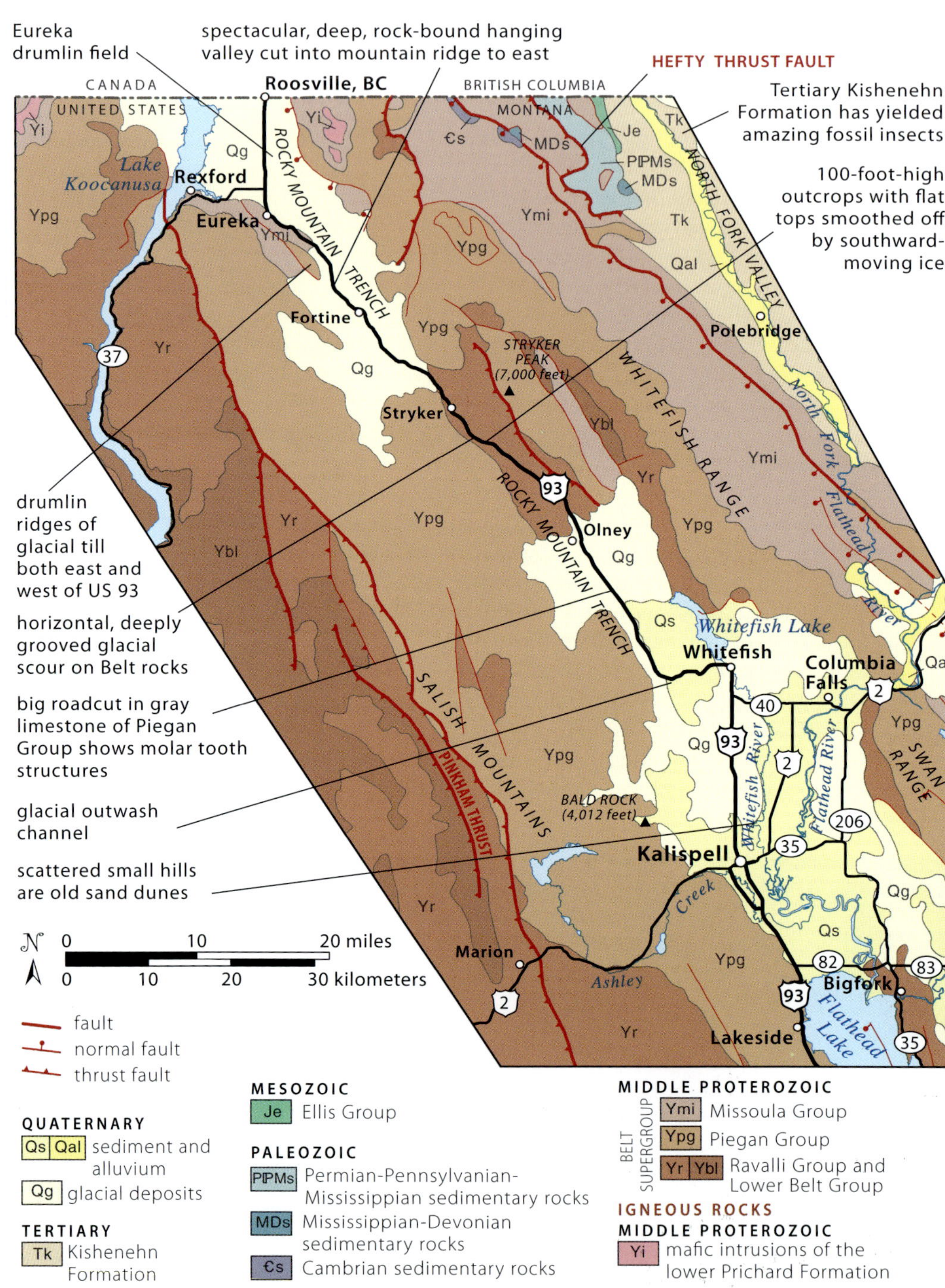

Geology along US 93 between Kalispell and the Canadian border.

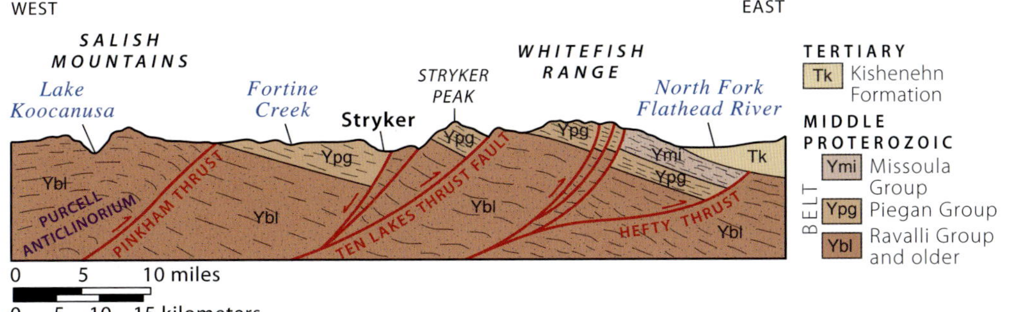

Cross section across US 93 at Stryker. Formations in the upper part of the Belt section are exposed along Fortine Creek and in the Whitefish Range.

of the glacier. They are elongated in a north-south direction, parallel to the glacier's flow. Raceway Park, just west of the highway 7 miles north of the US 2 junction in Kalispell, is on the crest of a drumlin. Scattered sand dunes along this stretch formed from winds blowing finer particles off the broad flats of glacial outwash near the Flathead River. Big cuts in till are exposed about 2 miles south of the MT 40 junction.

The south end of Whitefish Lake is blocked by a glacial moraine that formed as the Flathead lobe receded to the north. Five miles west of Whitefish, US 93 abruptly turns northward along the east side of the Stillwater River, which follows an old glacial outwash channel in the Rocky Mountain Trench. Like various parts of the trench, this section is generally low lying, with higher ground both east and west, but even here there are patches of hard bedrock. The trench here is marked by a group of steeply oriented faults in a zone about 5 miles wide, all trending northwesterly parallel to the trench. Most are buried under glacial outwash but are intermittently exposed to the north. Roadcuts in this area are in Proterozoic Belt limestone of the Helena Formation of the Piegan Group.

About 5 miles north of the turn, big cuts on the east side of the road are in southeast-dipping light-gray to slightly rusty limestone of the Helena Formation. The dark squiggly lines in the limestone are called "molar tooth structures." They were once thought to be related to vertical films of cyanobacteria that were compressed in the soft rock as it formed. More recently it's been proposed that earthquakes shook the lime mud, causing vertical cracks that then squiggled due to the dewatering and compaction of the still-fluid sediment. Others argue that gas produced by cyanobacteria got trapped in the wet, sticky sediment as the gas rose toward the surface, making open cracks. The open cracks then quickly filled with calcite, which precipitated from water in the cracks. Whatever their origin, the molar tooth structures must have formed very soon after the lime mud was deposited.

Low roadcuts in bouldery dirt near Olney are glacial till. About 7 to 8 miles north of Olney and 3 miles south of Stryker, outcrops of Ravalli Group sandstone along the east edge of the road show big, smooth horizontal grooves that look like they were left by a woodworker's giant shaper. These striations, and the 100-foot-high ledges, were smoothed flat by the relentless grinding of rocks embedded in overriding ice.

One mile farther north, 2 miles south of Stryker, there's another beautiful outcrop of Proterozoic Belt rocks on the east side of the highway. The broad, smooth horizontal

This limestone outcrop of Piegan Group Belt carbonates north of Whitefish is near milepost 136 on the east side of US 93. The inset shows squiggly gray molar tooth structures in the same outcrop.

grooves parallel the highway, the same direction the Flathead lobe moved. Mountaintops to the east were all completely rounded off by the overriding Cordilleran ice sheet, in stark contrast to the ragged crests just east of Whitefish, which protruded above the ice. Between Stryker and Fortine there are roadcuts in beige glacial lake sediments.

If you watch carefully for a gap in the trees east of the highway about 4 miles north of Fortine (near milepost 171), you can get a really nice view of a perfect U-shaped hanging valley. The regional ice sheet scraped off and rounded the top of the ridge above the valley as it moved south. The hanging valley formed later, during the Pinedale glaciation, when the Flathead lobe was confined to lower elevations near where the highway runs. The hanging valley glacier would have met this main valley glacier near the top of the ice filling the main valley. Two miles south of Fortine, 0.5 mile northwest of Murphy Lake, there is a cut in glacial lake sediments showing repeated light and dark layers known as varves. The darker layers contain the remains of vegetation killed off during winter. These sediments were deposited in a temporary lake at the toe of the receding Flathead lobe. Two miles north of Fortine and just west of the highway is a big drumlin strung out parallel to the highway.

At the south edge of Eureka is a large outcrop of dark-gray basalt with sparkling white veins of feldspar. The basalt is one of many such basaltic intrusions in the Belt rocks. Water dissolved the silica for the feldspar veins out of the basalt while the rock was still deep underground. The same thing happens in many limestones, except that the veins are composed of calcite dissolved from calcium carbonate of the limestone.

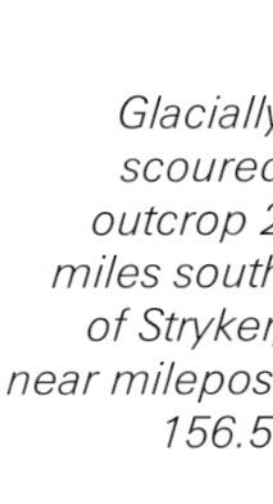

Glacially scoured outcrop 2 miles south of Stryker, near milepost 156.5.

This exposure of basalt, at Eureka, was injected into the adjacent Prichard Formation. White feldspar fills fractures in the rock.

At the north side of Eureka the landscape changes suddenly. The broad grass-covered flat-bottomed land, 8 to 10 miles across, is prominently laced with striking drumlins all lined up north to south, parallel to the highway. These peculiar hills are deposits of stream gravel and till that were molded into streamlined forms beneath flowing ice. As the ice melted back to the north and its thickness decreased, sand and gravel in glacial streams flowing in tunnels under the ice clogged their channels. The thinned ice flowed over the stalled sand and gravel, shaping it into long hills in the direction of ice flow. Thicker, heavier ice would have been too heavy to ride over the loose gravel; it would have merely pushed away and flattened the accumulated gravel. The high upstream end of each drumlin is blunt, facing into the oncoming ice; each drumlin trails off into a thin tail downstream. From the air, drumlins look like schools of giant tadpoles lined up with their big heads facing upstream. These drumlins continue to Roosville, at the Canadian border.

A short side trip of about 4 miles west from Eureka on MT 37 leads to the resort community of Rexford on Lake Koocanusa, a long reservoir impounded by the Libby Dam. A local resident coined the name Koocanusa in response to a contest. It's a

Huge drumlins like these are present between mileposts 180 and 182 north of Eureka. The northern ends are rounded and the southern ends taper. Ice flowed in the direction of the arrow.

combination of abbreviations: *Koo*tenai River, *Can*ada, and *USA*. About 1 mile east of Rexford is a spectacular exposure of black slates of the Prichard Formation (Lower Belt Group) showing very well-preserved, symmetrical, wave-generated ripple marks. Thin slabs break off from different layers to reveal varying ripple orientations, indicating that the wind blowing over the shallow water changed direction from time to time.

There are boat ramps at Rexford, where MT 37 turns south down the lake; they provide the best access to the lake. Across the bay to the north, thick layers of beige glacial sand break into blocks along vertical cracks. Continued erosion of the cracks leaves hoodoos, rock columns with sometimes interesting shapes, though these are best seen from the water to the north around the point from the public campground at Rexford.

MT 37 clings to the mountainside along the east side of Lake Koocanusa, mostly well above lake level. Huge roadcuts in gently dipping Proterozoic Belt mudstones are numerous along MT 37 because the Flathead lobe that scoured the Kootenai River valley during the last ice advance removed most of the soil and steepened the valley sides.

MT 83
Clearwater Junction—Bigfork
89 miles

MT 83 follows the length of the heavily forested Swan Valley, crossing glacial debris all the way. Belt rock crops out only around Salmon Lake and near Bigfork. The Swan Valley lies between the Mission Range on the west and the Swan Range on the east. Both ranges are great slabs of nearly identical Proterozoic Belt sedimentary formations, mostly of the Ravalli, Piegan, and Missoula Groups, each similarly tilted down to the east. Both ranges were pulled apart starting in Eocene time, around 50 million years ago; the valleys at their western bases dropped along steep listric normal faults,

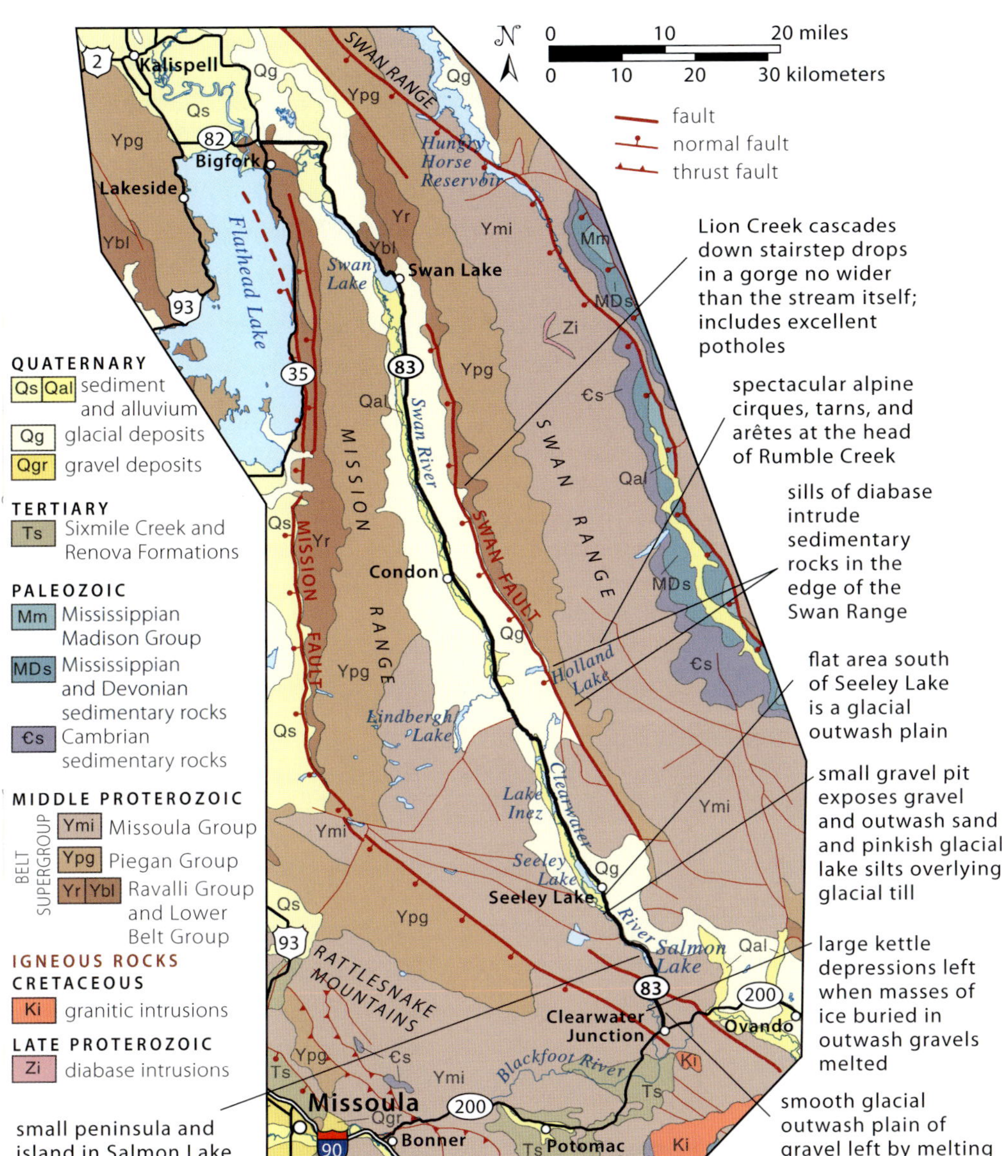

Geology along MT 83 between Clearwater Junction and Bigfork. See cross section on page 79.

forming half grabens. It seems likely that the faults are still active, based on earthquakes on other faults in the region, including a magnitude 7.5 earthquake on the Mission fault about 7,700 years ago. The lack of significant erosional deposits, such as alluvial fans, at the foot of the steep west-facing mountain fronts may be due to rapid basin subsidence, which traps valley-floor alluvial sediment against the active range fronts. It's also possible that ice filling the valleys during Pleistocene glaciation prevented alluvial fans from forming until after the ice retreated about 12,000 years ago.

During the maximum extent of the Bull Lake glacial advance, around 136,000 years ago, the Swan Valley branch of the Flathead lobe left moraines that form many of the low hills at the south end of the valley just south of Clearwater Junction. The road between Clearwater Junction and Salmon Lake crosses the smooth surface of an outwash plain, deposited by streams flowing from the Swan Valley glacier of the Pinedale glaciation, which left its terminal moraine near the lower end of Salmon Lake. Between 19,000 and 15,000 years ago, at the height of that last glacial advance, this smooth surface must have been a broad expanse of watery sand and gravel that flooded with torrents of glacial meltwater on warm summer days. Ice buried in the outwash eventually melted, and the surface collapsed to form Harpers Lake (1.5 miles north of MT 200) and several small ponds near the road.

Salmon Lake, at the southern end of the Swan Valley, lies north of the Pinedale terminal moraine, which formed a natural earth-fill dam. The Clearwater River eroded through the moraine right down to hard Proterozoic bedrock. Had that bedrock been a little lower in elevation, the river would have drained the lake.

After glacial ice finally melts, the gravel beds of the meltwater streams that flowed beneath the ice remain as long ridges of sand and gravel called eskers. One of those ridges winds for several miles parallel to the road in and beside Salmon Lake. A small public recreation area is on the southern end of the ridge, and its northern end forms several small islands, one occupied by a large house.

North of Salmon Lake, the road passes many large and small lakes that spangle the forested floor of the Swan Valley; all of them fill ice-melt depressions in the glacial deposits. As the Swan Valley glacier receded, leaving its unsorted piles of glacial till, Glacial Lake Missoula intermittently lapped onto those tills, in places leaving silt deposits. For example, at the Seeley Lake Ready Mix gravel pit, about 3 miles south of the town of Seeley Lake, silts overlie glacial till, which overlies Proterozoic Belt sedimentary rocks tilted at 45 degrees. The pit is private property; please don't enter without permission.

The rugged peaks of the Swan Range stood above the level of the Swan Valley glacier. Alpine glaciers with bottoms at the level of this main valley glacier eroded

The small peninsula protruding into Salmon Lake near its south end (near milepost 6) is an esker deposited by a stream that flowed under a glacier.

West from MT 83, a few miles north of Summit Lake, the ragged peaks of the Mission Range stood above the Swan Valley glacier, but the ice rode over the lower, snow-free mountains. The horizontal tree-covered surface midview is a moraine deposited by the glacier.

the smaller side valleys that now hang above the valley floor. After the Swan Valley glacier was finally gone, large glaciers continued to creep down these tributary valleys of the Mission and Swan Ranges. Those last lobes of ice left high moraines around their edges that now enclose Holland and Lindbergh Lakes, as well as several others. Holland Lake is about 20 miles north of Seeley Lake and 3 miles east of MT 83. Huge pale-green and red boulders at its campground beach are Missoula Group mudstones that came from the nearby Swan Range.

Near the big right-angle turn west toward Bigfork, MT 83 crosses lumpy moraines left as the Swan Valley glacier receded and descends to the smooth glacial outwash surface to the west. The east shore of Flathead Lake is marked by lumpy glacial moraines extending due north from Bigfork to just east of Creston and Lake Blaine. As the Flathead lobe melted back to the north, leaving these moraines, Glacial Lake Missoula silts were deposited on the top of a bluff south of Bigfork at an elevation of 3,000 feet. The postglacial Swan River cut a sequence of terraces down from this level to its current level at 2,884 feet at Bigfork. Kerr Dam (now Seli'š Ksanka Qlispe' Dam), built at the south outlet of Flathead Lake near Polson in 1938, raised the lake 10 feet above this to its current controlled level of 2,894 feet.

MT 200
Ravalli—Thompson Falls—Idaho Border
115 miles

Along MT 200, from its junction with US 93 at Ravalli to the Idaho border, most of the thickness of the Proterozoic Belt sedimentary rocks that dominate northwest Montana's bedrock are exposed. The highway follows much of the extent of the Lewis and Clark fault zone, the faults of which slice through both the Belt rocks and north-trending Cretaceous folds and thrust faults. Much of the Belt rock along this stretch

Geology along MT 200 between Ravalli and the Idaho border. Steeply dipping normal faults and strike-slip faults of the Lewis and Clark fault zone control the northwestward trend of the dominant valleys along MT 200.

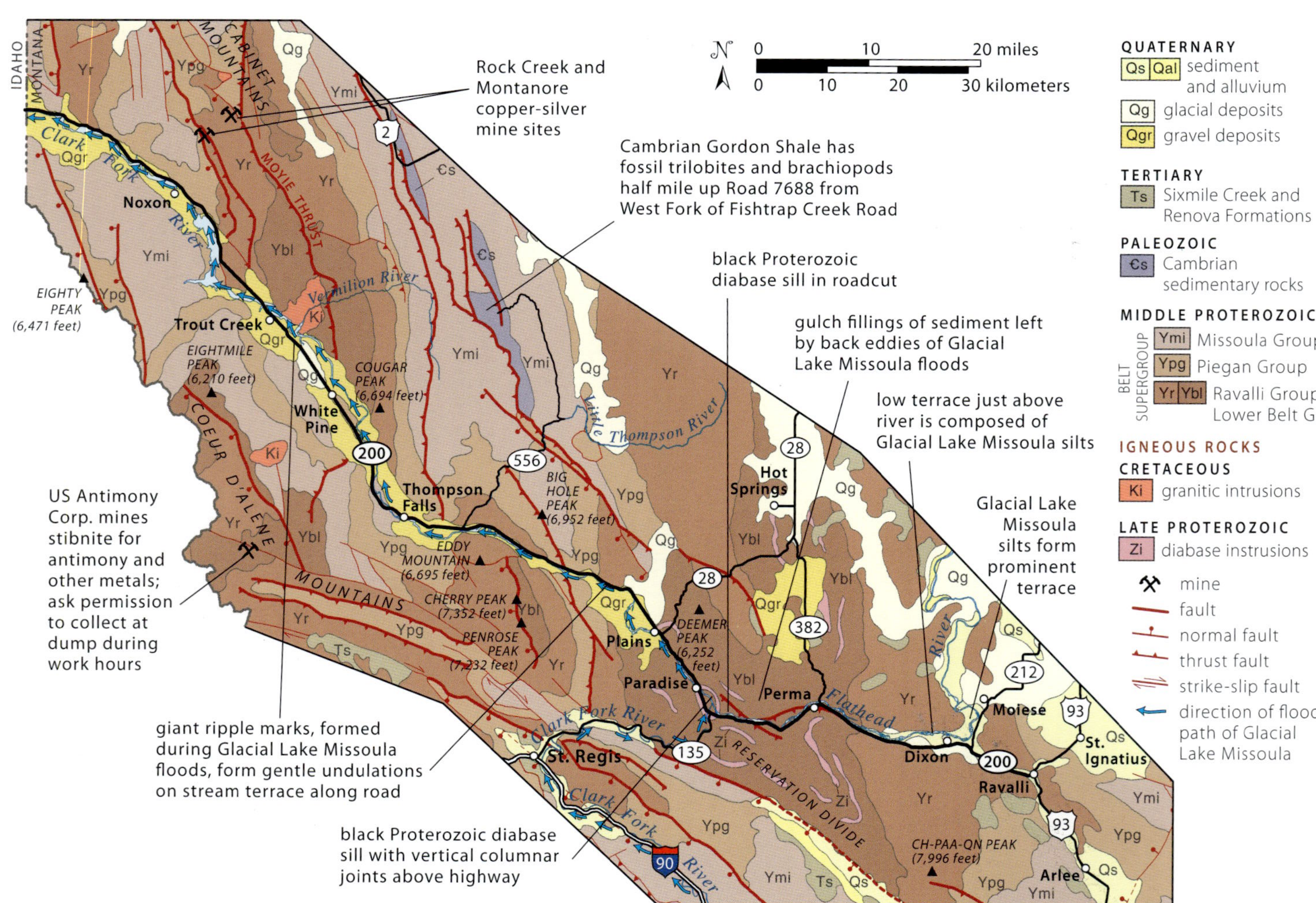

of highway is the Prichard Formation, thousands of feet of very dark-gray mudstone that weathers into reddish-brown outcrops because it contains tiny crystals of iron pyrite that oxidize to reddish iron oxide. The Belt formations in this part of Montana must have been deeply buried. They contain metamorphic minerals that formed as the rocks recrystallized at a temperature of 500 to 600 degrees Fahrenheit about 1.4 billion years ago.

The road also passes several big diabase sills (grainier versions of basalt that cooled more slowly) that intruded the sedimentary layers about 1.47 billion years ago, when the Nuna supercontinent was extending to form the Belt Basin. The diabase is black, and in places it split into distinctive columns when the magma shrank as it crystallized. Such columns always form perpendicular to the cooling surfaces of the magma body, in this case the top and bottom of the big sills. The younger Ravalli Group quartzite, such as that exposed about 8 miles east of Thompson Falls, tends to be richer in quartz and paler in color.

The Belt rocks were eroded for about 1 billion years before they were buried by beach sand, shelf mud, and limestone as an ocean rose and spread across the North American continent during middle Cambrian time. Cambrian rocks north of Thompson Falls are rich in tracks, trails, and fossils of crab-like trilobites, fossil evidence that the Proterozoic formations lack. The trilobites occur primarily in the Gordon Shale (Wolsey Shale in southwest Montana), a greenish shelf mud studied by Charles Doolittle Walcott, the pioneering paleontologist who discovered the famous soft-bodied fossils of the Cambrian Burgess Shale near Field, British Columbia, in 1909. Several quarries have worked the underlying Flathead Sandstone, the oldest Cambrian formation in Montana, for flagstone and building material. The rock comes out of the quarries in colorful slabs beautifully patterned in shades of red, brown, and yellow.

MT 200 follows the former long arm of Glacial Lake Missoula down the Clark Fork River, northwestward toward its glacial ice dam near the Idaho border. The road tracks the Jocko River between Ravalli and Dixon, the Flathead River between Dixon

Just north of the MT 135 junction a thick diabase sill with vertical columnar joints forms this cockscomb ridge. It intruded between layers of the Proterozoic Prichard Formation.

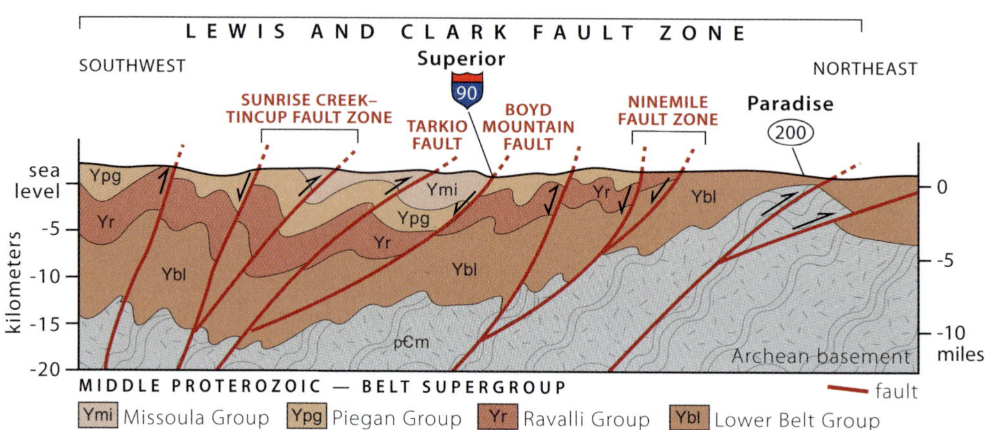

Faults in the Lewis and Clark fault zone are primarily steeply south-dipping thrust faults (with left-lateral shear that rotated structures counterclockwise); rocks on the south sides of most of the faults moved upward. The zone also contains tight folds, the axial planes of which parallel the faults.

and Paradise, and the Clark Fork River between Paradise and the Idaho line. Except for a few miles east of the Idaho border, no part of the valley contained glaciers, but this valley carried all the drainage of Glacial Lake Missoula. We see evidence of the giant floods everywhere, including where its ice dam rested at the valley's west end.

For about 3 miles east and west of Dixon, the low terrace at the base of the hills north of the highway consists of beige Glacial Lake Missoula silts. The same silts are visible in low hills to the south. In places, especially south of the highway between Dixon and Perma, very irregular hills just south of the highway are part of a large, blocky, slow-moving landslide that also underlies the highway, creating ongoing maintenance problems for the highway department.

Flood flow through the narrower stretches of the valley was very fast and highly erosive, especially between Perma and Plains, and for several miles east of Thompson Falls. The torrents scrubbed off most of the soil, leaving ragged bedrock on the lower valley walls and valley floor. Many of the rough bedrock knobs on the valley floor have smooth, grassy slopes on their downstream sides. These are flood deposits of sand and gravel that were swept into the lee side of the knobs. The scoured lower part of the valley contrasts sharply with the soil-upholstered slopes that were above the reach of the flood. Water rushing through the narrow reaches of the Flathead River valley eddied into the mouths of tributaries and dumped sediment there to form deposits that look almost like small earthen dams. Watch for these gulch-filling deposits between Perma and Paradise, east of Plains.

You can calculate the amount of water flowing through a channel by multiplying the area of its cross section by the speed of the flow. The gulch fillings between Perma and Paradise record the upper limit of the floodwaters and therefore make it possible for us to reconstruct the cross section of the largest floods. It's possible to approximate their speed by looking for the largest rocks they rolled along the valley bottom. In this case we're talking rocks almost the size of basketballs, which suggests a flow speed of about 45 miles per hour. And that is a minimum figure, because the size of

About 4 miles east of the MT 135 junction, the view north across the Flathead River shows sand and gravel filling in a side valley; a back eddy of the Glacial Lake Missoula floods left the fill. About 2 miles west of the junction is another such gulch filling.

the largest transported rocks may have been limited by the size of the largest chunks that broke free of the bedrock, and not by the actual speed of the flow. The tendency of torrential floods to bury the largest boulders they carry under smaller debris as the water subsides further minimizes the figure—in other words, larger rocks are likely to be buried out of sight.

With this in mind, it's estimated that between 8 and 10 cubic miles of water per hour surged down the valley between Perma and Paradise while the greatest fillings of Glacial Lake Missoula were draining. To put that into better perspective, consider that the greatest flood discharge ever measured on the Mississippi River at Memphis was 0.02 cubic miles per hour. The Glacial Lake Missoula flood was about 450 times that discharge—more water than the combined flow of all the rivers of the world! Gentle undulations on the surface of fields just west of Plains, and for about 5 miles both east and west of Trout Creek, are giant ripples left by the flood. Some are exposed in gravel roadcuts.

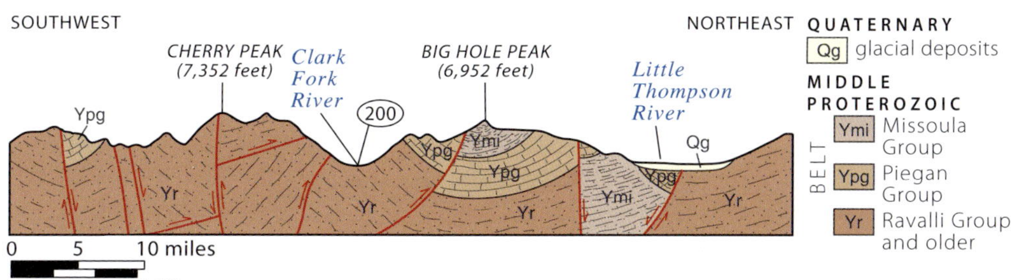

Cross section across MT 200 about 10 miles east of Thompson Falls. The structure near Cherry Peak, where a steep, near-vertical fault offsets an older thrust fault along which rocks were shoved east, is typical of the region.

West of Plains to Thompson Falls and to the Idaho border, Glacial Lake Missoula's high stand came to about halfway up the uppermost slopes on both sides of the valley. The Clark Fork River was dammed by the huge tongue of the Cordilleran ice sheet that filled the Purcell Trench of northern Idaho, covering Sandpoint and Lake Pend Oreille. The ice dam spread east up the Clark Fork drainage at least a few miles into Montana.

Thompson Falls contains several cold air springs long used as natural refrigerators. The town stands on a hill that looks like the debris dump of a big rockfall from many thousands of years ago. The ground is somewhat lumpy and full of angular boulders that project at odd angles from the soil. Air sinks into open spaces between those rocks only on winter days, when the air aboveground is colder and therefore denser than that underground. That, combined with the good insulating qualities of soil and rock, keeps food stored in any ground cavity in the hill well refrigerated.

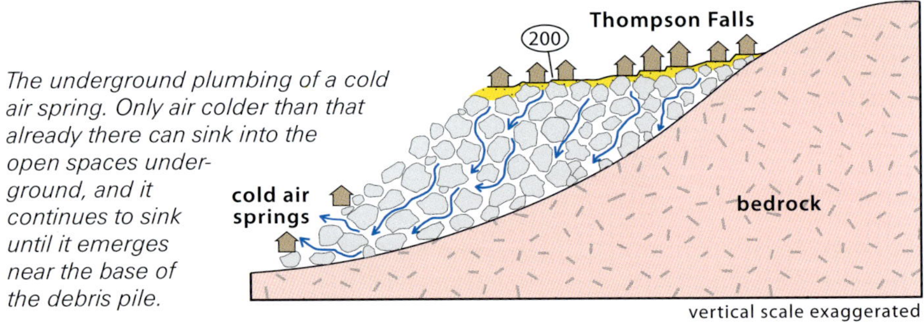

The underground plumbing of a cold air spring. Only air colder than that already there can sink into the open spaces underground, and it continues to sink until it emerges near the base of the debris pile.

MT 200
Missoula—Rogers Pass—Bowmans Corners (US 287 Junction)

116 miles

MT 200 east of Missoula follows the Blackfoot River to its headwaters on the Continental Divide at Rogers Pass. Between Missoula and Ovando, the road angles across the Garnet Range past steeply dipping sedimentary layers of Belt rocks that are tightly folded; the layers were originally horizontal. This is the north edge of the Sapphire detachment block, a great mass of rock that moved east into Montana off the top of the Idaho batholith between about 75 and 70 million years ago. The folds are the crumpled leading edge of the block.

Granite intrusions in the Garnet Range appear to have been emplaced along the big thrust faults that carried the mountains east as the Sapphire block. The Sapphire block probably began to move before the Idaho batholith crystallized, and it may have moved on a base of partly molten granite magma, the granite we now see in the numerous intrusions.

Rocks near the highway are mostly pink and gray sandstone and colorful red and green mudstone of the Missoula Group. Some of the sandy rocks have internal layers that lie at steep angles to the bedding. These are cross beds, and the layers are tilted in the direction that streamflow moved the sand grains. Exposures are especially good in the Blackfoot River canyon between Bonner and the Potomac Valley.

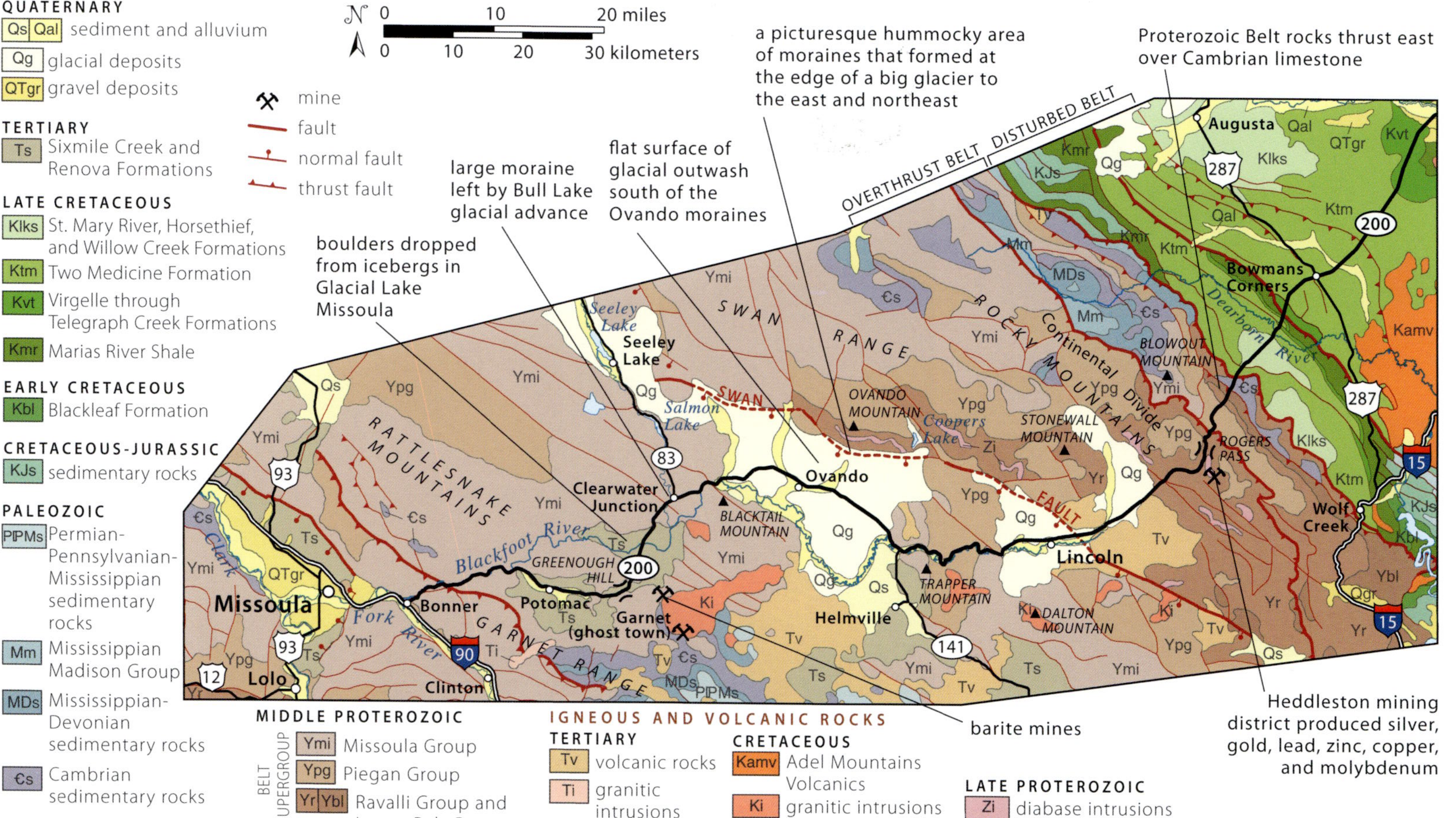

Geology along MT 200 between Missoula and Bowmans Corners.

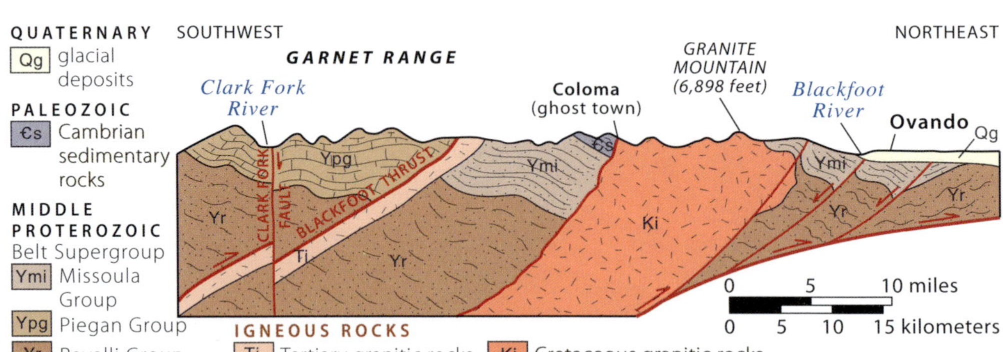

Cross section south of MT 200 between Missoula and Ovando.

Missoula Group sedimentary rocks east of Bonner near milepost 8.5.

The Potomac Valley and the broad valley around Clearwater Junction are both deep extensional basins filled with sediment, including the Renova Formation, which accumulated in lakes and streams from Eocene to Miocene time. The two valleys were one continuous basin until the ridge along the north side (informally known as Greenough Hill) was likely tilted up along a northwest-trending normal fault. Occasional earthquakes, some sharp enough to rattle windows in Missoula, come from the Greenough area. They may mean that Greenough Hill is still rising along the fault.

Roadcuts on the north side of MT 200 where it passes through the Potomac Valley expose soft pale silts and clays of the Renova Formation capped by a thin veneer of

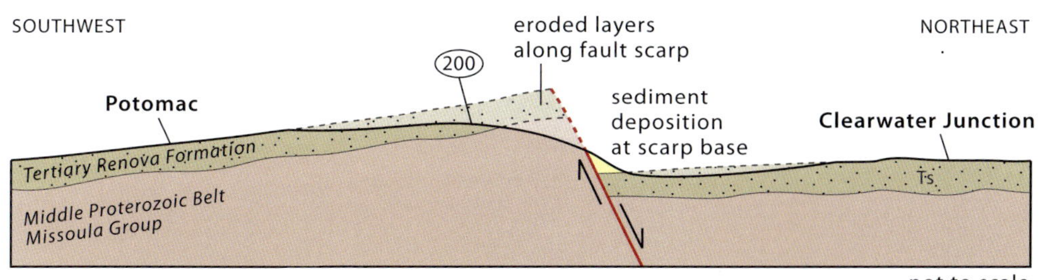

Cross section along the line of the road between Potomac and Clearwater Junction showing how the ridge at Greenough may still be rising along an active fault. The dashed line shows the ground surface before erosion.

the Sixmile Creek Formation, coarse Miocene and Pliocene gravel. Roadcuts on both sides of the road on the northeast side of Greenough Hill expose the same material.

At the top of Greenough Hill, about 6 miles east of Potomac (between mile markers 22 and 23), a scenic, well-maintained, mostly gravel road leads 11 miles south to the well-preserved ghost town of Garnet. Miners working gold gravels downstream near I-90 searched upstream for the bedrock source of the gold and discovered the ore. Underground mining for gold there occurred mostly between 1895 and 1908. A fire destroyed half of the town in 1912. When President Roosevelt jumped the price of gold from $16 to $32 per ounce in 1934, mining resumed, but it didn't last because dynamite use was restricted during World War II. The ores formed where Cretaceous granitic magma of the Garnet stock came into contact with Paleozoic limestones, reacting to form the garnet-rich rock called skarn.

The broad, flat area around Clearwater Junction, where MT 83 heads north, is a glacial outwash plain deposited by meltwater from the Swan Valley glaciers that deposited the low, lumpy glacial moraines largely along the north side of the outwash plain. From a few miles southwest of Clearwater Junction to Rogers Pass, the highway crosses several areas of lumpy landscape full of rounded humps and hollows—typical glacial moraine topography. The humps are liberally strewn with erratic boulders, and the hollows commonly hold marshes, shallow ponds, and a few small lakes. Glaciers from the big canyons in the mountains to the north reached down the Blackfoot Valley, where they spread out to form large ponds of nearly stagnant ice called piedmont glaciers. Most of those expanses of ponded ice were several miles across. Muddy meltwater pouring off them during the warm days of the ice-age summers spread sand and gravel across the ice-free parts of the valley to create large outwash plains. Moraines, such as those just east of Ovando, are strikingly lumpy because high lumps are places where large patches of rocks and dirt happened to accumulate on parts of the glacier. When the glaciers stagnated and melted, those patches remained as lumps of moraine.

Ovando stands on a smooth outwash plain between two expanses of hummocky morainal topography left by a piedmont glacier. Between Ovando and Lincoln, the road alternately crosses similar areas of smooth outwash and hummocky moraines, a pattern that's easy to recognize. Watch also for roadcuts in bouldery glacial till.

Between Lincoln and Rogers Pass, MT 200 follows a narrow valley cut through the high ridges of the Overthrust Belt. The ridges are hard Proterozoic Belt sedimentary

This hummocky topography near Ovando is composed of a glacial moraine. The flat surface in the foreground is a gravel outwash plain.

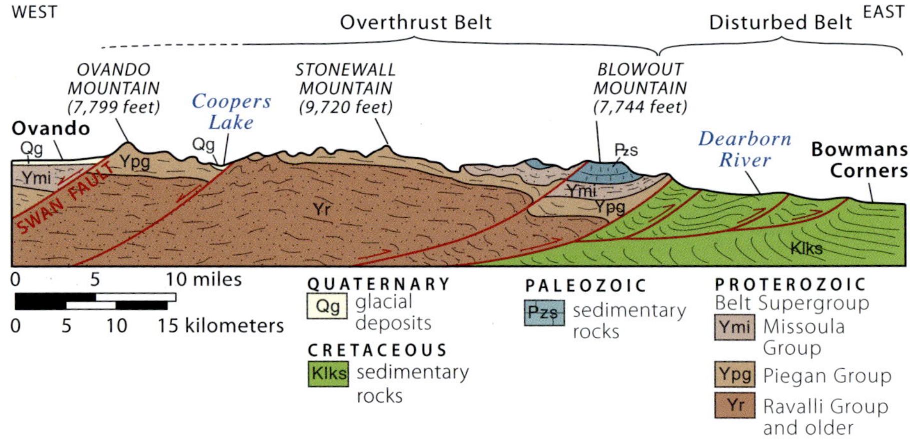

Cross section north of MT 200 between Ovando and Bowmans Corners.

rock that moved eastward in great slabs during Cretaceous time as the Farallon plate collided with the North American plate. Subduction of the ocean floor formed masses of molten granite magma, some up to several miles in diameter, that intruded the overthrust slabs, bringing small quantities of metallic minerals with them, including those at Garnet.

The Heddleston mining district between Lincoln and Rogers Pass includes several mines, of which the Mike Horse, just south of the highway, is by far the largest. Ore was discovered in 1898; the mine initially produced gold beginning in 1915, then silver, zinc, and lead, likely producing nearly $10 million worth of minerals before finally closing in 1955. Continued exploration after the mine closed revealed a large deposit of copper and molybdenum. The ore bodies are associated with a 500-foot-thick sill of diorite (a dark granitelike rock) that invaded the Overthrust Belt, probably about 35 million years ago. During the late 1960s, there was much talk—and controversy—surrounding plans to develop a large open pit mine to work the deposit. But the project was abandoned, because of both low copper prices and the potential for

Tilted layers of Cretaceous sedimentary rock in the Disturbed Belt east of Rogers Pass near milepost 105, 3 miles west of Bowmans Corners.

ongoing pollution of the Blackfoot River. A 65-foot-high tailings dam that partly collapsed in 1975 was finally removed by 2015.

Between Rogers Pass and Bowmans Corners (the US 287 junction), MT 200 crosses the Disturbed Belt, an area of tight folds and thrust faults in Mesozoic sedimentary rocks east of the Overthrust Belt. The structures are similar to those in the Overthrust Belt and formed at the same time, but the folds are smaller and the faults show less movement. They are harder to see because the Mesozoic rocks are not as resistant to erosion as the Proterozoic and Paleozoic rocks in the Overthrust Belt. The more flat-lying rock formations beneath the High Plains to the east are the same as those in the Disturbed Belt, but they are less deformed. The layers of brown sandstone and dark shales, relatively weak rocks, accumulated in shallow seawater that flooded this region during Cretaceous time, between about 145 and 66 million years ago.

SOUTHWEST MONTANA

ISOLATED RANGES, SPACIOUS VALLEYS

A swarm of ants crawling about on a painted board might conclude that the board consists of paint, unless they could see wood on a few bare spots. We are in a somewhat similar situation as we wander about the continents. Because we see mostly sedimentary and volcanic rocks of the continental crust at the surface, we conclude that the same exists below. Only in raised and eroded places do we see the basement rocks of the deep continental crust. Although the younger sedimentary and volcanic rocks are typically some thousands of feet thick, they represent no more than the paint on the board of the 25-mile-thick continental crust.

The Laramide orogeny, between about 75 and 50 million years ago, conveniently exposed large areas of Archean and earliest Proterozoic basement in several ranges in Southwest Montana, providing interesting windows into a planet very remote in time and character from the one we know. These basement rocks, ranging in age from about 3.5 to 1.75 billion years old, are mostly coarse-grained metamorphic rocks, often with streaky layers that look somewhat like those of sedimentary rocks but are discontinuous

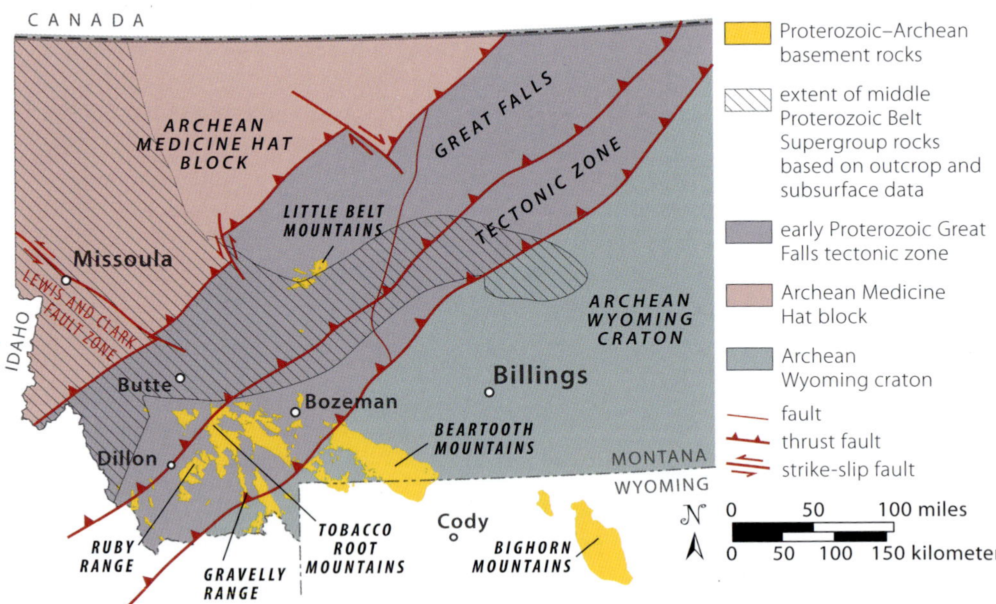

Three major northeast-southwest-trending faults delineate the Great Falls tectonic zone, the collision zone during the Big Sky orogeny. —Modified from Sims and others, 2004

or wavy. Some of the old rocks contain mineral grains eroded from even older rocks (nowhere exposed around here) that crystallized 4 billion years ago, making them some of the oldest mineral grains found on Earth. These old rocks were so cooked by high temperature, extreme pressure, and deformation that most don't look much like the original rocks that formed at or near Earth's surface. Just as fine snowflakes eventually turn to corn snow in the spring through repeated melting and freezing, deeply buried rocks "stew" and recrystallize for millions of years, becoming coarser grained. However, based on their composition, we can infer the original rock types, the deep crustal processes that affected them, and when they suffered those changes.

Less-deformed rocks show that surface water was occupied by single-celled (prokaryotic) organisms, some of which made laminated structures called stromatolites that likely formed much of the limestone (now altered to marble) found in southwest Montana. The atmosphere was rich in carbon dioxide, with very little free oxygen, imparting a light-green color to the oceans from unoxidized iron. As single-celled cyanobacteria flourished, they converted sunlight to food and produced oxygen that slowly rusted the iron, which settled to form the banded iron deposits found in Southwest Montana.

The Archean-age Medicine Hat block and Wyoming craton, both designated as terranes, collided about 1.85 billion years ago to form an early Proterozoic suture zone known as the Great Falls tectonic zone. This collision, known as the Big Sky orogeny, first formed a subduction zone that closed an ocean basin between the two terranes, making a volcanic arc along the margin of the Medicine Hat block. The collision crushed and reheated the old granites and metamorphic rocks of the two terranes. The older, colder Wyoming craton was jammed under the edge of the warmer Medicine Hat block, suturing them together and adding land to Laurentia. The three major faults in the suture zone trend northeast-southwest and dip down to the northwest, under the Medicine Hat block.

The Proterozoic Belt Supergroup sedimentary rocks, which dominate northwest Montana, are not quite as widespread in southwest Montana but do occur in major thrust slabs, such as the Grasshopper plate west of Dillon, and to the north, where coarse deposits define the southern edge of a large basin. The sediments were eroded from and deposited on top of the older Archean rocks in a huge intracontinental

A metamorphic rock from the Dillon area that was repeatedly deformed deep in the Earth by heat and pressure over millions of years.

extensional basin called the Belt Basin, which was active from 1.47 to 1.4 billion years ago. Although most of the Belt rocks are fine-grained, coarse sediment accumulated in several places along the southern boundary of the basin, where raised highlands of Archean-age metamorphic rock, such as the Dillon block, were eroded. The highlands shed debris flows with boulders and cobbles into the Belt Basin to form the LaHood Formation conglomerate. The coarse deposits grade to fine-grained sediments to the north. Hot water percolating along some of the normal faults that formed while the Belt rocks were deposited altered Archean dolomitic marble to form economic deposits of white talc in southwest Montana. (See the introduction to the northwest Montana chapter for more information about the Belt Supergroup.)

Archean and Proterozoic rocks in western Montana were deeply eroded, and then during Paleozoic and Mesozoic time, they were buried under thousands of feet of marine and nonmarine sedimentary rocks. The contact between the Paleozoic sedimentary rocks and the old rocks is called the Great Unconformity due to the great span of time—nearly a billion years—it represents. In Late Cretaceous time, the sedimentary pile was crumpled and shoved eastward, and Archean basement blocks were raised during the Sevier and Laramide orogenies. These events occurred when North America collided with the oceanic Farallon plate to the west to form mountains, the foreland basin east of the mountains, and abundant granitic magma.

Boulder Batholith

One of the largest intrusions in Montana is the 80-to-70-million-year-old Boulder batholith between Butte and Helena. This batholith seems out of place, well east of the long northerly trend of granite batholiths—including the Idaho and Sierra Nevada batholiths—that parallels the old continental margin 100 miles to the west.

Many geologists view large areas of granite as having formed at great depths, typically intruding and lifting the older rocks around them. However, sedimentary rocks dip at 30 to 40 degrees under the north margin of the Boulder batholith near Helena—the opposite of what we might expect to see. The huge granite batholith seems to lie on top of older sedimentary rocks. US Geological Survey geologists

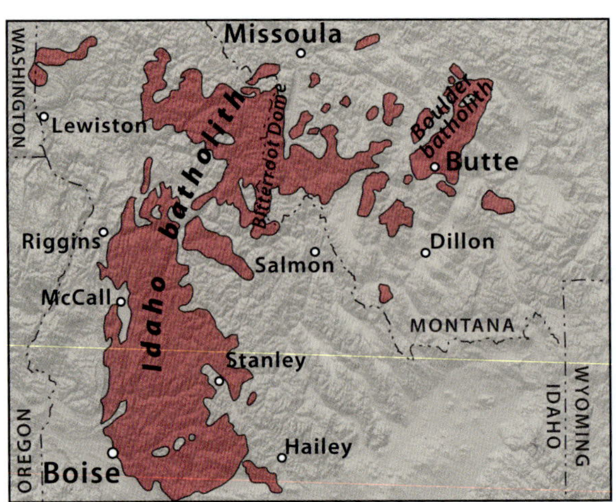

Distribution of Cretaceous granitic rocks of the Idaho batholith of central Idaho. Granitic rocks of about the same age, such as that of the Boulder batholith, between Butte and Helena, occurs farther east.

commented on this relationship in the early 1900s, and in the 1960s two other USGS geologists, Warren Hamilton and Bradley Myers, concluded that the only reasonable explanation was that the huge volume of granitic magma rose to the surface to spread out on top of the older sedimentary rocks as a gigantic, thick "lava flow." The initial magma quickly chilled at the surface to form the Late Cretaceous Elkhorn Mountains Volcanics, which are the same age as the granite directly underneath, about 76 million years old. The magma of the Boulder batholith spread under the volcanic rocks, crystallizing more slowly because of the insulating cover. The Elkhorn Mountains Volcanics appear to be only 1.2 to 2 miles thick, and the Boulder batholith only 3 to 4 miles thick. This early and imaginative interpretation has held up well over time.

It appears that the thin, broad plate carrying the Boulder batholith was later transported east. The granitic magma of the batholith may have been smeared out to the east from the Idaho batholith along the mylonite zones of the Bitterroot and Anaconda metamorphic core complexes.

Other granite intrusions of similar age in the Sapphire Mountains and Flint Creek Range were also emplaced at shallow, colder levels in the crust, and farther east from the main trend of granites of this age. Paleozoic and Mesozoic rocks surround these intrusions. The sedimentary rocks that extend a mile or less out from their contact with the granites were metamorphosed at low temperatures. All of this activity was in Late Cretaceous time, when dinosaurs grazed around water holes and along river floodplains and coastal marshes in the foreland basin. Though some may have scattered at the rumbling blast of the erupting Elkhorn Mountains volcano, herds of dinosaurs have been found buried under its ash.

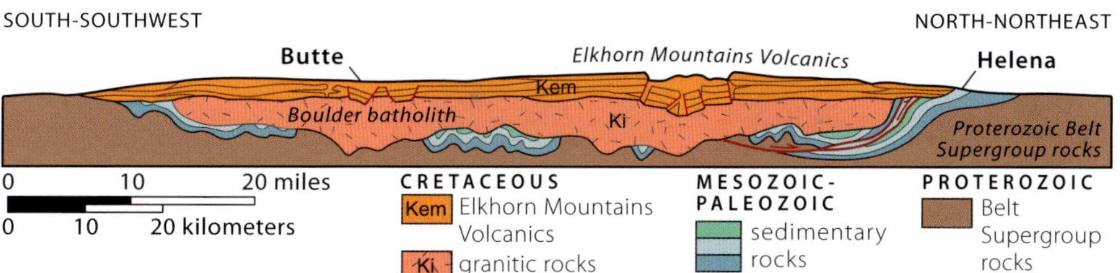

Schematic cross section through the Boulder batholith from south of Butte to north of Helena. —Modified from Hamilton and Myers, 1967

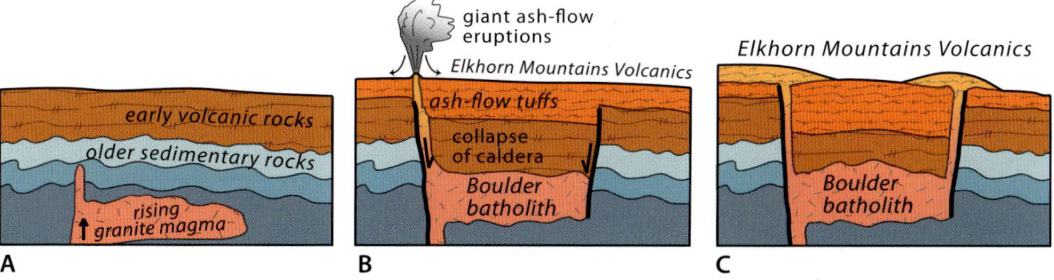

(A) Granite magma of the Boulder batholith rises toward the surface. (B) The roof over the magma chamber collapses, leading to a giant volcanic eruption and the initial deposition of the Elkhorn Mountains Volcanics. (C) Erosion happens over time, and the magma reintrudes the region, erupting and leaving younger deposits. Heat has melted the early volcanic rocks immediately above the magma, so they are thinner than in diagram A.

The prominent north-south-trending cracks, or joints, in the Boulder batholith granite may not all have the same origins. When magma crystallizes it shrinks about 10 percent, forming joints. However, tectonic movement as the magma crystallized or afterward can also form joints. In much of the Boulder batholith, one dominant north-south vertical set is about parallel to the eastern border of the batholith and to the axes of folds in older sedimentary rocks to the east. If the batholith moved over growing north-south folds, the solid granite may have flexed upward, stretched, and fractured to form the joints. These joints are responsible for the north-south blades of rock we see today. Although tectonic extension that occurred later in Cenozoic time could have produced the joints, some are filled with late-stage granite, suggesting they formed shortly after the pluton had crystallized, rather than millions of years later.

Most outcrops of granite have two or three persistent joint directions that run at about 90 degrees to one another, so the rock cracks into rectangular blocks. Rainwater, slightly acidic from dissolved carbon dioxide, slowly attacks the feldspar grains in the granite, turning them to clay. Weathering on two joint surfaces attacks the rock from two directions, causing rounded edges; weathering at three intersecting joints forms rounded corners. Over thousands of years, granite gradually rounds into bouldery outcrops, providing inspiration for the Boulder batholith's name.

Vertical jointing in the Humbug Spires, part of the Boulder batholith, south of Butte. Similar joints are prominent north of I-90, east of Butte.

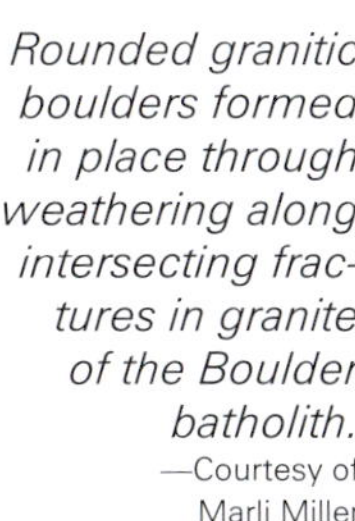

Rounded granitic boulders formed in place through weathering along intersecting fractures in granite of the Boulder batholith. —Courtesy of Marli Miller

EOCENE VOLCANISM

After about 20 million years of volcanic quiescence, volcanism returned with a vengeance during the beginning of Eocene time. Volcanic rocks, largely erupted from stratovolcanoes, cover large parts of the Absaroka and Gallatin Ranges and northwestern Wyoming (Absaroka Volcanics Supergroup), the area around Dillon (Dillon Volcanics), and Butte (Lowland Creek Volcanics), and stretch into central Idaho (Challis Volcanics). At about the same time, a broad area of scattered alkali-rich volcanoes erupted across central Montana. The specific cause of this outbreak of widespread volcanism is not clear, but it might be related to the shallow subduction of the Farallon plate, the beginning of crustal extension, or both.

This Eocene basaltic lava flow, north of Dillon on Block Mountain, fractured into columns as it solidified.

Cenozoic Extensional Basins

As volcanism slowly waned in southwest Montana around 50 million years ago, extensional topography began to develop in a north-south zone that roughly overlaps the area of thin-skin (Sevier) deformation that occurred during Cretaceous time. The crust that had experienced thick-skin (Laramide) deformation in that area remained stable. The cause of extension is debated, but it might be related to relaxing of the Sevier thrust faults as the subduction zone between the Farallon and North American plates moved west. Whatever the cause, extension thinned the continental crust, permitting deep rocks to rise as big domes of high-temperature, high-pressure metamorphic rocks. The hot boundary zone—tens of feet thick—between the deep, high-grade metamorphic rocks and the near-surface sedimentary rocks (including the Belt rocks) became a thoroughly smeared mylonite zone. As the deforming rock rose, it cooled, became brittle, and fractured, producing a thinner, thoroughly shattered breccia zone. Good examples in western Montana include the Bitterroot metamorphic core complex south of Missoula, and the Anaconda metamorphic core complex near Anaconda.

Extension along normal faults initiated the subsidence of complex basins, such as the Big Hole and Bitterroot Valleys, which filled with light-colored, ash-rich stream and lake sediments of the Renova Formation. Coarse-grained, angular gravels were shed off actively rising ranges, and water-saturated sediments disrupted by faulting and earthquake shaking show that the basins were tectonically active during deposition. East of the active zone of extension (east of Dillon), Renova sediments accumulated in a large, tectonically quiet basin with almost none of the tell-tale signs of active normal faulting

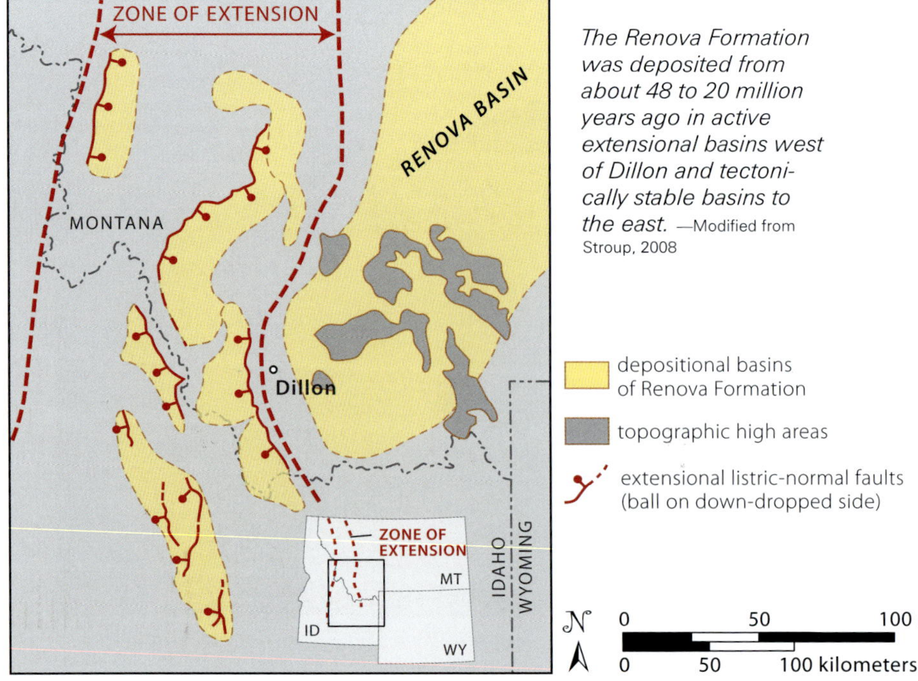

The Renova Formation was deposited from about 48 to 20 million years ago in active extensional basins west of Dillon and tectonically stable basins to the east. —Modified from Stroup, 2008

during deposition. This basin, the Renova Basin, developed on the stable Laramide-deformed crust that was subsiding under the weight of the accumulated sediments.

Few volcanoes were erupting near southwest Montana at the time, so much of the ash may have blown in from major eruptions in the then-active Western Cascades in Oregon and Washington, or rivers may have brought it from the giant caldera volcanoes of southern Utah and Nevada. That volcanic ash weathered into expandable clay that swells by as much as 40 percent of its original volume when wet, making it very slippery. Dirt roads built on the Renova Formation can simply be impassable when wet. With population growth, several western Montana towns are now spreading outward onto higher benches underlain by the Renova Formation, and landslides may become a serious problem in the future.

A landslide in Tertiary sedimentary rocks east of Dillon in the Blacktail Mountains.

The Renova Formation includes an endless variety of sedimentary rock types, including sandstone, conglomerate, limestone, clayey siltstone, and even coal. They preserve fossils of animals and plants that indicate the climate became progressively drier through Oligocene and early Miocene time. Globally, grasslands were expanding at this time, and a variety of mammals that depended on them were evolving. In Montana streams shriveled, and water pooled in valleys to form saline lakes. Mountains remained moist enough for dawn redwood trees, a close relative of modern sequoias, to grow. A short period of wet, humid climate around 20 million years ago

Oligocene-age dawn redwood, or metasequoia, preserved in the Renova Formation in southwest Montana.

formed red laterite soil on the Renova Formation, followed by the return of desert dryness throughout Miocene and Pliocene time.

In the zone of active extension, the Renova Formation was tilted continuously during its deposition as southwest Montana's valleys widened. As one side of a valley dropped along curved listric normal faults, the upturned edges of the Renova were eroded away. In the area of the Renova Basin, and elsewhere in southwest Montana where the Renova was deposited on Laramide-deformed crust, tilting did not start until Miocene time; the tilting was associated with the extension of the Basin and Range Province and the outbreak of the Yellowstone hot spot in northern Nevada. By Miocene time (around 16.5 million years ago), the deepened valley edges started to fill with sediment of the Sixmile Creek Formation, which was deposited over the tilted and eroded Renova Formation, resulting in what geologists call an angular unconformity, or gap in the rock record, between the two formations.

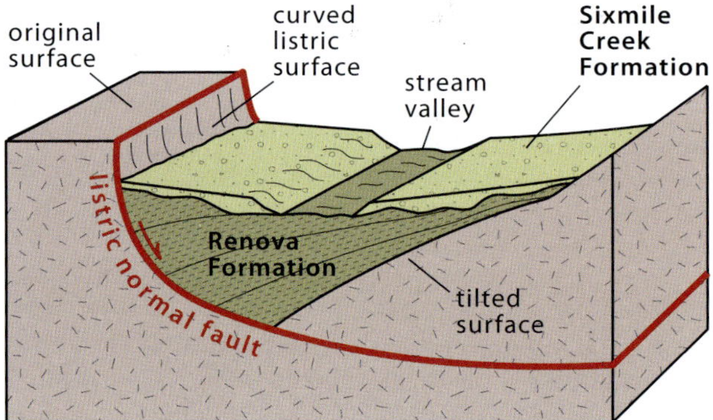

The Renova Formation in the active zone of extension west of Dillon was tilted continuously during its deposition as one side of a valley dropped along a listric normal fault. The Sixmile Creek Formation was deposited over this tilted and eroded surface.

Yellowstone Hot Spot

Approximately 16.5 million years ago, a gigantic thermal plume rose through Earth's mantle, bulging the continental crust near the intersection of Oregon, Idaho, and Nevada. Very hot basalt magma from the plume, or hot spot, rose into the crust, which melted to form granite magma. The overlying near-surface continental crust stretched over the bulge and cracked in a big circular ring, permitting the crust to sink into the huge mass of rising magma and leave a broad caldera basin like that in today's Yellowstone National Park. Erupting granite magma expanded, chilled, and blasted fine particles of hot ash and crystals that spread for tens of miles over the surrounding countryside.

As North America slowly continued southwestward over the stationary hot spot, successive Yellowstone-type caldera volcanoes developed along a path to the northeast, eventually forming today's Snake River Plain. As each caldera volcano passed beyond the underlying hot spot, volcanic eruptions ceased and the volcano cooled and shrank, lengthening the broad valley of the Snake River Plain.

Stretching of the crust over the thermal bulge created northeast-trending basins and ranges, the beginning of today's basin and range terrain in Montana; the basins filled with sediments of the Sixmile Creek Formation throughout the remainder of

Miocene time. Most of the initial Sixmile Creek basin fill was coarse sand and gravel shed as debris flows off mountain ranges adjacent to the basins. Over time, stream-transported ash from the giant caldera eruptions and far-traveled gravels filled the basins. The ancestral Missouri River formed in one of the valleys that drained radially outward from those volcanoes.

Since the Yellowstone hot spot reared its head nearly 17 million years ago, the North American plate has been moving very slowly (about 2 centimeters per year) over it to the southwest. The Yellowstone hot spot is stationary, so as the plate moves, the active eruptive center migrates to the northeast, leaving a volcanic scar behind it to the southwest. Imagine moving your hand over a stationary candle. The candle burns your hand as it passes over it, leaving burn scars behind. In this analogy, the burn scar is the Snake River Plain in Idaho; buried under it are many old calderas that formed during explosive eruptions of the hot spot. The ancestral Missouri River deposits in Montana show us that as the hot spot "migrated" northeastward toward southwest Montana, the river's gradient increased, resulting in larger and larger river cobbles and thicker ash deposits. Even a few basalt lava flows that erupted from the Snake River Plain reached Montana.

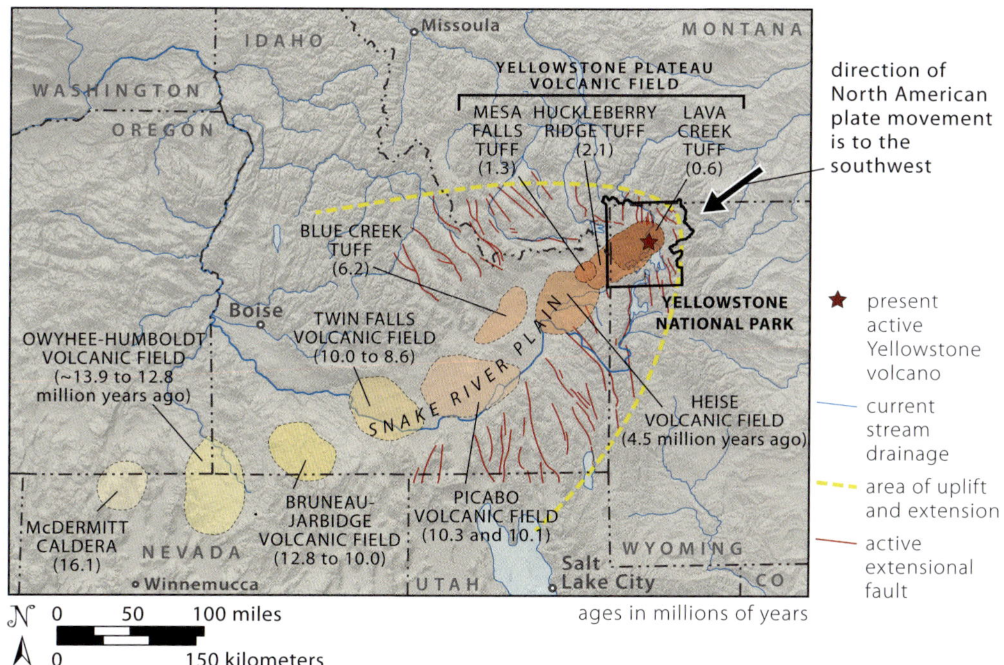

Map of the Snake River Plain showing the migration and timing (millions of years ago in parentheses) of the Yellowstone volcano's ancestral eruptions. The older, cooler calderas (yellow ovals) grade into the younger, hotter calderas (reddish ovals). The parabola-shaped area of uplift and extension around the active volcano is shown with a yellow dashed line, and active extensional faults are thin red lines. The Snake River Plain is the cooled and shrunken wake left behind the active hot spot. Streams drain away from the high ground of the present Yellowstone volcano (red circle). Modern basin and range topography trends northwest. —Modified from Pierce and Morgan, 1992

Around 4.5 million years ago, the stresses the hot spot placed on the crust in southwest Montana caused the crust to crack again, forming basins and ranges in a northwest trend, perpendicular to the trend that had existed since the hot spot emerged around 17 million years ago. This changed the flow pattern of the upper Missouri River drainage to the one we see today. Imagine an earthquake offsetting the ground 20 feet or more, instantly diverting the ancestral Missouri River water into a new valley. The old river went dry, with fish flopping around on the ground. Today, rivers like the Red Rock follow the most recent basins that started forming 4.5 million years ago; the ancestral northeast pathway can still be seen where the Beaverhead River cut a canyon south of Dillon through the mountains that were raised across it. Over the last 4.5 million years, some of the old deposits of the Missouri were lifted to an elevation of nearly 10,000 feet in the northwest-southeast-trending mountain ranges, preserving evidence of the old Missouri River drainage.

Crustal extension is still happening along the northwest-southeast-trending normal faults. On the evening of August 17, 1959, a 7.5 magnitude quake struck near Hebgen Lake west of West Yellowstone, burying nineteen people in a massive rockslide. On July 25, 2005, a 5.6 magnitude quake struck on a previously unknown fault at the base of the Pioneer Mountains north of Dillon. Southwest Montana is earthquake country, and the steep fronts of the Bridger Range near Bozeman and the East Ridge in Butte make us wonder when those faults might release their pent-up energy.

Flat-topped benches that slope gently toward the valley bottoms are common along the flanks of many of the ranges in southwest Montana. These desert pediment

Ash and pumice from the Yellowstone volcano in Idaho, exposed near Timber Hill in the Sweetwater Range east of Dillon, was transported by water down northeast-trending grabens of the ancestral Missouri River drainage. Note the cross beds.

A desert pediment slopes gently to the left in front of the Snowcrest Range. Gullies (foreground) were eroded into the Sixmile Creek and Renova Formations when moisture increased during Pleistocene time.

surfaces were eroded by sheetlike flows during intense rainfall. The pediments developed in places on the Sixmile Creek Formation, so they are at least as young as Pliocene in age. Old valley-floor surfaces in southwest Montana are everywhere dissected, probably due to continued faulting and increased moisture during Pleistocene time.

Yellowstone National Park is home to one of the largest and most violent volcanoes in the world. Over the last 2.1 million years, huge eruptions formed three calderas that each spewed out more than 1,000 cubic miles of ash. By comparison, Mount St. Helens ejected only 0.24 cubic mile of ash in 1980. The most recent eruption of the Yellowstone volcano ejected enough material from its magma chamber to form a caldera 45 miles long by 29 miles wide. A major eruption of Yellowstone would have dramatic effects on the surrounding region but not cause the global extinction popularized by some press.

The magma chamber beneath the Yellowstone volcano is big, some 50 miles long, 25 miles wide, and 5 miles thick. Two conduits of magma reach above this as close as 5 to 7 miles below the surface and dome the ground above them. The magma is gas-charged rhyolite that is only about 9 percent liquid, so it's very thick and pasty. The chamber expands and contracts on human timescales, but the movements do not yet appear to foretell an eruption.

An eruption will occur when magma rises high enough that overlying rock cracks and pressure is rapidly released. Dissolved gas will rapidly expand, shattering the magma into fine, glassy ash. This mixture of hot gas and ash will rapidly rise in giant columns to over 100,000 feet, spreading ash around the globe. Since the ash column will be relatively dense and heavy, some of the mixture will collapse in waves to flow across the ground at more than a 100 miles per hour and at a temperature greater than 1,000 degrees Fahrenheit. These pyroclastic flows, as they are called, will leave deposits up to 100 miles away from their vents, as happened in the Madison and Centennial Valleys in southwest Montana during the Huckleberry Ridge eruption 2.1 million years ago. When it's all over, the ground will have collapsed into the emptying top of the magma chamber, creating a caldera.

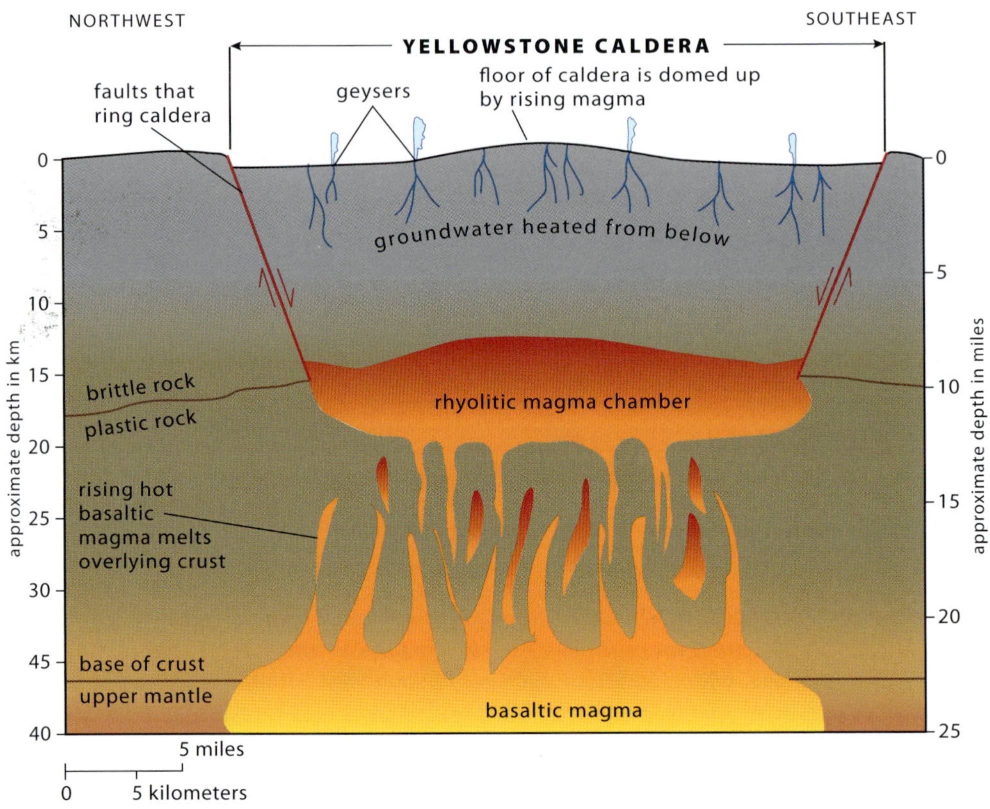

Inferred cross section of the Yellowstone magma system and caldera.
—Modified from Lowenstern and Hurwitz, 2008

When is Yellowstone going to erupt again? There is no doubt that it will, but the question is when and what type of eruption will occur. A supervolcanic eruption has happened at Yellowstone about every 600,000 to 700,000 years, and the last eruption occurred 640,000 years ago. But before you sell your house, the interval is based on only three eruptions, and that makes a poor statistical argument that we are "overdue." You are much more likely to die from lightning or slipping in the bathtub! However, small eruptions of hot gases or volcanic ash are likely within a generation or two. Now, lest that seem too comforting, a group of researchers at Arizona State University studying mineral changes in rocks from Yellowstone's last mega-eruption concluded that an influx of fresh magma from below the magma chamber triggered the eruption. They discovered that changes in temperature and composition seemed to build up over just a few decades, so when the next eruption does happen, we may have less warning than previously thought.

The last eruption of lava from the Yellowstone volcano was of the caldera-filling variety. These eruptions produce thick, slow-moving lavas that spread slowly away from the vent. The piles accumulate from the bottom up, building until they collapse under their own weight and spread outward. The last flow occurred about 70,000 years ago, so the magmatic system has been quiet for a very long time. Each year there

Huckleberry Ridge Tuff is a pyroclastic flow deposit from a caldera-forming eruption of the Yellowstone volcano 2.1 million years ago. This exposure, 5 miles south of Gardiner, shows (from bottom to top) baked Cretaceous sedimentary rocks (orange) overlain by airfall ash (white), volcanic glass (black) that formed when the hot flow quickly cooled as it first hit the ground, and the volcanic matter that settled in pulses from the flow.

The long white streaks are flattened pieces of pumice, which show that the Huckleberry Ridge Tuff was very hot when deposited. The light-gray groundmass is composed of ash and early-formed crystals that were blasted from the magma chamber.

are thousands of earthquakes, and the ground rises and falls slowly over a few months, which can make people concerned that the caldera is about to erupt. However, if it were to blow, we would see major uplift, perhaps feet per year, accompanied by many big earthquakes around the rising dome. If a supervolcanic eruption were to occur, Montana would certainly be hit hard. Computer modeling of the ash distribution during a month-long supervolcanic eruption show that Missoula might be covered in 4 to 12 inches of ash, but Billings could be buried under as much as 3 feet! That said, we don't expect to see a big eruption any time soon.

BUTTE
The Richest Hill on Earth

See map on page 126.

Butte started as a placer gold camp in 1864, but gold was almost as scarce as water to wash it out of the ground. The first miners hauled gravel down the hill in oxcarts to Silver Bow Creek, only to find that it contained very few nuggets and was too diluted with silver to bring the best price.

The earliest prospectors commented on the big quartz veins that made bold outcrops stained black as coal with manganese oxide and full of silver minerals. These were discovered at about the same time as the placer gold, mainly because they were too big and obvious to overlook, but the silver was so awash in copper that there seemed no hope of recovering it. In those days copper was not especially valuable. But the rapid rise of big electrical industries and the need for copper wire during the final two decades of the 1800s created an enormous market for copper, as well as a place in the sun for Butte—as a copper producer, with silver and gold as by-products. The copper mines of Butte thrived as Americans built big electrical transformers and motors, wired their houses, and wove a tight web of telegraph, telephone, and transmission lines across the continent. By 1898, the United States was producing about 60 percent of the world's copper, and Butte was producing 40 percent of that. Something like 211 billion pounds of copper metal had been produced from Butte through 2005. Butte also went on to become one of the few districts in North America to produce much manganese, totaling more than 3.7 billion pounds. No industrial economy can function without manganese, which is used for making steel alloys with different properties.

The ore minerals were deposited between 66 and 62 million years ago in the 80-to-70-million-year-old granite of the largest of several plutons that make up the Boulder batholith. Hydrothermal fluids (hot circulating water) deposited the minerals in cracks (veins) in the rock. These hot solutions either formed from the magma in the latest stages of the pluton's crystallization or developed by the exchange of fluids between the very hot magma and the surrounding water-bearing rock, or both.

Long after an igneous intrusion has cooled and the ore mineral veins have formed, warm near-surface water, often in a hot spring environment, dissolves many of the ore minerals near the ground surface and precipitates the metals as low-temperature minerals, such as chalcocite, a black, sooty copper sulfide, just below the water table. Miners, including those at Butte, exploit this enriched higher-grade ore zone. The leached zone above is waste rock that must be excavated to get at the good stuff.

There were two important sets of veins at Butte: silver-bearing quartz veins stained black with manganese minerals and veins full of copper minerals. The copper veins, the older and deeper of the two sets, were congregated in a tight bull's-eye that later

Old headframes, or "gallows frames," still stand on the hill upslope of the old part of Butte and above the Berkeley Pit. Headframes mark the access point to deep underground mines via crude elevator shafts. Most of the old headframes were abandoned when it became more profitable to mine large tonnages in open pits. Note the nearby houses at lower left.
—Courtesy of Richard Gibson

became the Berkeley Pit, and the silver veins, from a younger, shallower mineralization event, formed a halo around them, grading outward into higher concentrations of lead, zinc, gold, and manganese. Some of the many mines that started working during the 1880s produced mostly silver, others mostly copper, depending on the kind of veins they were following.

The Butte veins run broadly east to west and vertically; some are fairly thick, many thin. In order to mine them underground, an adit (tunnel) had to be excavated wide enough for a man to work and for a small rail car and its tracks to efficiently trundle the ore to a shaft; a lift (primitive elevator) brought it up to the ground surface. Even if a vein was high-grade, the surrounding waste rock had to be removed as well, so the average grade (concentration of the ore) of the full width of the adit dictated whether the ore was rich enough to mine. It all depended, of course, on the price of gold, silver, and copper, or some combination, and the cost of mining, milling, and smelting the ore to extract the metal. A vein was "worked out" when the value of the ore was not high enough to eke out a profit.

Depth also played a role in the viability of a mine. The deepest mine was the Mountain Con in uptown Butte, which reached a depth of 5,380 feet. Temperature in the Earth tends to increase with depth, so working conditions could be extreme. Workers in an area of the Stewart Mine called the "Chinese laundry," for example, faced temperatures of about 135 degrees Fahrenheit.

Many of the rich veins had been completely worked out by the 1950s. In all, about 10,000 miles of underground tunnels were excavated. Butte's mining industry then switched to producing enormous tonnages of low-grade ore from the Berkeley Pit. Open pit mining involves excavating an immense hole in the most valuable part of an ore deposit. The Berkeley Pit is more than 1 mile across and about 1,800 feet deep. Altogether, the pit produced some 316 million tons of rather lean ore. Three tons of waste rock were mined for every ton of ore. The highly fractured ore deposit had been altered to clay minerals, so digging was not difficult. The loose rock was scooped up

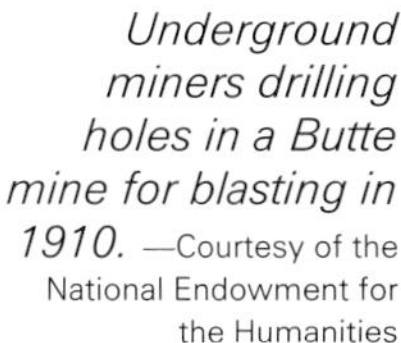

Underground miners drilling holes in a Butte mine for blasting in 1910. —Courtesy of the National Endowment for the Humanities

with giant power shovels, loaded into even bigger trucks, and taken to the mill for pulverizing and the separation of ore minerals from waste rock. The sidewall slopes of the pit posed the biggest danger; they were so steep that they could collapse without warning and bury workers.

The Berkeley Pit opened in 1955, and underground mining was gradually phased out over the following twenty years. At Butte, the average ore grade in the Berkeley Pit was less than 0.5 percent copper. It's almost impossible to see copper minerals in it, even with a strong hand lens. Nevertheless, the pit produced incredible quantities of metal. The pit itself became uneconomic, and mining ceased in July 1983.

Groundwater seeping into the Berkeley Pit had been kept at bay by pumping water out of nearby underground mines, even after they closed. After mining in the Berkeley Pit ended, pumping stopped and groundwater began flooding both the 10,000 miles of underground tunnels and the pit itself. The lake in the pit is now approximately 900 feet deep. For many years, its 45 billion gallons of toxic water had an acidity of around 2.5 (7.0 is neutral), similar to that of cola or lemon juice but full of deadly heavy metals. The water acidity has now been reduced to 4.0 because lime treatment sludge has been disposed of in it since 2003. The acidic pit water remains heavily laden with dangerous chemicals, such as copper, arsenic, cadmium, zinc, and lead, that leach from the rock. The water is so nasty that it poisoned more than three thousand migrating snow geese when they landed on it in December 2016. As of September 2018, after more than forty years, water in the pit had reached an elevation of 5,349 feet, just below the 5,410-foot elevation where it will contaminate the groundwater under the city itself. To deal with the toxic water, a treatment plant has been constructed that filters and neutralizes pit water so that it can be drained into Silver Bow Creek, the headwaters of the Clark Fork River. Water treatment began in the spring of 2019, and for the first time in many years the lake is not rising.

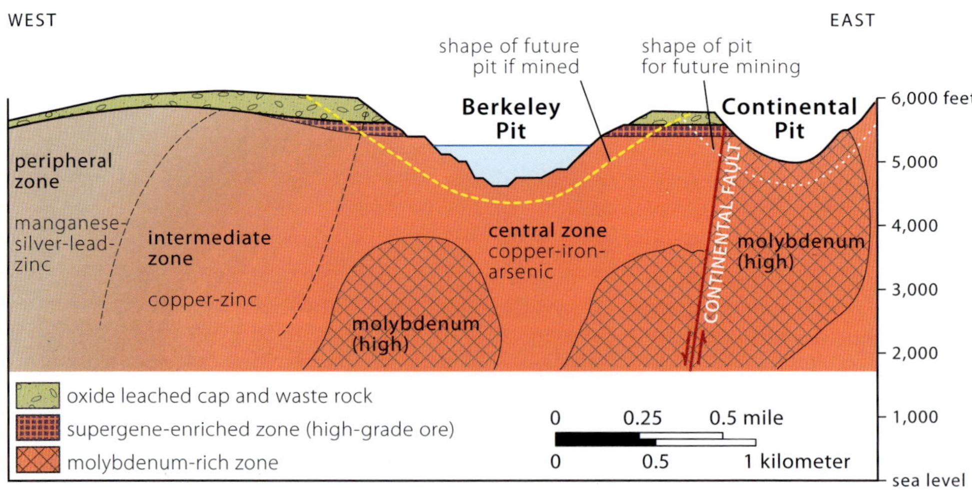

Simplified cross section through the Butte mining district. Note that deeper mining in the Berkeley Pit (yellow dashed line) would require the removal of much more waste rock, making it uneconomic. Supergene enrichment is a zone in which metals oxidized and dissolved near the surface and then percolated down to concentrate at a deeper level to become minable ore. —Modified from Gammons and others, 2006

A satellite view of the Berkeley Pit (full of black water) and the Continental Pit (upper right) on August 2, 2006. —Courtesy of National Aeronautics and Space Administration

In spite of the toxic nature of the Berkeley Pit water, it does have some value. Beginning in 1998, copper has been recovered from the acidic water by precipitating it on scrap iron, including old tin cans. Merely putting cans in the copper-rich water causes copper to precipitate, transferring iron from the cans back into the water. The process has provided the company with about 400,000 pounds of copper per month.

The pit water also hosts some thriving organisms, including a single-celled organism called *Euglena mutabilis*. It has evolved to produce highly toxic compounds that can fight certain cancer cells. Other organisms ingest metals and are being evaluated for use in cleaning toxic water.

After mining ended in the Berkeley Pit, Missoula industrialist Dennis Washington purchased the property, and in 1986 his company, Montana Resources, began mining the idled Continental Pit (East Pit), which lies about a half mile east of the Berkeley Pit and extends as far east as I-15. The deposit occurs in an area of copper and molybdenum mineralization exposed by extensional faulting along the Continental fault. About 0.35 percent of the Continental Pit ore is copper, with some molybdenum, and a little silver to sweeten the mix. "Moly" is a major component of high-strength steel. Electric cars could soon be an additional market for the copper since about 21 pounds of it are used in the construction of each, whereas only 3 pounds are used in a standard automobile. As of 2018, the company estimated that it had enough ore to continue mining for another thirty years.

After well over one hundred years of profitable mining, the whole area is a mining wasteland. Mining waste lies all around Butte Hill and downstream in Silver Bow Creek. Small smelters near the early mines created toxic air pollution all over Butte. When Marcus Daly built a big smelter in Anaconda in 1894, 25 miles to the west, ores were shipped there, where they were smelted—heated in the presence of a reducing (opposite of oxidizing) agent, such as carbon or charcoal—to remove oxygen in the ore minerals. This process produced carbon dioxide, which went up the smelter stack, along with other minor elements that vaporized: sulfur oxides, arsenic, and mercury, for example. Smelter stacks are tall in order to carry noxious gases away from smelter workers and towns, but that merely dilutes the gases that poison the air and soil downwind. The Anaconda smelter closed at about the same time as the Berkeley Pit.

Butte copper ores are mostly sulfide minerals—copper sulfide, copper-iron sulfide, and various sulfides containing iron, arsenic, cadmium, antimony, manganese, and zinc. Sulfide minerals sitting on tailings piles, in contact with water and air, oxidize. The iron oxidizes to a bright-orange rust, the stain on rocks and soil that often leads prospectors to dig deeper for ore. The sulfur oxidizes to form sulfuric acid, so it's no surprise that water downstream of Butte was severely polluted. On the Clark Fork River near Deer Lodge, downstream of Butte, a federal netting operation in 1991 failed to find a single fish. Butte's Berkeley Pit was declared a Superfund site in 1983, and the tailings along Silver Bow Creek were declared a Superfund site in 1985. Other related Superfund sites include the Anaconda smelter and the Milltown Dam reservoir a few miles upstream from Missoula. The dam was removed in 2008, and the area has been reclaimed.

North of the Berkeley Pit stands a giant earthen dam that impounds an immense pile of waste rock from the Berkeley Pit and the active Continental Pit. The dam stands over 650 feet tall, and the tailings pond is more than three times larger than the Berkeley Pit and filled with hundreds of millions of tons of fine-grained waste rock. With Montana Resources pumping 18,000 gallons of wet sand and silt per minute up the hill to the tailings pond, it grows substantially larger each day. Lime is added to the mix in order to reduce the acidity of the tailings slurry to avoid acid mine drainage from the site.

A large earthquake could jeopardize the stability of these water-saturated tailings. The nearby Continental fault is an active normal fault along which East Ridge has been rising and the Butte valley dropping for millions of years. Like the Hebgen Lake earthquake in 1959, this fault could generate a magnitude 7.0 or greater earthquake, which could rupture the dam and liquefy the tailings, producing a fast-moving slurry. Without

A giant dump truck (arrow), with wheels 10 feet high, is dwarfed by the scale of the Yankee Doodle Tailings Pond (upslope from the Berkeley Pit) impounded behind its earthen dam, as it looked in 2007. —Courtesy of Marli Miller

an earthquake, an iron-ore tailings dam in the state of Minas Gerais, Brazil, failed in 2018, killing 19 people downstream. Another tailings dam in the same area failed in January 2019, killing more than 237 people. The Continental fault has not ruptured the ground for more than 10,000 years, but that is not necessarily a good sign, because stress along the fault may be building during its dormancy.

Dust is an additional problem. When mining operations were suspended between 2000 and 2003, the tailings dried out and clouds of toxic dust blew into Butte from the tailings pond. To mitigate the problem, Montana Resources spread about 1.5 million tons of rock about 18 inches deep over 506 acres of the tailings-impoundment site.

Butte's mining legacy is highly visible still. You can get close to the old underground headframes by going north from I-90 at exit 126, all the way through the old part of Butte to the top of the hill. An overlook offering views of the acidic lake in the Berkeley Pit is east on Park Street, from Montana Street, in the old uptown. Montana Tech, with an extensive library of geology and mining literature and an outstanding collection of mineral specimens from Butte and around the world, is about 1 mile west along Park Street to the top of the hill. Less than 0.5 mile farther west, and downslope, is the World Museum of Mining. It preserves early shops, a fascinating collection of mining equipment from Butte's early days, a headframe, a fine collection of mineral samples, and part of one of the old underground mines that you can visit.

I-15
Butte—Helena
68 miles

Montana is full of historic mining districts that had their moment in the sun but were quickly abandoned for better opportunities elsewhere. Butte was one of the few major discoveries that permitted profitable mining for more than a century. No one knows why there is such an amazing concentration of gold, silver, copper, and other ore minerals in this particular place. There are lots of small mineral deposits and old mines scattered around the Boulder batholith, but nothing high-grade enough or in sufficient volume to justify opening a big project that might peter out after only a few years. Going forward, it appears that geologic and economic factors will continue to keep Butte the vibrant mining district it has been since the mid-1800s. (See the previous section for more on Butte's mines.)

About a half mile north of the junction of I-15 with I-90 at the southeast edge of Butte, a scenic overlook on the southbound lanes of I-15 looks down on Butte's Continental Pit and across it to the northwest over the Berkeley Pit and the old parts of Butte. On the north side of the Continental Pit are, from east to west, the headframes of the old Leonard, Kelly, and Mountain Con underground mines. The numerous horizontal terraces in the walls of the pit are haulage roads that spiral down to the bottom of the pit. Waste rock above the ore is blasted into fragments, hauled up the sides of the pit in gigantic dump trucks, and dumped to form a large hill away from the pit. Ore deeper in the pit is handled the same way but taken to a mill for further crushing and separation. The ore is typically mixed with a chemical solution that floats either the ore mineral or the waste rock, which is then skimmed off the solution. The ore mineral is then sent to a smelter to separate the copper metal from sulfur in the ore mineral.

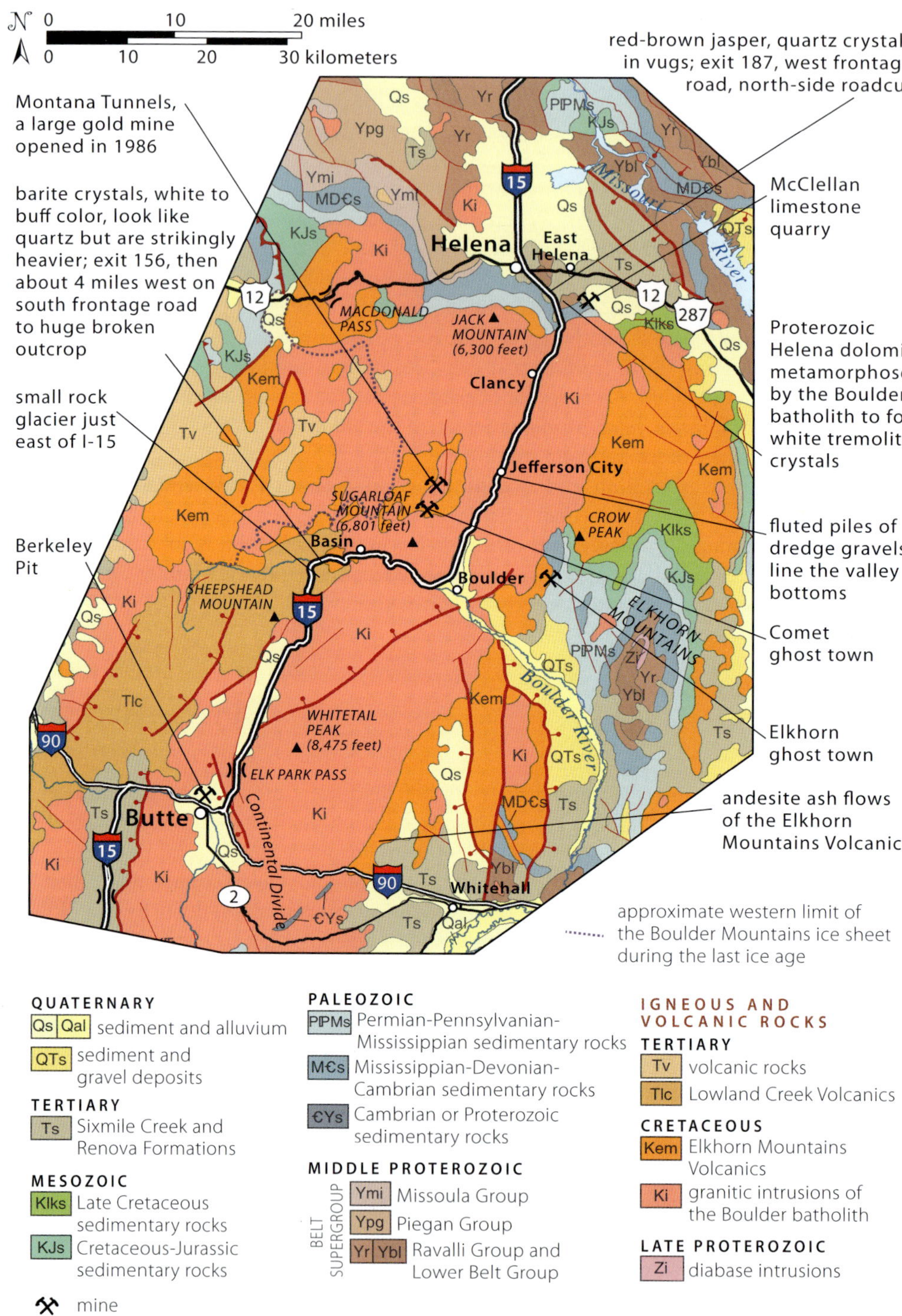

Geology along I-15 between Butte and Helena.

A panoramic view to the southwest from East Ridge shows the city of Butte, the Continental Pit in foreground, the Berkeley Pit full of water, and the Yankee Doodle Tailings Pond to the right. I-15 and the Continental fault span the whole view at the base of the near slope and out of sight. Most of the old headframes disappeared with the growth of Berkeley Pit, along with much of the oldest parts of downtown Butte. —Courtesy of Marli Miller

About 3 miles north from the east edge of Butte, the highway crosses the Continental Divide at Elk Park Pass into Elk Park, a flat-bottomed valley about 11 miles long and 1 to 2 miles wide. Cold, heavy winter air sinks into this valley, making it one of the coldest places in Montana. On March 4, 2019, the temperature hit −46 degrees Fahrenheit, the coldest March temperature ever recorded in Montana! Rocks on both sides of the interstate are granite of the Cretaceous Boulder batholith. At the pass, the Continental fault crosses I-15 and follows the base of the ridge to the west. It appears that Elk Park once drained southward into the Clark Fork drainage, but fault movement beheaded the drainage, reversing the flow, so Bison Creek now drains north into the Missouri River drainage.

About 7 miles north of the pass, the Lowland Creek Volcanics of Eocene age cap the Boulder batholith west of the interstate. Look for the dark-red rocks rising above the trees, especially on Sheepshead Mountain. About 3 miles south of the junction with Boulder River Road, I-15 crosses a normal fault that has dropped the pale-gray-to-red, angular-fractured Lowland Creek Volcanics down to road level. The road continues along these rocks for a mile or so east of the junction, where they rest on Cretaceous Elkhorn Mountains Volcanics. Granite of the Boulder batholith is exposed about 1 mile west of Basin and continues into Boulder.

The old mining town of Basin developed in the 1890s as a gold and silver mining center. Remains of the Glass Brothers smelter, prominent on the north side of I-15 in Basin, was built to process ores from the Hope Katy Mine, across the highway to the south, along with other ores shipped in by rail from other mines. Flooding and fires thwarted the operation even before it began. While the mill at the mines operated, excess waste tailings were dumped into the river. A century later, in 1999, the EPA designated the Basin mining area a Superfund site.

More recently Basin has been noted for promoting the supposed health benefits of sitting in an old mine to deliberately irradiate yourself with the radioactive element radon. Some granites have an unusually high uranium content, which decays to radon, a colorless, odorless gas that is considered a health hazard. Research shows a clear correlation of breathing high radon concentrations and lung cancer.

Between Basin and Boulder, outcrops of the Boulder batholith granite show prominent vertical jointing with a north-south-trending orientation. North of I-15 between Basin and Boulder, a zinc, copper, silver, and gold discovery in 1874 became

Lowland Creek Volcanics 3 or 4 miles southwest of Basin. This heavy, iron-rich breccia (broken rock) is highly silicified, forming a very hard rock called chert.

the largest mine outside of Butte; it finally closed in 1941. The ghost town of Comet features houses and ore-processing buildings.

In the Elkhorn Mountains east of Boulder, silver was discovered where the Boulder batholith was in contact with limestone, sandstone, and shale of Devonian age. Boulder, at the east edge of I-15, with its prominent hot springs, became an early stagecoach station and regional trading center. Like hot springs elsewhere, this one is along a stream where the groundwater intersects the land surface. The stream follows a northeast-trending normal fault that likely allows water to seep deep below the surface following permeable fault-fractured rock. If the circulating groundwater seeps deep enough, it is heated by the hot bedrock below. In the case of Boulder Hot Springs, the Elkhorn Mountains Volcanics that originally covered the Boulder batholith, or the Boulder batholith itself, are the likely heat sources. Heat in igneous rocks is primarily generated by the radioactive decay of the element potassium, which is fairly abundant in such pale igneous rocks. Potassium has a radioactive half-life of 1.2 billion years, meaning that half of the original radioactive potassium will be gone in that time. Since these rocks are roughly 76 million years old, much of the radioactivity is still there and the rocks are still generating heat.

I-15 continues north along the bottom of Prickly Pear Creek, wholly within Boulder batholith granites exposed on both sides of the valley, and through Clancy (milepost 182) along the old abandoned railway line. Like many old towns around the batholith, Clancy began as a silver mining camp in the 1870s and continued to the 1890s. Clancy was also the location of the T. Kain and Sons Quarry that supplied the granite for the Montana State Capitol building in Helena.

Rounded, bouldery-looking granite outcrops west of I-15, 1 or 2 miles south of Montana City, almost look river transported, but they actually weathered in place from the Boulder batholith. White Mississippian-age limestone just south of Montana City sparkles in the sun. The impure limestone was heated and metamorphosed by nearby

Boulder-like outcrops of granite of the Boulder batholith near I-15 north of Clancy.

granite of the Boulder batholith; calcium carbonate of the limestone reacted with impurities and silica in hot solutions from the granite to form sparkling blades of tremolite, a white calcium-magnesium silicate mineral. The Montana City Quarry at Clancy uses the limestone to make portland cement. (See the US 12: Helena—Garrison Junction road guide for information about mining and earthquakes in Helena.)

I-15
Butte—Dillon
58 miles

From the interchange with I-90, I-15 follows a graben, a block that dropped down to the east along the Rocker fault, a normal fault that has raised the Highland Mountains, to the east. Tertiary-age sediments that filled the graben have an overall tilt down to the east. The graben is bounded on the west by Fleecer Ridge, a raised block at the northern end of the Pioneer Mountains. As you drive south, the relatively flat topography of the graben makes it seem odd that you are passing over the Continental Divide, which separates water flowing to the Pacific Ocean (north) from water flowing to the Gulf of Mexico (south). The location of the drainage divide has changed over geologic time due to topographic changes brought about by tectonics, erosion, and deposition. In the distant future, Divide Creek may erode northward and capture Silver Bow Creek, sending its water to the Gulf of Mexico instead of the Pacific Ocean.

Rounded boulders scattered along the flanks of the Highland Mountains 4 to 5 miles south of the I-90 interchange are Pliocene or Pleistocene debris flow deposits shed off the Cretaceous Boulder batholith. The boulders at the surface are aplite, a fine-grained version of granite that generally forms dikes in granite; it is a minor component of the batholith. Where are the more abundant, coarse-grained granitic boulders? They have likely weathered to granitic sand, called grus, because the granite's abundant feldspar slowly weathers to clay where it's exposed, causing the rock to disintegrate.

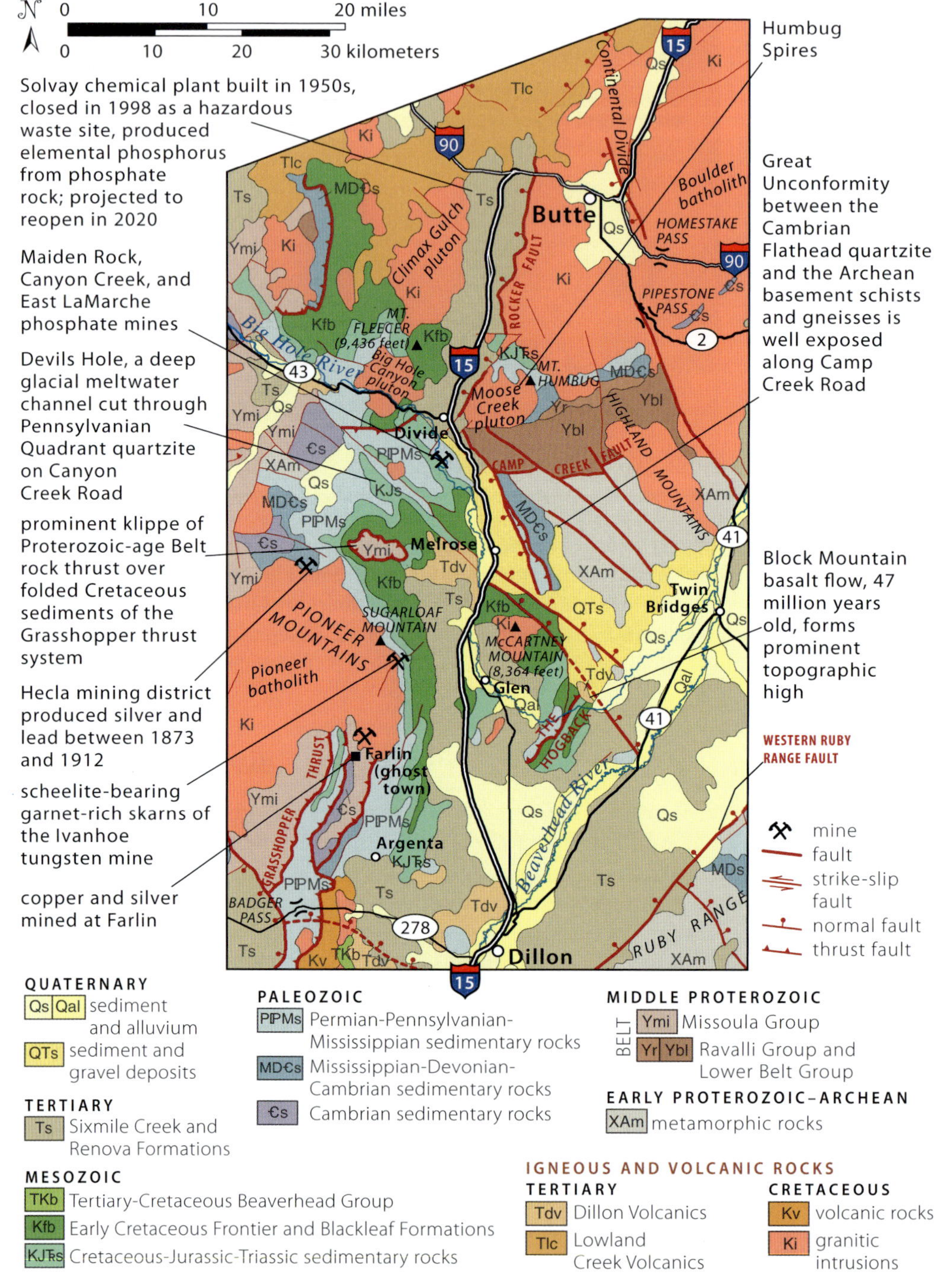

Geology along I-15 between Butte and Dillon. —Boulder batholith plutons from McDonald and others, 2012

Granitic plutons of the Boulder batholith, a Cretaceous granitic intrusion more than 62 miles long and 40 to 50 miles wide, are exposed in mountains on both sides of I-15. The batholith is composed of fifteen major separate intrusions, or plutons. The plutons have various shapes, and their emplacement mechanisms are diverse and complex. Recent data suggest that at least some of these plutons are tabular in shape and were likely injected along thrust faults. Evidence indicates that the batholith magma was not intruded where we see it today, but rather a minimum of 15 to 17 miles to the west. It was displaced to its present location during Eocene time on low-angle extensional, or detachment, faults. (See the introduction to this chapter for more about the Boulder batholith.)

Stop at the rest area south of the I-90 junction to take in the view. The prominent peak to the west is 9,436-foot Mt. Fleecer. The peak is composed of Cretaceous sedimentary rocks of the Kootenai Formation that were once coastal stream sediments deposited on the edge of the Western Interior Seaway, a shallow inland sea. Mt. Fleecer is sandwiched between two Cretaceous granitic intrusions, the Climax Gulch (76 million years) and Big Hole Canyon (85 to 75 million years) plutons, that baked the adjacent sedimentary rocks through contact metamorphism. The process made the rocks harder and more resistant to erosion, which may be why these normally erodible sedimentary rocks are able to hold up this prominent peak. Contact metamorphism can also concentrate economic minerals, such as gold, silver, and copper, by the diffusion of hot, watery fluids. In the 1800s, mining camps sprung up wherever mineralization was found along the contact between igneous and sedimentary rocks. The metamorphic rocks can be seen along MT 43 west of the town of Divide (see the MT 43: Divide—Lost Trail Pass road guide).

The light-colored bluffs along the west side of Divide Creek are river and floodplain deposits of the Eocene to Miocene Renova Formation. The Renova has yielded the bones and teeth of mammals such as camels and rhinos not found in North America today, as well as other fossils. The remains of three-toed horses,

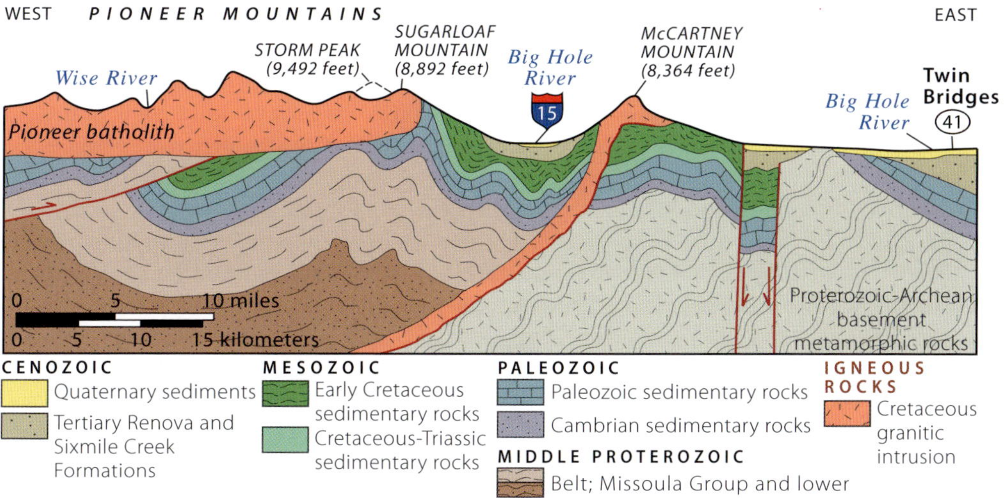

West-east cross section from the East Pioneer Mountains to Twin Bridges.

The Humbug Spires Wilderness Study Area as seen east from I-15.

carnivorous doglike animals, and a sloth-like mammal called a chalicothere (illustrated on an interpretive sign at the rest area) have all been found in the bluffs of Renova along Divide Creek.

East of exit 102 (MT 43 to Divide) are excellent views of the Humbug Spires, towering fins of granite. The fins formed in the 74-million-year-old Moose Creek pluton of the Boulder batholith. Some of the granite spires rise as much as 600 feet above the trees within the Humbug Spires Wilderness Study Area. The area is popular with rock climbers, who like the vertical cracks and flat surfaces that have developed in the hard granite.

The Humbug Spires formed when slightly acidic rainwater seeped into vertical joints, weathering minerals and widening and deepening the cracks; freezing water spreads them farther apart. Over time, vertical slabs of resistant rock became widely separated to stand out in relief from the surrounding weathered rock and soil. Much of the Boulder batholith weathers into spheroidal blocks that look like boulders, not spires. But in the Moose Creek pluton, the joints are more widely spaced, with few cross fractures, leaving some hard rock to form spires. To get a close-up view of the Humbug Spires, take exit 99 at Moose Creek and go east on the dirt road about 3 miles to the well-marked trailhead. The hike to the Wedge, a popular climbing spot, is about 4 miles on a good trail with moderate elevation gain.

Take MT 43 (exit 102) west toward Divide to see light-colored Tertiary sediments of the Renova and Sixmile Creek Formations. They are well exposed on the north side of MT 43 between I-15 and Divide, where they tilt to the northeast into the Rocker fault. The fault is easy to see to the east of I-15, and also where Moose Creek Road crosses it. At the fault, light-colored Tertiary sedimentary rocks butt up against granitic rocks of the Moose Creek pluton. The fault has raised the mountains; the valley has also risen, just not as much.

Exposed south of the Humbug Spires are weakly metamorphosed sedimentary rocks of the Proterozoic Belt Supergroup. These rocks were deposited from 1.47 to 1.4 billion years ago in the extensional-fault-bounded Belt Basin, which formed where the North American craton and another landmass were joined as a supercontinent (Nuna). The southern boundary of this extensional basin, the Camp Creek fault, is exposed in the southern Highland Mountains about halfway between Divide and Melrose. This east-west fault was active while the Belt sediments were deposited, lifting Archean metamorphic rocks to the south. Debris eroded from the Archean rocks was deposited in the basin to the north. In the Highland Mountains these deposits are more than 4,920 feet thick.

GREAT UNCONFORMITY AT CAMP CREEK

Cambrian through Mississippian sedimentary rocks dip gently to the west along the west flank of the southern Highland Mountains. The base of the Cambrian section is the Flathead Sandstone, deposited on a beach about 520 million years ago. It rests on igneous and metamorphic rocks that formed during a plate collision about 1.8 billion years ago (Big Sky orogeny). The contact of the units is an unconformity, or gap in the rock record, marking a time during which erosion occurred or no sediments were deposited. This one is called the Great Unconformity because it represents over 1 billion years of time. Gaps in the rock record are as important as the represents itself because they can help geologists identify areas of the Earth that were involved in mountain building, uplift, and erosion. To see the Great Unconformity up close, take Camp Creek Road off I-15 east of Melrose for about 3 miles and park at a small pull-out on the north side of the road.

The underside of a Cambrian trilobite carapace from Camp Creek. View is about 2 inches across.

The Great Unconformity at Camp Creek. The geologist's right hand is on 1.8-billion-year-old igneous and metamorphic rocks that formed during a collision of continents (Big Sky orogeny). Her left hand is on the Flathead Sandstone, 520-million-year-old beach sand that was deposited after the ancient mountains had been eroded down to sea level. —Courtesy of Marli Miller

A walk through the Cambrian section above the unconformity provides a great example of a marine transgression, the landward migration of depositional environments produced by a relative rise in sea level. The Flathead Sandstone was deposited directly on the eroded igneous and metamorphic basement rock as ocean water flooded across most of North America. The overlying Wolsey Shale is an offshore shelf deposit of mud that buried the beach sand as sea level continued to rise and the oceanic environment shifted landward. Continued sea level rise buried the Wolsey Shale beneath the Meagher Limestone, which formed on the outer reaches of the continental shelf, where calcite-producing animals and plants tend to live. Sea level fluctuated throughout the remainder of Cambrian time, resulting in alternating deposits of shale and limestone, which you can see by walking west from the Great Unconformity past the small reservoir. The rocks contain fossils of trilobites, brachiopods, and other ancient creatures that can be found if you have the patience to crack open the limestone or peel apart the shale. It's worth the effort to liberate a trilobite fossil that has not seen the light of day in more than 500 million years!

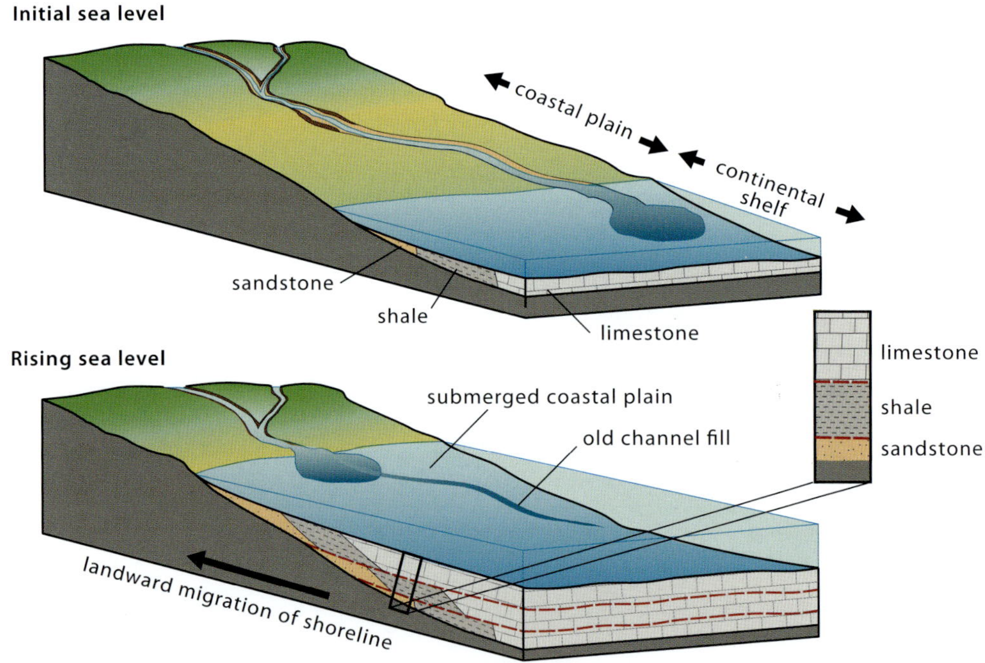

Sequence of rock types deposited during the transgression of a sea onto a continent.
—Courtesy of Marli Miller

South of Divide and west of I-15 are complexly folded and thrust-faulted Proterozoic through Mesozoic sedimentary rocks that were intruded by granitic plutons of the Pioneer batholith, a roughly 75-million-year-old intrusion produced during the same plate collision that formed the Boulder batholith. The collision, the Sevier mountain-building event, caused the layered sedimentary rocks to detach from the underlying Archean metamorphic rocks, producing large slabs of sedimentary rocks (thrust sheets) that were shoved as much as 100 miles to the east along flat faults called décollements. Where the rocks could not easily slide eastward, they were forced to cut up through the pile of sedimentary rocks along high-angle reverse faults, or ramps, putting older rocks over younger rocks.

A large thrust sheet called the Grasshopper plate spans the length of the East Pioneers. It consists of a pile of folded and thrust-faulted rocks that can be seen from the interstate west of Melrose. Proterozoic Belt Supergroup sedimentary rocks form a klippe, an isolated patch of older rocks resting on much younger and intensely folded sedimentary rocks, in this case mostly of Cretaceous age.

Granitic magma generated during plate collision intruded the Sevier folds and thrusts and is now well exposed along the crest of the range. The granitic magma of the Pioneer batholith and the 74-million-year-old McCartney Mountain pluton to the east (north of Glen) was injected along thrust faults. The magma appears to have followed the faults, pooling at the top of major thrust ramps to form plutons. These ramps are like steps along which older rocks are thrust up and over younger rocks along flat faults until stress forces another step to form. The top of a ramp appears to be a good place for magma to collect to form a pluton. You can actually see the rough outline of the McCartney Mountain pluton from I-15 because trees and larger scrub brush preferentially grow on the granite, which holds moisture better than the shale of the surrounding Cretaceous Blackleaf Formation.

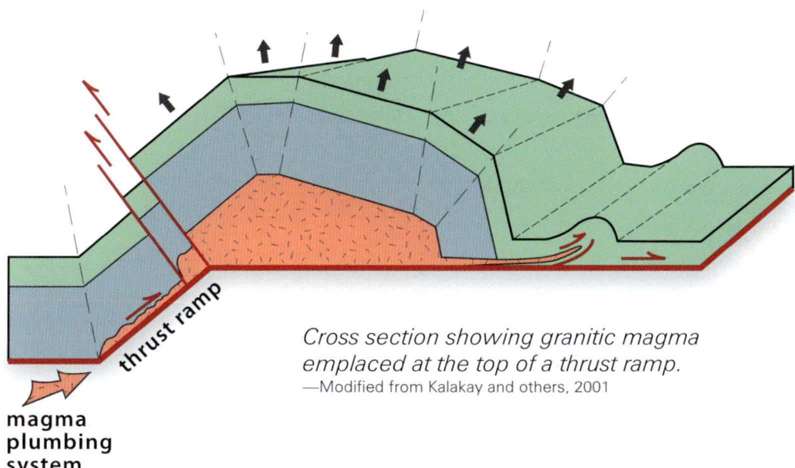

Cross section showing granitic magma emplaced at the top of a thrust ramp.
—Modified from Kalakay and others, 2001

The East Pioneer Mountains are much younger than the rocks and faults exposed within the range. Over the last 16.5 million years, active basin and range extension has raised the range to its lofty heights, with recent uplift concentrated along northwest-trending faults. On the evening of July 25, 2005, residents of the Beaverhead Valley abruptly discovered that these faults are active when a 5.6 magnitude earthquake shook buildings and rattled nerves. Analysis of the earthquake hypocenter, the subsurface location of fault rupture, shows that the earthquake occurred about 3 miles below the surface on a northwest-trending normal fault that is raising the East Pioneers and tilting the Beaverhead graben down to the southwest.

Two side trips into the East Pioneers offer examples of contact metamorphism. At exit 85, north of Glen, turn west onto Brownes Bridge Road, then turn west (left) on

Tungsten-bearing garnet skarn (dark rock above and below) in the Snowcrest Range Group (light-colored rock).

Rock Creek Road to Brownes Lake and the quarry at the west end of the lake. The pit, the Ivanhoe tungsten mine, was the largest producer of tungsten in Montana from 1954 to 1956. The tungsten formed where the Pioneer batholith intruded limestone of the Mississippian Snowcrest Range Group; the granitic magma altered the limestone with hot fluids to form a garnet-rich skarn with the tungsten mineral scheelite. Scheelite is nondescript and off-white in daylight but fluoresces bright to pale-blue under short-wave ultraviolet light.

To get to the ghost town of Farlin, take exit 74 (Apex) and turn right (west). Proceed for 6 miles on Birch Creek Road. Mining in Farlin started in 1864 and reached its peak in the late 1800s, when the town boasted about five hundred citizens and a post office, smelter, and one-room schoolhouse, which still stands north of the road. Its copper and silver deposits were produced by contact metamorphism and fluid exchange with the Pioneer batholith, which intruded the Mississippian Madison Group limestone during Cretaceous time. The ore deposits are also loaded with iron-bearing minerals and copper carbonates, such as malachite (green) and azurite (blue). The ore was smelted and the waste product, or slag, was dumped into Birch Creek. The pile of glassy black slag, south of the road next to the stream, shows ropy flow features (pahoehoe), some with large gas bubbles.

HOGBACK RIDGE AND BLOCK MOUNTAIN

To see the geologically famous structures of the Sevier orogeny at Block Mountain, take exit 85 to Glen and turn east on Burma Road, a dirt road less than 1 mile south of Glen. Rocks along this road are primarily Mesozoic and late Paleozoic sedimentary rocks that were folded and thrust faulted during the Late Cretaceous Sevier orogeny. About 8 miles east, at the Notch Bottom fishing access on the Big Hole River, an interpretive sign explains the geology of Hogback Ridge, the linear ridge stretching south of the Big Hole River. The hogback is a steeply overturned anticline cored here with the Pennsylvanian Quadrant Formation. The anticline was thrust faulted over Cretaceous sedimentary rocks to the east. In Sandy Hollow, north of the Big Hole River, the thrust fault plunges beneath a pair of folds that formed when the fault plane rose through the sedimentary rocks (think raising the edge of your hand through a pile of papers). For thousands of years, native peoples used the north-plunging syncline east of the anticline as a topographic funnel where they drove bison off a cliff.

Block Mountain, the dark-colored, flat-topped mountain looming to the north, is capped by 47-million-year-old basaltic lava that flowed down a stream drainage cut into the folded and thrust-faulted Paleozoic and Mesozoic sedimentary rocks. The basalt is resistant to erosion, so what was once a topographic low is now a topographic high. Beautiful columnar joints, fractures that developed as the magma contracted while it crystallized, are well exposed at road level east of the buffalo jump. Look for glassy grains of quartz and light-colored fragments of sandstone and siltstone in the basalt that it picked up from the Cretaceous Blackleaf Formation.

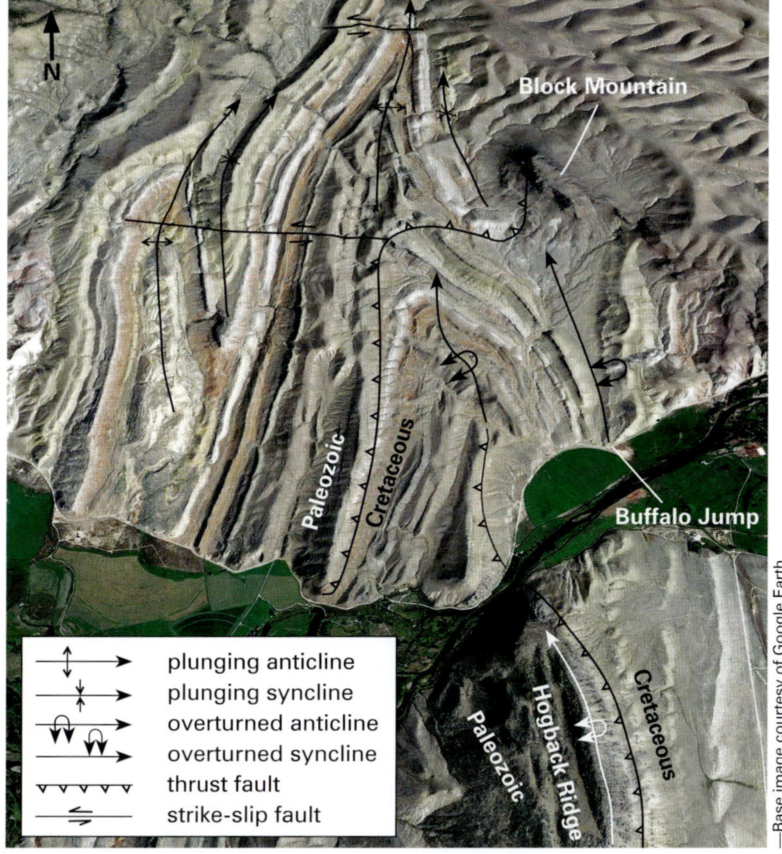

—Base image courtesy of Google Earth

I-15
Dillon—Monida Pass
64 miles

The Union Pacific Railroad established Dillon due to its proximity to gold fields, such as at Bannack, the first territorial capital of Montana Territory. But talc is now the mineral of value in the region, especially because deposits here are free of asbestos. A plant south of Dillon processes it for use in ceramics, paints, plastics, cosmetics, and nonstick coatings on chewing gum. Dillon is home to the University of Montana Western. Founded in 1893 as the Montana State Normal School, it's now an experiential-learning campus that uses immersion scheduling to teach students through hands-on practice. Dillon is also the summer home for up to thirty geology field camps. Participants from around the globe come to Dillon for the tremendous geological diversity and access to public lands.

East of Dillon is the east bench of the Ruby Range, a pediment (erosional surface) that developed during a climatic dry period in the Pliocene Epoch. Basement rocks are exposed in the Ruby Range and the Sweetwater Range (the northwest-trending, southern end of the Ruby Range), both of which are active extensional fault blocks along their northwest trends. The Regal talc mine on Sweetwater Road east of Dillon is an open pit excavated into Archean dolomitic (magnesium-rich) marble. Talc is a magnesium-silicate clay mineral; here it formed when percolating hot water altered the marble along extensional faults that created the Proterozoic Belt Basin. The hot water may have been circulated by basaltic dikes injected into the Archean rocks during the formation of the Belt Basin. One of these dikes occurs in the west wall of the open pit mine.

Regal talc mine east of Dillon. The excavation pit is on the left and the tailings dump on the right. Inset shows talc from mine. —Aerial view courtesy of Julia Gwinn

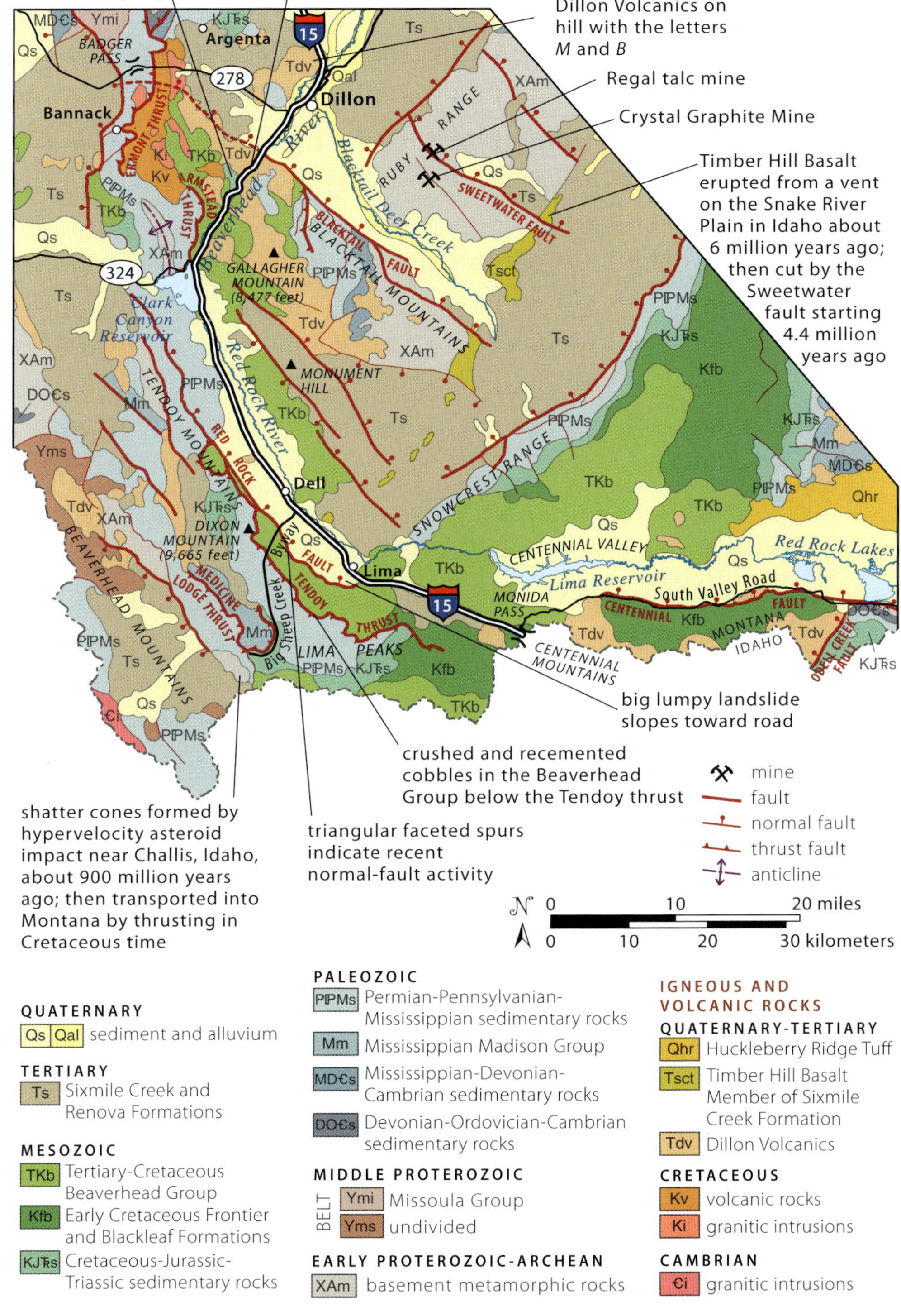

Geology along I-15 between Dillon and Monida Pass.

The basement rocks in these ranges have a diverse and complicated history of metamorphism. Sedimentary rocks were initially metamorphosed to marble, quartzite, metaconglomerate, schist, and gneiss around 2.5 billion years ago. Metamorphism occurred again around 1.8 billion years ago during a plate collision (Big Sky orogeny). The heat and pressure of metamorphism produced some unusual and valuable minerals. Single crystals of graphite several inches long have been found at the Crystal Graphite Mine; banded iron deposits have caught the attention of international iron-mining companies; and red garnets, which are currently mined for use as abrasives, are so abundant in some rocks that they gave rise to the name Ruby Range.

I-15 south of Dillon follows the canyon of the Beaverhead River across the Blacktail Mountains, which are rising as a result of northeast-southwest crustal extension on fault planes that dip to the northeast. The adjacent Blacktail Deer Creek valley tilts down to the southwest into the fault planes. This seesaw pattern, with ranges on the up side and valleys on the down side, can be traced from at least the Tobacco Root Mountains in the north to the Beaverhead Mountains in the south.

The very abrupt and steep front of the Blacktail Mountains shows the range's youth; erosional processes have had little time to modify the front. The Blacktail fault shows at least 3,400 feet of offset has happened in the last 4.5 million years. The most recent movement is uncertain, but sometime between 35,000 and 15,000 years ago an earthquake produced a 6.5-foot-high scarp. Such a ground rupture would only occur during a large-magnitude earthquake, and it's estimated that this fault is capable of producing an earthquake at least equal to the 7.5-magnitude Hebgen Lake earthquake of 1959. This is not good news for Dillon, with its many unreinforced turn-of-the-century brick and rock buildings.

Ash and pumice form spectacular rock towers along East Fork Blacktail Road southeast of Dillon. They were deposited 10 million years ago when volcanic mudflows went down the ancestral Missouri River drainage off the bulge of the Yellowstone hot spot of that time in Idaho. Since streams don't flow uphill, the current topography could not have been in the way at that time. Near the turnoff for East Fork Blacktail Road, the Blacktail fault offsets the Timber Hill Basalt. It erupted from the Heise volcanic field near Idaho Falls about 6 million years ago, again showing that the current mountains could not have been in the path of the flow at that time.

Notice that tree growth along the length of the Blacktail Mountains ends abruptly at about the same elevation near the base of the range. The trees mark the transition from bedrock exposure to alluvial fan gravel downslope. Alluvial fans are permeable, so any water that falls on the surface soaks in rapidly, sinking out of reach of tree roots; the surface remains dry and capable of growing only grass. In contrast, rocks are much better at growing trees in dry climates because they retain moisture in their cracks and provide protection for seeds and cones. This pattern of tree growth occurs throughout southwest Montana, one of many examples of how geology plays an important role in the distribution of animals and plants.

At the entrance to the Beaverhead Canyon, the Blacktail fault has raised and exposed Eocene-age Dillon Volcanics, a subdivision of the Renova Formation. These rocks range from rhyolite to andesite and basalt; they erupted from explosive volcanoes similar to those in the Cascades in Oregon and Washington. Like the Eocene Lowland Creek, Challis, and Absaroka volcanic fields around it, the Dillon Volcanics formed as the Farallon plate was subducted below the North American continent.

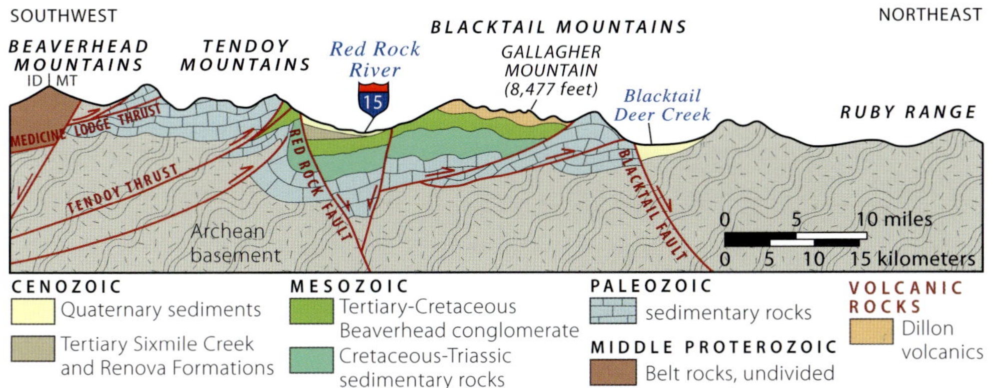

Southwest-northeast cross section from the Beaverhead Mountains to the Ruby Range, crossing the Blacktail fault southwest of Dillon.

Rhyolite (dark) and ashy volcanic mudflow deposits (light) of the Dillon Volcanics exposed by the Blacktail fault at the entrance to Beaverhead Canyon on I-15.

From Eocene to Miocene time, the Dillon Volcanics became more basaltic, probably because the Farallon plate began to sink, causing rising mantle to partially melt.

The volcanoes of the Dillon Volcanics are no longer visible because they were eroded away. Only the volcanic vent rocks remain in a few places to show where the volcanoes were located. Deposits of lava, ash, and volcanic mudflows were preserved in the stream valleys surrounding the volcanoes because they were exposed to less erosion. Because the hard volcanic rock resists erosion, down-dropping of modern valleys during recent crustal extension inverted many of the old stream valleys, leaving them standing higher than the surrounding rocks. Petrified wood is abundant on the undersides of many lava flows and within the volcanic mudflows that buried and carried a variety of trees, including spruce, redwood, and tropical hardwoods.

The Dillon Volcanics are accessible along 10 Mile Road to the hillside *M* west of Dillon, and at Barretts Park along the Beaverhead River on the frontage road south of exit 56. Variations in color and texture bring out the flow banding in the rhyolite. The

Flow banding in rhyolite of the Dillon Volcanics on the road to the M on the hillside west of Dillon.

layers probably formed when larger crystals in the flowing viscous lava segregated to form bands of crystals. Gas bubbles in the solidifying lava were later filled with fine-grained quartz to form geodes that easily erode from the rock. In 1805, Lewis and Clark had lunch on the rocks at the canyon entrance and named the spot "rattlesnake clifts" after the reptiles that live in the cracks.

The Beaverhead River occupies a 600-foot-deep canyon that its precursor cut across the Blacktail Mountains from Barretts Park to the Clark Canyon Reservoir. The path of the Beaverhead River in the canyon is older than the mountains. Starting some 16.5 million years ago, basin and range extension formed broad grabens with northeast-flowing streams that deposited gravel, ash, and lava from the Yellowstone thermal bulge when it was in southern Idaho. Then, around 4.5 million years ago, changes in crustal stretching brought on by the eastward-migrating bulge caused the Blacktail Mountains and other northwest-trending ranges in the area to rise across the path of the old rivers. Some of the northwest-trending grabens captured and diverted the old rivers, forming new pathways, such as that of the Red Rock River; others continued to follow their old pathways, eroding through the rising mountains and forming superimposed river canyons, such as the Beaverhead Canyon. The steep walls of this canyon are prone to slope failure, resulting in landslides that have blocked, narrowed, and diverted the river. Look for their hummocky topography on the lower slopes.

The columnar-jointed Pipe Organ Rock, at the west side of I-15 a few miles south of Dalys (exit 51), is Eocene Dillon Volcanics. The prominent rock west of I-15 is Dalys Spur, west-dipping sandstone of the Pennsylvanian Quadrant Formation deposited in a once-vast coastal dune field. The overlying Phosphoria Formation contains phosphate-rich rock, which was deposited when cold, nutrient-rich water was upwelling along an oceanic shelf edge during early Permian time. At the top of the sequence is the Triassic Dinwoody Formation, composed of brachiopod-rich limestone and shale deposited on a marine shelf several million years after the largest mass extinction event in Earth's history, at the end of Permian time.

These sedimentary layers were folded during Late Cretaceous time, resulting in the westward tilt. Dalys Spur must have been important to native peoples, because it

The Beaverhead River cut a canyon across the Blacktail Mountains as they rose across the river's path during the last 4.5 million years. The snow-covered Tendoy Mountains in the distance (to the southwest) rose at the same time; they blocked the old river and diverted it into a new, northwest-flowing pathway, forming the Red Rock River. Pipe Organ Rock rises above the interstate on the right (west).

Dalys Spur is a prominent rock composed of wind-deposited sandstone of the Quadrant Formation. It is exposed in the west-dipping limb of an anticline.

preserves lookout structures and pictographs. If you stop to look, please show respect for this legally protected artwork. The road scar up the north-facing wall of Dalys Spur was a recent, ill-conceived attempt to mine sand from this sacred rock for use in hydraulic fracturing, or fracking. The scar will undoubtedly outlast humanity.

South of Pipe Organ Rock, the rust-red and red-gray rocks on either side of I-15 are alluvial fan and lake deposits of the Beaverhead Group. These sedimentary rocks were shed off and deposited in front of the eastward-advancing fold-and-thrust belt during the Late Cretaceous Sevier orogeny. They were caught up in the folding and were overridden by the heavy thrust plates as these slabs of rock were pushed eastward, crushing, shearing, and deeply denting the cobbles, which may have acted like ball bearings under the thrust plates. Many of the broken cobbles were later cemented back together, preserving their tortured past. The west side of Clark Canyon Dam is anchored in Mississippian limestone loaded with fossils of tropical marine animals, such as crinoids, brachiopods, bryozoans, and corals. Many of these weather completely out of the rock, especially on the rock berm next to the spillway.

Fossils in the Mississippian-age Madison Group limestone at Clark Canyon Reservoir. Fossils include (a) crinoids, (b) brachiopods, and (c) rugose corals.

On August 17, 1805, near what is today an island in Clark Canyon Reservoir, Lewis and Clark met with Sacagawea's brother, Shoshone Chief Cameahwait, to barter for horses. Their "Fortunate Camp" was located at the base of the fossil-rich Mission Canyon Limestone, which is in the upper plate of the Armstead thrust fault. The thrust continues to the southeast as the Tendoy thrust fault, with Paleozoic rocks pushed over Late Cretaceous Beaverhead Group rocks in the Tendoy Mountains. The prominent red rock in the valley southeast of Dell is the Red Butte Conglomerate of the Beaverhead Group.

The Tendoy Mountains rose over the last 4.5 million years along the northeast-dipping Red Rock fault, a normal fault formed by crustal stretching around the Yellowstone hot spot. The steep range front exposes folds and thrust faults in Paleozoic and Mesozoic rocks that the normal fault has cut through over the last 4.5 million years. The Red Rock River valley is tilted down to the southwest into the fault; Monument Hill, east of the valley, is tilted up on the northeastern side of the fault block, forming impressive dissected surfaces and alluvial fans on Miocene and older rocks. Minor faults, such as those around Dell, have produced numerous small earthquakes

BIG SHEEP CREEK CANYON

To see Big Sheep Canyon, take Big Sheep Creek Back Country Byway, a gravel road on the west side of I-15 that is accessed from Dell via Westside Frontage Road. Big Sheep Creek, a superimposed stream that cuts across the Tendoy Mountains, was inherited from the northeast-trending basin and range topography that existed here from 16.5 to 4.5 million years ago. At the mouth of the canyon, the road bumps up to cross a fault scarp made by multiple ground ruptures of the Red Rock fault that cut glacial outwash. From here, the road passes through conglomerate of the Beaverhead Group that was overridden in Late Cretaceous time by the Tendoy thrust plate, which carried Mississippian-through-Permian sedimentary rocks in its upper plate. South of the extensional Muddy Creek basin, the byway winds through folded and faulted Madison Group limestone that's riddled with karstic holes dissolved out of the rock by slightly acidic groundwater. Many of these caves have yielded Pleistocene mammal bones and archaeological artifacts.

The road exits the canyon and enters a northwest-trending extensional graben that parallels the Beaverhead Mountains, which rose around 4.5 million years ago. Belt rocks in the Beaverhead Mountains preserve shatter cones from a major asteroid impact that hit near where Challis, Idaho, is today around 900 million years ago. Shatter cones are conical rock structures that high-pressure shock waves form on the bedrock below and outward away from an impact crater. Subsurface data show the crater to be about 65 miles in diameter, making it one of the largest impact craters on Earth. These structures are now exposed in Montana, in part because they were shoved here along thrust faults during Cretaceous compression.

Karstic caves in an exposure of intensely folded Madison Group limestone on Big Sheep Creek Back Country Byway.

The red rock southeast of Dell is Beaverhead Group conglomerate. Inset shows limestone cobbles in the Lima Conglomerate of the Beaverhead Group, which is well exposed southeast of Lima next to I-15.

in recent years. Trenches dug across the Red Rock fault show that it last ruptured the ground surface between 3,705 and 2,240 years ago. Single earthquake events on this fault have produced fault scarps that are up to 20 feet high. These events were similar in size to the 1959 Hebgen Lake earthquake, which measured 7.5 on the Richter scale. Some of the scarps on the Red Rock fault are as high as 131 feet but were produced by multiple events. Between Dell and Lima, triangular-shaped faceted spurs (between-gulley remnants of the fault plane) along the front of the Tendoy Mountains attest to the very recent movement on the fault. Landslides along the range front may be triggered by shaking during large earthquakes.

Northeast of Dell on Sage Creek Road are excellent exposures of the Timber Hill Basalt flow, a 6-million-year-old flow from the Heise volcanic field in Idaho that flowed more than 100 miles down the ancestral Missouri River before the mountains rose. Underlying the basalt flow are exposures of light-colored, ashy lake and stream sediments of the Eocene-to-Miocene Renova Formation.

South of Lima, the glaciated Lima Peaks reach to more than 10,000 feet, held up largely by hard Pennsylvanian Quadrant Formation sandstone and quartzite. Intensely folded Paleozoic rocks in the Lima Peaks are in the upper plate of the Tendoy thrust plate and rest on the Late Cretaceous Beaverhead Group. Basalt that flowed north (now uphill) into Montana from the Snake River Plain rest on the Beaverhead Group rocks in several places all the way to Monida.

View west from the Dell area shows triangular-shaped faceted spurs, preserved surfaces of the Red Rock fault plane.

CENTENNIAL VALLEY

If you're driving a vehicle with good tires, consider traveling east on South Valley Road (MT 509), a well-maintained gravel road through the Centennial Valley along the northern front of the east-west-trending Centennial Mountains. Take the Monida exit east from I-15 at the Montana-Idaho border. The Centennial Valley formed over the last 2.1 million years along the Centennial fault, a listric normal fault system. Extensional stresses placed on the crust by thermal bulging around the Yellowstone hot spot produced the fault. About 2.1 million years ago, before the valley existed, the Yellowstone volcano erupted a ground-hugging pyroclastic flow, a fast-moving, incinerating, red-hot flow of ash that deposited the Huckleberry Ridge Tuff. Parts of the once-continuous flow are preserved at the top of the Centennial Mountains and elsewhere in the Centennial Valley. Movement along the Centennial fault has offset the tuff at least 6,000 feet in the last 2.1 million years, and younger deposits by as much as 33 feet over the last 12,000 years. The recent fault scarps along the range front occur in glacial, alluvial, and landslide deposits of Pleistocene and Holocene age. By dating organic material obtained from trenches dug across the fault, scientists determined the time between earthquakes to be more than 3,000 years. Seismic data show that the fault is still active and capable of producing a large earthquake.

The western part of the Centennial Mountains exposes young volcanic rocks that rest on nonmarine Cretaceous sandstone and shale deposited by streams on the margin of the Western Interior Seaway. Around 95 million years ago, *Oryctodromeus*, a small plant-eating dinosaur, lived in burrows, like Montana's ground squirrels do today, and its remains are preserved as fossils in the Cretaceous rocks. Bentonite clay in the Cretaceous sedimentary rocks absorbs water; it has caused numerous landslides along the range front. The slides form bowl-shaped depressions with hummocky deposits below. Aspen trees grow in the wet bowl areas.

The eastern part of the range exposes much older rock, likely due to uplift on the Odell Creek fault, a normal fault that probably first formed as a reverse fault during the Laramide orogeny. On the ridge above the homestead near Battle Creek are Archean and Proterozoic rocks that were deeply buried and metamorphosed when continents collided some 3 billion years ago. The mountains that formed during the collision were slowly eroded down to sea level and eventually buried by thousands of feet of Paleozoic and Mesozoic marine and nonmarine sedimentary rocks that make up most of the steep face of the Centennial Mountains above Upper Red Rock Lake.

An artist's rendition of the burrow-dwelling Oryctodromeus *with its young.*
—Courtesy of Norm Dwyer

The geologist is standing on fine-grained lake sediments deposited in Lake Centennial during Pleistocene time. When the lake drained, Red Rock Creek deposited gravel on top of the lake bottom. The geologist's right hand is at the transition. —Courtesy of Ken Pierce

Today, the Centennial Valley drains west from the uppermost headwaters of the Missouri River into the Red Rock River. However, prior to the start of the Pinedale glaciation, the valley drained northeast into the Madison Valley through an old canyon now marked by a chain of steep-walled lakes, including Elk, Hidden, Cliff, and Wade Lakes. Sometime between 30,000 and 20,000 years ago, a landslide at the northeast end of Elk Lake dammed the old drainage and created Lake Centennial. Once dammed, the Pleistocene-age lake rose enough to release some water over a low area at the western end of the valley near the present-day Lima Dam to form the current path of the Red Rock River. At its deepest, the lake was about 65 feet deep, as shown by the old lake shorelines visible on the north side of the valley near Long Creek. Thinly bedded clay-rich sediments in the valley were deposited in the lake.

Lake Centennial's level fluctuated as it dropped, exposing shoreline sand that was piled into dunes on the northern side of Centennial Valley. The longitudinal dunes stand as high as 98 feet and are aligned with the prevailing southwest-to-northeast wind direction. Most of the sand accumulated during the late Pleistocene, and it appears that the lake was gone by 12,000 years ago.

Roaming the valley in Pleistocene time were small rodents, pronghorn, deer, bison, bears, beavers, coyotes, wolves, cats, horses, camels, and Columbian mammoths, a fur-covered relative of the modern elephant. The mammoths are particularly well-known from preserved remains of their teeth, tusks, and other bones excavated from the Merrell Site near the west end of present-day Lima Reservoir. Radiocarbon dates indicate that humans were in the valley by at least 10,500 years ago. They hunted the mammoths.

Tectonic tilting of the valley to the south against the Centennial fault, a listric normal fault, and constriction of Red Rock Creek by alluvial fans, together created Red Rock Lakes. These alluvial fans, built of sediment eroded from the Centennial Mountains, limit the outflow of water from the valley, creating lakes behind them. The lakes are only 5 or 6 feet deep and treacherously muddy. Don't try to stand on the bottom or you may sink into the sticky clay.

Red Rock Lakes are shallow and loaded with aquatic vegetation that supports a variety of bird species. The wetlands were instrumental in the recovery of the trumpeter swan.

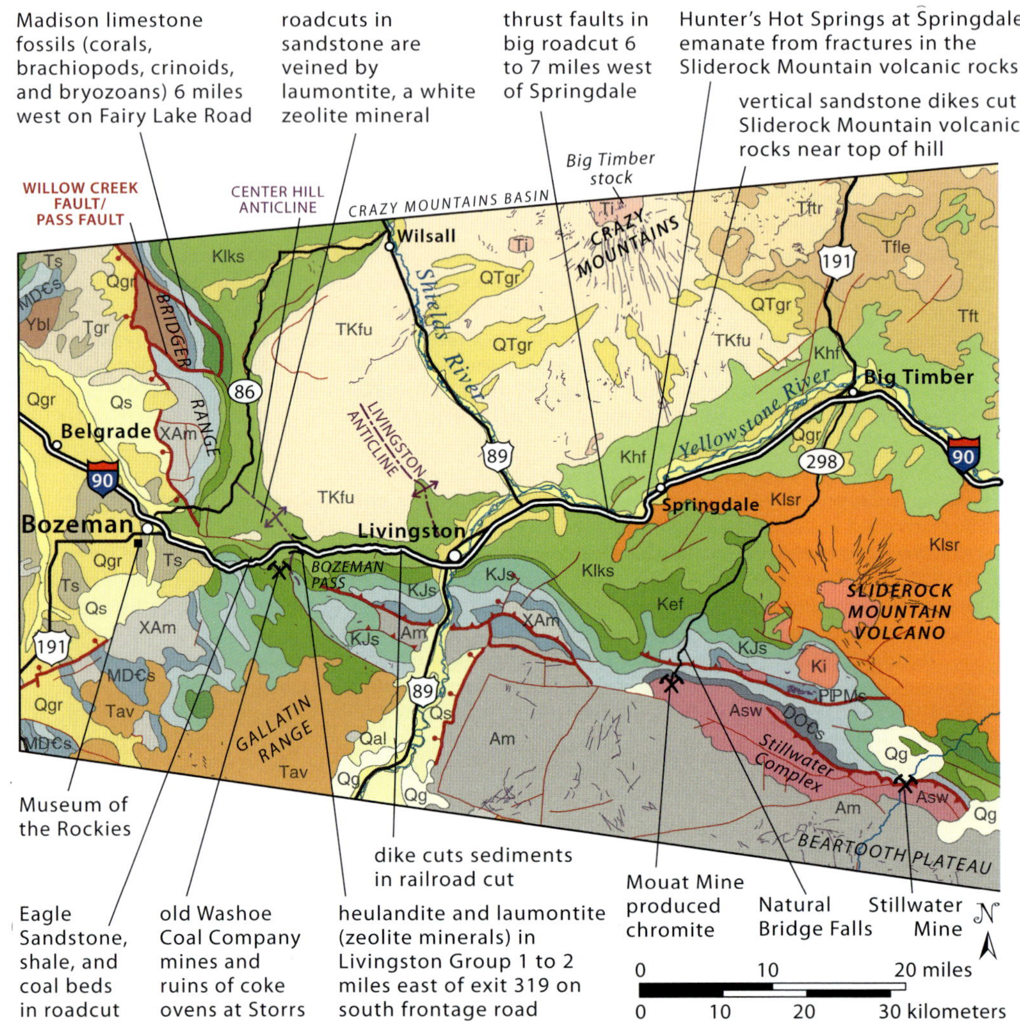

I-90
Billings—Bozeman
141 miles

For a discussion of the dramatic cliffs of Eagle Sandstone known as the Rimrocks, on the north side of Billings, see the I-94: Billings—Miles City road guide in the Central and Eastern Montana chapter.

Between Billings and Livingston, I-90 follows the Yellowstone River valley. At low water you can find agates, jasper, and petrified wood in the river gravels. They are primarily eroded from Eocene volcanic rocks upstream. Riverfront Park in Billings, Riverside Park in Laurel, and Itch-Kep-Pe Park (City Park) in Columbus are good places to search.

Much of the route from a few miles west of Billings to Columbus, and from there on to Big Timber, crosses forested hills eroded in Late Cretaceous Hell Creek

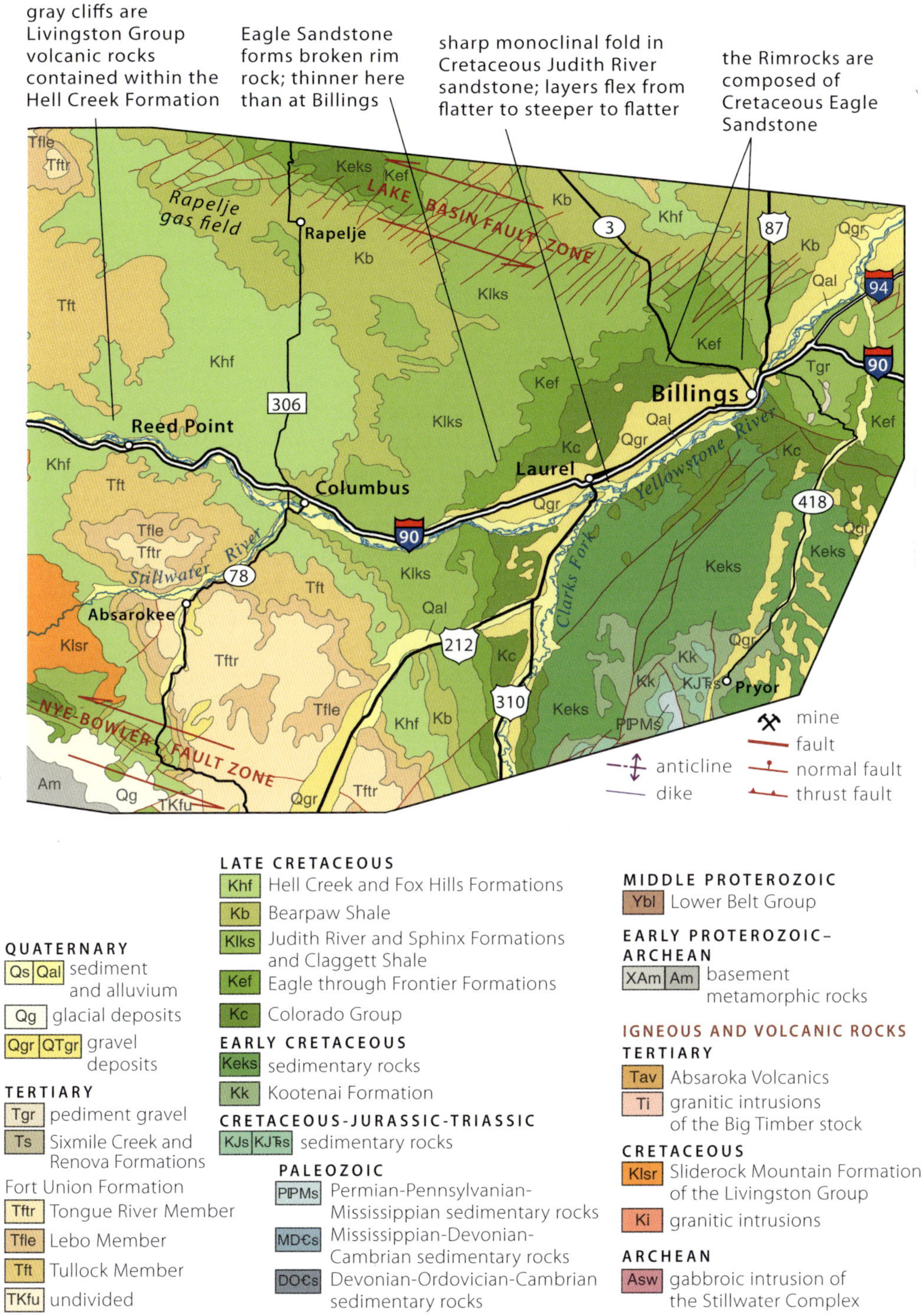
gray cliffs are Livingston Group volcanic rocks contained within the Hell Creek Formation
Eagle Sandstone forms broken rim rock; thinner here than at Billings
sharp monoclinal fold in Cretaceous Judith River sandstone; layers flex from flatter to steeper to flatter
the Rimrocks are composed of Cretaceous Eagle Sandstone
Rapelje gas field
LAKE BASIN FAULT ZONE
NYE-BOWLER FAULT ZONE
Rapelje
Reed Point
Columbus
Laurel
Billings
Absarokee
Pryor
Yellowstone River
Stillwater River
Clarks Fork
306
3
87
94
90
418
78
212
310
Tfle
Tftr
Tft
Khf
Keks
Kef
Kb
Klks
Kc
Qgr
Qal
Tgr
Klsr
Kk
KJTRs
PIPMs
Am
Qg
TKfu
mine
fault
normal fault
thrust fault
anticline
dike
QUATERNARY
Qs Qal sediment and alluvium
Qg glacial deposits
Qgr QTgr gravel deposits
TERTIARY
Tgr pediment gravel
Ts Sixmile Creek and Renova Formations
Fort Union Formation
Tftr Tongue River Member
Tfle Lebo Member
Tft Tullock Member
TKfu undivided
LATE CRETACEOUS
Khf Hell Creek and Fox Hills Formations
Kb Bearpaw Shale
Klks Judith River and Sphinx Formations and Claggett Shale
Kef Eagle through Frontier Formations
Kc Colorado Group
EARLY CRETACEOUS
Keks sedimentary rocks
Kk Kootenai Formation
CRETACEOUS-JURASSIC-TRIASSIC
KJs KJTRs sedimentary rocks
PALEOZOIC
PIPMs Permian-Pennsylvanian-Mississippian sedimentary rocks
MD€s Mississippian-Devonian-Cambrian sedimentary rocks
DO€s Devonian-Ordovician-Cambrian sedimentary rocks
MIDDLE PROTEROZOIC
Ybl Lower Belt Group
EARLY PROTEROZOIC–ARCHEAN
XAm Am basement metamorphic rocks
IGNEOUS AND VOLCANIC ROCKS
TERTIARY
Tav Absaroka Volcanics
Ti granitic intrusions of the Big Timber stock
CRETACEOUS
Klsr Sliderock Mountain Formation of the Livingston Group
Ki granitic intrusions
ARCHEAN
Asw gabbroic intrusion of the Stillwater Complex

Geology along I-90 between Billings and Bozeman.

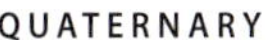

Formation, dinosaur-bearing sedimentary rocks laid down in a nonmarine environment of river channels (sandstone) and floodplains (mudstone). Watch for bouldery outcrops of pale-gray or tan rock among the pine trees, which flourish on moisture-trapping sandstone everywhere in central and eastern Montana. About the turn of the century, a quarry just north of the Columbus exit supplied handsome pale-gray stone for the older part of the Montana State Capitol building in Helena, as well as for many other buildings in several Montana towns.

South of I-90, the Beartooth Mountains dominate the skyline. The mountains contain more than twenty-five peaks that exceed 12,000 feet, with Granite Peak south of Nye, at 12,807 feet, the highest point in Montana. This massive block was first thrust upward from the south about 50 million years ago, during the Laramide orogeny. The core of the mountains consists of hard Archean-age granite and metamorphic rocks that are among the oldest rocks on Earth. They are flanked on the north by light-colored, steeply dipping Paleozoic and Mesozoic sedimentary rocks that were draped over the thrust block, and early Tertiary rocks eroded off the block as it was raised and deeply eroded. The mountains were lifted again beginning in Miocene time by regional crustal thinning and thermal bulging caused by basin and range extension and the Yellowstone hot spot. The mountains are distinctly flat on top (a plateau) due to the reexposure of an ancient flat surface that had formed after the hard Archean rocks had undergone billions of years of erosion prior to their burial by sedimentary rocks starting in Cambrian time. (See the US 212: Laurel—Wyoming Border on the Beartooth Plateau road guide for a trip to the spectacular top of the plateau.)

West of Columbus near milepost 409, pale-gray to beige shale and sandstone of the Hell Creek Formation form big roadcuts on the north side of the Yellowstone River. A couple of miles east of Reed Point, on the south side of the river, layers of pale-gray shale dip about 20 degrees south into the axis of the Reed Point syncline.

STILLWATER MINE

The Stillwater Mine, one of the world's largest platinum-palladium mines and the only one in the United States, lies at the base of the Beartooth Mountains near Nye, southwest of Columbus. The Stillwater Complex of rocks is one of the most unique and geologically famous Archean rock exposures in Montana. This layered ultramafic-mafic intrusion was emplaced 2.7 billion years ago and is now tilted up on edge and spectacularly exposed for 30 miles along the front of the range. The top of the intrusion was eroded away hundreds of million years ago, before Paleozoic sedimentary rocks were deposited on it, but the complex is still up to 21,000 feet thick.

The Stillwater Complex is divided into three distinct zones of peridotite (olivine and pyroxene) and gabbro (pyroxene and calcium-rich plagioclase). These minerals are low in silica but rich in iron and magnesium. The gabbro, a coarsely crystalline version of basalt, is layered like sedimentary rock. The layering in the gabbro is the result of a process called fractional crystallization. Minerals crystallize at different temperatures. Heavier crystals that form first sink to the bottom of the magma chamber, followed by other minerals, such as plagioclase feldspar, as the magma cools, forming layers of different minerals. Alternating layers of light and dark minerals may also form as slurries of one mineral sink in the convecting currents in a magma chamber. In layered mafic intrusions, olivine tends to crystallize at the highest temperatures, followed by pyroxene and then plagioclase, to form distinct concentrations of these minerals.

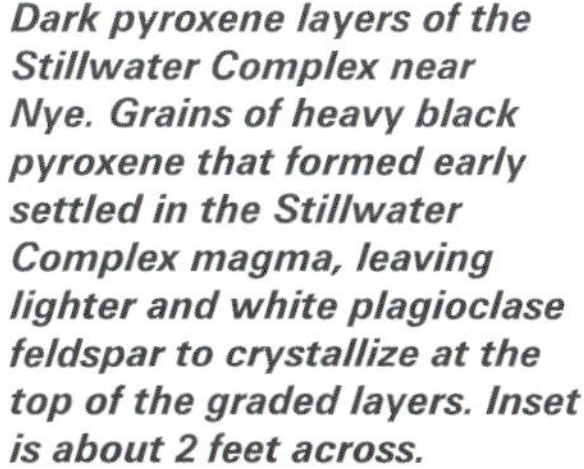

Dark pyroxene layers of the Stillwater Complex near Nye. Grains of heavy black pyroxene that formed early settled in the Stillwater Complex magma, leaving lighter and white plagioclase feldspar to crystallize at the top of the graded layers. Inset is about 2 feet across.

The origin of the intrusion and the processes that caused the layering are debated, but the magma certainly came from melting in Earth's mantle, and such rocks tend to contain valuable ore bodies. Chromite was mined during World War II for making steel, and platinum and palladium are mined today for use primarily in catalytic converters in cars. The Stillwater Mine began extracting platinum and palladium from a very thin zone near the upper part of the banded zone in 1986, a smelter was constructed in Columbus in 1990, and a sister mine, the East Boulder Project, opened in 2001.

Late Cretaceous Hell Creek Formation shale and sandstone continue to the west. The slightly older volcanic rocks of the Late Cretaceous Sliderock Mountain volcano lie under the uppermost Hell Creek beds. The Sliderock Mountain volcanics, part of the Livingston Group, only reach I-90 along its south side about 3 miles east of exit 377 and near Springdale (exit 354), west of Big Timber. The main mass of this small volcano and its diorite magma chamber lie 20 miles south of Big Timber. Its eruptions built a stratovolcano from 78 to 75 million years ago, mostly of basaltic andesite ash, broken volcanic rubble, lava flows, and volcanic mudflows. Ash was blown out of the volcanic vent, and still-soft volcanic chunks were blown directly out of the vent, or solid lava broke up and collapsed as rubble down the volcano's flanks during

Agglomerate of Sliderock Mountain volcano 7 to 8 miles southwest of Big Timber. The dark-colored, subangular rock fragments are andesite suspended in volcanic ash. This rock probably formed as rubble on the flank of the volcano or as a mudflow deposit. View is 3 to 4 feet across.

eruptions. Volcanic mudflows most commonly form on a volcano's flanks when rainstorms saturate loose material, causing it to flow downslope as a fast-moving slurry. Dikes radiate outward from the central intrusion. The volcano was active while Hell Creek sediments were being deposited because its volcanic rocks interfinger with that formation.

Almost all of the sedimentary rocks along I-90 between Billings and Springdale have been minimally deformed and lie essentially flat. West of Springdale sedimentary layers have been flexed into folds and sliced by faults. If you follow a group of nicely parallel sedimentary layers with your eye to where another adjacent layer intersects at an angle, you're likely looking at a fault surface. A small angle suggests a thrust fault, where one section of rock was shoved up and over another section. A large intersecting angle may indicate a strike-slip fault, where rocks slid sideways along the fault, or a normal fault, along which rock on one side of the fault moved down relative to that on the other side.

Layers dip at various angles, and the crests (or axes) of folds are equally variable, but most trend generally east-west. The massive Absaroka Range and Beartooth Plateau are on the skyline about 10 miles south of I-90. Their Archean-age rocks were shoved up during Laramide compression, so maybe these folds along I-90 are related to the same deformation, or possibly to the Late Cretaceous compressional mountain building that happened farther to the west.

Between Big Timber and Livingston, I-90 crosses the southern part of the Crazy Mountains Basin, a deep accumulation of sediments centered on the high Crazy Mountains north of Livingston; the basin extends from about US 89 on the west to US 191 on the east, and north to US 12. Most of the rocks exposed in roadcuts and streambanks south of I-90 erupted from the late Cretaceous Sliderock Mountain volcano, centered south of I-90. Watch for generally greenish or gray roadcuts full of angular chunks of dark andesite.

NATURAL BRIDGE AND FALLS

For a side trip into the high country of the Absaroka Range that includes a spectacular waterfall, travel southwest on MT 298, along the Boulder River from Big Timber, approximately 27 miles. MT 298 is a paved road for about the first 20 miles before becoming a good gravel road. For the first 12 or 13 miles the high cliffs on both sides of the valley are rocks of the Late Cretaceous Sliderock Mountain volcano. The gently dipping, indistinct layers were laid down as ejected volcanic chunks or slide rubble and mudflows on the volcano's flanks during and shortly after eruptions. Regional deformation of the last 75 million years, since the volcano was active, may have tilted the layers out of their original orientations. Scattered, very rounded boulders on nearly flat surfaces near the road are also parts of mudflows or debris flows but are not related to volcanic eruptions. They probably came out of the mountains to the south during flood events of the last ice age.

Farther up-valley, around McLeod, the mountains to your right (to the west) expose the Late Cretaceous Judith River Formation, still nearly horizontal cliffs about 22 miles southwest of Big Timber, about 3 miles downstream from the falls. In places, distant views to the southeast show the broad, nearly flat crest of the Beartooth Plateau, a huge raised block of ancient Archean metamorphic rocks. Just downstream from Natural Bridge Falls, bouldery deposits near the road include lots of granite that was carried north by streams and glaciers from the distant Beartooth Plateau. Even-topped piles with more-rounded boulders are likely mudflow deposits in which the jostling and grinding of boulders rounded them.

For the last few miles to the falls, the road dives deeper into the sedimentary rock section draped over the Beartooth uplift during the Laramide orogeny. The road passes into older and older, steeply dipping, folded and thrust-faulted rocks, including those deposited during Jurassic, Permian, and Pennsylvanian time. At Natural Bridge Falls, the rocks are sandy pale-gray limestone of the Mississippian Madison Group. Limestones are noted for being easily dissolved in slightly acidic rainwater, so finding caves and caverns in this area is not a big surprise. Surface water that has drained down from the stream into the caverns exits them farther downstream, where erosion has exposed the caverns and created a waterfall. During low water, the stream flows only through the lower cavern. The natural bridge that was here collapsed in 1988, but the falls are spectacular.

The river and two underground caverns at different levels feed the falls at high water. During low water, the falls go dry and all the water exits via the lower cavern.

Hell Creek shales and sandstones on the north side of the Yellowstone River just west of Springdale.

About 1 to 2 miles west of Springdale (exit 354) and across the Yellowstone River north of the highway, a big bluff close to the river nicely exposes prominently layered, finer-grained sedimentary rocks ranging in color from dark gray to pale gray to rusty. These are Hell Creek Formation shales and sandstones deposited in the last stages of Cretaceous time, before the sudden extermination of the dinosaurs. These rocks lie right under the similar-looking Fort Union Formation. Both fill the Crazy Mountains Basin.

On the south side of the river 4 or 5 miles west of Springdale, the road leaves the Sliderock Mountain volcanics, and the big roadcuts are in sandstone of the Cretaceous Judith River Formation and Eagle Sandstone. Two miles farther west, greenish-gray, ash-rich sandstone and pebbly conglomerate of the Cretaceous Sedan Formation are exposed south of I-90. These rocks are considered equivalent in age to the finer-grained Livingston Group.

Looming on the horizon north of I-90 is the big dome of the Crazy Mountains, composed of Cretaceous to early Tertiary sediments of the Crazy Mountains Basin. They were pushed upward by the Eocene-age Big Timber stock. The ragged crest of the range consists of igneous rocks of the central intrusion with a big swarm of vertical dikes radiating outward. The intrusions were injected into Paleocene Fort Union Formation sediments, which must have been at rather shallow depths because the intrusions are almost the same age as the sediments. (See the US 191: Big Timber—Lewistown road guide in the Central and Eastern Montana chapter for more information on the Crazies.)

West of US 89 and extending north along US 89 around the west side of the Crazy Mountains Basin, the Fort Union is underlain by volcanic-ash-rich sediments of the Livingston Group rather than the Hell Creek. This ash is about the same age as that erupted from the Sliderock Mountain volcano south of I-90 but is finer grained. It erupted from a source to the west, perhaps that of the Elkhorn Mountains Volcanics around the Boulder batholith. The fine-grained volcanic tuff contains well-preserved impressions of leaves that range from 80 to 70 million years old.

Livingston is built on Yellowstone River gravel that covers Late Cretaceous and early Tertiary sedimentary and volcanic sedimentary rocks exposed in the hills to the north. The most conspicuous is the Paleocene Fort Union Formation, a nonmarine deposit of sandstone, siltstone, and conglomerate as thick as 10,000 feet and loaded with fossil plants and animals.

Livingston was founded when the Northern Pacific Railway established railroad shops to service steam-powered trains before their steep climb over Bozeman Pass. Because of the excellent rail access of a branch line through the Paradise Valley to Gardiner, Livingston became the first gateway town to Yellowstone National Park. The unfortunate legacy of the rail shops has been contamination of soil and groundwater with chlorinated solvents, hydrocarbons, and asbestos.

Underground mines began producing coal around Cokedale, west of Livingston on Cokedale Road, during the 1870s. Early production was sold for domestic use

Late Cretaceous sedimentary rocks exposed along I-90 about 7 miles west of Springdale show clear-cut deformation, including thrust faults. The deformation likely occurred during Laramide compression.

The prominent, continuous cliff capping the slope on the north side of the Yellowstone River 9 to 10 miles west of Springdale is a thick syenite sill. It's a thin, broad laccolith that was injected between horizontal layers of the Fort Union Formation.

in Livingston and Bozeman, but ore smelters and steam engines soon became the largest consumers. Coal from the Livingston field was especially suited for smelting because it could be roasted into coke. Many of the hundreds of coke ovens that operated during the last century survive as ruins in the area a couple of miles south of the highway. After about 1910, Cokedale went into rapid decline. The most easily minable coal was gone by then, more efficient smelter designs had diminished the demand for coke, and cheap oil was becoming readily available. Production diminished further after World War I and finally ceased about 1942. Far more coal remains in the ground than the mines produced.

The Cokedale mines worked several seams near the top of the Eagle Sandstone, beach deposits laid down along the shore of the Western Interior Seaway that still flooded much of inland North America about 70 million years ago. The coal probably began as thick deposits of peat in broad tidewater marshes that may have resembled those along the west coast of Florida today. The weight of younger sediments deposited on top of the peat compressed it, and over time it became coal.

Between Livingston and Bozeman Pass, I-90 follows Late Cretaceous sedimentary rocks; many of the cliffs north of the highway are Livingston Group, beige to dark-gray shales and volcanic sandstones. Approaching Bozeman, I-90 cuts through the northern tip of the Gallatin Range, a western continuation of the tightly crumpled Paleozoic and Mesozoic sedimentary rocks that were plastered against the Archean metamorphic rocks along the northwestern edge of the Beartooth Mountains. The road cuts across the northern tips of several north-tilted anticlines and synclines, so formations are repeated going west.

At Bozeman Pass, the road crosses an anticline in the Livingston Group rocks, green, blocky, fractured volcanic sandstone with brown-stained surfaces, and then a syncline-anticline pair, still in the same formation. Two miles west of the pass, I-90 enters a series of tight turns where Meadow Creek from the south joins the main valley of Rocky Creek.

For the next 2 miles I-90 twists down a very narrow cliff-bound valley across the east limb of a north-tilted anticline through older and older rocks, including Late Cretaceous light-brownish-gray, fine-grained Eagle Sandstone, Mowry Shale, reddish-brown to olive Early Cretaceous Kootenai mudstone and sandstone, greenish to reddish-gray Jurassic Morrison shale, Pennsylvanian Quadrant quartzite and red Amsden limestone and sandstone, and finally the tight axis of the anticline in light-gray Mississippian Madison Group limestone. On the north side of I-90 about halfway between mileposts 316 and 315, the limestone is tightly crumpled, heavily fractured, and full of caves and caverns; their growth has probably been facilitated by water flowing through the fractures. Giant pinnacled cliffs south of the highway are the same Madison Group limestone.

Between the limestone and the valley floor at Bozeman, I-90 passes west up through the west limb of the anticline, through most of the same younger sedimentary rock formations, and then onto Tertiary valley-fill silt, sand, and gravel. As you exit the narrow canyon headed west, the Bridger Range comes into view several miles to the north, trending slightly west of north. This big, eastward-tilted slab of Earth's crust was raised up from the Gallatin Valley along the Bridger fault, an active listric normal fault that lies at the base of the range's steep western front. The Bridger Range is near the eastern extent of basin and range spreading, which extends from Nevada

The tightly folded and faulted, ragged, weathered Madison Group limestone west of Bozeman Pass, is full of caves and caverns.

and Utah, north through Yellowstone and southwest Montana, and on north through Helena to Canada. Vertical offset on the Bridger fault is several thousand feet, and it is likely still capable of generating large earthquakes.

Several small faults cross the range, chopping it into segments. One of the more important is the Pass fault (also called the Willow Creek fault), which cuts across the range in Ross Pass, the deep gap midway along the crest of the range. Around 1.47 billion years ago, movement on this fault raised and exposed Archean basement rocks to the south; these were eroded, and their sediment was deposited in the Belt Basin to the north as a thick pile called the LaHood conglomerate. Paleozoic formations now lie on these Archean metamorphic rocks south of the fault, and on younger Proterozoic Belt sedimentary rocks north of the fault. Mesozoic sedimentary formations cap the rock layers in both the Archean and Proterozoic parts of the range.

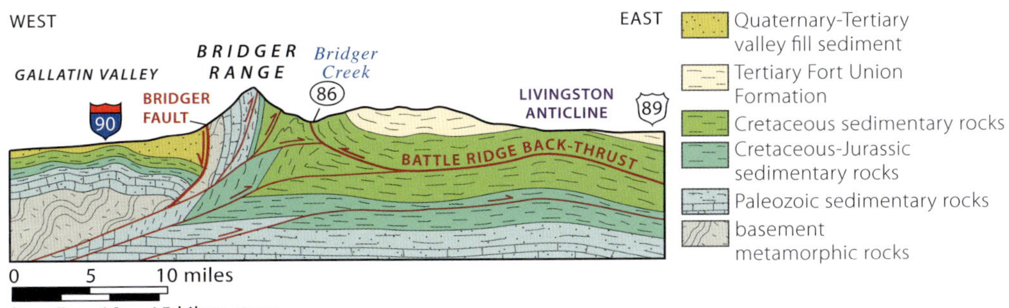

Cross section across the Bridger Range north of Bozeman. Note that the range is a big extensional fault block tilted up to the east, with the Gallatin Valley dropped on the west. The Bridger normal fault may be a reactivated, older thrust fault. It is likely active, posing an earthquake threat. The Battle Ridge back-thrust is a thrust fault whose movement direction is opposite the dominant thrust direction. —Modified from Skipp and others, 1999

I-90
Bozeman—Butte
85 miles

While in Bozeman, be sure to visit the Museum of the Rockies, which has a world-class collection of dinosaur fossils. South of Bozeman and I-90 is the Gallatin Range, a block that most recently rose due to tectonic extension; it exposes Laramide structures with Archean and earliest Proterozoic basement rocks partly overlain by younger sedimentary rocks. Much of the eastern part of the range is covered by Absaroka Volcanics, which erupted about 50 million years ago, in Eocene time.

The wide-open Gallatin Valley west of Bozeman is filled with nearly flat-lying, poorly cemented sediments ranging from Eocene through Miocene age. These are mostly covered by Quaternary alluvial-fan gravels eroded from the surrounding mountains; some of these fans are well exposed in huge gravel pits near the highway. Halfway between Manhattan and Three Forks, I-90 cuts across the southern tip of the Horseshoe Hills, which consist of Proterozoic, Paleozoic, and Mesozoic sedimentary rocks that were folded and thrust faulted during the Sevier orogeny. Well-exposed rocks just north of I-90 near Logan are northwest-dipping layers of Cambrian through Permian sedimentary rocks that form the east limb of a tightly folded syncline with Early Cretaceous Kootenai Formation in its core.

The rocks and prominent folds are broken by northeast-trending thrust faults that dip at shallow angles to the northwest. Tectonic forces caused by the collision of the Farallon plate with the North American continent compressed the rocks in a northwest to southeast direction, forming the folds. When the rocks could not fold, stresses caused them to break along faults that pushed older rocks eastward up and over younger rocks below. I-90 crosses the best known of these faults, the Lombard thrust fault, 2 miles west of Three Forks. It was active sometime after 77 million years ago, when movement on it deformed igneous sills of that age. Although it displaced rocks by no more than a few miles, some geologists reason that the Lombard thrust fault should be considered the geological eastern boundary of the northern Rocky Mountains.

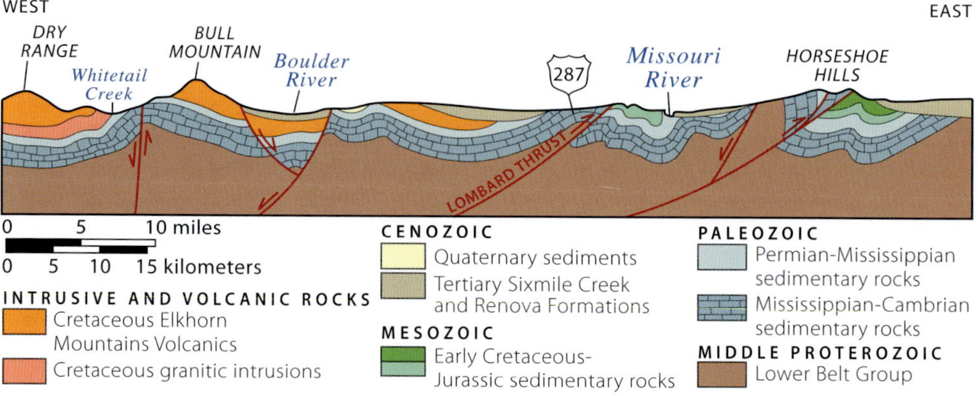

West-east cross section from the area north of Whitehall to north of Manhattan.

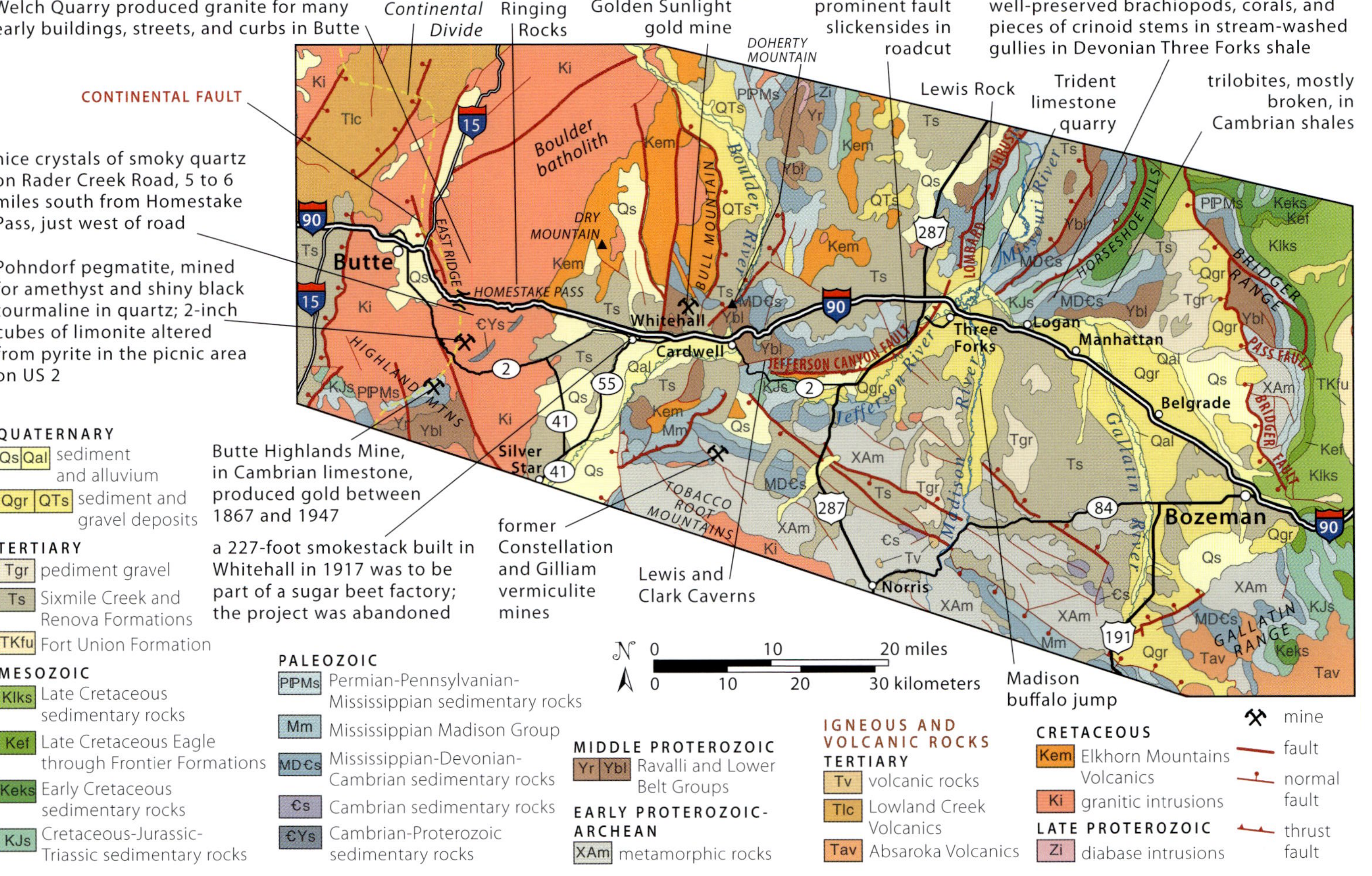

Geology along I-90 between Bozeman and Butte.

View looking east at the Madison buffalo jump.

About 7 miles south of Logan on Buffalo Jump Road is Madison Buffalo Jump State Park. Native Americans stampeded herds of bison off the massive *pishkun* (cliff) for two thousand years. This activity ended with the introduction of horses sometime after 1700. The resistant cliff is a pebbly Miocene-age sandstone that overlooks the Madison River valley to the west. Buffalo bones are abundant in the sediment at the base of the cliff, and archaeologists have located evidence suggesting an extensive village existed in the area.

The town of Three Forks is so named because it lies near the confluence of the three major tributaries of the Missouri River—the Madison, Jefferson, and Gallatin—where they join to form the main Missouri. In late July of 1805, the Lewis and Clark Expedition made its way up the Missouri River from the Gates of the Mountains north of present-day Helena. They were assured by their Shoshone guide, Sacagawea, that the three forks of the Missouri were "at no great distance." She was correct, and on July 26 they camped at the confluence. While at the site, Lewis unknowingly noted the basin and range topography in his journal: "The country opens suddenly to extensive and beatifull plains and meadows which appear to be surrounded in every direction with distant and lofty mountains." On the morning of July 27, 1805, Lewis surveyed the area from a cliff of west-dipping Mission Canyon Limestone of the Madison Group in the limb of a fold formed during Sevier deformation. The cliff is now called Lewis Rock.

The community of Trident, several miles north of Three Forks, was founded in 1908 with the building of a portland cement plant. Operating off and on for more than one hundred years, the lime quarry in the Madison Group limestone provided cement for notable projects, such as Fort Peck Dam, Holter Dam, and the Grand Coulee Dam in Washington State. Eventually, the town dwindled as people built their own houses elsewhere and commuted to work. The Trident plant still produces cement from the limestone and distributes it throughout the Northern Rockies.

LEWIS AND CLARK CAVERNS AND THE JEFFERSON CANYON FAULT

To visit Lewis and Clark Caverns, head south and west from Three Forks (exit 274) on US 287 and then follow MT 2 through the deep Jefferson River canyon. This spectacular route will return you to I-90 and adds only 5 miles to the trip. For the first 3 miles (at the east end) US 287 follows the flat valley bottom next to the Jefferson River with a steep mountainside to the west. Roadcuts are primarily in Cambrian shale and limestone, with younger Cambrian and Devonian limestones upslope and cliffs of massive Mississippian Madison Group limestone forming rugged peaks at the top.

A major fault zone, the Jefferson Canyon fault (formerly called the Willow Creek fault), traces an east-west line just north of the highway and west almost to Cardwell, where MT 2 rejoins I-90. This steeply dipping fault, first active during the formation of the Belt Basin more than 1.47 billion years ago, places deep-crustal Archean rocks to the south against Proterozoic LaHood Formation mudstone and conglomerate (eroded from the Archean rocks) to the north. A fault surface is nicely exposed along the highway about 4.5 miles southwest of I-90 from exit 274. Here is one of best exposures of slickensides anywhere. These conspicuous, steeply oriented, straight, shallow grooves were gouged into Cambrian sandstone by rock as slippage occurred along a fault plane in the rock. The direction of motion (up or down) is not known for sure, because the grooves only provide a trend. The gently dipping thrust faults in the area, including the Lombard, which carried the sedimentary rocks here from the west, are cut by younger, near-vertical normal faults.

Slickensides gouged into the rock by movement along a fault in the Jefferson River canyon. This steep exposure is a couple of feet across.

Two or three miles west of Sappington (where MT 2 and US 287 part ways), the Jefferson River swings north through a narrow canyon in the London Hills and comes close to the highway. These east-trending hills were gradually raised across the path of the river, slowly enough that the river was able to continue cutting its channel downward as the ridge continued to rise. Geologists call such a stream **antecedent**—that is, the stream preceded the hills.

Four to eight miles farther northwest, the river meanders through another narrow, deep canyon—through another, still-larger anticline ridge in Mississippian limestones that was gradually raised across the river's path, presumably at the same time. The terrace gravels that follow the river's meander may be as old as late Miocene, so the river may have begun downcutting then. Several faults in this region have moved as recently as 1.6 million years ago, and at least one moved between 15,000 and 13,000 years ago. Broad floodplain deposits from Quaternary time that are stranded above the present river channel suggest that the uplift is continuing.

North-tilted Paleozoic sedimentary rocks exposed in the Jefferson River canyon. View looking west.

Lewis and Clark Caverns are in Mississippian-age Madison Group limestone, deposited between 360 and 325 million years ago, then folded during Late Cretaceous through earliest Tertiary compression. From the upper end of the road at the upper visitor center, the mile-long paved trail climbs 300 feet to the cavern entrance. Look for fossils of tropical marine animals, such as button-shaped crinoids, clam-like brachiopods, and lacy-looking bryozoans in the limestone. The course of the two-hour tour through the caverns drops a total of about 300 feet in stages, returning you to about the same elevation as the visitor center. Wear an extra layer and good shoes because the temperature in the caverns is only 50 degrees year-round, and sloping surfaces inside are sometimes wet.

Although the caverns are 1,300 feet above the river at 5,500 feet of elevation, they likely formed near river level, where the water table was located. Caverns everywhere tend to form near the water table, where slightly acidic water can dissolve the limestone. Rainwater dissolves carbon dioxide in the air to form weak carbonic acid that percolates down to the water table. Over time, individual joints or cracks in the limestone enlarge into wider channels or caverns. As the mountains here slowly rose across the path of the river over thousands of years, the caverns ended up far above the river.

Once the caverns are open to air, calcium-rich rainwater seeping through fractures in the roof of the caverns evaporates, loses its carbon dioxide, and precipitates calcium carbonate—calcite. Each droplet at the cavern roof partially evaporates, depositing calcite as it drops from a growing, icicle-like stalactite. Where water drops to the floor of the cavern and evaporates, stalagmites grow upward from the floor. The dripping water leaves red iron oxide stains in parts of the cavern; the iron may come from minute amounts of iron-rich clay left after the limestone dissolves.

The same big cliffs of pale-gray Madison Group limestone come down to road level on MT 2 about 1.2 miles west of Lewis and Clark Caverns. The limestone is tightly folded, as if it was pushed over to the right (to the east). To the west, the younger Amsden Formation is limestone that was exposed shortly after deposition. It weathered for millions of years, and much of the limestone dissolved away, leaving behind minute amounts of insoluble iron-bearing red clay that colors the limestone.

About 1 mile farther west of Lewis and Clark Caverns, look for gray Madison Group limestone capped with red limestone of the Amsden Formation on the south side of the Jefferson River. Above are distinctly layered younger sedimentary rocks of Permian and Jurassic age.

Stalactites hanging down from the roof and stalagmites growing up from the floor of Lewis and Clark Caverns.

Between 3.6 and 3.9 miles west of Lewis and Clark Caverns are big roadcuts in dark-gray conglomerate of the LaHood Formation. Up close you can see that it's full of distinctive round pebbles and cobbles that range from less than an inch to many inches across. Most of the rocks are streaky, lighter-colored gneiss and schist (high-temperature metamorphic rocks) or streaky metamorphosed granite from very deep in Earth's crust, the same rocks that make up parts of the isolated mountain ranges just south of here. About 1.47 billion years ago, the ancient Archean terrane called the Dillon block slowly rose along the south side of the Jefferson Canyon fault, which runs east-west through the canyon, and began eroding. Mountain streams carried pebbly mudflows north into the Belt Basin. Near the fault, where stream channels were steep, larger rocks dropped first; finer-grained sediment collected farther north. This fault marks the southeastern extent of sedimentary deposits of the Proterozoic Belt Basin, as does the Pass fault in the Bridger Range.

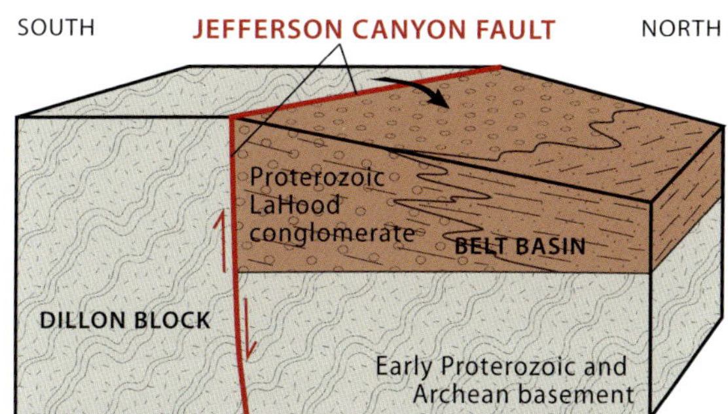

Block diagram showing how the gravel that became the LaHood Formation conglomerate was shed north in middle Proterozoic time off the steep fault scarp of the Jefferson Canyon fault.

Conglomerate in the LaHood Formation in the Jefferson River canyon is full of distinctive metamorphic pebbles and cobbles eroded from the ancient Archean terrane that rose south of the Jefferson Canyon fault.

West of Three Forks (exit 274), I-90 crosses light-gray to grayish-brown mudstones of the Renova Formation deposited in Eocene time, between about 38 and 36 million years ago. The layers are nearly horizontal for 14 or 15 miles as the highway climbs gently to the west, except where faults cut the formations. Fragments of petrified wood are present locally, as are leaf fossils on split surfaces of light-gray shale. Near the top of the rise, the road lies on Miocene sediment of the Renova Formation. West of this high point the terrain becomes very rugged, especially in the deep canyon north of the highway. Distinctly layered and tightly folded sandstone, shale, and limestone formations, ranging from Cambrian through Devonian and Mississippian ages, are exposed. The prominent light-colored cliffs in Doherty Mountain, to the north, are Cambrian limestone formations wrapped around tight folds formed by Sevier thin-skin compression. Occasional greenish stripes in the roadcuts are deeply weathered sills of andesite sandwiched between the sedimentary layers. Less than 1 mile farther west these Paleozoic sedimentary rocks give way to much older, yellowish Proterozoic LaHood Formation sandstones and mudstones. South of I-90, these rocks grade rapidly into the pebbly LaHood Formation conglomerate that is so well exposed along MT 2 in the Jefferson River canyon.

Between Cardwell and Whitehall, the high glaciated peaks of the Tobacco Root Mountains rise directly south of I-90. At their core is a large granite batholith, probably injected along a thrust fault that punched up through Archean basement rock about 70 million years ago. An unpaved but generally passable road goes south from Cardwell to the old gold-mining town of Mammoth, at the headwaters of the South Boulder River in alpine meadows high in the range. The road passes southward through a nice cross section of the range, first passing a nearly complete section of successively older Mesozoic and Paleozoic formations until it enters Archean basement rock and finally gets onto the granite batholith. The most southern parts of the route get into spectacular glaciated landscapes.

The huge rusty scar on the hills north of I-90 between Cardwell and Whitehall is the Golden Sunlight gold mine. Gold was discovered here in 1892, and the mining of it limped along for nearly a century until large-tonnage production from the low-grade ores began in 1983. The gold is finely disseminated both in strongly shattered sedimentary rock of the Proterozoic LaHood Formation, and in the Late Cretaceous volcanic dikes intruding it. The ore was crushed, and the gold was dissolved out in

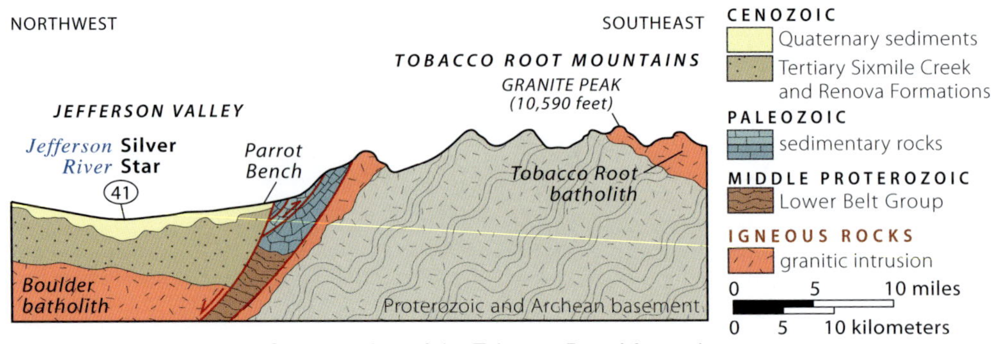

Cross section of the Tobacco Root Mountains.

The Golden Sunlight gold mine. —Courtesy of Chris Boyer

giant cyanide tanks. To keep it economically viable, tailings from local historic mines were trucked in for additional processing, but operations were slated for closure in 2019. Since 1983, more than 3 million ounces of gold were recovered, making Golden Sunlight the largest gold producer in Montana. In 1998, Montanans voted to ban the use of heap or vat leaching with cyanide ore-processing reagents, but Golden Sunlight was exempted because the law applied only to new mine operations.

The low, forested Bull Mountain near the Golden Sunlight gold mine consists of Late Cretaceous andesite of the Elkhorn Mountains Volcanics, which erupted from the Boulder batholith to the west. The volcanic rocks rest on tightly folded Belt and Paleozoic sedimentary rocks.

West of Whitehall, I-90 crosses a broad area of yellowish sand and gravel of late Miocene age and tan to orange sands of the older Renova Formation. Around Pipestone Hot Springs (exit 241), the Renova is rich in small-mammal fossils and coprolites (fossilized dung) of late Eocene age.

Ringing Rocks, a little known geologic site north of the highway between Whitehall and Butte, is an adventurous side trip for those with 4-wheel drive. Take exit 241, drive 1 mile east on the frontage road, then about 3 miles north and northwest on a rough, winding track not suitable for passenger cars. This fascinating hill is composed of a very fine-grained, very hard, dark igneous rock called gabbro that weathers reddish-brown and intruded the area's rocks 78 million years ago. If you strike it with a hammer, it rings like a bell. Striking different-sized boulders provides distinctly different tones. Freezing and thawing probably broke the once-solid rock into a boulder field. It's uncertain what causes the ringing, but it seems to be related to how the fine-grained minerals grew together so tightly and the way the rocks are piled. If a boulder is removed from the pile, it doesn't ring!

North-south-trending vertical joints in weathered granite of the Boulder batholith east of Homestake Pass.

As I-90 climbs toward Homestake Pass through the Boulder batholith, huge vertical blades of granite form a striking landscape north of the highway. These narrow blades mark one set of prominent north-south-trending vertical joints. (See this chapter's introduction for more about joints in the batholith.)

The pale-gray granite, exposed in outcrops and roadcuts, consists of milky-white feldspar, glassy dark-gray grains of quartz, glossy black flakes of biotite mica, and stubby black needles of hornblende. The granite encloses round patches of darker rock that has the same minerals but a greater proportion of dark minerals. Although these are sometimes called xenoliths (pieces of intruded country rock), most are rounded and sometimes show rounded indentations, indicating that they're more likely patches of dark magmas that mingled with and chilled in the paler, lower-temperature granite magma.

Granite magma forms in the lower continental crust, which is heated and melted by very hot basaltic magma rising from deep in the Earth's mantle. As the two magmas mingle, the basalt solidifies because it crystallizes at much higher temperatures than granite, which remains molten as the magma rises toward the surface. The basalt is preserved as the dark rounded blobs, a few inches across, that we see in the granite.

West of Homestake Pass, I-90 descends a steep grade where it crosses the Continental fault, a young and potentially active extensional fault near the base of the grade. Old maps show marshland along Silver Bow Creek and in the flats south of I-90. The wetlands dried out as the Butte mines pumped the groundwater level down; buildings now cover much of the formerly marshy ground. When the last pumps went silent, the water table began slowly returning to its natural level. If that were to continue, the rising water would again flood the old marshland, along with the buildings. Water-saturated sediment has a tendency to liquefy when shaken, so any future earthquakes on the Continental fault could pose a serious risk to buildings on "the flats." Continued movement on the Continental fault tilts the valley further down to the east, compounding the water problem. (For more information on Butte's geology and mining history, see the section on Butte.)

I-90
Butte—Missoula
119 miles

(For more information on Butte's geology and mining history, see the section on Butte.) The prominent knob on the west side of Butte, marked by the white *M*, is Big Butte, an eroded volcanic neck that intruded the roughly 76-million-year-old Boulder batholith about 50 million years ago. Montana Tech, the Montana Bureau of Mines and Geology with its great mineral museum, and the World Museum of Mining are at the south base of the knob.

Between Butte and the MT 1 junction at exit 208, I-90 crosses weathered granite of the western edge of the Boulder batholith. The granite tends to weather into big, rounded, pinkish-brown boulders due to the intersecting joints in the rock. Around the junction with I-15 to the south, the granite is covered by light-gray Tertiary valley-fill sediments; west of Ramsay it's covered by the Lowland Creek Volcanics, which erupted 50 million years ago—no boulders here! This rhyolite is similar in composition to the granite but erupted onto its surface 25 million years later. It forms relatively soft rocks, but it's preserved here because it accumulated in a north-northeast-trending graben.

As I-90 gradually turns north, it drops into the broad Deer Lodge Valley, a large graben tilted down to the west. Granite outcrops in the rugged hills to the east belong to the Boulder batholith; the hills are capped by Cretaceous Elkhorn Mountains Volcanics. West of the junction with MT 1, south of Warm Springs, the giant Anaconda smelter stack, all that remains of the century-old smelter, is visible from several miles away. Ore was shipped to it from Butte mines to extract the copper, leaving black piles of slag as waste. In 1908 massive flooding distributed the mine waste from Butte and Anaconda to Milltown Reservoir near East Missoula.

Water used in smelting came from hills to the west, and fine-grained waste was collected in broad settling ponds on the broad valley floor on both sides of I-90. The water was treated with limestone to neutralize the sulfuric acid created while smelting the copper-iron sulfide minerals, to precipitate the remaining metals, and to make the water slightly alkaline. The treatment was moderately effective but didn't remove noxious metals, and it wasn't used until many years after the smelting had begun. High concentrations of copper, arsenic, and cadmium collected in the sediment in Milltown Reservoir. Those sediments persisted until Superfund remediation began in 2006.

The gently sloping benches at the valley sides are pediments, erosional surfaces that formed in the desertlike conditions of Pliocene time. These are partly covered by younger, arched alluvial fans. During the last ice age, the wetter climate provided year-round stream flow that eroded gullies, leaving the pediments standing as benches above the new, flat-bottomed floodplain next to the river. North of Deer Lodge, the river sweeps against the outside bends of meanders, eroding steep banks into the benches and widening the Clark Fork River floodplain over time.

Rocks in the Flint Creek Range, west of the Deer Lodge Valley, range from Proterozoic Belt formations through Paleozoic and Mesozoic sedimentary rocks, all tightly crumpled and thrust faulted. In Late Cretaceous time, large granite plutons, similar in age to the roughly 76-million-year-old Boulder batholith, intruded that deformed mass, probably late in its deformation. Some intruded along thrust faults. Mt. Powell,

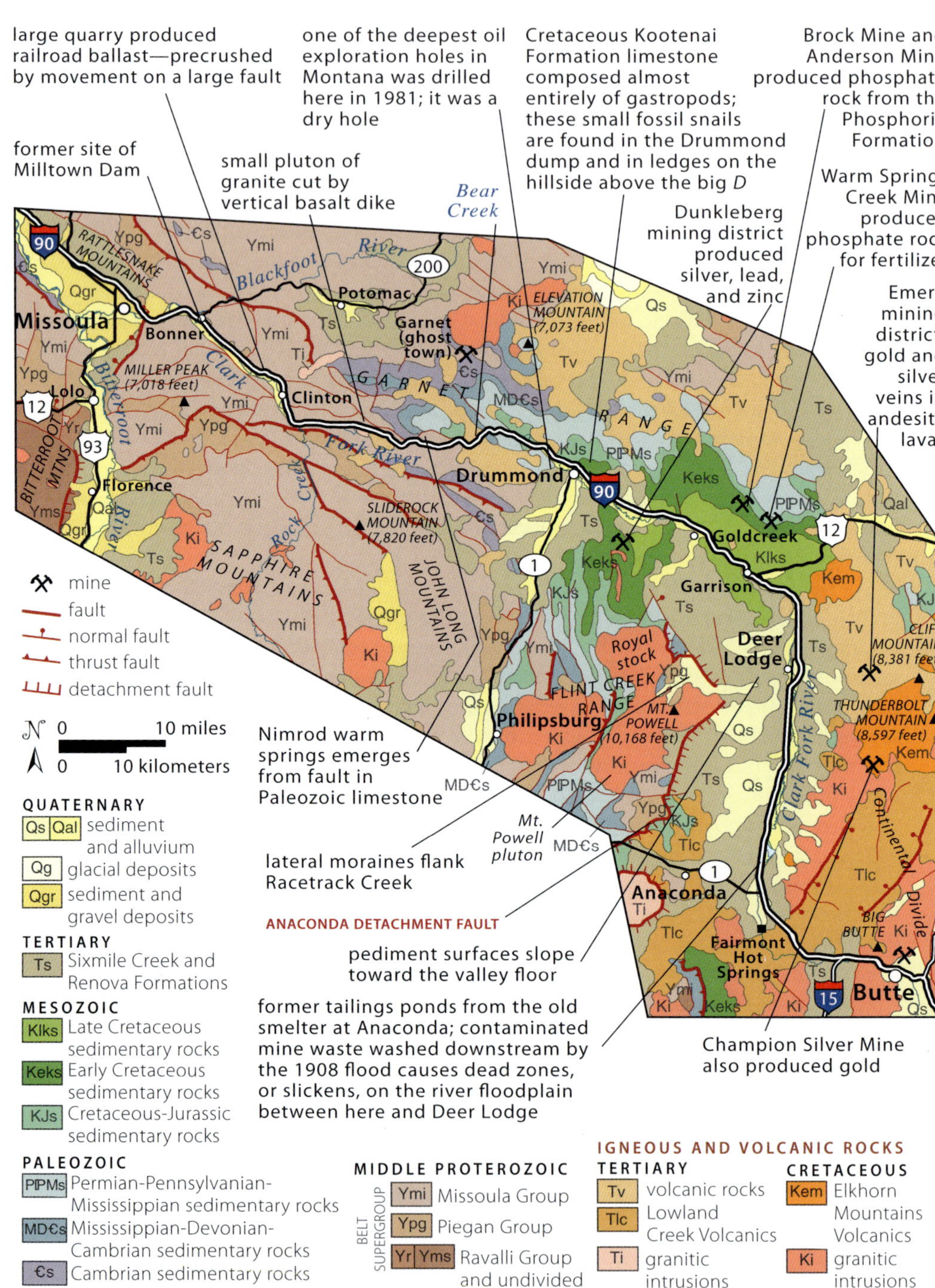

Geology along I-90 between Butte and Missoula.

A gently sloping Pliocene pediment surface on the Renova Formation northwest of Deer Lodge. The old valley-floor surface (pediment) was dissected by streams, forming the new valley floor largely after the climate turned wet in Pleistocene time.

10,168 feet, the prominent pyramid-shaped peak on the skyline southwest of Deer Lodge, is part of the Mt. Powell pluton.

Pleistocene mountain glaciers carved the rugged topography of the Flint Creek Range. About 7 miles south of Deer Lodge, west of the Racetrack exit off I-90, a large valley glacier that filled the Racetrack Creek valley left a pair of long, high ridges flanking the creek. These glacial moraines, left by the melting glacier, show that it reached down-valley to within 5 miles of the interstate. Racetrack Road west of I-90 winds through the hummocky, bouldery terrain of the moraines.

Rocks flanking the east side of the Flint Creek Range are part of a big Eocene-age detachment zone that dips gently east; rocks above this zone of the sheared rock called mylonite moved down to the east away from the range along the Anaconda detachment fault. Metamorphosed rocks below the detachment make up the underlying

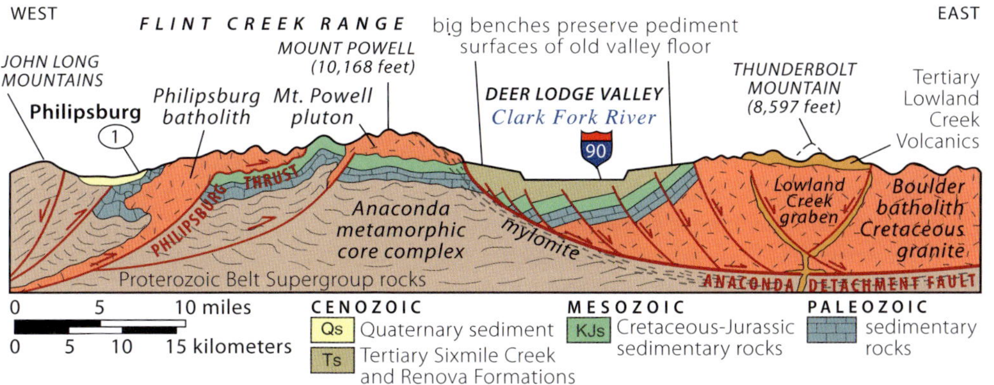

A schematic cross section across the Deer Lodge Valley. Granite intrusions were injected up and eastward in Late Cretaceous time, many along thrust faults. The Anaconda detachment fault dips eastward under the Deer Lodge Valley and under the Boulder batholith, having carried the batholith with it. Big benches preserve parts of the old desert valley floor of Pliocene time.

metamorphic core complex. Petroleum exploration wells drilled through Tertiary sediments on the west side of the Deer Lodge Valley intersected the big detachment, which is well exposed in the Anaconda Range to the southwest. It appears to flatten under the Boulder batholith east of the valley. The batholith and its surrounding rocks must have moved at least 15 miles east on the detachment surface in Eocene time. The metamorphic core complex and detachment zone are comparable to and of similar age to the Bitterroot metamorphic core complex, farther west.

Between Deer Lodge and Garrison, and farther west toward Drummond, are soft gray Cretaceous shales. Rocks in the hills north of Garrison and Goldcreek become progressively older, with Paleozoic limestones holding up the higher hills. Brock Creek, 3 to 4 miles northwest of Garrison, was the site of phosphate mines for years beginning in 1929. The phosphate rock in the Permian Phosphoria Formation reportedly contained more than 600 million tons of ore with more than 31 percent phosphorus oxide. The phosphate was shipped to Trail, British Columbia, where its sulfuric acid was used to make fertilizer. A small plant was built at Garrison in 1963 to use sulfuric acid to roast fluorides out of the phosphate rock to make a cattle feed additive. Severe air pollution from the plant spread high concentrations of hydrogen fluoride and sulfur dioxide for miles around. Grass and trees died, cattle developed severe deformities, and children suffered, while the plant owner and political supporters denied responsibility. The plant was finally forced to close in 1967.

In 1858 a couple of prospectors discovered gold in what's now called Gold Creek, and they returned with equipment in 1862. Others flocked to the area, but interest soon waned as would-be miners drifted off for richer opportunities. The creek is about a mile east of Gold Creek Road and the now almost-vanished community of Goldcreek. The site, near milepost 171, is also near the location of the final spike driven on the Northern Pacific Railway on September 8, 1883.

Between Drummond and Rock Creek (exit 126), I-90 passes a series of southwest-dipping thrust faults along the north edge of the Sapphire Mountains. Belt sedimentary rocks of the Missoula Group were pushed northeast along the faults over Paleozoic and older Missoula Group sedimentary rocks in the Garnet Range. North of I-90 and west of Drummond, most of the rocks near the road are tightly folded Paleozoic sedimentary rocks with fold axes about parallel to the highway. Pale-gray Madison Group limestone forms the cliffs both north and south of I-90. Big cuts a few miles west of Drummond expose steeply dipping, green to black Cretaceous sandstone and shale sliced by thrust faults, all active during the Sevier orogeny.

About 6 miles west of Drummond, right at the north edge of I-90, is a beautiful near-vertical streambank cut in soft horizontal layers of beige to buff sediments. These are Glacial Lake Missoula sediments; the main part of the lake was far downstream around Missoula, but the lake was deep enough to flood the Clark Fork River valley upstream almost to Deer Lodge. It was deep enough here to deposit silt a few tens of feet thick over a number of years and probably from several separate fillings of the lake. The alternating light and darker layers are likely varves. During the melting season, streams deposited light-colored sand and silt. When the lake was frozen, dark-colored organics and clay settled to the bottom. (See the book's first chapter and the relevant road guides in the Northwest Montana chapter for more on the Glacial Lake Missoula story.)

Cuts in steeply dipping, folded, and thrust-faulted nonmarine sandstone and shale of the Cretaceous Kootenai Formation 5.4 miles west of Drummond, between mileposts 149 and 148.

The Clark Fork River cut a beautiful exposure in beige to buff, horizontally layered Glacial Lake Missoula silt between mileposts 148 and 147.

South of I-90 between Drummond and Missoula, the rolling, forested Sapphire Mountains consist mainly of sedimentary and lightly metamorphosed Proterozoic Belt rocks that a number of good-sized granite plutons intruded in Late Cretaceous time. At one time all of these rocks stood atop the Idaho batholith, along the Idaho-Montana border. In Late Cretaceous time, 80 or 70 million years ago, the huge mass now forming the Sapphire Mountains crept several tens of miles down and to the northeast on an intensely sheared mylonite zone on the east flank of the Bitterroot Mountains. The granite plutons were emplaced at about that time, and they may have been either carried with the mass or injected as it moved. It's possible the magma helped lubricate the movement. (See the US 93: Missoula—Lost Trail Pass road guide for more information.)

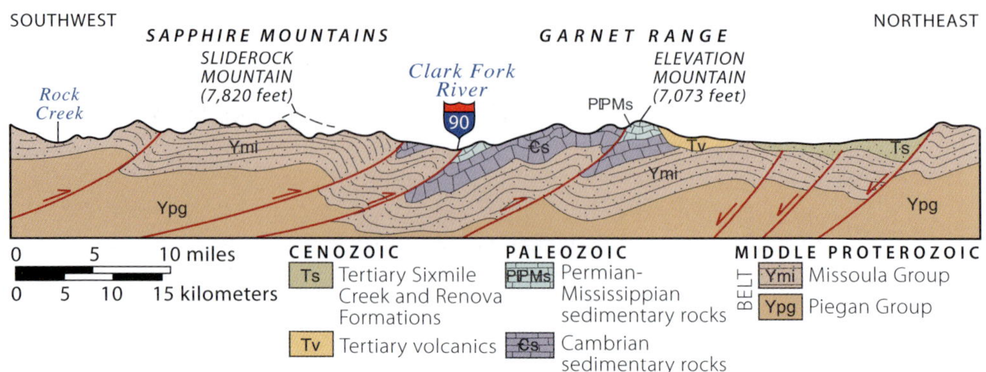

Cross section across I-90 near Bear Creek shows the thrust faults and folds along the north edge of the Sapphire Mountains, all of which were pushed to the northeast.

Hoodoo-like spires of Madison Group limestone on the north side of I-90 near milepost 145.

Look for a large overturned fold in the Mississippian Madison Group limestone in a big cliff south of I-90 at milepost 145. Just west of milepost 145, I-90 makes a sharp turn around the end of a high cliff of light-gray limestone that's full of large solution cavities and marked by tall, hoodoo-like spires. The whole mass was intensely folded and thrust faulted during the Sevier orogeny. Splashes of red in the limestone are from iron oxide formed by the deep weathering and dissolution of the limestone, which left behind minute concentrated amounts of red clay.

Bear Creek reaches the Clark Fork River at Bearmouth. Just upstream and east of the interstate rest areas, prospectors found gold in gravels in Bear Gulch in 1863, leading to the usual stampede of miners expecting to strike it rich. Thousands of miners worked the gravels in Bear Creek. They recovered an estimated $30 million—an impressive haul in the nineteenth century. They dug pits as deep as 70 feet in the gravels, down to bedrock, and hoisted the gravel out to the surface to wash out the gold—an extraordinarily dangerous practice in a collapsible gravel deposit, yet they had no alternative since gold, being seven times as heavy as gravel, sinks to the bottom of gravel fill. Many years later, between 1933 and 1939, an old gold dredge operated near the mouth of the creek and left the usual gravel ridges, but recent recycling of the gravel for construction has removed most of those.

Miners found gold as they headed upstream, and some searched for gold-bearing veins in bedrock around the margins of granite intrusions. Gold-bearing veins were found in 1866, but it was tough going and not profitable. With the erection of a stamp mill in 1895, miners could finally crush the rock mechanically to release the gold, and underground mining took off. Gold, silver, and copper were found next to granite that had intruded along thrust faults. Much was in the Madison Group limestone, which reacted with the hot fluids of the granitic magma to form skarns that consist mostly of garnet, hence the town of the same name. After 1900 mining tapered off. The well-preserved ghost town of Garnet is 8 miles north of Bearmouth (exit 138) via Bear Gulch Road. This road up from I-90 is steep, narrow, and not suitable for trailers or some passenger cars. The road in from MT 200, 24 miles east from the I-90 junction at Bonner, is excellent.

About a half mile east of the exit to Beavertail Hill State Park, on Bonita Station Road, there's a big roadcut in Tertiary white granite. A vertical, dark-gray to brown dike of basalt, also cooled from magma in early Tertiary time, cuts the granite pluton, showing that it is at least a bit younger than the granitic intrusion.

A big, dark diabase (grainy basalt) sill south of the highway near the junction of I-90 and MT 200 is much older. It was intruded in late Proterozoic time. Milltown Dam, formerly anchored in the diabase sill, was removed in 2015. That hydroelectric dam was built in 1908 to supply power to lumber mills in nearby Bonner. A record-breaking flood on the Clark Fork River a few months later washed large quantities of mining waste downstream from the upper Clark Fork drainage, depositing it in the shallow reservoir behind the dam just upstream from Missoula. The sediments contained large concentrations of arsenic, copper, lead, zinc, and other metals. A routine study of water quality in the area found that heavy concentrations of arsenic had contaminated the wells of thirty-five homes in Milltown. Declared a Superfund site, the dam and the

A prominent vertical, dark-gray to brown basalt dike cuts through a Tertiary-age granite intrusion on the north side of I-90 between mileposts 132 and 131. The dike is younger than the granitic rock and therefore more than a half billion years younger than a similar-looking sill at Milltown Dam.

This huge dark-brown diabase sill was injected along layers in pale-gray Belt Supergroup sediments at Milltown State Park in late Proterozoic time, around 780 million years ago, during the initial breakup of the supercontinent Rodinia. View to the south from the I-90 bridge over the Clark Fork River.

contaminated sediments were removed and transported upstream for burial south of Deer Lodge. After removal of the dam and sediments, stumps of trees that grew on the river floodplain before 1908 became visible at low water.

US 12
Helena—Garrison
44 miles

Helena in the late 1800s was the hub of a lot of mining activity, starting with placer gold and followed by underground workings. Heat and hot, watery fluids along the north edge of the Boulder batholith baked and chemically altered the surrounding rocks, depositing ore. Four gold prospectors returning empty-handed from the new gold discoveries in the Kootenai country, to the west, decided to take one last chance with prospecting; in 1864 they discovered gold in stream gravels in the southwestern part of today's Helena. They hit it big-time and named their find Last Chance Gulch. The strike brought in thousands of new prospectors. By the summer of 1865, three thousand people in the area needed food and supplies. The four initial entrepreneurs added to their growing wealth by building support services for all the newcomers. They provided mining tools, food stores, cabins and then hotels, and other services—the beginnings of the new town of Helena. A lack of water slowed mining success, so one of the four entrepreneurs remedied this by organizing the building of a ditch to bring in water. Placer discoveries in Grizzly Gulch and other nearby locations kept the place going.

Underground lode mines at Grizzly Gulch and the Whitlatch-Union Mine extended the mining activity and lasted longer than the surface mining. Most of the underground activity was centered around Unionville, almost 4 miles southwest of Last Chance Gulch. The Whitlatch-Union veins were mined from 1864 to 1872, and again from 1905 to 1942, recovering 17,000 ounces of gold and 8,000 of silver.

Last Chance Gulch, long since abandoned, is now a major hub of commercial activity in Helena. It's covered with stores, restaurants, and office buildings. To a geologist, the location of this commercial district is a bit concerning, because although the gulches are dry most of the time, they drain the mountains above the city and have the potential to cause flash flooding during rare heavy storms.

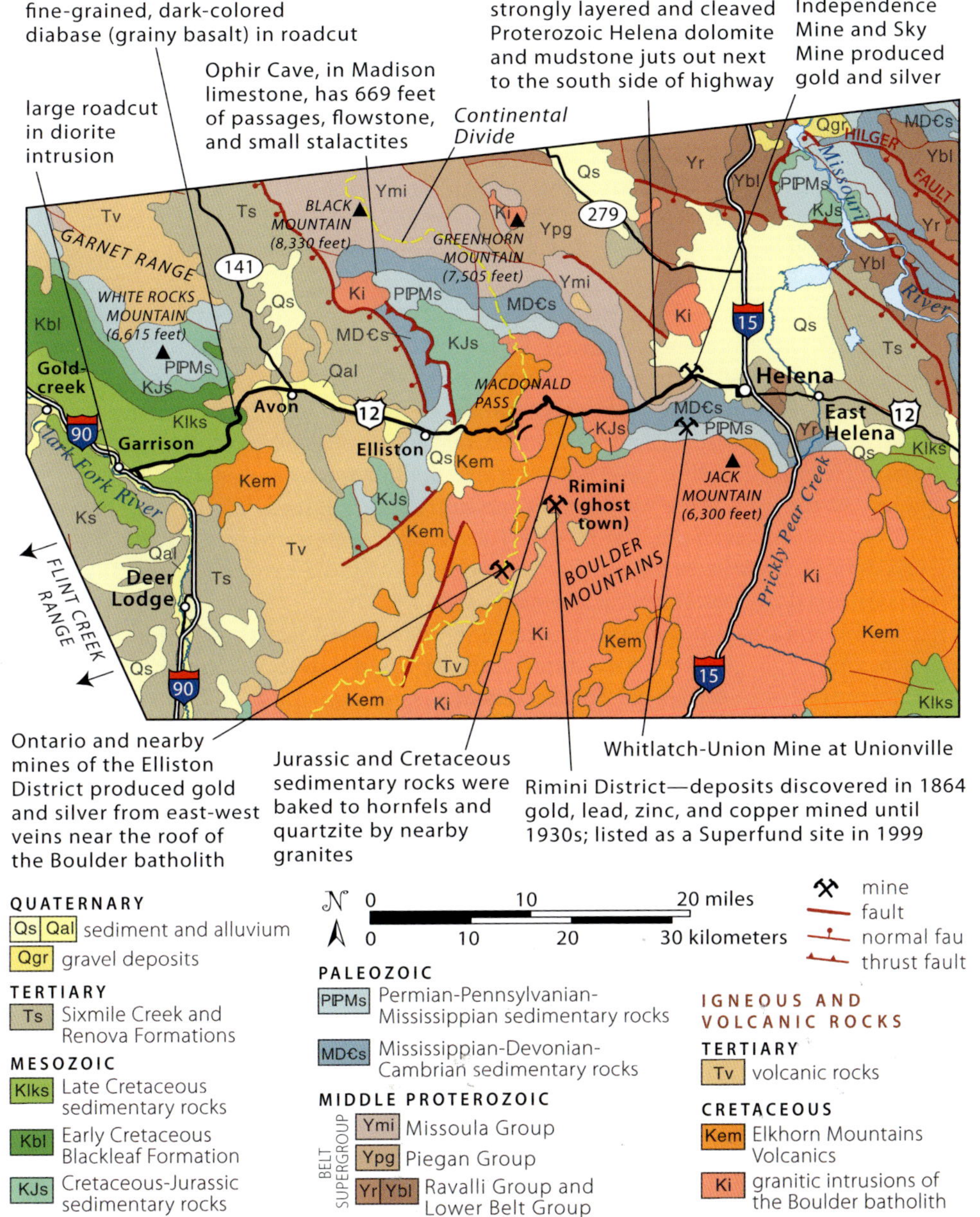

Geology along US 12 between Helena and Garrison.

Early wooden buildings in Helena were destroyed by fire five times between 1869 and 1874, but stone and brick buildings mostly survived. Stone and brick needed mortar for construction, so lime was made from local limestone. The remains of the old Grizzly Gulch lime kilns are in the Last Chance Gulch area on South Park Street, just southwest of the Bureau of Land Management buildings. Some of that lime was used for mortar in constructing the Montana State Capitol building here. Granite for the building came from the T. Kain and Sons Quarry at Clancy on I-15, about 10 miles south of Helena, and sandstone came from a quarry at Columbus, Montana.

Earthquakes can also be a concern for buildings in Helena, being in the northern part of the Intermountain Seismic Belt and near the Lewis and Clark fault zone, an ancient zone of faults that runs from Helena through Missoula and all the way to Spokane. Old newspapers report of a moderate earthquake in 1889 and more than sixty others between 1903 and 1935. Most faulting in the zone reflects ongoing crustal spreading and slip on north-trending normal faults, along which the crustal blocks that form today's valleys drop. On October 18, 1935, just before 11 p.m., Helena was shaken by a severe magnitude 6.2 earthquake, with an aftershock of 6.0 on October 31. Two people died in each of those events. In the following three months, more than 1,200 earthquakes were felt, but few were violent enough to cause significant damage. Brick walls and chimneys collapsed, and many old brick buildings in Helena still show cracks or other damage. Structural brick used in Montana in the late 1800s to early 1900s was often poor quality, sometimes covered by a better-quality brick facade imported from out of state. It's estimated that more than 60 percent of the structures in Helena, including 3,000 homes, were either damaged or destroyed during the quakes. At the time, there was little recognition that earthquakes could even happen in Montana. For example, the relatively expensive and newly constructed Helena High School, built in 1935, was not engineered to handle earthquake shaking and was destroyed two weeks after it was dedicated.

The Helena area is still shaking and has a significant seismic risk going forward. On July 6, 2017, a magnitude 5.8 earthquake struck about 5 miles southeast of Lincoln, 34 miles northwest of Helena. This earthquake occurred within the Lewis and Clark fault zone on a strike-slip fault, the type where one crustal block slides past another, such as on the San Andreas fault in California.

Brick walls collapsed in Helena in the October 18, 1935, magnitude 6.2 earthquake. Note the poor-quality brick of the inner walls (lower left), common construction in the late 1800s and early 1900s. —Courtesy of Montana Historical Society

Helena is especially vulnerable to earthquakes because most of the town is built on soft valley-fill sediments that amplify the shaking of earthquakes. Think of a large bowl full of soft pudding, the rigid bowl simulating bedrock and the pudding the soft sediments. If you sharply tap the side of the bowl to simulate an earthquake, the pudding will move far more visibly than the bowl.

Helena rests in the shadow of the Boulder Mountains which dominate the view to the south. Almost every rock in the Boulder Mountains from a couple miles south of Helena to Butte is Boulder batholith granite. Helena itself is in Proterozoic Belt Supergroup sedimentary rocks, overlain to the south by progressively younger sedimentary rocks—Cambrian, Devonian through Permian, Jurassic, and Cretaceous. The granite magma heated the older rocks, saturating them with superheated steam and water long enough that their mineral grains reacted with others and grew to larger mineral grains. The metamorphic rock halo extends for a mile or more north from the granite contact. The heated groundwater dissolved minor elements in the rocks, finally precipitating them at lower temperatures in pore spaces and in cracks in the rocks. Those crack fillings are the veins that we see in many ore deposits, including gold, silver, and copper ores that were exploited in the Butte mines and around Helena.

A landmark at the west edge of Helena denotes the location of the Broadwater Hotel and Hot Springs Natatorium, a luxury hotel and bath house built in 1889 by a local businessman and lauded by locals. The hot springs water, which likely migrated from deep in the Earth to the surface through porous and permeable rocks along a fault, came out at about 150 degrees but was cooled for the pools. The hotel's huge indoor pool was the largest in the world. The hotel closed in 1924 and was briefly revived for a few years, but the 1941 crackdown on gambling was the last straw. Badly deteriorated, it was finally torn down in 1974.

Along US 12 about 2 miles west of where the road takes a distinct bend to the south (between mileposts 37 and 38), a big roadcut on the south side of the road exposes thinly layered Helena Formation dolomite and quartzite, rocks of the middle part of the Proterozoic Belt Supergroup that were heated and metamorphosed by the Boulder batholith. Another 1 to 2 miles farther west are outcrops of white granite. For 3 miles after the granite the rocks are again Belt sedimentary rocks. The road then

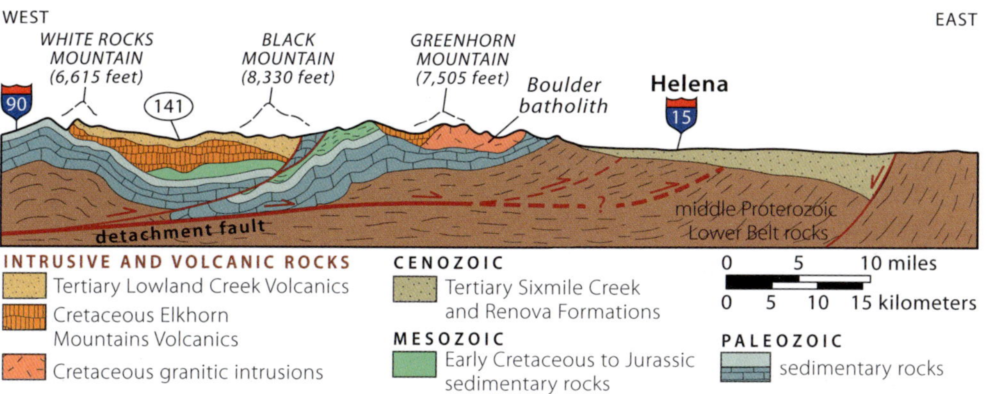

West-east cross section just north of US 12 between Garrison and Helena.

climbs into Paleozoic sedimentary rock before getting into the main mass of Boulder batholith granite 5 or 6 miles east of MacDonald Pass.

One mile east of the big hairpin turn on the east side of MacDonald Pass, a patch of Cretaceous Elkhorn Mountains Volcanics is preserved above the granite, only a couple of hundred feet above the highway; the same volcanic rock is exposed 2 miles west of the pass. This is a remnant of volcanic rocks that once covered the batholith; they are the same age as the granite and grade downward into it. The Elkhorn Mountains Volcanics erupted first, and the granite magma moved into place and solidified more slowly below the volcanics because it was insulated by them.

About 5 miles west of MacDonald Pass (1 mile east of Elliston), the Elliston Lime Company quarried Mississippian Madison Group limestone to make quicklime (calcium oxide) from the early 1900s to 1965, when the buildings were destroyed in a fire. Alternating layers of wood and limestone were burned in kilns for four to six days. Adding a little water to the quicklime produces mortar, the active ingredient in cement and plaster.

Between Elliston and Garrison, and on west to Lookout Pass on the Idaho border, US 12 follows the Little Blackfoot River and the Clark Fork River, as well as Mullan Road, a military wagon road originally constructed by Captain John Mullan between 1859 and 1860. It was intended to open land to settlement and to prevent any territorial claims by France and England. Within a year of completion it was used by some twenty thousand people, including many drawn by newly discovered gold deposits.

Between Elliston and Avon, at the junction of MT 141 to Helmville, are big, dark-gray roadcuts in massive fine-grained volcanic rocks, part of the Lowland Creek Volcanics, which erupted about 50 million years ago, during Eocene time. About 1 mile west of Avon, a few hundred yards north of US 12, these rocks form huge, dark-gray, glacially scoured smooth-topped cliffs that have weathered into tall hoodoo-like spires. These same rocks continue for at least 3 miles west of Avon.

Old lime kilns from the early 1900s above the north side of US 12 near milepost 23.5.

Hoodoo-like spires of the Eocene Lowland Creek Volcanics to the north across a hayfield west of Avon near milepost 12.

About 4 or 5 miles southwest of Avon are big roadcuts in folded and thrust-faulted Jurassic and Cretaceous shale and sandstone. These sedimentary rocks were derived from the erosion of mountains to the west that were built during the collision of the oceanic Farallon plate and the continental North American plate during Mesozoic time. During this event, sedimentary rocks above the Archean basement were shoved eastward and crumpled into mountains. Their eroded sediment was deposited in a foreland basin adjacent to the rising mountains. Over time, thin-skin Sevier deformation spread into the foreland basin, folding and thrust faulting the sedimentary rocks now exposed in the roadcuts. These same gray sedimentary rocks continue for the remaining 8 or 9 miles to Garrison and I-90. At Garrison, if you continue under the interstate onto the frontage road, you'll curve around a large roadcut in a 76-million-year-old diorite intrusion.

US 12
Helena—White Sulphur Springs
74 miles

Five miles east of Helena, US 12 passes the site of the now-infamous Helena lead smelter, which was built in 1888 and operated by ASARCO until it closed in 2001. Soil and groundwater were contaminated by lead, other heavy metals, and arsenic, so the site was declared a Superfund site. The buildings, stacks, and soil were removed, but the 16-million-ton black slag heap at the south edge of the highway remains. American Chemet Corporation sources copper and zinc oxides from it for use in paint and fungicides.

The reddish landscape northeast of US 12 about 10 miles southeast of East Helena is the Spokane Hills, marked by the red mudstones of the Spokane Formation of the Proterozoic Belt Supergroup. The hills are part of an intact extensional fault block of

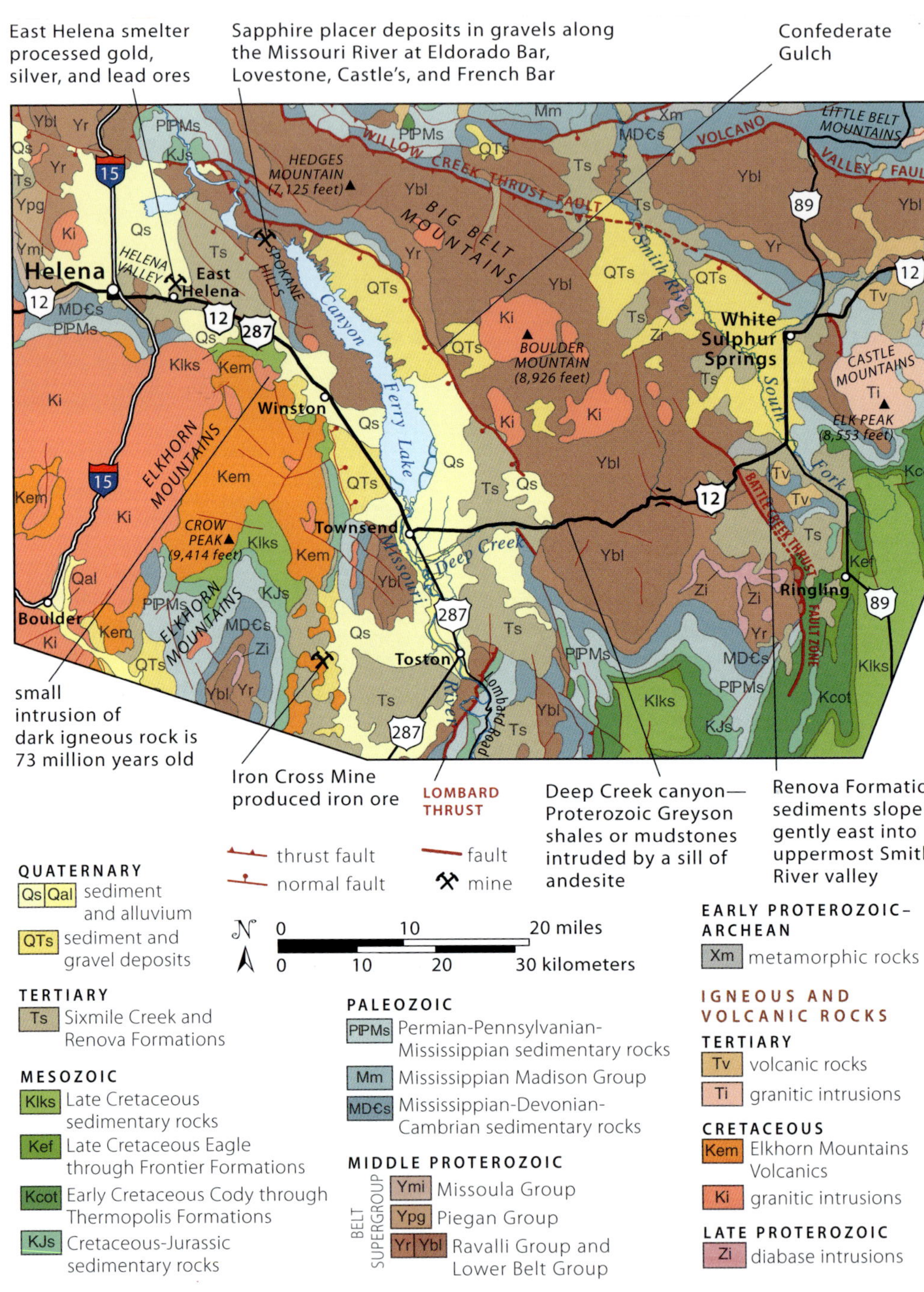

Geology along US 12 between Helena and White Sulphur Springs.

sedimentary rocks tilted to the east. The Belt rocks are on the west edge of the block, with progressively younger formations all the way up to Cretaceous age to the east. The mountains southwest of the highway are made of the Cretaceous Elkhorn Mountains Volcanics that were intruded by many andesite sills related to the volcanic rocks.

The 27-mile-long Canyon Ferry Lake, a dammed section of the Missouri River, lies in an extensionally faulted swale between the Elkhorn Mountains on the west and the Big Belt Mountains on the east. A long normal fault runs the length of the lake but is not visible at the surface. The broad, gently sloping surfaces along both sides of the lake are pediments that were eroded by sheetlike flows during intense rainfall events in the desert climates of Pliocene time. The pediments were likely dissected when moisture dramatically increased during Pleistocene time.

Light-colored lake deposits of the Renova Formation occur along the west side of Canyon Ferry Lake. The Renova here was deposited in extensional basins from 48 to 20 million years ago. Lake deposits dated at 32 million years have yielded some of the most exquisitely preserved fossils of plants and insects found anywhere in the world. The preservation is so good that some insect fossils are fully intact, retaining color patterns and delicate feeding and reproductive structures. One extraordinary discovery preserved a pair of caddis flies that died while mating. Preservation of this quality is rare and suggests that the chemical composition of the lake was unusual, possibly so alkaline from volcanic ash that it prohibited the growth of microorganisms that would ordinarily break down dead insects. The fossils are mostly of nonaquatic insects, strongly supporting the notion that the alkaline water did not support much aquatic life.

In 1955, prospectors discovered uranium in the Renova Formation along the shore about 5 miles northeast of Winston. Here, as in many places, the volcanic ash deposits in the Renova contain abundant plant remains. Uranium dissolved in groundwater tends to precipitate in low-oxygen environments with organic matter, such as that found on a lake bottom, mostly as the yellow uranium oxide mineral carnotite. The deposits near Winston were too small and lean to mine.

Thorium is found across the lake from Winston in small, high-grade deposits in and around small granite intrusions. In the 1950s it was demonstrated to be viable in nuclear reactors to generate power but abandoned because it was a less-efficient fuel than uranium and plutonium and could not be used to make bombs. However, it has several advantages. Thorium is more abundant, it poses zero risk for reactor meltdown, the by-products cannot be used to make nuclear weapons, and it produces far less nuclear waste. Many countries are researching its potential use, and China and India are building prototype thorium reactors that should be on line before 2025.

Two former Confederate soldiers, paroled prisoners of war, found placer gold in Confederate Gulch near the center of the Big Belt Mountains in 1864. If we can believe the reports of the people who were there, the gravels in the gulch were among the richest ever washed anywhere—freakishly rich. Single pans are said to have yielded as much as $1,000 worth of gold, at a time when an ounce of gold was worth less than $20. Several accounts say that no less than 2.5 tons of gold were recovered in a glorious final cleanup of the sluice boxes. Total production from Confederate Gulch may have been worth about $16 million—in late-1860s money!

The gold-bearing gravels of Confederate Gulch existed only within the 2 acres of a gravel bar, the Montana Bar, and nearly everyone left as soon as that ground had been worked. By 1870, the population of Confederate Gulch had declined from more than 10,000 to 255. Hardly a trace now remains of the roaring town of Diamond City, and the gulch yields grudging little to occasional prospectors who come seeking an overlooked scrap of the bonanza. Despite the incredible placer bonanzas, bedrock sources of the gold in narrow quartz veins in diorite and shale near Confederate Gulch proved uninspiring.

Miners used large-scale hydraulic mining in Confederate Gulch. The technique involves bringing water to the site in a high flume, and then sending it down through several hundred feet of pipe to huge nozzles that look like muzzle-loading cannons. The massive jets of high-pressure water were said to have enough force to smash a brick building with a single pass at a range of several hundred feet; they certainly had enough force to wash down high gravel banks and flush their contents through sluice boxes. Hydraulic mining was especially effective in places like Confederate Gulch, where the gold-bearing gravels were in terraces well above stream level.

For 5 miles east of Townsend, US 12 crosses gently sloping pediment gravels and underlying valley-fill deposits of the Sixmile Creek and Renova Formations. The overlying, coarser-grained Sixmile was deposited by streams flowing to the northeast in valleys formed by basin and range extension, starting around 16.5 million years ago. From 16.5 to 4.4 million years ago, mega-eruptions of the Yellowstone volcano, then well southwest of its present location, sent incredibly large volumes of ash and pumice down these valleys, choking the stream channels and spreading water-deposited ash across the valley floodplains. About 3 miles east of Townsend on US 12 are thin beds of gently west-dipping ash and gravel of the Sixmile Creek Formation. The ash beds were transported down the ancestral Missouri River Valley all the way from south-central Idaho, probably around 10 million years ago.

For 10 miles east of Townsend, US 12 mostly follows Deep Creek, which has cut a narrow slot through the Renova Formation. The road crosses numerous normal faults that drop the blocks west of them. Once you see the hard, layered rocks of the Proterozoic Belt Supergroup, you've crossed the range-bounding normal fault that

Trough-shaped cross bedding in ashy sandstone of the Sixmile Creek Formation east of Townsend. In this photo's perspective, the stream that deposited the sediment flowed from right to left.

dropped the Missouri River Valley and raised the Big Belt Mountains. (As an aside, the Belt Supergroup was named after the Big Belt Mountains, one of the areas where they were first recognized.) This fault has likely been active for the last 16.5 million years, forming the basin that the ancestral Missouri River flowed through as it deposited the Sixmile Creek Formation.

East of the fault, Deep Creek cuts through a broad crustal arch of Proterozoic Belt rocks that were deformed into big folds. First it goes through 1 mile of pale-gray shales or mudstones, then 9 miles of gray to off-white Newland Formation limestones, and finally a couple more miles of shales or mudstones, mostly covered by Renova Formation, to the crest of the range. The bedrock layers west of the crest of the big arch all dip to the west; those east of the crest all dip to the east. Note that the crest of the fold is not at the pass but about 7 miles west of it. East of the pass US 12 crosses a syncline in younger Spokane Formation shale for about 3 miles.

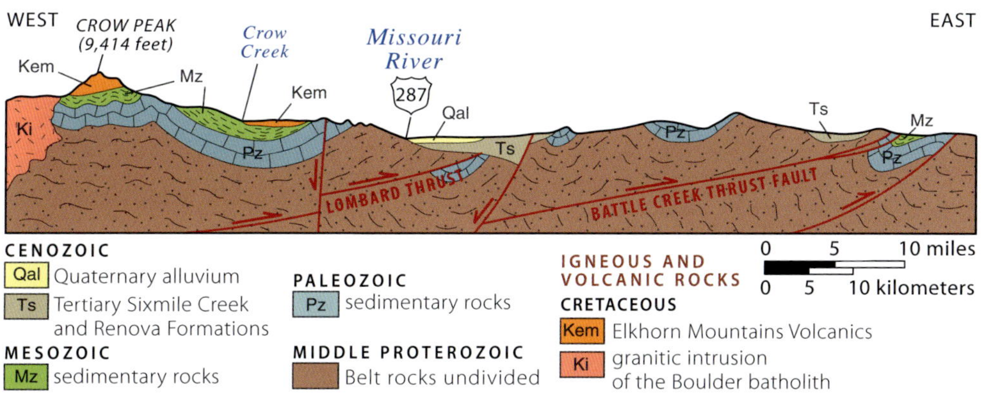

West-east cross section drawn along a line south of US 12 from west of Townsend to north of Ringling on US 89. The big thrust faults moved rocks east several miles from the area of the Boulder batholith during Cretaceous Sevier deformation.

About 8 miles east of Townsend the road crosses the trace of the Lombard thrust fault, which some geologists argue is the eastern edge of the northern Rocky Mountains if they're defined by rocks deformed in the Sevier orogeny, simply because it's the easternmost noteworthy thrust fault. This structural definition of the Rocky Mountains should not be confused with the existing landscape, which consists mostly of extensional mountain ranges and basins. The west-to-east displacement on the Lombard thrust fault is probably no more than a few miles, quite modest by the standards of many northern Rocky Mountain thrust faults.

The Proterozoic rocks show a well-developed platy structure known as slaty cleavage, which is typical of strongly deformed and slightly metamorphosed rocks in compressional tectonic environments. As the rock was compressed, its minerals recrystallized in a direction perpendicular to the compression, not parallel to the original sedimentary bedding. Here in the Big Belts, thin slabs of slate spall off nearly every exposure along slaty cleavage. In some places the mica-rich sedimentary layers they cut across glint in the sun to make ribbons of light and dark gray on their flat surfaces.

Quartzite and slate of the Proterozoic Belt Supergroup in Deep Creek canyon 12 miles east of Townsend. The close-up shows slaty cleavage that splits across the original sedimentary layering along surfaces parallel to one that would slice the fold in half.

Patches of Renova lake sediments on top of the range show that these Proterozoic rocks were under a valley floor as recently as 20 to 16 million years ago and were raised into mountains sometime after. The fault under Canyon Ferry Lake and a group of six parallel normal faults, including the fault along the front of the Big Belt Mountains, all dropped down on their west sides, are active extensional faults that account for that rise.

Descending the east side of the pass toward White Sulphur Springs, the road enters open country and continues down a broad Pliocene pediment surface eroded into the Renova Formation. The road crosses into the broad, uppermost Smith River valley to connect with US 89. The highway continues 8 miles straight north on grayish-orange sand and volcanic ash of Tertiary age, with layers of gravel. The gravel has lots of fragments of fossil mammals and turtles from Oligocene and Miocene time.

LOMBARD THRUST FAULT

To see the Lombard thrust fault, head south from Townsend on US 287 and turn east at Toston. Follow Lombard Road southeast for about 7 miles into the canyon cut by the Missouri River. In the hills on the left (north), dark-brown Proterozoic Belt rocks rest on red rocks of the Cretaceous Kootenai Formation, with the Lombard thrust fault between them. This older-over-younger relationship happened during Sevier deformation about 77 million years ago as a result of the collision of the oceanic Farallon plate and the continental North American plate. The cliff-forming rocks below the Kootenai Formation are the Permian Phosphoria and Pennsylvanian Quadrant Formations, shelf and coastal sediments deposited in a shallow sea that covered the area during Paleozoic time.

The Missouri River's sinuous pattern developed when it was meandering over a low-gradient surface composed of soft rocks or sediment; the channel maintained its path as the mountains rose during basin and range extension. As the river eroded downward, its channel could no longer migrate back and forth, so it remained stuck in the meander that it cut over millions of years.

The Lombard thrust exposed in the canyon of the Missouri River about 7 miles upstream from Toston. This view to the northwest shows Proterozoic Belt rocks shoved northeastward over Cretaceous rocks.

Gold in the Elkhorn Mountains from ores high in iron sulfides such as pyrite led to the construction of a smelter in Toston in 1885. Designed by William L. Austin, the smelter was the first of its kind in the United States to utilize pyrite-containing sulfide ores. Ore came primarily from Radersburg, west of Toston, but some from as far away as Coeur d'Alene, Idaho. The smelter closed in 1891, but it left a legacy of slag waste that had been dumped into the Missouri River. In 2009, the Montana Department of Environmental Quality completed the reclamation of a 9.4-acre site by removing heavy metal contamination. An interpretive sign with samples of the slag is located in Toston.

US 89
Livingston—Gardiner
55 miles

South of Livingston, US 89 enters a narrow canyon that the Yellowstone River cut through a large fold in Paleozoic and Mesozoic sedimentary rocks. At the entrance to the canyon, these rocks dip north but bend over to dip south as the road exits the canyon. The fold, the Canyon Mountain anticline, formed in Late Cretaceous time as a block of Archean crystalline rock rose under the sedimentary rocks during the Laramide orogeny. The Suce Creek thrust fault that caused the fold is exposed west of US 89, where older Cambrian limestone rests on sedimentary rocks of Jurassic and Cretaceous age. Elsewhere along the fault, Archean metamorphic rocks were shoved over Paleozoic and Mesozoic sedimentary rocks, a diagnostic feature of a Laramide structure. Livingston Peak, south of Livingston in the Absaroka Range, is erosion-resistant amphibolite and gneiss of Archean age above the Suce Creek fault.

In the 1940s, the US Army Corps of Engineers considered the canyon a good place for a dam. Opposition to the dam addressed the desire to preserve agricultural land and scenic beauty, but the geology of the canyon finally killed the project. The abutment for the dam would have been placed in Madison Group limestone that is full of karst cavities. Karst forms when limestone is dissolved by slightly acidic groundwater, causing fissures, caverns, and sinkholes. In addition, it is now known that the area is prone to earthquakes related to the Yellowstone volcano and the extensional faulting that has dropped the Paradise Valley relative to the Absaroka Range. To top it off, construction activities for US 89 in 1967 caused a large landslide in slick Cambrian-age shale that blocked the road, demonstrating that the canyon is prone to landslides.

South of the canyon, US 89 follows the Paradise Valley. Bounded on the west by the Gallatin Range and the east by the Absaroka Range, the valley is the easternmost major graben, or down-dropped block, of the Basin and Range Province in Montana. Extensional faulting began in Miocene time around 17 million years ago, but the

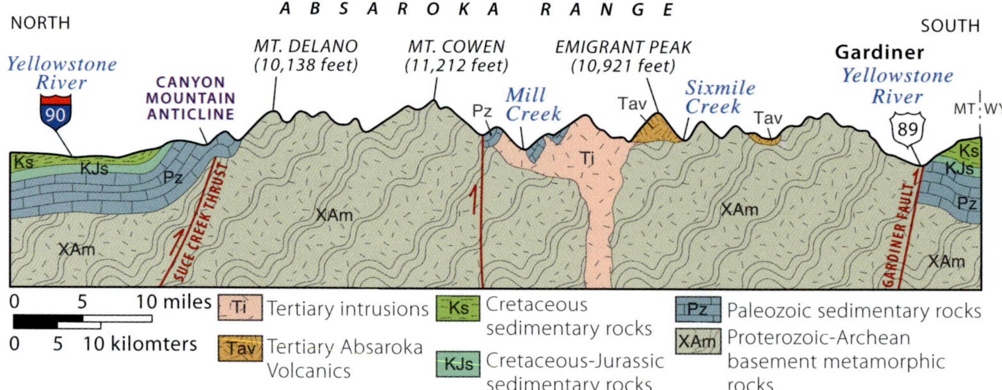

North-south cross section parallel to US 89 through the Canyon Mountain anticline south of Livingston. Layered Paleozoic and Mesozoic rocks were folded over crystalline Archean rocks that were pushed up along the Suce Creek thrust fault during the Laramide orogeny.

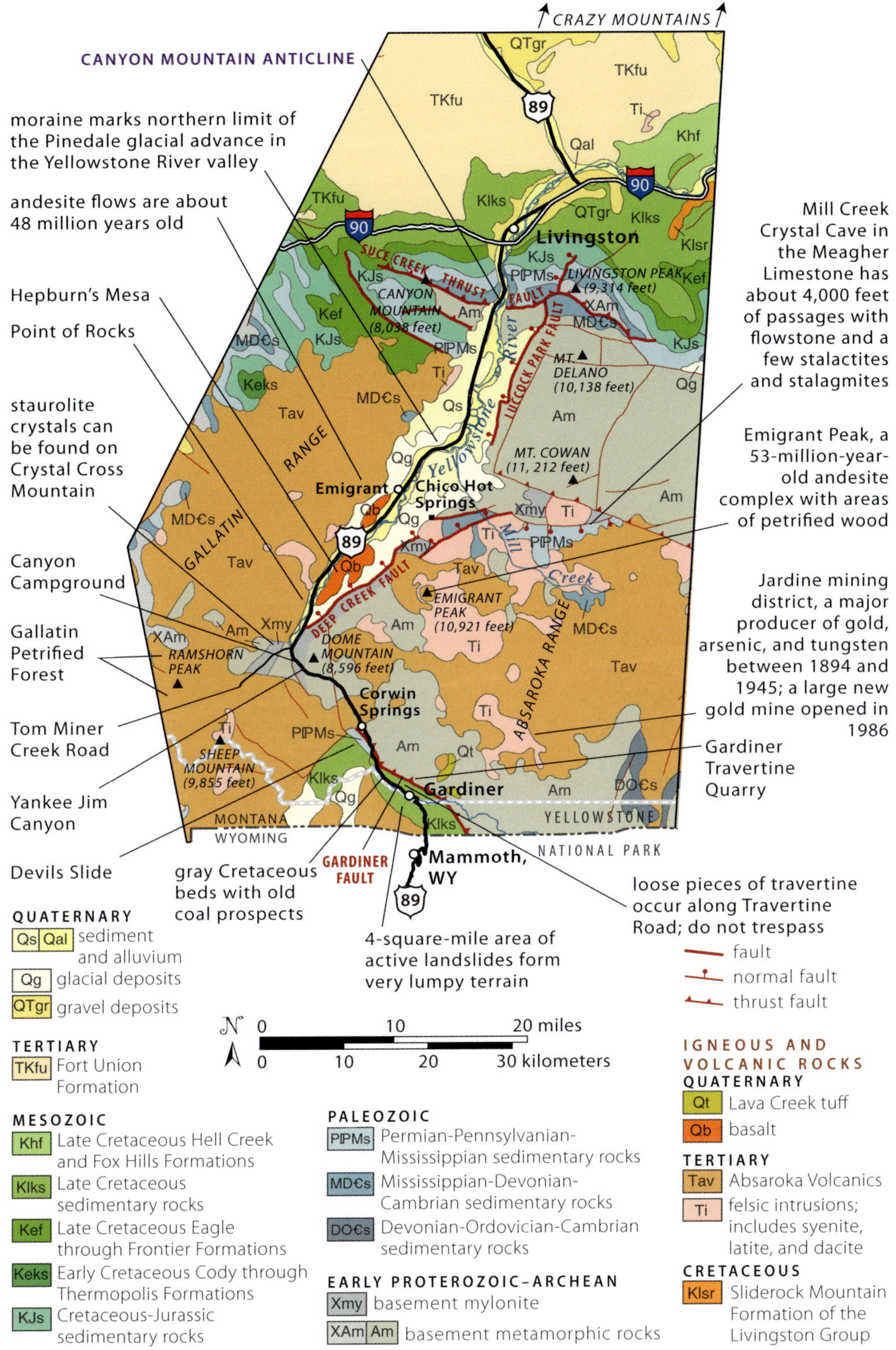

Geology along US 89 between Livingston and Gardiner.

presence of fault scarps cutting recent valley deposits shows that the faults have moved recently, dropping the valley relative to the rising mountains. Fault movement occurs predominantly along the Absaroka Range side of the valley, where fault scarps mark the northwest-dipping Deep Creek and Luccock Park faults. These scarps are up to 164 feet high from multiple ground ruptures; as much as 39 feet of surface rupture has occurred in the last 11,700 years.

At least 18,000 feet of vertical displacement has occurred on the Deep Creek fault near Chico Hot Springs since Miocene time. Groundwater circulates deep below the surface in fractured rock along this fault, then surfaces as hot springs. Clearly, large-magnitude earthquakes occurred here in the past and likely will again.

The landscape of the Paradise Valley owes much of its form to several large glaciers that flowed northward from the Yellowstone region, leaving glacial outwash and moraines composed of glacial till. Moraines that form east-west-trending ridges across the Paradise Valley mark the ends of glaciers; those that parallel the valley mark the former sides of them. Not all glacial deposits in the Paradise Valley came from the Yellowstone region. About 10 miles south of Livingston, on the east side of the valley, a large moraine extends several miles from the mouth of Pine Creek onto the valley floor. Here the glacier was able to spread onto the valley floor because there wasn't a large glacier to block its advance. South of Mill Creek, however, the large glacier from Yellowstone likely blocked ice descending from the Absaroka Range.

Near the Mill Creek Road turnoff, US 89 crosses the terminal moraine left by the last glacier to flow down the Paradise Valley from Yellowstone. It was active during the Pinedale glaciation, which started 30,000 years ago and peaked between 19,000 and 15,000 years ago. The ice reached its maximum extent near here and deposited its till about 16,500 years ago. Glaciers of the Bull Lake glaciation, which started 160,000 years ago and ended about 130,000 years ago, also flowed down the valley, but the Pinedale glaciers erased most of the evidence left by the older glacial advance.

At Hepburn's Mesa, about 7 miles south of Emigrant, a rest area east of the highway provides a good view across the river of a 2.2-million-year-old columnar-jointed basalt flow erupted from the Yellowstone region just prior to a caldera-forming eruption 2.1 million years ago. The basalt caps gravel that's less than 4 million years old and was likely deposited by the ancestral Yellowstone River flowing off the Yellowstone thermal bulge. Below the gravel is light-colored siltstone of the Hepburn's Mesa Formation, a local deposit that accumulated in alkaline lakes and mudflats during Miocene time. These deposits have yielded fossils of rodents, moles, camels, tortoises, and horses buried and preserved between 16.8 and 13.7 million years ago. The climate at that time was arid to semiarid, like much of Montana today.

Hepburn's Mesa is an excellent example of topographic inversion. The basalt flowed down the old Yellowstone River channel 2.2 million years ago. As the region rose and the river cut downward, the hard basalt resisted erosion to leave the old river channel standing as a mesa. The dropped blocks of basalt along the margins of the mesa to the south make it look like there are two flows, but they are large landslides that formed as the softer sediments under the basalt failed and destabilized the slope. This is a good reason to not build below or too close to the edge of any mesa!

About 1 mile south of Hepburn's Mesa and west of the road is an eye-catching dike of dacite, a volcanic rock intermediate in composition between rhyolite and andesite that was injected into a crack in layered volcanic conglomerate. Because the hard dike

Light-colored lake deposits exposed at Hepburn's Mesa have yielded abundant fossils. They are capped by gravel likely deposited by the ancestral Yellowstone River and basalt erupted from the Yellowstone volcano around 2.2 million years ago.

resists weathering better than the softer conglomerate, it stands out in relief. This dike and most of the area west of the Paradise Valley in the Gallatin Range consist of Archean through Mesozoic rocks intruded by and covered with Eocene Hyalite Peak Volcanics of the Absaroka Volcanics Supergroup. From around 53 to 43 million years ago, subduction of the Farallon plate to the west produced several northwest-trending chains of explosive stratovolcanoes near here. They produced mostly andesitic magmas that intruded older rocks and erupted as lava flows and ash. These deposits later produced lahars, torrents of water and rock debris that flowed down the volcano flanks into the surrounding valleys. The mudflows carried and buried trees that became fossilized stone replicas as silica-rich fluids slowly replaced their organic material. The volcanoes rose to great heights, so a variety of trees that grow at different elevations—such as pines, firs, redwoods, and sycamores—were preserved. The Gallatin Petrified Forest Interpretive Trail, at the Tom Miner Campground on the Tom Miner Creek Road about 19 miles south of the Hepburn's Mesa rest area, is a

A petrified tree buried in conglomerate of the Absaroka Volcanics Supergroup exposed along the Gallatin Petrified Forest Interpretive Trail.
—Courtesy of Chris Pace

great place to see petrified trees in the Absaroka Volcanics Supergroup. Near the top of Ramshorn Peak are some magnificent petrified stumps, including a redwood over 12 feet in diameter. Please leave the petrified wood you see along the interpretive trail for others to enjoy. You can collect petrified wood in the gravels of the Yellowstone River, as well as agates and jasper, all eroded from the volcanic rocks.

About a half mile south of the turnoff to Tom Miner Basin, the road enters Yankee Jim Canyon, a major obstacle to early travel into Yellowstone National Park. The mouth of the canyon marks the southernmost trace of the Deep Creek fault. Uplift along this fault and bulging of the crust around the Yellowstone volcano forced the Yellowstone River to cut a deep canyon through the Archean igneous and metamorphic rocks. The rocks are complexly folded and sheared granitic gneiss and amphibolite, the latter a rock rich in black, rectangular grains of hornblende. In places they are intruded by dark basaltic dikes and pale, very coarse-grained pegmatite dikes.

Most of the Absaroka Range is made up of these rocks, some of which are more than 3 billion years old. Between Mill Creek and Yankee Jim Canyon, the Absaroka Volcanics rest directly on Archean metamorphic rocks. The Paleozoic and Mesozoic rocks that once covered the Archean rocks must have eroded away prior to Eocene time, likely the result of compression and uplift during the Late Cretaceous Laramide orogeny.

During the last ice advance the canyon contained as much as 3,200 feet of ice. Around 5,000 years ago, well after the glacier had melted, and probably facilitated by its scouring of the canyon, a landslide in the canyon dammed the river, forming a lake that deposited silt that has yielded human artifacts. Slides are still a hazard in the canyon. About 2 miles south of Tom Miner Creek, picturesque Canyon Campground is nestled among scattered giant boulders that fell from the cliff walls. Might additional big rocks tumble down that slope, perhaps due to freeze-thaw processes? Earthquakes in this area could certainly shake things enough to further facilitate rockfalls.

Igneous and metamorphic rocks of Archean age exposed in the walls of Yankee Jim Canyon.

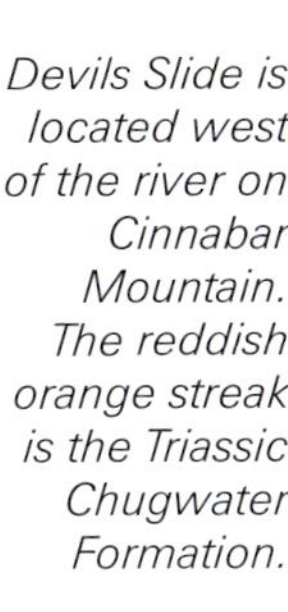

Devils Slide is located west of the river on Cinnabar Mountain. The reddish orange streak is the Triassic Chugwater Formation.

The vertical reddish orange streak on Cinnabar Mountain west of the road is Devils Slide. It is not a landslide, but rather an exposed layer of the Chugwater Formation, a 200-million-year-old Triassic tidal-flat deposit. The reddish orange color formed from the oxidation of iron-bearing minerals in the Chugwater's soft siltstone, which erodes faster than the more resistant "rails" on either side of the slide. The originally horizontal layers were tilted by Laramide compression and reverse faulting along a splay of the Gardiner fault. That major Late Cretaceous compressional fault lifted Archean igneous and metamorphic rocks over the younger sedimentary rock.

North of Gardiner, several quarries mine 29,000-year-old travertine for ornamental building stone. Travertine commonly forms around hot springs, as at Mammoth Hot Springs in Yellowstone National Park, when acidic, carbon dioxide–rich water percolates through limestone deep below the surface and dissolves the rock. The water, saturated with calcium carbonate, surfaces through cracks, and the resulting pressure and temperature drop releases carbon dioxide as a gas, decreasing the acidity and allowing the calcium carbonate to precipitate from the water as layers of travertine. Thermophiles, heat-loving microorganisms, live on and aid in the precipitation of the travertine, helping produce the intricate lacy layering that makes the rock a beautiful building material.

Near Gardiner, glaciers carved steep walls in weak Eocene-age conglomerate. After the ice melted, many of the cliffs slumped into the valley in giant landslides that are easy to recognize by their chaotic, hummocky surfaces, numerous springs and aspen groves, sharp toes on the valley floor, and steep scarps where the slides broke away from the slopes. There is an especially large slide southwest of Gardiner, at the entrance to Yellowstone National Park.

In Gardiner, piles of boulders near Gardiner High School, west of the Yellowstone River, attest to the torrents of water released after a glacially dammed lake failed far to the southeast in the Lamar Valley. As the ice cap on the Yellowstone Plateau began to melt and stagnate near the end of Pleistocene time, active ice continued to flow off the Beartooth Mountains and into drainages, such as Slough Creek in the northeast part of the park. The ice flowed down this drainage and into a deglaciated segment of the Lamar Valley, damming it to form a deep lake. When the water rose high enough to float the ice dam, water rushed toward Gardiner, carrying the boulders in a surge of water estimated to have been several hundred feet high. It happened at least twice. (For information about Yellowstone National Park, see *Roadside Geology of Yellowstone Country*.)

US 93
Missoula—Lost Trail Pass
94 miles

Between Missoula and Lolo, US 93 follows the Bitterroot River past roadcuts in red and green mudstone in the Missoula Group of the Proterozoic Belt Supergroup. Watch for colorful bedding surfaces covered with mud cracks and ripple marks. From Lolo to south of Darby, US 93 follows the lush Bitterroot Valley between the spectacular Bitterroot Mountains to the west and the lower Sapphire Mountains to the east. The bulk of the Bitterroot Mountains is granite of the immense Cretaceous-age Idaho batholith, raised into a huge arch called the Bitterroot dome. The arch is beautifully displayed in profile when viewed from Missoula, where its smooth eastern flank curves from the range crest down to the Bitterroot Valley on the east. The northern end of the Bitterroots consists of highly metamorphosed rocks; the peaks from Stevensville south are granite. Forest Service trails lead up every canyon to pristine alpine lakes huddled in glacial cirques at the Montana-Idaho divide. Thirty-two peaks in the range are over 9,000 feet high, and Trapper Peak in the southern part reaches 10,157 feet.

The eastern flank of the Bitterroot Mountains viewed southwest from Missoula shows the smooth curve of the Bitterroot dome.

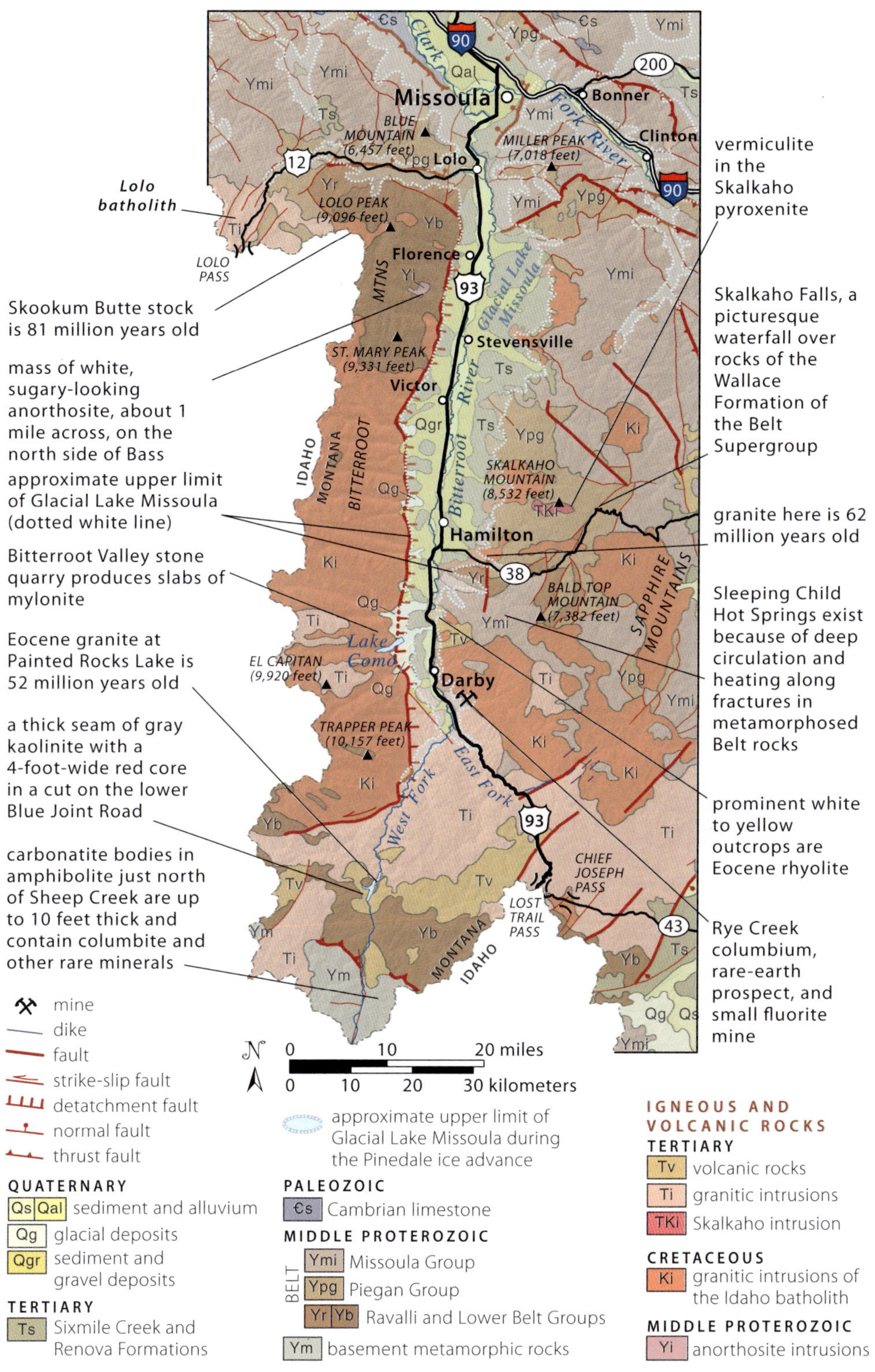

Geology along US 93 between Missoula and Lost Trail Pass.

The ancestral Bitterroot Mountains must have been about 20,000 feet high, about twice their present height. They were buoyed up by continental crust that had thickened as additional crust was shoved under this area during subduction of the Farallon plate. Melting of the overlying crust formed the Idaho batholith granite between about 90 and 70 million years ago.

In the northern end of the range, that magma intruded gneiss and schist, high-temperature metamorphic rocks that decrease in grade to the north. The age and origin of these basement rocks is uncertain, but some age dates suggest they may be 1.47 billion years old, similar to the age of the older Belt rocks. Compositions and sequences of these rocks also closely resemble early Belt formations, except that they're highly metamorphosed.

If these metamorphosed rocks are in fact Belt rocks, the heat that cooked them was probably not imposed by the huge Idaho batholith. More likely they are early Belt rocks that were tightly crumpled in once much thicker continental crust, which would have placed them deep in the hot crust. There they would have been heated by rising basaltic magma above the active subduction zone that existed at the margin of the North American continent when the batholith was emplaced. Where highly metamorphosed continental mudstones and sandstones are in contact with basalt magma, they are likely to melt. When sedimentary rocks such as these melt, they produce granitic magma. That granitic magma, being lighter than the surrounding solid metamorphic rock, would float upward in the crust to solidify as a granite intrusion consisting of quartz, feldspar, and mica. Thus, the Idaho batholith is probably the product of high-temperature metamorphism rather than the cause of it.

Almost every vantage point that gives a good view of the Bitterroot Mountains also reveals that most of its eastern front was originally a smooth surface that curved down to the east at an angle of about 25 degrees. A zone of distinctly platy and streaky-looking rock more than 1,000 feet thick, called the Bitterroot mylonite, lies on the surface of the range front. Mylonite is a distinctive rock that forms where rocks deep below the surface are strongly sheared in a fault zone. The easiest way to see the Bitterroot mylonite at close range is to drive to the mouth of any of the canyons south of Stevensville (for example, Sweathouse Creek near Victor). Look for off-white platy rock with a streaky lineation oriented downslope on the surfaces of the slabs.

Such domed areas of intensely metamorphosed and sheared mylonite rocks surrounded by lower-grade metamorphic rocks are called metamorphic core complexes. These originally deep crustal rocks occur in a long string from southern Arizona to northern British Columbia. Adjacent rocks east of the core complexes were either pulled off or slowly slid off, unloading the domes and causing them to rise. The smooth surface on the lower elevations of the east flank of the Bitterroot front is so well preserved because it was buried for many millions of years beneath valley-fill sediments of the Renova Formation, which have now eroded off.

The Bitterroot mylonite is the deep fault zone that carried the slightly metamorphosed Belt rocks of the Sapphire Mountains east over the rocks beneath it. The distinct lineation on the slabby surfaces invariably points just south of due east—the direction the overlying rocks moved. Two explanations have been proposed to explain how the Sapphires moved. The first explanation, based on relationships between rocks observed in the field, suggests that they slowly slid to the east off the rising Idaho batholith during mountain building at the time of its intrusion, and

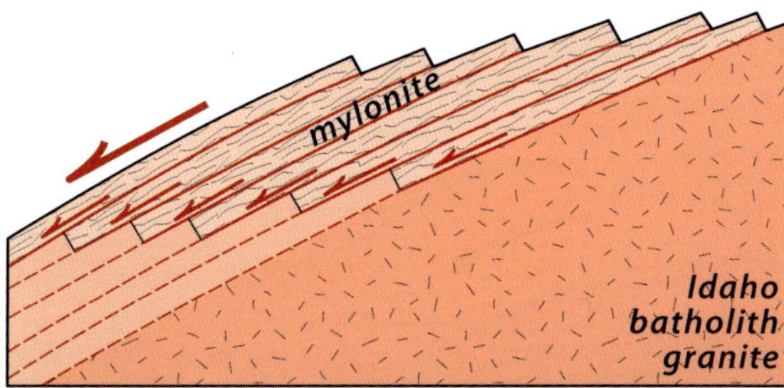

Sheared mylonite detachment zone along Sweathouse Creek in the Bitterroot Mountains. View facing south. The Bitterroot mylonite probably moved like a sliding deck of cards.

Sheared mylonite surface in the detachment zone at Sweathouse Creek.

remnant magmas, still weak enough to shear easily, lubricated the movement as the batholith solidified. That would have happened 80 or 70 so million years ago.

The second explanation, based more on radiometric age dating, suggests that the Bitterroot Mountains were pulled out from under the Sapphire Mountains during crustal stretching in an earlier stage of basin and range extension. This would have been about 50 million years ago, after the Idaho batholith had crystallized but before younger, shallower granites in the region were intruded.

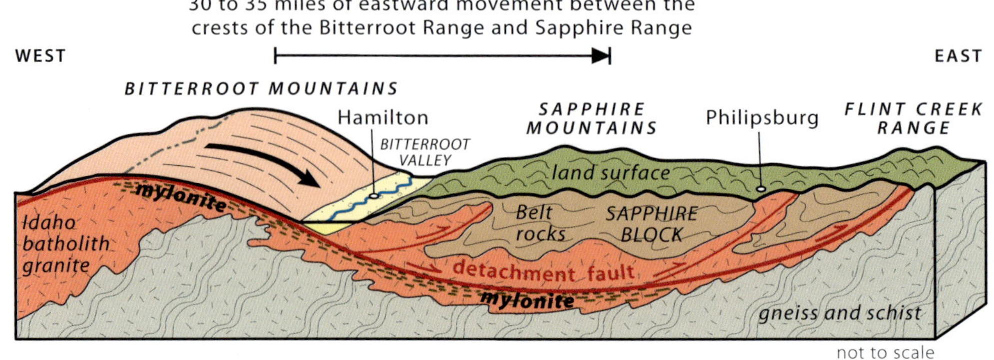

Cross section across the Bitterroot dome in the Idaho batholith and the Sapphire Mountains. The Sapphires moved 30 to 35 miles off the crest of the Bitterroot dome.

Both explanations could also be correct: the intrusion of the Idaho batholith during mountain building 70 million years ago caused the Sapphire Mountains to slide off the top of the Bitterroots, and some of that motion recurred (or never stopped) as late as 50 million years ago as the giant mountain range spread and collapsed. The old, still-hot batholith could have remelted as it rose to shallower levels in the crust, generating the 50-million-year-old granites.

The serrated skyline of the Bitterroot Mountains speaks eloquently of heavy glaciation during Pleistocene time. Most glacial valleys in other mountain ranges drain in a variety of directions, but here they line up directly east-west. The smooth east-sloping surface of the Bitterroot dome directed streamflow straight downslope, and during the ice age those valleys filled with glaciers that scoured them wider and deeper. Only the highest jagged peaks protruded above the glaciers. Most of the glaciers descended to an elevation near 4,000 feet before they finally reached a climate warm enough to melt the ice as fast as it advanced. North of Hamilton, where the floor of the Bitterroot Valley is lower, the glaciers melted before they could emerge from the mountains, so a large fan of glacial outwash spreads down toward the highway from the mouth of every canyon. South of Hamilton, the glaciers emerged from the mountains onto the floor of the Bitterroot Valley. They left large moraines, the hummocky expanses of landscape between US 93 and the mountain front. Notable examples of glacial moraines are at Bear Creek and Mill Creek south of Victor, Blodgett Creek near Hamilton, and Roaring Lion to its south. Lost Horse Road, between Hamilton and Darby, heads west, climbs across a moraine, and enters the big glaciated canyon of Lost Horse Creek, providing easy access to this amazing terrain almost all the way to the Idaho border at the crest of the range.

A large arm of Glacial Lake Missoula reached about 9 miles south of Darby to just beyond the community of Connor, deep into the Bitterroot Valley. At its maximum height, the lake reached up to 4,150 feet. Darby, with an elevation of 3,885 feet, was under 265 feet of water. When at its fullest, the lake would have lapped against the toes of valley glaciers descending from the Bitterroots. Boulders that fell on the surface of the ice up-valley were entrained in the ice and carried to the glacier ends, where icebergs calved into the lake and dropped them as they melted. These boulders are now scattered around the Bitterroot Valley. Lone Rock School northeast of Hamilton preserves a big one, although it was moved here from a nearby ranch. Wave-eroded shorelines left at high stands of the lake are also preserved along the east side of the valley, though you have to look carefully for them on grassy slopes undisturbed by houses.

US 93 takes a big bend to the west 9 or so miles south of Hamilton, around a large area of rusty-looking yellowish to whitish volcanic rocks that are exposed in ragged cuts along the highway but are more prominent low in the hills to the east.

SKALKAHO MOUNTAIN

The Sapphire Mountains southeast of Hamilton, from Skalkaho Road south for 5 or 6 miles, including the valley of Sleeping Child Creek, are made of medium-grade metamorphic rocks, including quartzite, quartz-feldspar-rich gneisses, and granitic gneisses. They appear to be derived from Belt sedimentary rocks, but from which formations is unclear.

The top of Skalkaho Mountain, in the Sapphire Mountains about 10 miles directly east of Hamilton, exposes an extraordinary igneous intrusion almost identical to the Rainy Creek stock near Libby. The lower, largest part of the Skalkaho intrusion consists primarily of a wild assortment of strange pyroxenites, black rocks composed mostly of pyroxene. The white upper part of the intrusion is syenite, a rock composed almost entirely of potassium feldspar. These rock types indicate that the original magma must have formed deep in Earth's mantle.

For many years these bodies of pyroxenite were considered valuable by miners because they contain black mica heavily altered to vermiculite, a mineral used for home insulation. Unfortunately, in the last few decades it has become clear that the vermiculite mined from the Rainy Creek stock contained a needlelike form of asbestos that has killed hundreds of people in and around Libby. (We elaborate in the US 2: Kalispell—Idaho Border road guide in the Northwest Montana chapter.) Several early attempts to mine vermiculite in the Skalkaho intrusion went poorly. While at one time that seemed unfortunate, now it's clear that we narrowly escaped having another major environmental disaster.

Metamorphosed rock of the Belt Supergroup along Skalkaho Road, southeast of Hamilton.

The canyons were eroded straight east down the east flank of the Bitterroot dome.

This Eocene rhyolite caps slightly older granites that are exposed intermittently in valleys. In some places metamorphosed Belt rocks cap higher hills above the rhyolite or granite, so the igneous rocks may form a sill (or sills) that was injected into the Belt rocks along a nearly flat thrust fault. If so, both the rhyolite and massive to slightly foliated granites seem to form a thin, nearly horizontal sheet lying on top of metamorphosed Belt rocks. The bluffs east of Darby are granites, perhaps those that fed the same volcanic eruptions. Those granites form most of the bedrock between Darby and Lost Trail Pass.

Most of the broad benches sloping down toward the Bitterroot River on both sides of the whole valley are alluvial fan deposits of sand and gravel eroded from the mountains above. The bulk of these deposits are of Tertiary age, probably including the Renova Formation, covered in many areas, especially on the west side, by Quaternary sand and gravel.

In 2017 Montana Bureau of Mines and Geology geologists spotted a fault scarp on a radar image along the foot of the Bitterroot Mountains from Victor south to the Lake Como–Darby area. The earthquake that caused it appears to have occurred 15,000 to 10,000 years ago, while glaciers were active in the Bitterroots. That is recent enough to consider the fault active, but it's not known if or when it might again move.

US 191
Bozeman—West Yellowstone
89 miles

West of Bozeman, US 191 crosses Tertiary sedimentary rocks and Quaternary sediment in the Gallatin Valley, a relatively young and complex extensional graben, or down-dropped block. The oldest sedimentary deposits belong to the Renova Formation, a fine-grained ashy silt and sand deposit with some pebbly conglomerate and airfall ash deposited in a broad basin that existed well before today's landscape.

The overlying Sixmile Creek Formation consists of gravel deposited by streams that were also likely active before the present topography existed. How do we know?

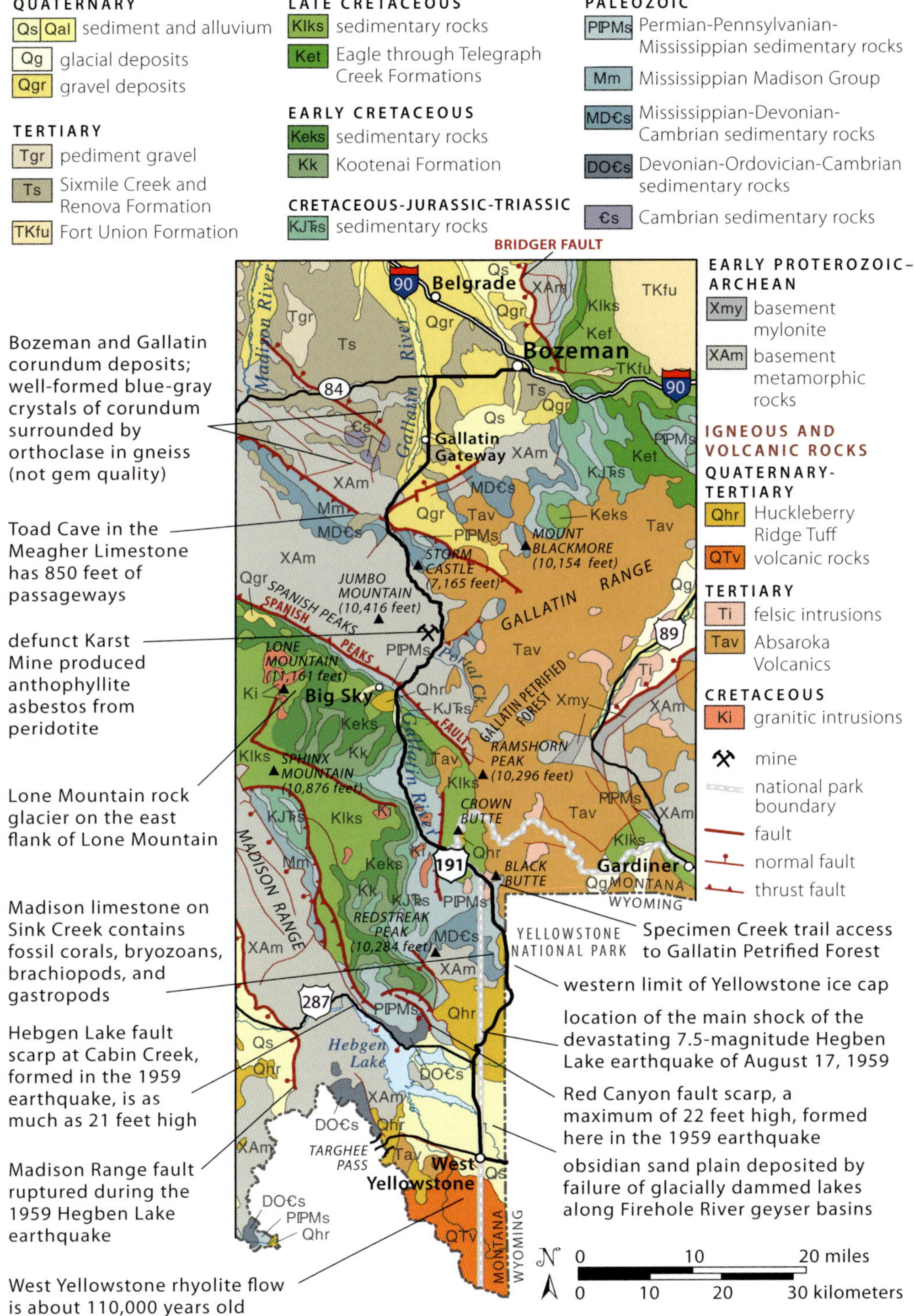

Geology along US 191 between Bozeman and West Yellowstone.

Because the Sixmile Creek gravel near the Bridger Range, north of Bozeman, is composed of cobbles eroded from the Eocene-age Absaroka volcanic field, yet none of the rocks related to these volcanoes occur within the Bridger Range. Since the gravel had to be eroded off higher ground, the Bridger Range, at least in its present form, probably didn't rise until relatively recently. Datable ash deposits show us that it may be as young as 4 million years. Similar evidence suggests that this is also true of the Gallatin Range.

The most recent deposits, Quaternary-age sand and gravel, which clearly eroded off the surrounding mountains, show that the modern topography was developing around 2 million years ago. Variations in the thickness of these sediments show that they were deposited over irregular topography, probably created by faulting, but the absence of fault scarps in the youngest valley-fill deposits shows that there haven't been any recent ground-rupturing earthquakes. However, the youth of the Bridger and Gallatin Ranges should be of concern to county planners as they consider future development in one of Montana's fastest-growing areas. The faults are not completely bad news for locals, given they provide conduits for hot water to percolate to the surface, forming hot springs, such as Bozeman Hot Springs west of Bozeman and Chico Hot Springs south of Livingston.

Even though there's a lack of evidence for recent fault movement, there is abundant evidence for uplift in the Gallatin Valley, the result of basin and range extension, which forms high-angle normal faults that raise mountains relative to valleys. River terraces along the Gallatin River show that the valley is rising also but not as fast as the mountains, and the river is excavating through its own, older deposits.

US 191 heads south toward the mountains at Four Corners (the junction with MT 84) on Quaternary sediments carried out of the mountains by mudflows and streams. The road crosses stream terraces of the Gallatin River, glacial outwash, and even some older Tertiary sedimentary rocks. Somewhere buried under that sediment is an active extensional fault along which the mountains you are about to enter are rising.

US 191 crosses primarily Archean metamorphic rocks between the canyon mouth and Big Sky. Folded Paleozoic and Mesozoic sedimentary rocks that overlie the old rocks are visible along the canyon walls in places, such as in Storm Castle, a peak that exposes Cambrian through Mississippian rocks that were deposited in a tropical ocean. The basement metamorphic rocks in the Gallatin Range to the east are covered more extensively by the Eocene-age Absaroka Volcanics, erupted from a volcanic arc that likely formed during the shallow subduction of the Farallon plate beneath the North American plate.

Gallatin Canyon has the narrow, V-shaped profile characteristic of stream-cut valleys, evidence that no glaciers flowed down this main drainage. Glaciers did flow down many tributary valleys, as seen in their glacially scoured, U-shaped profiles. Large piles of unsorted glacial till mark the down-valley limit of these glaciers.

Most of the metamorphic basement rocks are gneiss, a layered rock with light and dark banding called foliation. The dark layers contain black amphibole, mica, and other iron-rich minerals. The light-colored layers consist of quartz and feldspar. In places, magma was injected into the gneiss, leaving black fine-grained basaltic dikes and white coarse-grained pegmatite dikes.

At Portal Creek, gneiss in the roadcut has been age dated to about 2 billion years. An outcrop of 2.1-million-year-old Huckleberry Ridge Tuff overlies the gneiss,

A rock climber on Archean gneiss with white pegmatite dikes in the Gallatin Canyon. —Courtesy of Parker Webb

marking the northern limit of the rhyolitic pyroclastic-flow deposits erupted during the first of three eruptive cycles of the Yellowstone volcano. The boundary between the two rocks is an unconformity representing a time gap of nearly 2 billion years.

Just north of the turnoff to the Big Sky Resort, near Dudley Creek, primarily Archean metamorphic rocks overlie sedimentary rocks of Paleozoic age. This older-over-younger relationship was caused by the Spanish Peaks fault, a high-angle Laramide reverse fault. This fault was active during Late Cretaceous and early Tertiary time as a result of the collision of the oceanic Farallon plate to the west with the continental North American plate. Forces were directed to the southwest, and this large block of

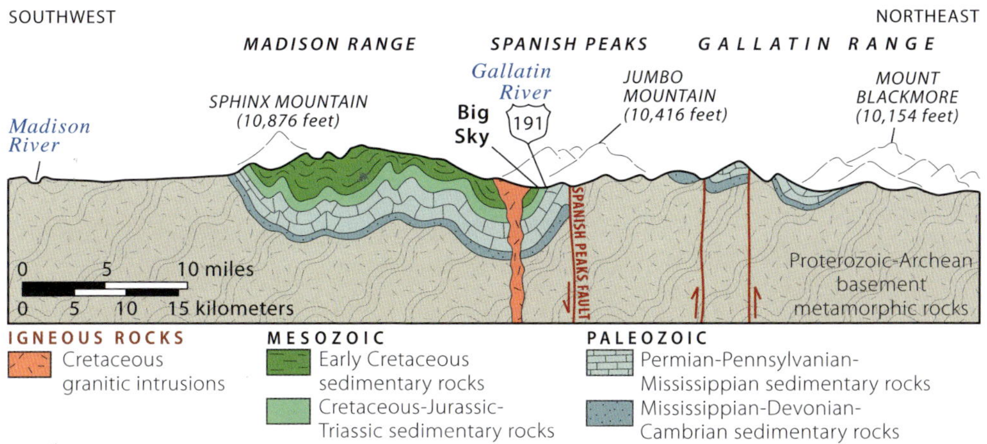

Cross section showing the Spanish Peaks fault, along which Archean basement rocks were shoved over Paleozoic sedimentary rocks during Laramide compression.

continental crust was raised a total of 13,500 feet, or 2.5 miles. The uplift resulted in the erosion of the overlying sedimentary rocks, exposing the once deeply buried, older metamorphic rocks. Today, compressional uplift is no longer active, and extension of the crust is now lifting these rocks, creating the mountain range we see today.

At 11,166 feet, Lone Mountain is visible from US 191 and dominates the skyline near the Big Sky Resort. The peak is a glacial horn that was eroded by several glaciers that flowed down its side valleys. The peak stands alone because its core is a hard dacite (igneous) of Late Cretaceous age that resists weathering more than the surrounding, easily eroded sedimentary rocks of the Cretaceous Frontier Formation. Shale belonging to this formation is notoriously unstable and forms landslides throughout the Big Sky area. Look for hummocky or irregular topography, a telltale sign of a landslide.

From the Big Sky turnoff south, US 191 passes through Paleozoic sedimentary rocks. The prominent grayish-white ledges are fossiliferous Madison Group limestone, a marine sedimentary rock that formed in shallow seawater about 300 million years ago.

At Porcupine Creek, 3 miles south of the Big Sky turnoff, a large landslide complex appears to the west where the valley floor broadens out. Low rounded hills in the valley are made of black, Cretaceous-age marine shale. When clay minerals (smectite) in the shale absorb water, they become very slippery and slide. The dirt road up Taylor Creek provides an interesting short side trip to see a huge landslide that came off the cliffs to the south and blocked the valley.

Crown Butte is on the boundary of Yellowstone National Park about 1 mile south of Taylor Creek. The top of the peak is a remnant of 2.1-million-year-old Huckleberry Ridge Tuff resting on Cretaceous-age shale. The tuff solidified from a pyroclastic flow that traveled down a valley that now has become a topographic high point. The hard caprock protects the underlying soft shale from erosion, inverting the topography. Lava Butte just to the south is composed of the same rocks.

The top of Crown Butte is an inverted stream valley composed of resistant Huckleberry Ridge Tuff on softer, Cretaceous-age shale.

Black Butte, a volcanic hill between Black Butte Creek and Specimen Creek, has a very different origin. Black Butte is an andesite stock, or plug, that cooled in the throat of an Eocene-age volcano 50 million years ago. It is part of the Absaroka Volcanics Supergroup, a predominantly volcanic sedimentary deposit that buried fossil forests and produced abundant petrified wood.

The Specimen Creek Trailhead, about 25 miles north of West Yellowstone, provides access to petrified trees buried in the Absaroka Volcanics. Its many upright stumps and prostrate logs can be seen about 2.5 miles up the trail. One tall redwood near the top of the first ridge is at least 12 feet tall. Please remember that exposures along Specimen Creek are within Yellowstone National Park, and it's illegal to collect specimens within the park.

South of Specimen Creek, US 191 is built on layers of pyroclastic flow deposits, some of them welded, from the Yellowstone volcano, as well as Archean schist, Paleozoic and Mesozoic sedimentary rock, and Eocene conglomerate and andesite of the Absaroka Volcanics. As the road descends a hill just north of the junction with US 287, a fine panoramic view of the Yellowstone Plateau opens up to the southeast. (See the introduction to this chapter, as well as *Roadside Geology of Yellowstone Country*, for more information about the Yellowstone volcano.)

Much of the uppermost Madison River valley in the West Yellowstone area is a gently sloping plain made up of obsidian sand. During the Pinedale glaciation, glaciers in the geyser basins along the Firehole River dammed several lakes that suddenly drained and flooded into the West Yellowstone basin, depositing the sand. At the airport turnoff on the west side of the highway, the ground is covered with small granules of black obsidian.

US 212
Laurel—Wyoming Border on the Beartooth Plateau

70 miles

US 212 between Laurel and the Wyoming border first follows the Clarks Fork Yellowstone River and then Rock Creek through Red Lodge, passing over a section of the Beartooth Highway. Late CBS correspondent Charles Kuralt called the Beartooth Highway "the most beautiful drive in America," but heavy snowfall only allows that beauty to be enjoyed from mid-May through mid-October. To the geologist, this route provides the greatest cross-sectional views of a Laramide structure in all of Montana. The Beartooth Mountains, once an active Laramide uplift, expose some of the oldest rocks on Earth and comprise a textbook of alpine glacial features. The descent from Beartooth Pass into Cooke City provides a longitudinal view of the Heart Mountain detachment, the largest terrestrial landslide on Earth. (See *Roadside Geology of Yellowstone Country* for the section of US 212 between Beartooth Pass and Cooke City.)

South of Laurel, US 212 follows the lower Clarks Fork River valley across gently west-dipping Cretaceous sedimentary rocks, many of which were deposited in the Western Interior Seaway. The seaway here teemed with life, including fish that left behind fossil teeth, bones, and scales in the Mowry Shale, exposed east (left) of the river.

West of US 212 is the White Horse Bench, a gravel-covered terrace at the confluence of the Yellowstone and Clarks Fork Yellowstone Rivers. The bench is a remnant

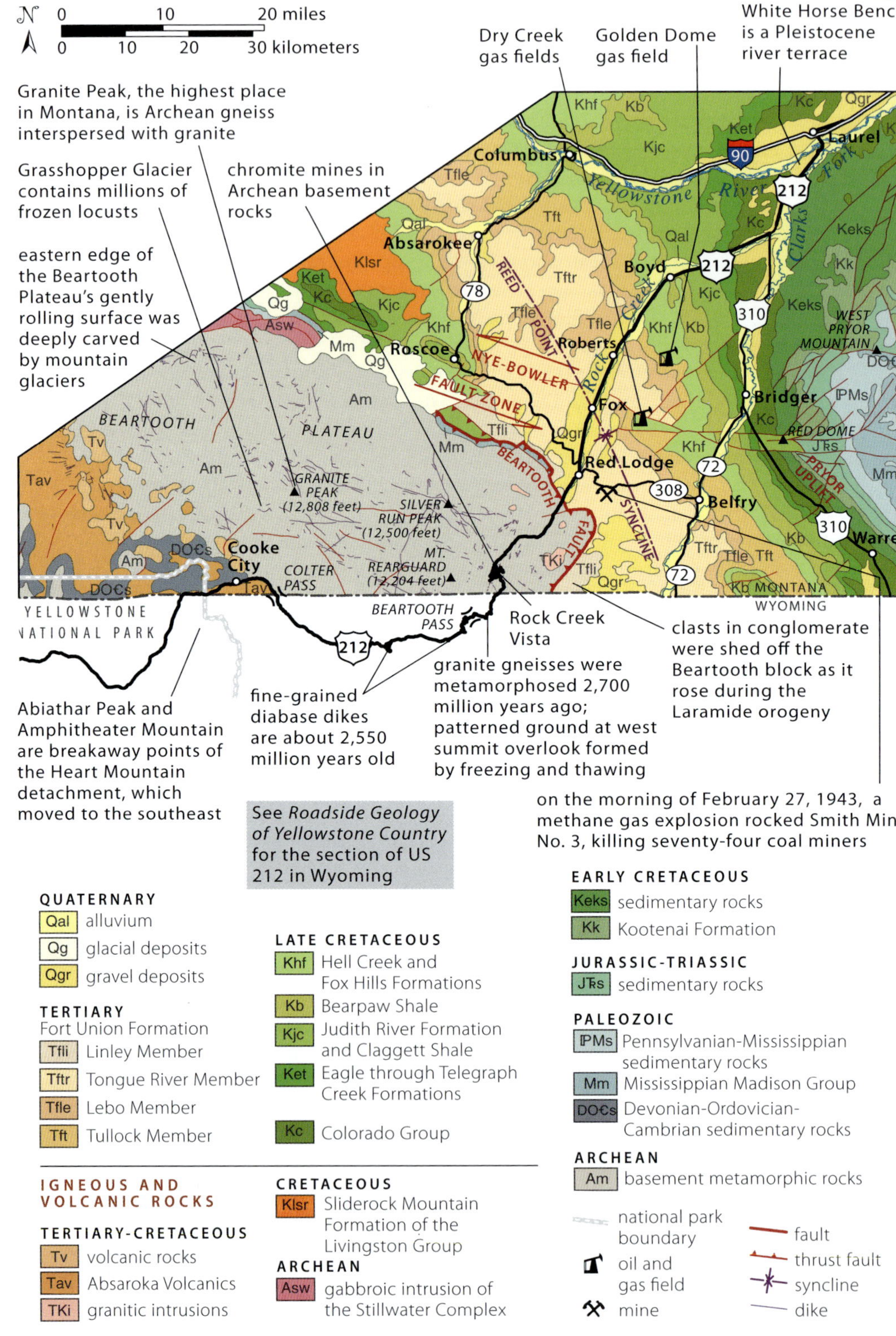

Geology along US 212 between Laurel and the Wyoming border on the Beartooth Plateau.

RED DOME AND THE PRYOR MOUNTAINS

Thirteen miles southwest of Laurel, US 310 heads south from US 212 and follows the broad valley of the Clarks Fork Yellowstone River between the Beartooth Plateau and the Pryor Mountains, through the north end of the Bighorn Basin. The latter is a 4-mile-thick, oil-rich sedimentary basin that extends into Montana from northwestern Wyoming. To the south as far as Fromberg, the subdued topography is eroded brownish-gray shales and some muddy sandstones of the Late Cretaceous Niobrara Formation. Almost flat lying, they dip less than 5 degrees west, so the formations become younger to the west. East of the highway dark-gray to light-brownish-gray Belle Fourche Shale, which is slightly older and also dips less than 5 degrees west. The sedimentary rocks are draped over the Pryor uplift, a chunk of Archean basement that was raised during the Laramide orogeny along west-dipping, high-angle compressional faults called reverse faults. Mississippian Madison Group limestone draped over the raised block tilts gently to the west and is raised almost 2,000 feet on the east flank of the mountains.

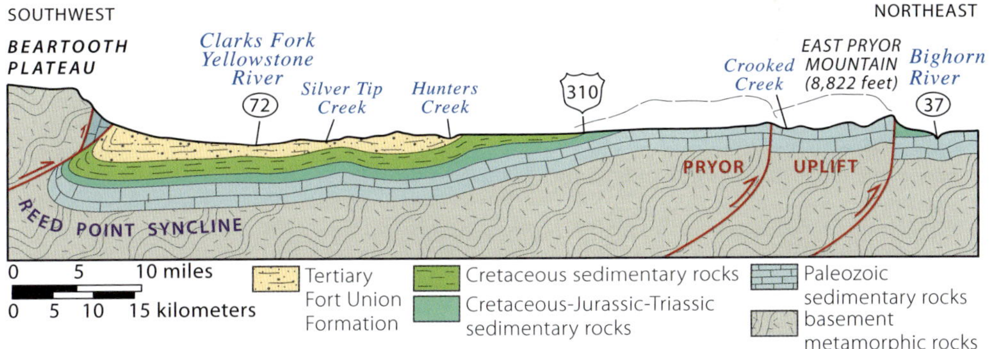

Southwest-northeast cross section from the Beartooth Plateau to the Pryor Mountains, showing the west-dipping faults that raised the Archean basement during compression related to the Laramide orogeny. **—Modified from Lopez, 2001**

Several large ice caves in the Madison Group limestone attract large numbers of visitors, although access via Pryor Mountain Road, which heads east from US 310 about 2.5 miles south of Bridger, requires a high-clearance vehicle. These caves, which fill with ice, open vertically to the surface. Air can only enter them by sinking, and that happens only when the air outside is colder, and therefore denser, than that inside. So air sinks into the caves on very cold days and stays cold all year. The Forest Service regulates visiting hours to the caves to conserve the ice.

Pryor Mountain Road passes through a strange complex of east-west faults and associated folds that brought Jurassic and even Triassic rocks to the surface at Red Dome. The faults are part of the Nye-Bowler fault zone, a northwest-trending group of structures. Pryor Mountain Road passes near Red Dome about 6 miles southeast of US 310, but you also can see the small but conspicuous hill from US 310. The Chugwater Formation's reddish-brown sandstone, siltstone, and mudstone are so typical of sedimentary rocks that formed in Triassic time that many geologists simply call them Triassic red beds. These sediments were probably deposited by rivers and in short-lived lakes on their floodplains. They contain fossils of star crinoids, oysters, snails, clams, and ammonites.

During the 1960s, some important and interesting dinosaurs called *Deinonychus* were found in Early Cretaceous rocks near Red Dome. At the time, the popular conception of dinosaurs was of giant, plodding, cold-blooded reptiles that laid eggs

Red Dome, seen here from Pryor Mountain Road southeast of Bridger, exposes the Triassic Chugwater Formation.

and let their young fend for themselves. Studies of this dinosaur ignited a debate about whether dinosaurs were warm- or cold-blooded animals, leading to a dinosaur renaissance in the paleontology world. Through the detailed study of many aspects of their bones, paleontologists such as John Ostrom at Yale University showed that *Deinonychus* was more like a big, nonflying bird than a lizard. He noted the small body size, posture, and spine similar to those of flightless birds; the flexible tail; and the enlarged raptor-like claws on the feet—all of which suggested an active, agile predator.

of an older (Pleistocene) floodplain left behind as the river cut downward to create new floodplains at lower elevations, perhaps due to uplift of the land surface, changes in water volume due to climate change, or a combination of these.

About 12 miles south of Laurel, Rock Creek is flanked by Pleistocene gravel terraces derived from meltwater of large glaciers that covered the Beartooth Plateau and flowed down the Rock Creek canyon. East of Red Lodge the horizontal terrace gravel rests on south-dipping beds of sandstone of the Paleocene Fort Union Formation (Tongue River Member). This originally horizontal sandstone was deformed, tilted, and eroded during early Tertiary Laramide deformation, long before the terrace gravels were laid on top.

North of Red Lodge is the Nye-Bowler fault zone, a potentially active west-northwest-trending zone of northeast-trending folds and left-lateral faults. Beneath this fault zone is a major zone of shearing in the deep Archean crust. This fault zone has been intermittently active for a very long time.

In 1866, rich coal deposits were discovered near Red Lodge; coal shipments began in 1889 with the construction of a rail line. Settlers from all over the world quickly moved into land that had belonged to the Crow Nation, and by 1892 the population of Red Lodge had reached 1,180. It peaked in the early twentieth century at 5,000 and then plummeted with the development of strip mines at Colstrip in 1924. With the closing of many mines during the Great Depression, coal mining faded away, but tourism associated with the opening of the Beartooth Highway in 1936 resurrected the town. The coal occurs in the Tongue River Member of the Fort Union Formation, deposited in a swamp in a foreland basin that accumulated abundant plant material in Paleocene time. This gigantic downward bend in the Earth's crust was the result of crustal thickening that occurred during the plate collision that formed the Rocky Mountains at that time. The Fort Union Formation and underlying older rocks are all gently folded north of the north-facing front of the Beartooth Mountains, forming a broad syncline called the Reed Point syncline at the north end of the Bighorn Basin.

Mining coal underground is a dangerous proposition, as was made all too clear on the morning of February 27, 1943, at the Smith Mine, just east of Red Lodge along MT 308. A methane gas explosion rocked Smith Mine No. 3, killing seventy-four miners. The explosion was so large that it knocked a 20-ton locomotive off its tracks a quarter mile from the blast. It remains the worst coal mine disaster in Montana.

About 4 miles south of Red Lodge, US 212 enters the Beartooth Mountains at Point of Rocks, the outcrop of gray Madison Group limestone that looks a bit like a thumbs-up on the right (west) side of the highway. Its overturned layers are draped over the Beartooth uplift, a massive block of Archean crystalline rock that was raised, along with its cover of Paleozoic and Mesozoic sedimentary rocks, northeastward along steep compressional faults between 65 and 57 million years ago, during the Laramide orogeny.

The Beartooth uplift eroded as it rose, depositing sediments of the Paleocene Fort Union Formation in the foreland basin to the northeast. Most of the Paleozoic and Mesozoic rocks were eroded off the top of the Beartooth Mountains, but those remaining along the front of the range are tilted on edge, forming walls, or palisades, of sedimentary rocks. The erosion, or unroofing history, of the Beartooth uplift is recorded in the Fort Union Formation. As with deconstructing a building from the top down, the sediment pile that accumulated in the foreland basin consists of debris from the youngest formations at the bottom and the oldest at the top. Because the Fort Union is folded, the Laramide compression must have extended into Paleocene time in this part of Montana. In a few places on the plateau, such as at Beartooth Butte, in Wyoming, down-faulting preserved remnants of some of the rocks above the basement.

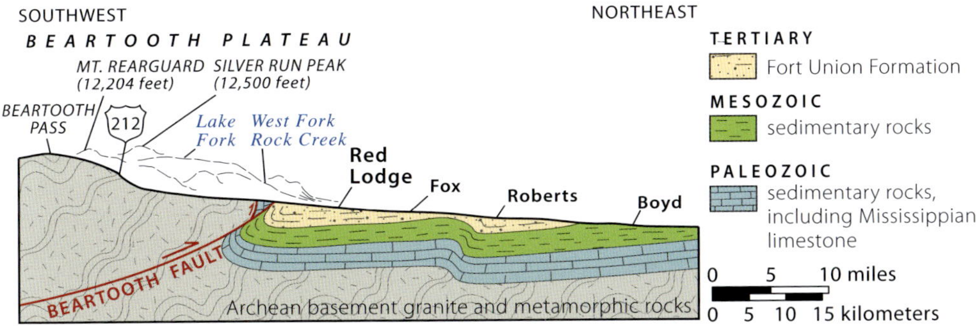

Cross section of the Beartooth Mountains front near Red Lodge showing the Archean basement and steeply tilted Paleozoic sedimentary rocks (Beartooth uplift) thrust faulted by the Beartooth fault over upturned, younger sedimentary rocks.

Palisades of Ordovician Bighorn Dolomite on the Beartooth front at North Fork Grove Creek, southeast of Red Lodge on Meeteetsee Trail Road. The horizontal sedimentary rocks in the foreground to the left are Cambrian in age.

We know that earthquakes occurred as the Beartooth uplift was actively rising in Paleocene time because the Fort Union Formation contains paleoseismites, sedimentary structures that form when water-saturated sediments are shaken and disrupted. Contorted beds and sand dikes are the most common, the latter forming when liquefied sand is injected into overlying sediments during earthquakes. It's estimated that between 4.5 and 7.5 miles of uplift occurred on the Beartooth fault when it was active. By Oligocene and Miocene time, weathering and erosion outpaced uplift, and much of the Beartooth uplift was worn down. In Miocene time, beginning around 20 to 17 million years ago, basin and range extension and crustal thinning began raising the Beartooth uplift once again—by as much as 2.5 miles. Over the last few million years, the Beartooth Mountains have moved very close to the Yellowstone hot spot due to the slow, relentless movement of the North American plate to the southwest; the heat of the hot spot causes the crust to rise like a loaf of bread, raising the Beartooth Mountains even more.

Most of the rock south of Red Lodge along the Beartooth Highway is incredibly old Archean granitic gneiss with intermittent bands of mafic (dark) gneiss, originally basalt dikes that were metamorphosed. Most of the pale rocks along the highway appear to be an ancient granite batholith that was later metamorphosed and deformed during a series of collisional mountain-building events. These events formed a part of the old core of the North American continent (Laurentia) between 3.5 and 2.5 billion years ago. Metamorphosed sandstone from the Hellroaring Plateau, west of US 212, contains zircon crystals eroded from even older rocks that date to nearly 4 billion years, making the source rocks some of the oldest rocks in North America. The old

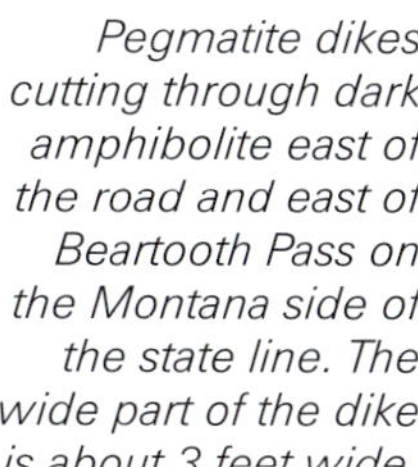

Pegmatite dikes cutting through dark amphibolite east of the road and east of Beartooth Pass on the Montana side of the state line. The wide part of the dike is about 3 feet wide.

metamorphic rocks are intruded by dark- and light-colored dikes of various ages, some of which are nearly as old as the rocks they intrude. East of Beartooth Pass, dark amphibolite (formerly basalt) east of the road was intruded by pale pegmatite dikes with very large crystals that crystallized from very water-rich magma.

The Beartooth Highway is a natural textbook on glacial landforms and deposits. About 12 miles south of Red Lodge, the road crosses a large terminal moraine composed of boulders of crystalline Archean rocks; it marks where a glacier from the Beartooth Plateau stopped and deposited its load. The glacier in the valley was part of a much larger ice sheet of the Pinedale glaciation that entirely covered the Beartooth Plateau with more than 3,000 feet of ice between 30,000 and 12,000 years ago.

Evidence for glaciation is striking from the Rock Creek Vista, about 20 miles from Red Lodge, near the top of the switchbacks; it provides a great view of the U-shaped profile of Rock Creek valley and the flat surface of the Beartooth Plateau. Rock Creek valley likely had a stream-eroded V-shape profile prior to glaciation, but it was widened and deepened by glacial scouring. Talus fallen from the valley sidewalls since the ice melted accentuates the curvature of the profile.

The flat surface of the plateau is a curious feature. Alpine glaciation usually erodes rolling, stream-dissected land into jagged topography with pointed peaks and sharp-crested, serrated ridges. So, what happened here? The most logical answer is that this flat, hard surface is difficult to erode. The plateau began developing in Archean time with the erosion of mountains that had formed from plate collisions. A few billion years later the mountains were gone, and the land was a flat surface of hard, crystalline rocks. During Cambrian time, around 520 million years ago, oceans flooded this surface, burying and preserving it under thousands of feet of sediment. In Cretaceous time, that surface rose due to Laramide compression, and much of the sedimentary cover was eroded away. Renewed uplift during Miocene time resulted in additional erosion, yet the flat surface on the hard Archean crystalline rocks persisted. When ice started building up on the high-elevation flat plateau in Pleistocene time, the low gradient and the hard rocks made it difficult for the ice to dig in and erode the plateau. Glaciers flowing off the edges of the plateau left behind bowl-shaped cirques and U-shaped valleys.

The nearly flat upper surface of the Beartooth plateau is interrupted by the deep, U-shaped Rock Creek valley (left center).

A few glaciers remain in the high elevations of the Beartooth Plateau, all of which formed during a temporary period of cooling called the Little Ice Age, from around AD 1300 to 1870. The most notable, Grasshopper Glacier, may actually be a snowfield because it may not be thick enough to move—a defining attribute of a glacier. The "grasshoppers" the glacier is named for were actually a species of migratory locust that's been extinct since the late 1800s; they were entombed by winter snow about three hundred years ago.

US 287
Three Forks—West Yellowstone
127 miles

US 287 south of Three Forks follows the route of Lewis and Clark for a dozen miles along the Jefferson River. Folded and thrust-faulted Cambrian sedimentary rocks are well exposed in the Cretaceous Lombard thrust sheet on the right (northwest) side of the road. If you look closely, you can find 500-million-year-old fossils of crab-like trilobites, clam-like brachiopods, and traces of wormlike animals that lived in a tropical ocean that covered this part of North America. MT 2 leads west to Lewis and Clark Caverns (see the I-90: Bozeman—Butte road guide).

Downstream (east) of the US 287 bridge over the Jefferson River is a small canyon called Sappington water gap. The river once flowed freely on gravel that covered the folded and faulted Paleozoic rocks, but as the land rose, the river eroded downward and got stuck in the harder Paleozoic rocks, eroding the canyon. Between the bridge and Harrison, US 287 crosses lake and stream sediments of the Eocene-Miocene Renova Formation. In places, such as in the low hills west of Norris, it was

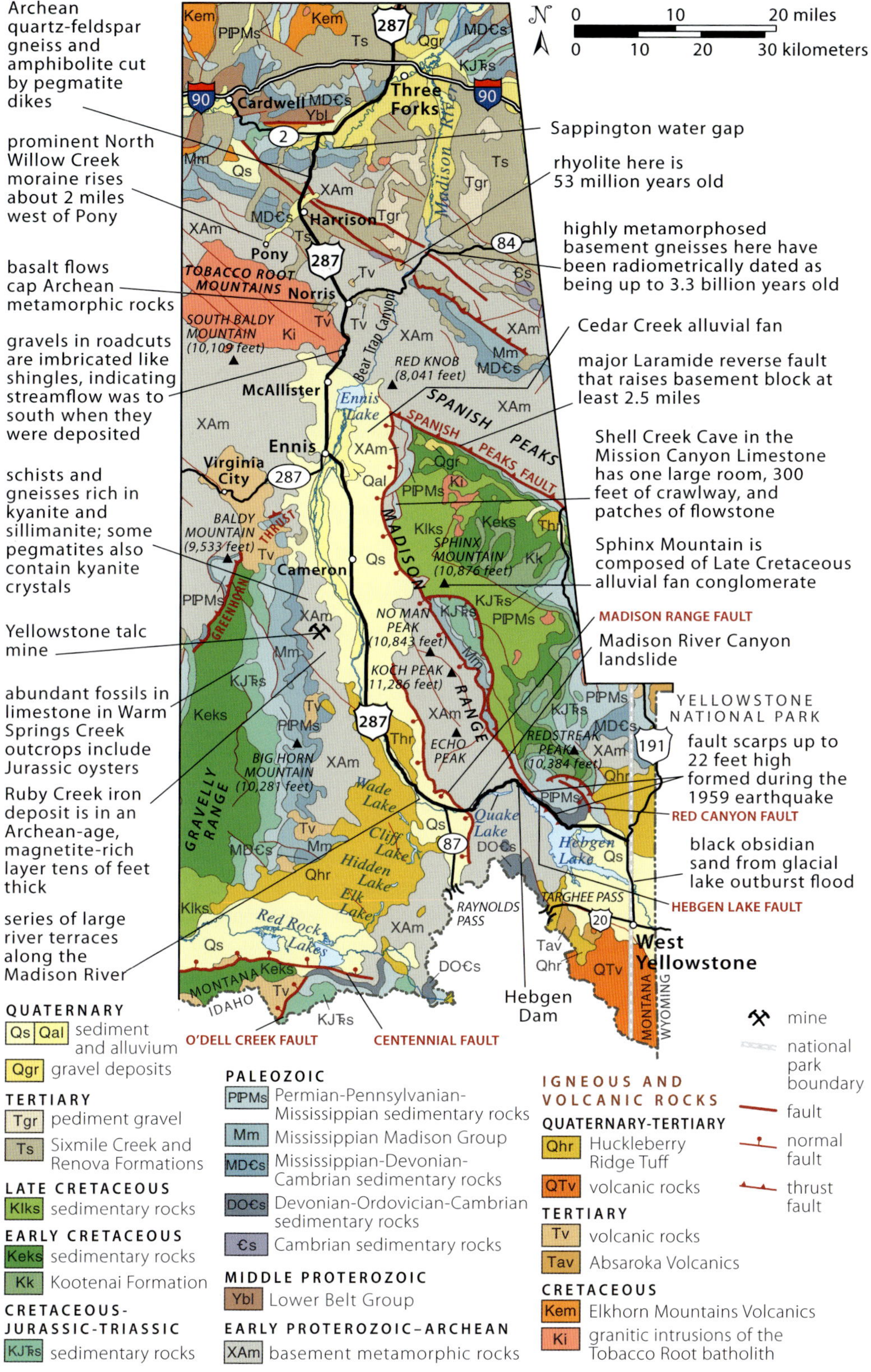

Geology along US 287 between Three Forks and West Yellowstone.

Wormlike Cambrian animals left these trace fossils while feeding in the marine mud of the Wolsey Shale.

deposited directly on Archean metamorphic rocks, a gap in the rock record of at least 2.5 billion years!

All along the western skyline are the spectacular, high, glaciated peaks of the Tobacco Root Mountains, which boast an impressive forty-three peaks above 10,000 feet in elevation. The core of the range consists of Archean metamorphic rocks and granitic rocks of the Tobacco Root batholith, which intruded the metamorphic rocks at a depth of about 10 miles around 75 million years ago. Mineral-rich fluids from the granite formed gold-bearing quartz veins that caught the attention of early prospectors; mining began in the 1870s in the Pony mining district, about 6 miles west of Harrison.

Between 16.5 and 4.5 million years ago, the northeast-flowing ancestral Missouri River deposited gravel and ash of the Sixmile Creek Formation on either side of the mountains. In order for the river to have flowed through here, the mountains could not have existed. So the northwest-trending range we see today probably began forming around 4.5 million years ago. The Sixmile Creek Formation can be seen in the bluffs around Three Forks and near Madison Buffalo Jump State Park.

The modern mountains of southwest Montana are forming because the crust is rising and being pulled apart. As the crust extends, steep normal faults drop the valleys relative to the mountains; this is called basin and range extension. The valleys drop about twice as much as the mountains rise, but the whole landscape is rising because the crust thins as it's pulled apart and the hot mantle rises. River terraces of former floodplain surfaces have been left behind, in part because the rise causes the rivers to erode downward relatively rapidly. The active normal faults provide conduits for hot water to reach the surface as hot springs, such as those at Potosi Hot Springs, near Pony, and Norris Hot Springs.

About 3 miles south of Norris and on the east side of the road are dark Archean metamorphic rocks riddled with intrusions of pale pegmatite, a granitic rock with big crystals (larger than 0.4 inch in diameter). The metamorphic rocks display foliation, thin layering caused by heat and shearing that align minerals as they're deformed and segregate them into light and dark compositional bands. This gneiss was likely metamorphosed by multiple stages of plate collision that buried the rocks to depths of as much as 25 miles during Archean and early Proterozoic time.

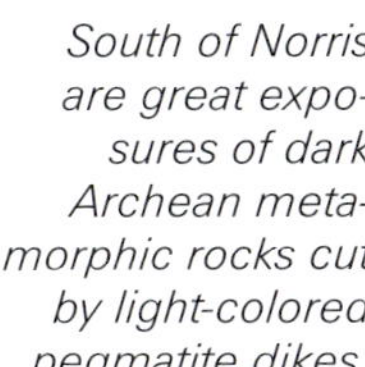

South of Norris are great exposures of dark Archean metamorphic rocks cut by light-colored pegmatite dikes.

Close-up view of the Cretaceous Tobacco Root granodiorite that intruded Archean metamorphic rocks about 3.5 miles south of Norris.

The pegmatite cuts across the metamorphic foliation, so it must be younger than the metamorphism. However, it may not be much younger, because during high-grade metamorphism high temperatures often melt the rock that's being deformed to produce granitic magma. Note that the intrusions are always lighter-colored rocks composed mostly of white feldspar and glassy pale-gray quartz. This is because light-colored minerals melt at a lower temperature than dark ones.

About a half mile farther south is an exposure of the Tobacco Root batholith, which intruded the Archean metamorphic rocks. The 75-million-year-old granodiorite of the batholith differs from granite in that it has more plagioclase (sodium-calcium) feldspar than orthoclase (potassium) feldspar. The magma must have crystallized slowly below the surface, because all of the mineral grains are large enough to be visible. The dark crystals are amphibole and biotite. Small, darker rocks within the granodiorite are probably remnants of basalt magma that rose into the crust and melted the basement rocks to form the granodiorite.

A great place to see Archean metamorphic rocks is in Bear Trap Canyon between the dam that impounds Ennis Lake and MT 84. You can reach the upstream end of the canyon by taking the road east from McAllister. You can also see the rocks at the Madison River boat launch (5.5 miles east of US 287) on MT 84, near the downstream end of the canyon. The rocks are 3.3 billion years old, three-quarters the age of the

Earth itself. Look for banded gneiss and layers of dark rocks loaded with black amphibole. Many of the larger white grains are plagioclase feldspar crystals that have been stretched into eye-shaped features called augen. The foliation, or mineral layering, was folded and contorted into some crazy shapes because the hot rock was very soft.

Unlike the lazy river in the Madison Valley upstream, the Madison River is a torrent in Bear Trap Canyon, making it a popular place for kayaking and rafting. A few million years ago the river flowed on gravel that covered the Archean rocks but got trapped in the hard rock and cut the canyon as basin and range extension raised the Madison Range.

Ennis lies at the MT 287 junction, which crosses a divide between the Tobacco Root Mountains to the north and the Gravelly Range to the south. The Gravelly Range is named for Late Cretaceous conglomerate (gravel) eroded from mountains built during Cretaceous folding and thrust faulting. The conglomerate was raised and exposed along the crest of the range by basin and range extension, starting around 17 million years ago, likely enhanced by crustal bulging associated with the Yellowstone hot spot. Based on topographic offset seen in the Huckleberry Ridge Tuff, a ground-hugging pyroclastic flow erupted from the Yellowstone volcano 2.1 million years ago, there has been as much as 3,600 feet of offset on normal faults since that time! The Gravelly Range is likely still rising, and those faults pose an earthquake hazard to the region.

Low hills west of US 287 south of Ennis are Archean metamorphic rocks, including dolomitic marble that was hydrothermally altered to talc. Like other talc deposits in the area, those in the Gravelly Range contain no asbestos. The purity of the talc enables area mines to be competitive despite the high shipping costs to reach markets. The Yellowstone Mine 18 miles south of Ennis is the largest talc mine in the United States.

Along the eastern skyline, the Madison Range rises abruptly from the Madison Valley floor, also due to active extensional faulting. Much of the range consists of Archean metamorphic rocks that were shoved over Paleozoic and Mesozoic rocks along reverse faults during Cretaceous time. The most prominent of these faults is

Black Butte in the Gravelly Range is a basaltic volcanic vent that was active 23 million years ago. The butte is visible from US 287 and can be reached via Gravelly Range Road.
—Courtesy of Debbie Hanneman

Stream terraces along the Madison River south of Ennis.

the Spanish Peaks fault, a classic thick-skin Laramide structure that trends northwest across the range. Signs of extensional uplift, including fault scarps, steep alluvial fans, and terraces, are everywhere in the Madison Valley south of Ennis. Prominent stream terraces rise from the Madison River like giant steps. They are remnants of former channels and floodplains of the Madison River; the highest terraces mark the oldest positions of the river. The terraces formed, in part, as the valley rose under the river, forcing it to cut downward. We tend to think of mountains rising and valleys dropping, but the terraces suggest that the Madison Valley is rising as well, just less so than the mountains. The MT 287 junction in Ennis lies at the edge of a terrace 50 feet above the present floodplain of the river.

The Cedar Creek alluvial fan, east of Ennis at the base of the Madison Range, is world famous. Alluvial fans form where mountain streams exit steep, narrow canyons onto broader gently sloping valley sides and fan out to deposit sediment as the water slows. As the Madison Range continues to rise, the Cedar Creek fan tilts and steepens, causing its streams to cut into the fan surface. As a result, the old fan surface no longer

The Cedar Creek alluvial fan looks like a textbook alluvial fan, but uplift has caused its streams to erode gullies in the fan, indicating that it is now inactive and that water no longer spreads out across the fan. View is eastward across the Madison Valley.

floods, but new fans are forming at the bottom of the Cedar Creek fan where the new channels exit the fan.

Between Ennis and Cameron, the most eye-catching peaks in the Madison Range are Sphinx Mountain and its companion, the Helmet. These glacially sculpted peaks are composed of the Sphinx Conglomerate: river gravels, sands, and silts deposited on alluvial fans at the base of mountains that rose to the west during Cretaceous time. Large slabs of crustal rocks were shoved eastward over the active alluvial fans, folding and faulting the deposits. The flat, ramping surface at the top of the Helmet is the exhumed plane of a thrust fault that cut through the conglomerate.

Approximately 27 miles south of Ennis, turn right (west) onto the road to the Palisades Campground to see a prominent northwest-trending normal fault on the west

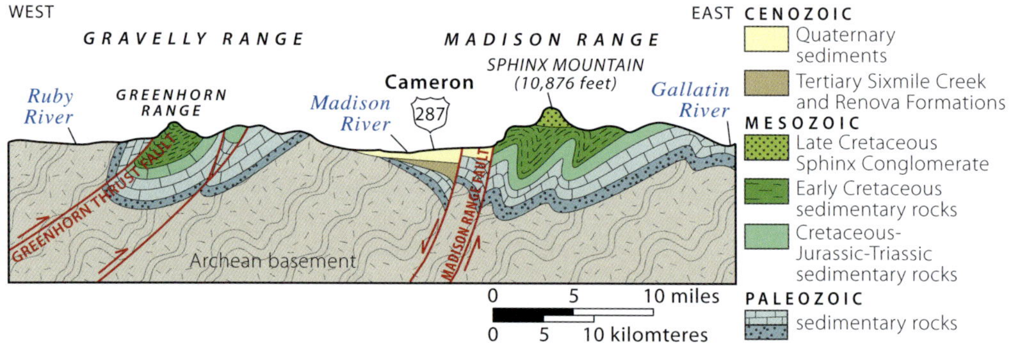

West-east cross section south of Cameron.

Sphinx Mountain (right) and the Helmet (left) are alluvial fan deposits shed off rising mountains during Cretaceous time. Inset shows the Sphinx Conglomerate, the alluvial fan deposit.
—Inset courtesy of Rod Benson, Bigskywalker.com

side of the Madison River. The fault is very young, cutting the Huckleberry Ridge Tuff, which was a ground-hugging pyroclastic flow that erupted from the Yellowstone volcano about 2.1 million years ago, when the volcano was about 50 miles to the southwest. This fault was likely caused by changes in extensional stress directions in the crust due to the slow migration of the North American plate over the Yellowstone hot spot.

To see the Huckleberry Ridge Tuff up close, turn right (west) onto Sundance Bench Road about 7.5 miles south of the Palisades turnoff, cross the river, and follow the dirt road south less than 2 miles to the large, fallen blocks of tuff. The tuff is composed of crystals and pumice in a matrix of glass, all of which were flattened and fused together because the tuff was hot and molten when deposited. The Yellowstone volcano erupted a column of superhot gas, silica-rich ash, crystals, and rock fragments as high as 100,000 feet into the atmosphere. When this eruption column collapsed, the pyroclastic flow, a surge of hot gas followed by a dense mixture of tephra (rock fragments and ash up to 750 degrees Fahrenheit) and gas, rushed down the Madison drainage at speeds up to 450 miles per hour, finally solidifying as this deposit.

US 287 leaves the wide Madison Valley and enters the narrow, steep-sided Madison River Canyon about 3 miles east of the turnoff for MT 87 to Ashton, Idaho. The abrupt change in topography happens where the road crosses the Madison Range fault, an extensional fault that lifts the range and drops the valley. This fault ruptured south of here in 1959, and older fault scarps from earlier ruptures can still be seen on the hillside to the northeast. Fault scarps don't always form during earthquakes, but they're common during large-magnitude earthquakes that originate at relatively shallow depths of less than 12 miles. Active faulting here occurs within the Intermountain Seismic Belt, an area of active extension from southern Nevada and Utah through western Montana. Earthquake activity is particularly intense near the Yellowstone hot spot because heated crust bulges, stretches, and cracks to create valleys and ranges.

Over the next 20 miles a series of interpretive signs tell the story of the Earthquake Lake Geologic Area, the site of the largest earthquake ever recorded in the Rocky Mountains. At 11:37 p.m. on Monday, August 17, 1959, an earthquake measuring 7.5 on the Richter scale jarred southwest Montana. It killed at least twenty-eight people, many of whom were buried under the massive rockslide that created Earthquake Lake. Felt as far away as Seattle, it changed the behavior of geysers and other geothermal features in Yellowstone National Park.

The Earthquake Lake Visitor Center, at the west end of the Madison River Canyon, is located atop debris from the slide, the scar of which looms to the south across the valley. This slide was a disaster waiting to happen. The upper part of the mountain is schist, a foliated metamorphic rock with platy minerals that here dip downslope toward the canyon bottom. A dolomitic marble buttress held up the unstable rocks, but river erosion had undermined that support. Earthquake shaking shattered the buttress, reduced friction between surfaces in the schist, and triggered sudden slope failure.

When the mountainside collapsed into the canyon, it compressed the air ahead of it, creating hurricane-force winds that sent 2-ton cars sailing through the air and blew people away. The slide moved as an intact mass at over 100 miles per hour on a cushion of air, riding 400 feet up the opposite canyon wall and burying nineteen people in the Rock Creek Campground. It all happened in less than 30 seconds, leaving behind 37 million cubic yards of slide debris to a depth of 225 feet. Gigantic boulders of dolomitic marble carried to the top of the slide mass remain as prominent

The landslide scar and debris of the Madison slide are visible from the Earthquake Lake Overlook. Dead trees in Earthquake Lake remain from the forest that was drowned as water rose behind the slide-debris dam. View to the west.

reminders of the force of the event. A memorial plaque with the names of the known victims is attached to one of the boulders.

The slide debris blocked the river to create a lake that drowned several people. Others escaped to Refuge Point, near Beaver Creek. Fear of a catastrophic failure of the landslide dam and the possible undermining of Hebgen Dam upstream prompted the US Army Corps of Engineers to cut a spillway through the landslide in October of 1959. The small trees along the south shore of the lake show how high the water reached before the spillway was completed.

The Madison Range, Red Canyon, and Hebgen Lake faults all ruptured the ground surface on the evening of August 17. The Hebgen Lake fault is well exposed as a 21-foot scarp at Cabin Creek, about 3 miles east of the visitor center. A trench dug across the scarp by the US Geological Survey showed that the fault had ruptured before. Over the last 10,000 to 14,000 years, the Hebgen Lake fault has ruptured three times. The last rupture before August 17, 1959, occurred between 1,000 and 3,000 years ago and produced about 10 feet of offset; some of the scarp we see today is likely inherited from this older event.

Hebgen Dam, about 1 mile east of Cabin Creek, dropped 10 feet during the earthquake. The concrete core and spillway cracked but did not fail. The entire Hebgen Lake basin instantly tilted down about 20 feet to the northeast into the Hebgen Lake and Red Canyon faults, drowning the northeast shore of the lake and leaving the southwest shore high and dry. Sudden movement of lake water produced a standing wave that swashed back and forth, topping Hebgen Dam three times with up to 4 feet of water. Partially submerged buildings can still be observed at the Building Destruction Site on Hebgen Lake, about 1 mile southeast of the dam.

The Red Canyon fault scarp is northeast of the lake on Red Canyon Road about 7.6 miles from the damaged buildings. The scarp has begun to fade but is easy to find

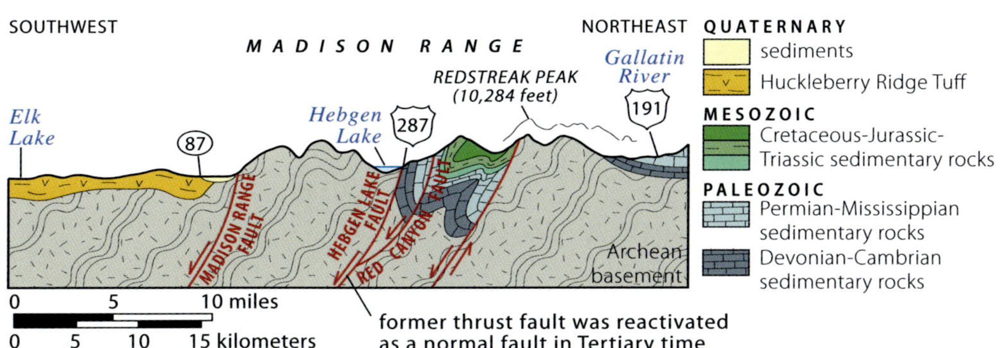

Cross section from Elk Lake to the Gallatin River through Hebgen Lake.

about a half mile by trail from the parking area for the Red Canyon Creek trailhead. The fault plane dips toward the lake (southwest) at an angle of 45 to 50 degrees, and seismic data show that the fault merges with the Hebgen Lake fault at a depth of about 5 miles. Two large earthquakes were produced on the evening of August 17, 1959, on this fault at depths of 6 to 9 miles, causing shaking that lasted between 30 and 40 seconds for each quake. Many aftershocks, smaller earthquakes that occur after the main shock, rocked the area, with four measuring more than 6.0 on the Richter scale. It must have been a terrifying night for people in the area. (For a more in-depth discussion of the earthquake, see *Roadside Geology of Yellowstone Country*.)

MT 1 (ANACONDA-PINTLER SCENIC HIGHWAY)
Drummond—Anaconda
56 miles

Drummond is at the north end of the Flint Creek valley, where Flint Creek joins the Clark Fork River. High peaks of the Flint Creek Range are Late Cretaceous granite plutons that intrude tightly crumpled and thrust-faulted sedimentary rocks ranging in age from Proterozoic Belt mudstones to Paleozoic limestones, sandstones, and shales to Mesozoic sandstones and shales. Early Tertiary sediments of the Renova Formation were deposited on the lower slopes on the eroded, steeply folded edges of the older rocks and were eroded into flat pediments before Pleistocene time. Low terraces along the Clark Fork River and Flint Creek are Pleistocene in age and younger.

About 12 miles south of Drummond are big outcrops of Paleozoic sedimentary rocks, including the distinctive white quartzite of the Pennsylvanian Quadrant Formation. For about 6 miles south of the small community of Maxville, MT 1 passes more Paleozoic and Mesozoic sedimentary rocks, including weathered gray to buff limestone and sandstone. About 1 mile north of Maxville, along both sides of the road, big rounded boulders of granite rest on a low, grassy rise. These came down Boulder Creek in the Flint Creek Range during the last ice age as part of a big mudflow. The mudflow passed through the present location of Maxville and on north down Flint Creek, parallel to the highway. The mudflow boulders resist erosion better than other stream deposits, so the mudflow deposit stands slightly higher.

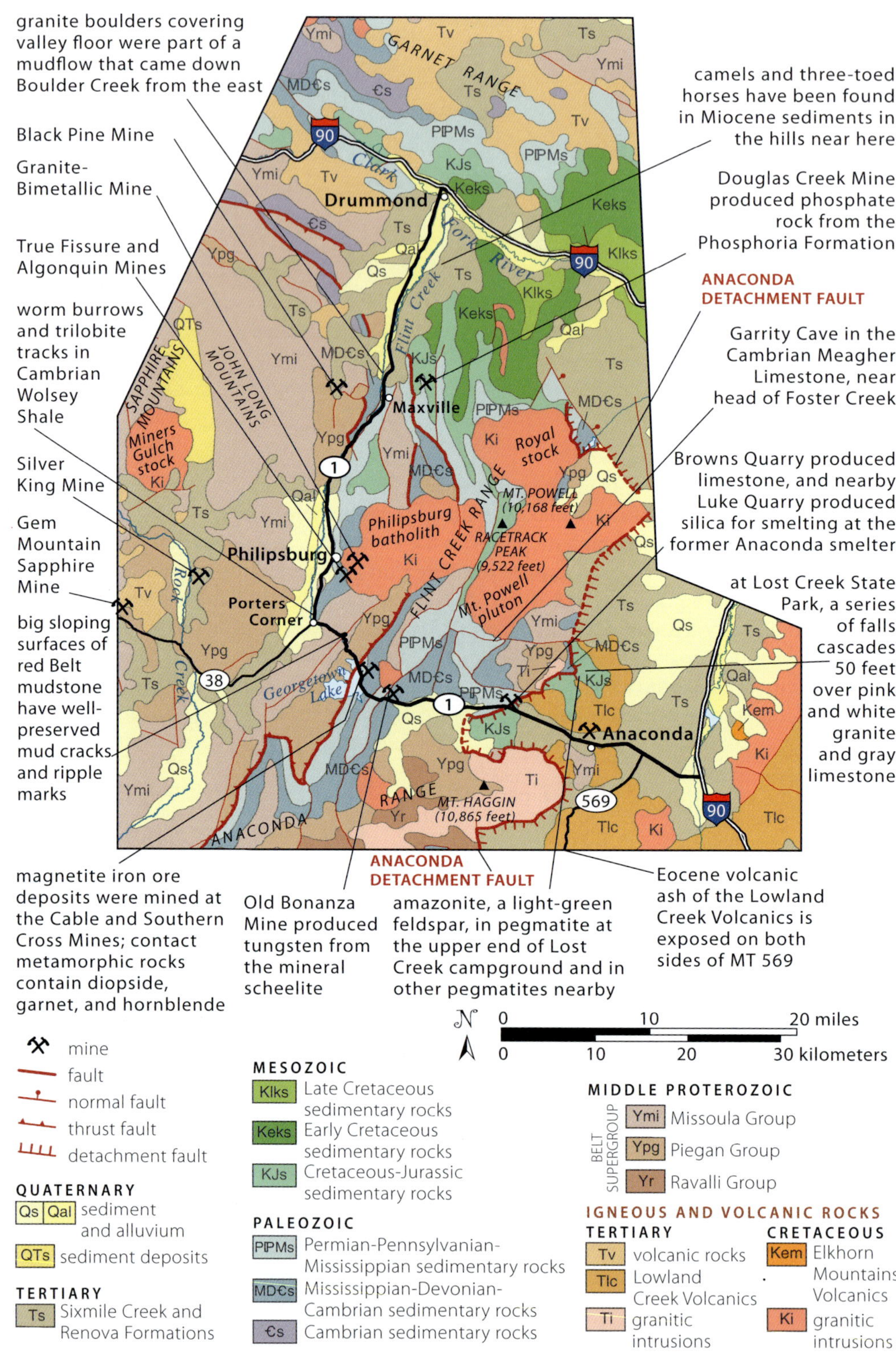

Geology along MT 1 between Drummond and Anaconda.

The Sapphire Mountains, Belt mudstones and sandstones, moved eastward off the Bitterroot dome along the Montana-Idaho border about 80 to 70 million years ago, in latest Cretaceous time, bulldozing the Paleozoic and Mesozoic sedimentary rocks of the Flint Creek Range into their steep, tight folds. The Sapphire block moved eastward on a deep mylonite zone that dives under the Bitterroot Valley south of Missoula; it splays upward into thrust faults in Belt mudstones and sandstones in the Sapphire Mountains and in the western Flint Creek Range.

Granite in the northeastern Idaho batholith appears to have been partly molten at the time of intense shearing that formed the mylonite and carried the Sapphires to the east off the batholith. Some 20,000 feet of the metamorphosed roof rocks, along with an unknown amount of the uppermost batholith, moved east in the Sapphire block. Partly molten granite under the Sapphire Mountains rose along thrust faults in a few places in the central part of the range, and also in the Flint Creek Range at the leading edge of the moving slab. (See cross section for I-90 on page 171.)

West of Maxville is the former Black Pine Mine. Its mineralized quartz veins, discovered in 1885, are in Proterozoic Missoula Group quartzite and red shale. The production, most of which was before 1893, then sporadically until 1985, consisted of 2,135,000 ounces of silver and 1,400 ounces of gold from underground mines. A couple of head frames above mine shafts remain, along with remains of the mill and building foundations. So much mercury was used to leach valuable metals from the ore in the early years that the toxic waste at this site was included in recent remediation of ASARCO's East Helena smelter.

Philipsburg was registered as a town in 1867, in the same time period as Granite and other nearby ghost towns. The only real difference is that P-burg, as it's affectionately known to locals, is still home to 820 people. The Granite-Bimetallic Mine at Granite ghost town lies about 7 miles east of Philipsburg via a rough track mainly suited to four-wheel drives with adequate clearance. The mine produced more than $50 million in gold and silver between 1882 and 1905 from the 72-million-year-old Philipsburg batholith. The remains of the mine, including a stamp mill constructed

The remains of the stamp mill at Granite ghost town.

in 1887, are remarkably well preserved, including the remnants of dozens of stamp machines that crushed the rock with steel-sheathed posts to facilitate the separation of ore minerals. A few wood buildings remain; unfortunately, many burned in a fire in 1958. Several houses and the remains of the Bimetallic mill, built of local granite, with smokestacks and chimneys, are preserved at Kirkville, a much smaller ghost town just southeast of Philipsburg.

Montana is home to several sapphire deposits. Except for Yogo Gulch near Lewistown, most are placer deposits in stream gravels. The best known is the Gem Mountain mine, on the Skalkaho Road southwest of Philipsburg. The sapphires occur in old stream gravels, but they were probably eroded from Eocene-age dikes. Large sapphires are highly valued as gemstones, and small grains and poor-quality material are used as an abrasive because of the mineral's hardness; sapphire is 9 on the hardness scale, next to the hardest of all minerals, diamond, which is 10. Both sapphire (ideally blue) and ruby (red) are the same mineral, otherwise known as corundum, an aluminum oxide.

To get to the Gem Mountain mine from Philipsburg, follow MT 1 south about 7 miles to Porters Corner, and then go west for 16 miles on MT 38 (Skalkaho Road), which is paved as far as the mine. For a modest charge at the Gem Mountain Sapphire Mine, you can pan for sapphires from buckets of gravel dug out of nearby deposits. The mine provides sieves, gold pans, wash water, and sorting tables, and you get to keep any sapphires you find. Though a few gems up to a few carats in size can be found, most are small. They are almost always transparent and very pretty, coming in all colors of the rainbow—blue, green, pink, yellow, and colorless. For a small fee, you can have your finds heat treated to bring out their colors, a process that's particularly effective for yellows, browns, and blues.

Steeply tilted cross-bedded sandstone and red shale of Cambrian age are well exposed east of MT 1 at Porters Corner. A rising sea that transgressed west to east across the continent around 520 million years ago deposited these marine sedimentary rocks. A close look at the bedding surfaces of the red Wolsey Shale can reveal tracks of crab-like trilobites and burrows of worms and other bottom-dwelling animals that thrived in the Cambrian ocean.

A giant roadcut 3 miles east of Porters Corner on a steep highway grade exposes sloping surfaces in red Missoula Group mudstones. These spectacular exposures are

This specimen of the Wolsey Shale reveals trilobite tracks with the scratch marks their many legs made as the animals plowed through the mud. The various tube-shaped fossils are burrows made by worms feeding in the mud.

full of perfectly preserved mud cracks and ripple marks that look like they just dried up yesterday—but "yesterday" was nearly 1.5 billion years ago! The mud cracks formed in two stages. The smaller polygons with thin cracks penetrated to shallow depth first; the larger ones formed later when larger cracks penetrated deeper into the drying mud. The symmetrical ripple marks formed when wind gently blew across shallow standing water, creating currents that moved sediment back and forth, building evenly spaced, symmetrical wave ripples. The asymmetrical ripple marks formed when directional flow from currents caused the sediment to hop along the bottom, building ripples with steep sides that slope in the direction of flow. If you run your finger over the ripples, you might be able to figure out which process formed the ripples.

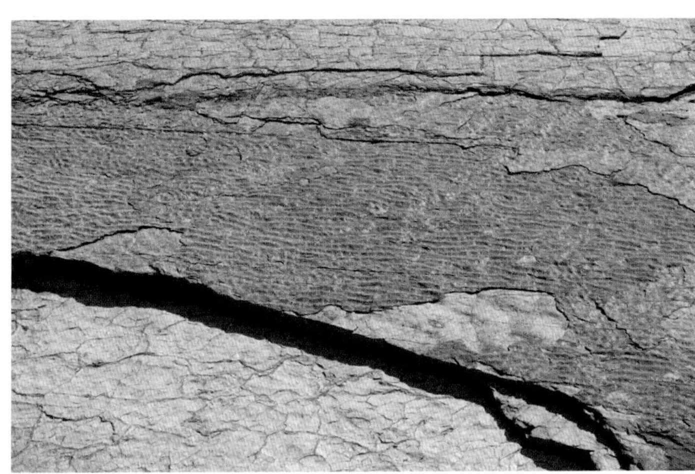

Ripple marks in Missoula Group bedding surfaces in the Flint Creek canyon near milepost 28.5 on MT 1. Note the mud cracks at the top and lower left.

Georgetown Lake is a reservoir created in 1885 by the damming of the North Fork Flint Creek to provide electric power for Philipsburg and area mines. Twenty years later it provided power to the smelter in Anaconda, as well as water for the smelting process.

East of the lake, placer deposits discovered in 1866 led to the discovery of gold vein deposits, which led to the building of the Cable Mine. The mine had its ups and downs, but by 1911 its total gold production was estimated to be between $3 and $4 million. The last exploration activity was in 1947, and the mill burned down in the 1960s. The deposits formed due to contact metamorphism between the Cambrian Meagher Limestone and a nearby granite. Ore minerals that formed include tetradymite, an unusual bismuth-tellurium sulfide, and pyrite, along with gold.

Mountain glaciers were widespread in the Flint Creek Range, north of MT 1, and the Anaconda Range, to the south. The larger of these glaciers reached almost to the valleys of the Clark Fork River on the east and Flint Creek on the west. Between Georgetown Lake and Anaconda, MT 1 follows a glacial valley in which mountain glaciers of these ranges merged. Watch for lumpy topography littered with boulders left by the melting ice.

Big roadcuts along the 12-mile stretch east of Georgetown Lake to 5 miles west of Anaconda expose Paleozoic limestones. The Anaconda smelter used limestone from the very rusty quarry a few miles west of Anaconda in the smelting process. Although not well exposed along the road, the eastern edge of these Paleozoic limestones marks

The Paleozoic limestone in this roadcut just east of Georgetown Lake was tilted by the same deformation that crumpled most of the rocks in the Flint Creek Range.

the gently east-dipping Anaconda detachment zone that carried roof rocks of the Flint Creek Range at least 15 miles eastward, taking the big Boulder batholith with them. (See the I-90: Butte—Missoula road guide for more information.)

An original 1883 smelter in Anaconda, the Old Works, lasted only about 3 years before it had to be replaced by a new, bigger smelter to treat 15,000 tons of Butte ore per day. Its 300-foot smokestack, completed in 1903, was intended to disperse the noxious fumes and to placate people living nearby. When the pollution lawsuits continued, the company erected the even larger stack, which is hard to miss as you approach Anaconda from any direction. It stands alone on a cluster of low hills just south of the highway. The 585-foot-tall stack is all that remains of the huge smelter that treated the copper ores from Butte from 1918, when it was built, until 1981, when the smelter closed. The inside diameter of the enormous steel-reinforced stack is 75 feet at the base and 60 feet at the top. At 58 stories, it's big enough to enclose the Washington Monument.

Like most industrial smokestacks, this one was built very high to better disperse hazardous fumes emitted from the mill. **Nonetheless, high concentrations of lead, copper, cadmium, arsenic, antimony, and zinc in the ore were released in the stack gases, which were hazardous to residents of Anaconda and beyond. The smelter area became a Superfund site in 1983, and the smelter buildings were dismantled, millions of cubic yards of waste were removed, almost 8 square miles of the site and surrounding contaminated ground were capped and revegetated, groundwater wells were tested and remediated, and more than 5 miles of stream channel were restored.**

The Old Works Golf Course, designed by Jack Nicklaus, was built on capped waste from the smelter, and ground-up smelter slag was used in place of sand in the sand traps. A huge pile of black smelter slag was dumped next to MT 1 just east of town. Dust from the slag pile contains dangerous levels of arsenic and lead. US Minerals reprocesses the slag into iron-silicate roofing granules and abrasives.

MT 55 and MT 41
Whitehall—Dillon
55 miles

The drive from Whitehall to Dillon follows the path taken by the Lewis and Clark Expedition during the summer of 1805. Tailings from the Golden Sunlight Mine can be seen on the skyline northeast of Whitehall. (See the I-90: Bozeman—Butte road guide for a discussion of the mine.)

A few miles south of Whitehall are good views to the east of Parrot Bench, a gently sloping pediment surface extending from the Tobacco Root Mountains into the Jefferson River valley. Sheetlike flows during intense rainfall events eroded and formed the pediment in a desert climate between 5 and 2.5 million years ago, during Pliocene time. Parrot Bench is mostly eroded into the Renova Formation here, an Eocene to early Miocene stream and lake deposit loaded with fossils of terrestrial animals, plants, and insects. These rocks were first described from outcrops in the former town of Renova, just west of the bench.

The Jefferson River valley is a graben that formed due to extension of the crust along high-angle normal faults. It initially formed along *northeast*-trending basin and range normal faults between 16.5 and 4.5 million years ago, filling with sediments of the Sixmile Creek Formation, some of which were derived from as far away as the Snake River Plain in Idaho. Then, around 4.5 million years ago, changes in extensional stresses related to the migration of the North American plate over the Yellowstone hot spot caused *northwest*-trending normal faults to develop; these cut the older northeast-trending structures and are now modifying the landscape into a complex combination of these two episodes of extension.

Large triangular-shaped rock faces called facets, running in a northeast direction, mark the front of the northern end of the Tobacco Root Mountains. The facets are the eroded remnants of a fault plane from an active north-northwest-trending normal fault that has ruptured the ground in geologically recent times. To the south, several northwest-trending normal faults cut Quaternary deposits in the Jefferson River

Triangular rock facets at the north end of the Tobacco Root Mountains illustrate that recent fault movements occurred on this north-northwest-trending fault. The bases of the triangular facets are at the bottom of the ridge. The fault slip surface of one triangle is outlined in white.

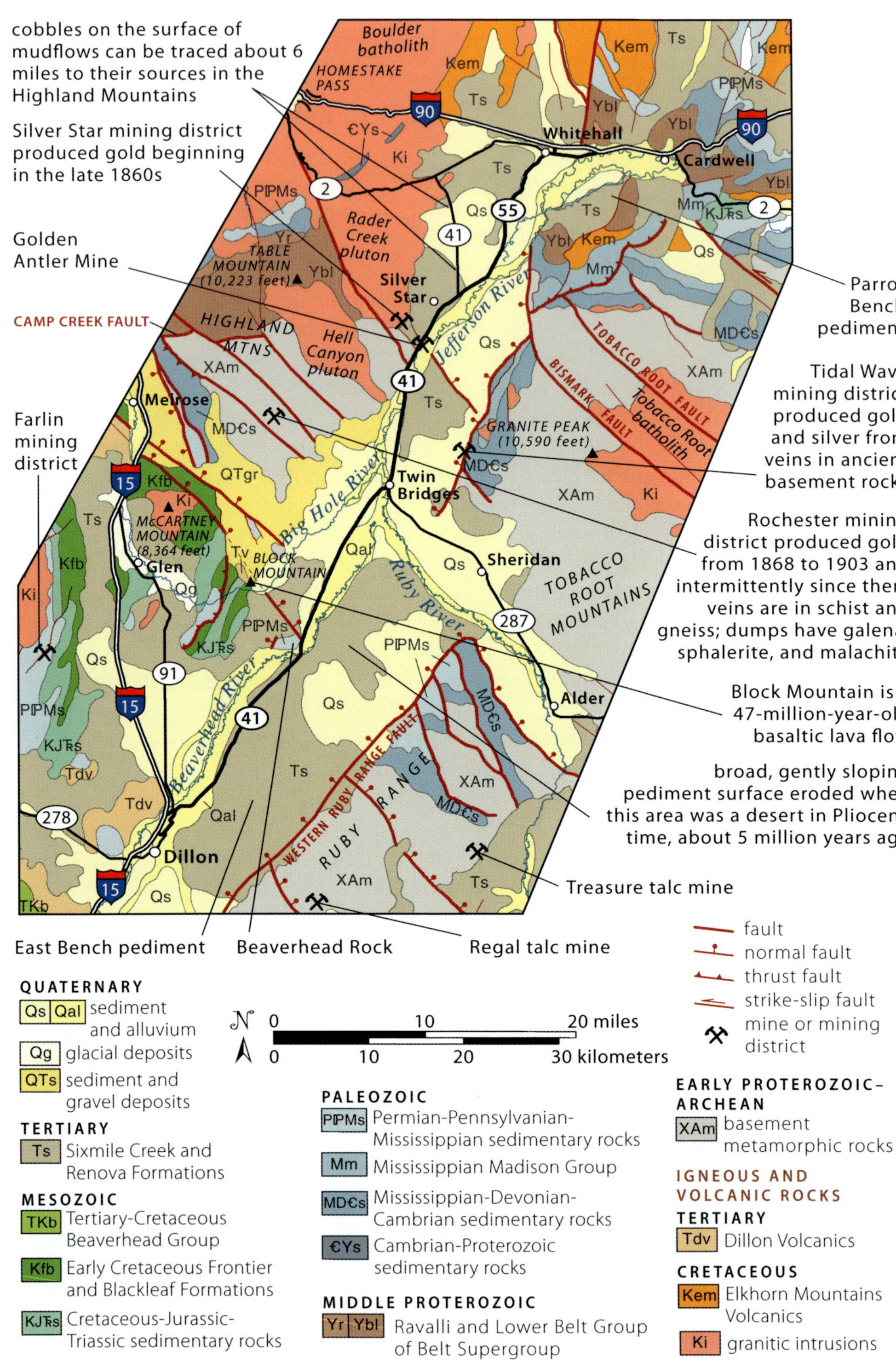

Geology along MT 55 and MT 41 between Whitehall and Dillon.

valley, and at least one fault south of Dry Boulder Creek ruptured the ground and left a fault scarp, a visible step in the topography. Seismic data show that the northwest-trending normal faults in this range are the ones that have produced earthquakes in historic times, meaning they are the active faults in the region.

The Tobacco Root (east) and Highland Mountains (west) on either side of the Jefferson River valley expose Archean through Mesozoic rocks that were intruded by Late Cretaceous granitic rocks. The Boulder batholith, which extends from the Highland Mountains to Helena, was intruded in three phases, from 80.7 to 73.7 million years ago, with each phase becoming richer in light-colored feldspar and quartz. The Rader Creek granodiorite, a more mafic (darker) pluton high in iron and magnesium-rich minerals, was intruded around 80.7 million years ago and is exposed along the southeastern flank of the Highland Mountains as far south as Silver Star. The Tobacco Root batholith is of similar age and composition and thought to be related to the Boulder batholith. To see the darker plutons of the Boulder batholith, take a side trip via MT 41, MT 2, and Pipestone Pass. The turnoff for this beautiful drive through the mountains is MT 41 (west toward Butte), about 12.5 miles south of Whitehall.

Paleozoic and Mesozoic sedimentary rocks once covered the Archean and early Proterozoic crystalline rocks and the Late Cretaceous igneous rocks exposed in the core of the Tobacco Root and Highland Mountains, but now they mostly occur along the flanks of these ranges, in places forming palisades, or fins, of steeply dipping light-colored sedimentary rocks.

Along the lower flanks of the Tobacco Root Mountains and resting on the Pliocene pediment surface are alluvial fan sediments that melting glaciers deposited during the wetter climate of Pleistocene time. Gravels from the fans gave Lewis and Clark no end of trouble as they traveled up "Jefferson's River." According to Lewis, the gravel caused the river to shoal, forcing them to drag their dugouts along the shore, where they frequently fell on the slippery stones, making their feet "tender and soar."

Silver Star, about 35 miles south of Whitehall, was once a rip-roaring mining town that developed around the gold, silver, lead, and copper that had formed in the contact zone between Paleozoic limestone and magma of the Rader Creek pluton. Hot magma in contact with the sedimentary rocks baked and soaked them with hot fluids to form new minerals. The large wheels on the west side of town are rope-drive compressor wheels from the Leonard Mine in Butte. They were used from 1903 to 1964 to make compressed air, which was piped to the mines to operate drills and power hoists. To see the granitic rocks of the 74-million-year-old Hell Canyon pluton of the Boulder batholith, turn west just north of the MT 41 bridge over the Jefferson River about 4 miles south of Silver Star, and head up Hells Canyon.

A mile down the road from Silver Star is an exposure of chloritic gneiss that was mined from 1977 to 1999 at the Golden Antler open pit mine, up the hill to the west. Chlorite, a mineral similar to talc, was mined for use as a paint filler and extender. The hydrothermal alteration of Archean gneiss formed the deposit. From Twin Bridges the flat-topped Table Mountain (10,233 feet) is visible 16 miles to the northwest. Because the surface is so flat, it was probably eroded by streams before the range was raised. A thin veneer of bouldery gravel resting on top of the flat surface on nearby East Peak may have been left behind by the streams that eroded the flat surface.

About 12 miles south of Twin Bridges, Beaverhead Rock comes into view. On August 8, 1805, Meriwether Lewis recorded in his journal that their Shoshone guide

View of Beaverhead Rock to the north from MT 41.

and interpreter, Sacagawea, recognized this prominent rock as the "beaver's head," a landmark she remembered from childhood as being close to the place where her people crossed the Continental Divide. Beaverhead Rock is a prominent exposure of gently west-dipping Mission Canyon Limestone of the Madison Group that rises above the Beaverhead River floodplain. The outcrop is the easternmost exposure of a stack of folded and thrust-faulted rocks deformed during the Sevier orogeny. The near-isolated nature of the exposure is, in part, due to northwest-trending extensional faults that cut through the east end of the stack of rocks, dropping them down to the northeast of the faults by approximately 4,000 feet. The limestone is filled with small caves and holes dissolved by slightly acidic groundwater. Many are filled with beautiful crystals of calcite. Please remember that Beaverhead Rock is a state park, so collecting minerals is prohibited.

Just south of Beaverhead Rock, the Beaverhead Valley opens up to provide a stunning view of the high Pioneer Mountains to the west. The valley likely began forming around 16.5 million years ago as a graben with streams that drained to the northeast off the Yellowstone thermal bulge in Idaho. Water-deposited ash beds in the Sixmile Creek Formation can be traced to their sources in southern Idaho, where they formed during mega-eruptions of the Yellowstone volcano. About 9 miles south of Beaverhead Rock, the ash is exposed east of the highway in a cut through a pediment that forms the East Bench along the flanks of the Ruby Range. If you have a hand lens or other magnifier, you can see that the sparkling ash is made up of hollow particles of glass. They likely formed as gas expanded in the sticky magma of the Yellowstone volcano, creating vesicles, or tiny holes. When the magma explosively erupted, it shattered the foamy magma, which then rapidly cooled and hardened to glass in the air, preserving the vesicles. The delicate particles were then transported down the graben as a thick slurry of ash and water (lahar), protecting the glass from breaking into smaller pieces.

On August 10, 1805, Lewis commented on what is now the Beaverhead River in his journal. He wrote, "I do not believe that the world can furnish an example of a river running to the extent which the Missouri and Jefferson's [Beaverhead] rivers do through such a mountainous country and at the same time so navigable as they are." Lewis was describing basin and range extensional topography, which had produced rugged mountains surrounding broad, low-gradient valleys with placid rivers. Much

Water-deposited ash in the Miocene Sixmile Creek Formation exposed in a cut in the East Bench pediment about 9 miles south of Beaverhead Rock.

to his dismay, steep topography with narrow canyons and rapid rivers west of the Great Divide would forever eliminate the notion of a navigable Northwest Passage.

The Ruby Range, south of Twin Bridges, is essentially the southern extension of the Tobacco Root Mountains, without the Late Cretaceous intrusion. The ranges are separated by an active, northwest-trending graben followed by the Ruby River. The oldest rocks in the range are Archean, and some may be as old as 3.3 billion years. Gneiss, schist, marble, quartzite, metaconglomerate, pods of peridotite, and banded iron formation are just a few of the common rock types that can be found by taking side trips into the range, especially east on Stone Creek Road about 6 miles north of Dillon, and on Sweetwater Road directly east of Dillon. These rocks are loaded with interesting minerals, such as corundum, graphite, magnetite, talc, and garnet. The talc is mined at the Treasure Mine in the Stone Creek drainage and the Regal Mine on Sweetwater Road. Ruby-colored garnets, some the size of softballs, are so abundant in many of these rocks that the mountains were named the Ruby Range.

MT 43
Divide—Lost Trail Pass
81 miles

The community of Divide is seated where the Big Hole River emerges from a northwest-trending canyon cut across the Pioneer Mountains. The river turns sharply south to follow the Divide valley (followed by I-15), a Tertiary extensional graben that preserves its fill of the Renova and Sixmile Creek Formations. These sediments are exposed on the north side of MT 43 just west of I-15, where they are tilted to the southeast into the active extensional Rocker fault. South of Divide, the Big Hole River was diverted to the west by extensional uplift, causing it to cut through folded and thrust-faulted Paleozoic and Mesozoic rocks to form the spectacular Maiden Rock Canyon. Gravel and sand deposits exposed along the frontage road east of the Big Hole River mark the older pathway of the river.

On the south (left) side of MT 43 about 3.5 miles west of I-15 is a prominent dark-and-light-striped outcrop of near-vertical rocks. These fine-grained sedimentary rocks

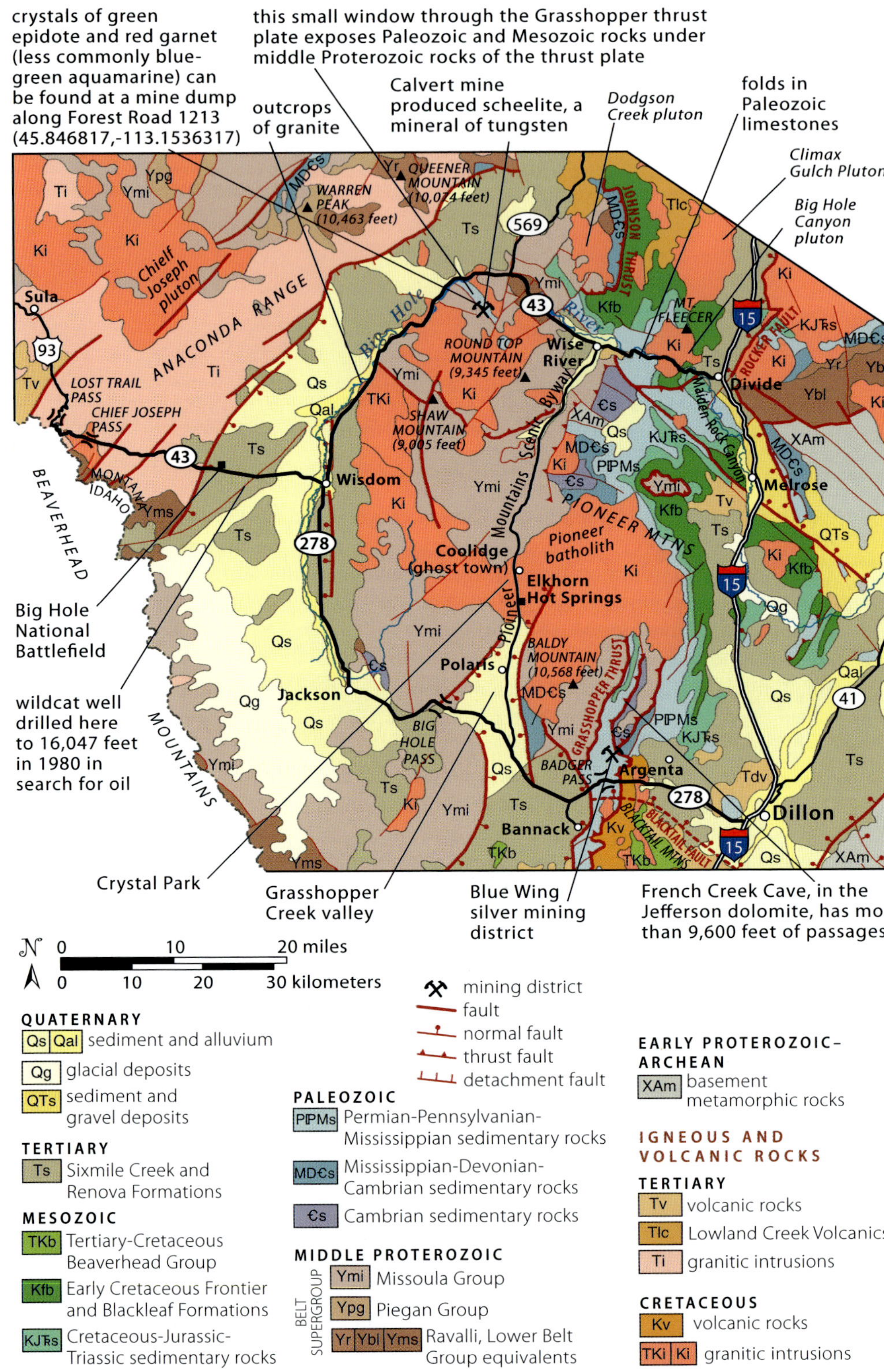

Geology along MT 43 between Divide and Lost Trail Pass.

The Renova and Sixmile Creek Formations are tilted to the southeast into the Rocker fault on the east side of the Divide valley graben in a roadcut on the north side of MT 43, just west of the I-15 exit to Divide. Mt. Fleecer looms in the background.

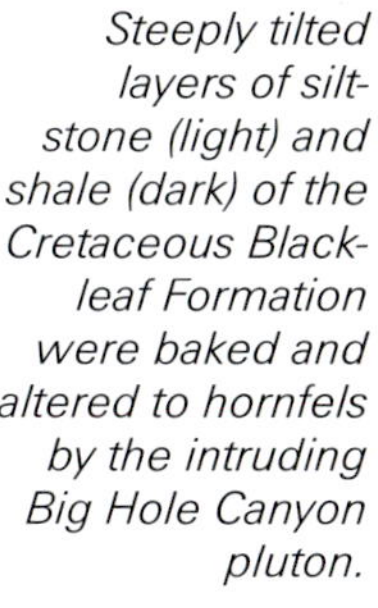

Steeply tilted layers of siltstone (light) and shale (dark) of the Cretaceous Blackleaf Formation were baked and altered to hornfels by the intruding Big Hole Canyon pluton.

of the Cretaceous Blackleaf Formation were metamorphosed when hot magma of the Big Hole Canyon pluton intruded them during Cretaceous time. The heat turned layers of siltstone and claystone into hornfels, a very hard, fine-grained metamorphic rock that sounds and looks a bit like a broken porcelain plate when struck with a rock hammer. Many of the broken pieces show conchoidal fractures that are typical of hornfels.

The Big Hole Canyon pluton can be seen in contact with the Blackleaf Formation across the river at the fishing access off Pumphouse Road. Turn right (north) on

Pumphouse Road, cross the bridge, and park in the gravel parking area on your left (west). Scramble up the rocks north of the parking area and walk west (left) on the abandoned railroad grade to a tunnel carved through the granitic rock. The narrow-gauge railroad was completed in 1919 to provide access to the then-booming mining camp of Coolidge in the Pioneer Mountains. The tunnel exposes granodiorite of the pluton, a granitic rock with a lot of black hornblende and biotite crystals. The hornblende crystals are rectangular, the biotites hexagonal. Their well-formed shapes show that these minerals crystallized early, while they were still surrounded by liquid magma. The granodiorite is riddled with angular mafic (dark) inclusions of older basalts that were broken up by the magma. This pluton formed during Cretaceous time as the oceanic Farallon plate was being subducted beneath the North American plate, the same tectonic event that formed the other plutons of the Boulder batholith.

Big Hole Canyon provides great exposures of the Big Hole Canyon pluton, but there are very few safe places to pull off. Look for a large pullout on the north side of the road about 2 miles west of the Pumphouse Road fishing access. The evenly spaced, near-vertical cracks in the granitic rock north of the river are joints that formed when the granodiorite either contracted as it crystallized or was later pulled apart by tectonic stresses. Across the highway are large inclusions of black basalt within the granodiorite. Some of the fragments have sharp, very angular edges, suggesting they were solid and brittle when broken up and engulfed by the magma. Looking downstream from here it seems logical that the west-to-east pathway of the Big Hole River was likely established before the mountains were raised in its path. The uplift history of the Pioneer Mountains is complex, but some geologists think that streams drained to the southeast off highlands west of the Pioneer Mountains long before the range existed, and that the Big Hole River cut down through the rising mountains starting around 17 million years ago.

Large inclusions of black basalt within the granodiorite of the Big Hole Canyon pluton.

In the hills south of the road, just west of the George Grant fishing access, is the contact between the Big Hole Canyon pluton and the Mississippian Mission Canyon Limestone. Most of the cliff-forming rocks on both sides of the road from here to Wise River are intensely folded and thrust-faulted Paleozoic and Mesozoic sedimentary rocks. These rocks were deformed by the same Late Cretaceous tectonic plate collision that produced the magma for the plutons.

For the adventurous, Quartz Hill Road, east of Wise River, heads south and provides access to the East Pioneers, where compressional faults and overturned anticlines and synclines are all well exposed. North of Wise River and visible on Granulated Mountain, these rocks are structurally buried beneath the Johnson fault, which is the basal thrust fault, or décollement, of the huge Grasshopper plate. Slightly metamorphosed sedimentary rocks of the Proterozoic Belt Supergroup were shoved tens of miles eastward along the fault over Paleozoic and Mesozoic sedimentary rocks, which are exposed at river level. Granitic rocks of the Dodgson Creek pluton intruded the thrust fault at about the same time the Big Hole Canyon pluton intruded rocks to the east. On the north side of the river, numerous caverns in the bright-reddish-orange Paleozoic limestones formed when slightly acidic groundwater dissolved the limestone. Such caverns are important archeological and paleontological sites throughout southwest Montana.

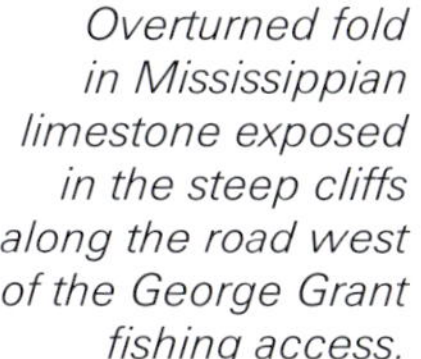

Overturned fold in Mississippian limestone exposed in the steep cliffs along the road west of the George Grant fishing access.

The town of Wise River is located at the northern end of an active extensional graben that marks the boundary between the East and West Pioneer Mountains. The Pioneer Mountains Scenic Byway south from town provides access to alpine scenery and historic mining towns, as well as opportunities to dig for quartz crystals at Crystal Park and soak at Elkhorn Hot Springs. (See the Pioneer Mountains Scenic Byway side trip in the MT 278: Dillon—Wisdom road guide.)

Between Wise River and the turnoff for MT 569 to Anaconda, the highway primarily passes fine-grained, slightly metamorphosed sedimentary rocks of the Proterozoic Belt Supergroup in the Grasshopper plate. The southwestern end of the Anaconda Range (also known as the Pintlers) comes into view at the turnoff for MT 569. The 50-mile-long, northeast-trending range is a raised extensional block that

Karst dissolution features in Mississippian-age limestone near Wise River.

exposes tightly folded sedimentary rocks predominantly of the Belt Supergroup. They were shoved east during the Sevier orogeny along big thrust faults and intruded by granitic rocks of Late Cretaceous and early Tertiary age. The Continental Divide follows the crest of the range and includes numerous peaks rising to well over 10,000 feet in elevation. The high peaks of the Anaconda Range were deeply glaciated by at least two advances of ice during Pleistocene time, leaving behind impressive glacial horns such as Warren Peak, which is carved into carbonate rocks of the Piegan Group of the Belt Supergroup. Many of the stream valleys extending out of the range are filled with glacial outwash or stream gravel and glacial till. The ridges of till show how far the ice extended down into the Big Hole Valley. MT 569 follows a topographic low point of an extensional graben over the northeastern end of the range to Anaconda. It seems illogical that the Big Hole River didn't flow through this gap and into the Clark Fork River on the other side of the Continental Divide, rather than cut its deep canyon. Some geologists speculate that it once did, but the evidence is inconclusive.

The Big Hole Valley is a complex, active extensional graben that forms a crescent around the West Pioneer Mountains. The northern end of the valley dropped along a normal fault on the southeast side of the Anaconda Range. This listric normal fault dips to the southeast and curves to a shallower angle deep beneath the surface, causing the Tertiary Renova Formation valley fill to tilt northwest into the fault. These rocks are well exposed north of MT 43 about 11 miles west of the MT 569 turnoff to Anaconda. Although there has been no major recent activity on this fault, fault scarps in recent stream sediments suggest the fault is active.

It's hard to reconstruct the history of the graben using only surface outcrops, so geoscientists rely on subsurface information from seismic waves and drilling. In the early 1980s, an exploratory oil well penetrated more than 14,000 feet of valley fill—including a thin veneer of glacial and recent stream deposits overlying the Tertiary Renova Formation—before reaching bedrock of the original valley floor. At the base of the well are Challis Volcanics of Eocene age. The valley began forming in Eocene time when all that rock that had piled up during the Late Cretaceous Sevier orogeny

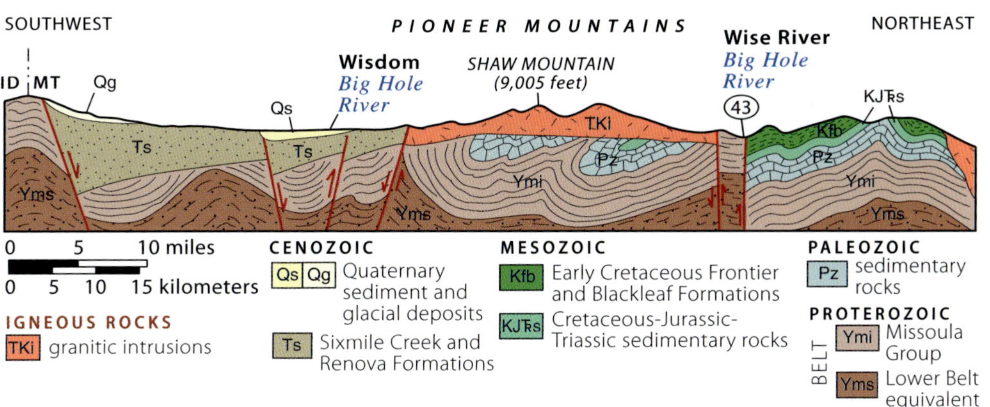

Cross section of the Big Hole Valley and Pioneer Mountains.

began to collapse. Imagine that this tectonic compression is like a giant bulldozer that has pushed up a large pile of dirt. When the bulldozer backs away from the pile, the dirt subsides. As the subduction zone of the Farallon plate shifted westward, the folded and faulted rock pile it had created collapsed to form a north-trending, 60-mile-wide zone of extension, which includes the ancestral Big Hole Valley and many other grabens.

Some of the sediment derived from the surrounding highlands of this mountain-building event didn't just fill the Big Hole Valley but was transported far to the east by streams. Studies of distinctive muscovite-mica-rich granitic sand that was eroded off the Late Cretaceous Chief Joseph pluton near Lost Trail Pass show that streams bypassed the Big Hole graben and the Pioneer Mountains to deposit their sediment load in the Renova Basin, to the east. This suggests that the Pioneer Mountains were not a significant topographic barrier at that time, and that the mountains we see today did not form until much later, probably during the basin and range extension that began regionally around 17 million years ago. The absence of the Sixmile Creek Formation—a widespread volcanic ash and gravel deposit—in the Big Hole Valley indicates that this graben was likely not physically connected to the Yellowstone hot spot, as were the Beaverhead and Ruby grabens to the east.

Between Wisdom and the Big Hole National Battlefield, 10 miles to the west, MT 43 crosses the floodplain of the Big Hole River and then jumps up onto a pediment surface eroded on the Tertiary Renova Formation. Like other pediment surfaces in southwest Montana, it likely formed in the desert climate of Pliocene time when intense rainfall produced sheetlike flows that eroded sloping, flat surfaces, or benches, across the underlying rocks. Much of the bench on both sides of the road is covered by peculiar, low, circular dome-like mounds about 30 feet in diameter. Different vegetation, typically sage, grows on the mounds than in the spaces between them, where grass typically grows, thus accentuating the mounds. These Mima mounds, or prairie mounds, commonly form on relatively flat, poorly drained surfaces, such as those of the alluvial fans of the Big Hole Valley. These curious features have been attributed to everything from burrowing by pocket gophers to earthquake shaking, although none have been observed to form during earthquakes. Many geologists suspect they form as

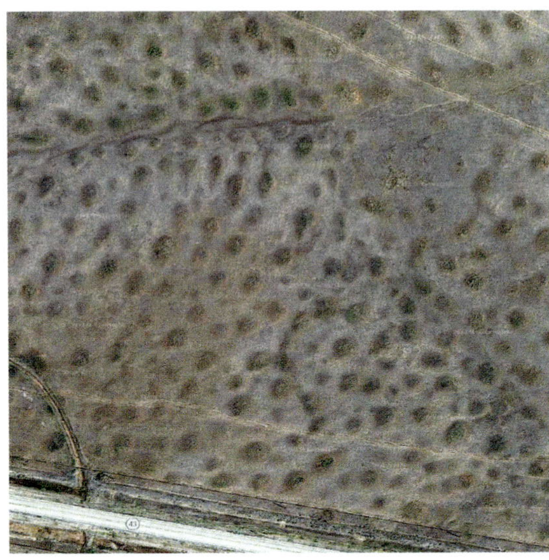

Aerial view of dome-like mounds on the pediment bench near the Big Hole National Battlefield. MT 43 runs across the lower left corner. —Imagery from Google Earth

Typical granitic rock found along MT 43 to Lost Trail Pass.

plants capture windblown sediments that then accumulate around them, or possibly by the repeated freezing and thawing of groundwater in glacial-margin areas.

The Big Hole National Battlefield commemorates a Nez Perce battle. On their way to Canada to avoid being forced onto a reservation in Idaho, Chief Joseph and a group of eight hundred men, women, and children were attacked on the morning of August 9, 1877, by US military forces. It's estimated that ninety Nez Perce and thirty-one US soldiers were killed. The Nez Perce trapped the soldiers on a hillside above the camp, while the women and children buried the dead and fled the battlefield. The hillside is composed of sandstone and conglomerate of the Renova Formation, here preserved on the side of the normal fault that has raised the Anaconda Range and dropped the northern end of the Big Hole Valley.

West of the battlefield MT 43 crosses the big normal fault. It's visible southwest of the highway, where it places younger Tertiary Renova Formation sediments against older Eocene granodiorite, which crystallized miles below the surface. Once you cross over the main range-front fault, Late Cretaceous and Eocene granitic rocks are exposed all the way to Lost Trail Pass. They intrude folded and thrust-faulted metamorphosed sedimentary rocks of the Belt Supergroup. The granitic rocks consist primarily of light-colored granodiorite loaded with black biotite and yellowish-white muscovite mica. The distinctive two-mica granitic sands derived from the erosion of this rock are found in the Renova Formation all across Southwest Montana, and are the sands mentioned earlier that lead geologists to conclude that the Pioneers rose relatively recently.

MT 278
Dillon—Wisdom

65 miles

See map on page 232.

West from Dillon, MT 278 climbs across a huge glacial outwash fan built by braided streams that carried gravel and sand off a melting glacier in the Rattlesnake Creek drainage above the old mining town of Argenta. This Pleistocene-age fan is an important aquifer (a source of groundwater) in the Beaverhead Valley. Based on mapping of glacial erosional features and deposits, it appears that the Pioneer Mountains were encased in a 127-square-mile ice field. The glacial outwash shows that the glaciers didn't extend into the Beaverhead Valley beyond the mountain valleys, as glaciers did in the Flathead Valley of northwest Montana. Instead, the Beaverhead Valley's upland areas were filled with lush grasses, and willow, cottonwood, and birch grew along the river. Roaming the valley were bison, Columbian mammoths, dire wolves, saber-toothed cats, camels, horses, and even humans, who arrived at least by 10,500 years ago.

The light-colored dirt road to the north about 11 miles west of Dillon leads to Argenta. Placer miners discovered this historic mining area in 1862, and it was later mined for lode deposits of lead, silver, and, more recently, gold. Since no smelters existed in Montana in 1862, the silver ore was shipped all the way to Swansea, Wales. The operation barely broke even, so the St. Louis and Montana Mining Company built the first smelter in Montana Territory in Argenta in 1866. Brick charcoal kilns used to make fuel to smelt the ore still survive west of town. At its peak, Argenta boasted a population of 1,500, with three hotels, six saloons, two grocery stores, and a dance hall. Most of the ore formed in a contact metamorphic zone where Cretaceous granitic magma reacted with Mississippian limestone, but some was concentrated in veins that at the surface are up to a mile from the granitic rocks, though unexposed granite beneath the surface may be closer to the veins. Northeast of Argenta in Frying Pan Basin, the Paleozoic and Mesozoic sedimentary rocks that were folded into anticlines and synclines during the Cretaceous Sevier orogeny are used to teach field-mapping techniques to geology students from around the world.

Trees abruptly appear along the skyline near Badger Pass. They are growing in water and soil trapped in cracks in the Pennsylvanian Quadrant Formation, a wind-blown sand deposit from a coastal dune field that existed here 310 million years ago. The Pleistocene outwash gravels and Tertiary sediments east of the pass are too porous to retain water, so only grasses and sage grow on them. The Quadrant and other Paleozoic formations are folded and thrust faulted at Badger Pass due to the compressive forces of the Sevier orogeny. A thrust fault carried these rocks here from the west, pushing them over Cretaceous igneous rocks that were intruded along the fault at shallow depths. The magma erupted onto the surface near the eastern edge of the thrust fault, producing lava flows and fast-moving pyroclastic flows of hot gas and volcanic debris. The volcanic rocks are similar in age to the Elkhorn Mountains Volcanics near Butte but are composed of dacite, a rock similar to andesite and rhyolite but higher in silica (quartz) and lower in potassium. A light-colored pyroclastic flow deposit exposed below the fault serves as a durable material for surfacing dirt roads all over the Dillon area and can be seen in a quarry south of MT 278 just east of the roadcut through the Quadrant Formation.

Seen from the quarry area, the abrupt front of the Blacktail Mountains rises to the southeast. These mountains are rising as a result of northeast-southwest crustal extension on the Blacktail fault, a normal fault that dips to the northeast. The adjacent Blacktail Deer Creek valley is tilted down to the southwest into the fault plane, which is mostly covered by alluvial fan deposits. The abrupt and very steep range front reveals the range's youth; erosional processes have had very little time to eat away at it. Also, stream deposits derived from the Yellowstone hot spot to the south can be found atop the range, and they are less than 4.5 million years old. Based on the elevation change from the valley floor, a minimum of 3,400 feet of offset has occurred on the Blacktail fault in the last 4.5 million years. The actual offset of rocks across the fault must be even greater, but they remain buried under valley fill. The most recent movement on the fault is uncertain, but between 35,000 and 15,000 years ago an earthquake was large enough to produce a 6.5-foot-high scarp. The fault can be traced to the northwest through the Badger Pass area, where the offset is much smaller. The Blacktail fault eventually merges to the west with other normal faults.

Approximately 21 miles from Dillon is the road to Bannack State Park, which preserves a ghost town that was the site of the first major gold discovery in Montana Territory. Founded in the late summer of 1862 when prospectors found gold in Grasshopper Creek, Bannack swelled to a city of more than ten thousand people

The Blacktail Mountains, as seen looking southeast from near Badger Pass. The Blacktail fault (dashed line) dips steeply to left (northeast).

and served as the territorial capital for a time in 1864, before the capital was moved to Virginia City. It was a raucous and dangerous town known for its outlaw sheriff, Henry Plummer, and the vigilantes who hung him without trial. The placer gold diggings not only included what could be panned from the bed and banks of the creek, but what could be washed from the terrace gravels above the creek. It was said that gold could be panned out of the sagebrush, because miners would pull it up, shake the gravel from the roots into a pan, and obtain between 25 cents and $1 (about $18 to $70 per pan today) in small pieces of gold.

A present-day view of Bannack, the first territorial capital of Montana and site of the first major gold discovery in Montana Territory.

From the beginning, the Grasshopper Creek diggings suffered from chronic water shortages. Miners constructed a series of ditches, but they were expensive to dig and didn't adequately solve the problem. In the late 1860s, hydraulic mining was introduced, creating a need for even more water. To meet the need, a 30-mile-long ditch was dug in 1867 from Coyote and Painter Creeks, followed by three more ditches in 1870. By the mid-1870s, most of the placer mining operations had ended, and the population had dwindled to just a few hundred people. Then, in 1895, the first gold dredge in the country was built on Grasshopper Creek. These floating barges had conveyor-belt bucket lines to scoop up gravel, which was run through sluice boxes to separate the gold. Three more dredges followed within the year, including the famous *Maggie A. Gibson*. Bannack prospered until the dredges washed all the gravels they could reach. Left behind are piles of gravel tailings and dredge ponds that today are used for ice-skating in the winter months.

From the beginning, lode (hard-rock) deposits were a part of Bannack's mining operations. The bedrock sources of the gold were primarily in the contact metamorphic zones around two small Cretaceous intrusions of granodiorite. They were located on opposite sides of Grasshopper Creek on the high ridge of Mississippian limestone above Bannack. Large quantities of garnet and small amounts of silver, lead, and copper are also found in the contact zones. Bedrock mining started in 1863 after a stamp mill was built to pound the rock to a powder from which the small particles of gold could be washed out. By 1870 four stamp mills were operating, but so little ore remained by 1914 that a cyanide mill was built to dissolve the gold out of the crushed rock. Total production from Bannack is difficult to know for sure because much of the gold that was mined was never reported, but estimates suggest about $12 million between 1862 and 1930—easily $100 million at today's values.

Baldy Mountain, north of MT 278 in Grasshopper Valley, is a prominent, treeless peak in the Pioneer Mountains. It's shaped like a volcano, but it's actually glacially sculpted Proterozoic quartzite of the Missoula Group. The quartzite is part of the Grasshopper plate, a huge slab of rock that was thrust eastward up and over Paleozoic and Mesozoic rocks during the Sevier orogeny. However, the quartzite's history goes back even further. More than 1.4 billion years ago, braided streams drained off southern highlands composed of Archean crystalline rocks and built large alluvial fans along the southern margin of the vast Belt Sea to the north. Because the Earth was more like Mars at that time, with no vegetation or soil to stabilize sediment, the fans were much larger than modern alluvial fans and had few channels. The cross bedding, formed of layers of migrating ripples, was beautifully preserved due to high sedimentation rates and a lack of multicellular animals or plant roots to churn up the sediments. Near the summit of Baldy, the highly fractured quartzite forms circles of broken rocks. The patterned ground develops when water freezes, expands, and pushes large particles away from smaller ones, sorting them by size. The fine-grained sediment patches make it possible for grasses to get a foothold among the boulders.

View to north across the Grasshopper Valley about 3.7 miles west of the turnoff to Bannack State Park. The red spot on the slope below Baldy Mountain is a laterite paleosol near the base of the Renova Formation.

Baldy Mountain is made up of cross-bedded, slightly metamorphosed quartzite of the Missoula Group (Belt Supergroup). Pocket knife for scale.

Rock circles of sorted quartzite along the trail near the summit of Baldy Mountain.

A prominent red spot in the foothills below Baldy Mountain is an ancient Tertiary soil. During several stretches of Tertiary time, Montana's climate was tropically warm and wet. The soils of some of these rocks were leached of all but the most chemically resistant common elements, such as iron and aluminum, forming rusty-red-orange laterite. Discontinuous layers of this paleosol (ancient soil) occur in Eocene and Miocene deposits throughout southwest Montana, reminding us that the climate was dramatically different at times. We know that Eocene time (early Tertiary) was extraordinarily warm and wet in Montana, because the undersides of many lava flows of this age yield petrified wood from a variety of tropical hardwood tree species.

These tropical conditions may very well be evidence for the globally recognized Paleocene-Eocene thermal maximum, a period of global warming when average temperatures were as much as 14 degrees Fahrenheit warmer than today. The warming was related to a spike in atmospheric carbon attributed to everything from volcanism, burning peat, methane released from the ocean floor, or changes in ocean circulation to even a comet impact. Whatever the cause, the thermal maximum has become a focal point of considerable research because it likely provides the best analog by which to understand the impacts of global climate warming and the massive amount of carbon we are releasing today.

West of the turnoff for the Pioneer Mountains Scenic Byway, MT 278 climbs over Big Hole Pass, also known as Carroll Hill. Several intersecting normal faults in Missoula Group rocks produced this topographic low point in the West Pioneers. From the pass is a breathtaking view to the west of the Beaverhead Mountains rising above the Big Hole Valley. The mountains are a raised extensional block of metasedimentary rocks of the Proterozoic Belt Supergroup. These rocks originated as sand deposited on immense alluvial fans on the southwestern edge of the Belt Sea.

As elsewhere in southwest Montana, these old rocks were folded and pushed to the east along thrust faults when North America collided with the oceanic Farallon plate

PIONEER MOUNTAINS SCENIC BYWAY

The Pioneer Mountains Scenic Byway, the southern junction of which is approximately 32 miles west from Dillon on MT 278, should not be missed. The 49-mile paved route to Wise River bisects the Pioneer Mountains, which were glaciated at least twice in the last 200,000 years, first during the Bull Lake glaciation, from 160,000 to 130,000 years ago, and then during the Pinedale glaciation, from around 30,000 to 12,000 years ago.

Sawtooth Lake, a tarn that sits in a glacial cirque below Sawtooth Ridge (an arête, at left), and Highboy Mountain (a glacial horn at right) are accessible from the Sawtooth Lake Trail off Clark Creek Road.

The scenic byway begins in the Grasshopper Valley, a graben that preserves lake and stream deposits of the Tertiary Renova Formation, which is overlain by younger glacial outwash and alluvial fan deposits. The Renova forms the light-colored bluffs exposed along the east side of Grasshopper Creek as you enter the valley. Tertiary alluvial fan deposits are well exposed along the road just south of the Polaris School. The raised fault blocks of the West and East Pioneer Mountains, located on either side of the Grasshopper graben, expose slightly metamorphosed Proterozoic sedimentary rocks of the Missoula Group, which have been broken into angular blocks of quartzite by the freezing and thawing of water in the cracks. These sedimentary rocks are intruded by granitic rocks of the roughly 75-million-year-old Pioneer batholith, which is well exposed in the high, glaciated ridge of the East Pioneers, east of the scenic byway.

The hot water at Elkhorn Hot Springs rises along normal faults on the west side of Grasshopper Valley. North of the hot springs, the scenic byway climbs out of the valley into a high, grassy plateau covered in glacial till. Large granitic boulders dumped during the Pinedale glacial advance are strewn across the landscape. The numerous small ponds and bogs are likely places where icebergs in the till melted, leaving depressions that later filled with water. Mosquitos love all that standing water during the summer.

Crystal Park, on the west side of the road about 17.5 miles from MT 278, is a great place for families to spend a day collecting quartz crystals. The crystals formed in cracks in the Cretaceous granodiorite of the Pioneer batholith. Hydrothermal fluid, superheated water containing silica in solution, permeated open spaces in the granodiorite and precipitated crystals of vein quartz as it cooled. The open spaces allowed the quartz crystals to grow into their ideal shape of six-sided prisms with a pyramid, typically at one end. The crystals range from smaller than your little finger to several inches in diameter. They can be clear, white, gray (smoky), or even purple (amethyst); many have tiny inclusions of other minerals or small bubbles. Some are stained red or orange from iron oxide in the groundwater. Crystals that have the shadowy outline of a crystal within a crystal (phantoms) are rare prizes. Many visitors come loaded

An 0.8-inch-long pale amethyst quartz crystal from Crystal Park.
—Courtesy of Richard Gibson

with picks and shovels, but the crystals are so abundant that nice specimens can easily be found at the surface. Chemical and mechanical weathering of feldspar in the granodiorite has caused it to disintegrate into granitic sand, or *grus*, liberating the quartz crystals.

A little more than 6 miles north of Crystal Park, Forest Road 484 heads southeast to the old mining town of Coolidge in the Elkhorn mining district, a silver-rich area discovered in 1872. The ore occurs in veins in the Pioneer batholith and includes quartz, pyrite, galena, sphalerite, chalcopyrite, and molybdenite. The crash of silver markets in 1893 closed the mines temporarily, but in 1917 construction of a narrow-gauge rail line brought the mines back to life. Coolidge boasted a population of 350 people at its peak, with electricity, telephone service, a post office, a school district, a restaurant, and a company store, but no saloon or church. In 1922, a mill was built to service the mines, but it never really prospered due to the low price of silver. By the mid-1920s, more than 24,000 feet of underground tunnels had been excavated and 50,000 tons of gold, silver, copper, lead, and zinc ore had been processed by the mill. By the early 1930s, the town was largely abandoned.

Today the abandoned tunnels produce a toxic reddish acid-mine drainage, which forms when the sulfide-rich ores are oxidized by contact with air and water to form sulfuric acid. The discharge contains measurable concentrations of iron, aluminum, cadmium, copper, nickel, lead, zinc, and arsenic, which reach Elkhorn Creek, a tributary to the Wise River.

North of the turnoff to Coolidge, the road descends into the Wise River valley, which was the outlet for a large glacier that flowed off the ice field that covered both the West and East Pioneer Mountains during Pleistocene time. The glacier's curved terminal moraine, visible from the road near Moose Creek, is about 200 feet high and formed a dam that enclosed a lake as the glacier retreated. The current pathway of Wise River follows the lake's spillway over a low point on the east end of the moraine, where it was deposited against hard rock of the Missoula Group. From Sheep Creek, about 5 miles north of Moose Creek, the Wise River valley is filled with glacial outwash, well rounded and sorted gravel deposited by braided streams coming from the melting glacial ice.

The terminal moraine of the Wise River glacier where the valley narrows, about 1 mile north of Gold Creek. The turnoff for Lacy Creek Road (Forest Road 1299) is 3 miles south of the moraine. The current path of the river was formed when a lake dammed by the moraine spilled over a low point located where the till overlaps the bedrock.

in Late Cretaceous time. About 50 million years ago, tectonic forces reversed, and the crust began spreading to form basin and range topography. The older compressional faults reactivated to become extensional faults. As the Beaverhead Mountains rose along these normal faults, erosion of the mountains filled the Big Hole Valley, an actively subsiding graben, with miles of sediment. Drilling and subsurface imaging show that more than 14,000 feet of sediment fills this graben, mostly lake, stream, and alluvial fan deposits of the Renova Formation. Distinctive deposits of gritty sandstone with flakes of both light-colored muscovite and dark-colored biotite mica, eroded from the Late Cretaceous Chief Joseph pluton exposed west of the Big Hole National Battlefield on MT 43, have allowed geologists to reconstruct the old river system that carried these sediments. Because some of this sediment was transported as far east as the Renova Basin east of Dillon, we know that at least some of the Pioneer Mountains were not a topographic barrier when the Renova was forming. Most of the range we see today likely formed during the last 17 million years of extension. Excellent exposures of two-mica, gritty sandstone beds of the Renova, some loaded with fossil wood, occur north of the highway from Big Hole Pass into Jackson.

Jackson is best known for its hot springs, which Captain William Clark's party used to cook venison in 1806. Since the party no longer had a thermometer at this point on their return trip, Clark estimated the temperature of the spring by how long a man could keep his hand in the water (3 seconds) and by how long it took to cook different-sized chunks of meat. Hot rocks at depth heat Jackson Hot Springs water. As water descends into the Earth along faults and fractures, it's heated; the deeper it goes, the hotter it gets. This geothermal gradient is quite robust in southwest Montana due to basin and range extension and the thinning of the crust, which brings hot mantle rocks closer to the surface. The heated water migrates back to the surface along active faults to form hot springs. We know that the extensional faults in the area are active because there's a prominent fault scarp just northeast of Jackson on the ridge with the letter *J*. Water at Jackson may travel more than 2 miles below the surface, reaching a temperature of about 260 degrees Fahrenheit. It loses some heat on the trip back up, but the outlet of the spring is still nearly 140 degrees Fahrenheit, warm enough to cook a meal.

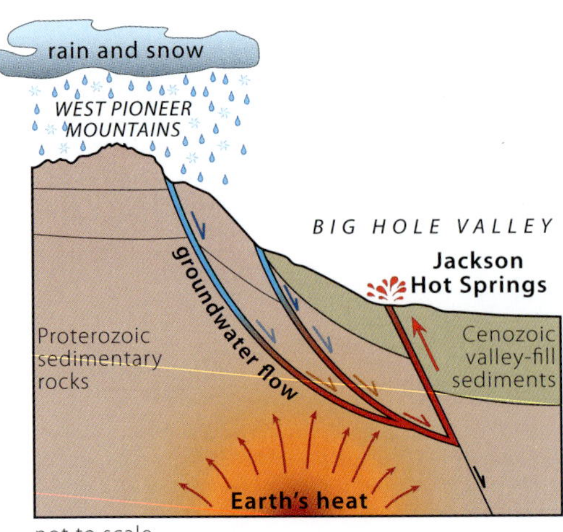

Generalized cross section of the Jackson Hot Springs system.
—Modified from Thomas and Roberts, 2000

The Beaverhead Mountains were glaciated during Pleistocene time, leaving glacial erosional features in the high country, an impressive blanket of glacial till along the northeastern flank of the range, and extensive glacial outwash in the valley. At least one glacier remains in the range, although it probably formed during the Little Ice Age, a climatic cooling event between AD 1300 and 1870. One of the more impressive glacial features in this range is a rock glacier, consisting of angular rock debris frozen in ice. Rock glaciers move downslope very slowly, creating distinctive flow features such as curved ridges and troughs.

A rock glacier along the Continental Divide in the Beaverhead Mountains.

MT 287
Twin Bridges—Ennis
43 miles

Twin Bridges is situated at the confluence of rivers and trails used by native peoples for thousands of years. On August 7, 1805, Lewis and Clark camped near the "two forks" of "Jefferson's River," naming the eastern fork "Philanthropy" (now the Ruby River) and the northern fork "Wisdom" (now the Big Hole River) in "commemoration of two of those cardinal virtues, which have so eminently marked that deservedly selibrated character [Thomas Jefferson] through life." Lewis noted that the Wisdom River was flooding, something that typically occurs in early June today. He was likely witnessing late-season melting as a result of cooler climates during the Little Ice Age, a period of cooling from about 1300 to 1870.

To the northeast loom the glaciated Tobacco Root Mountains, a raised block of Earth's crust bounded by normal faults. Exposed in its core are Archean and early Proterozoic metamorphic rocks and Cretaceous granite, most notably the Tobacco Root batholith. Paleozoic and Mesozoic sedimentary rocks were folded over and eroded off this block of basement when it rose during the Late Cretaceous Laramide orogeny. Remnants of the sedimentary rocks form the light-colored cliffs and palisades (vertical walls) along the flanks of the range visible from Twin Bridges.

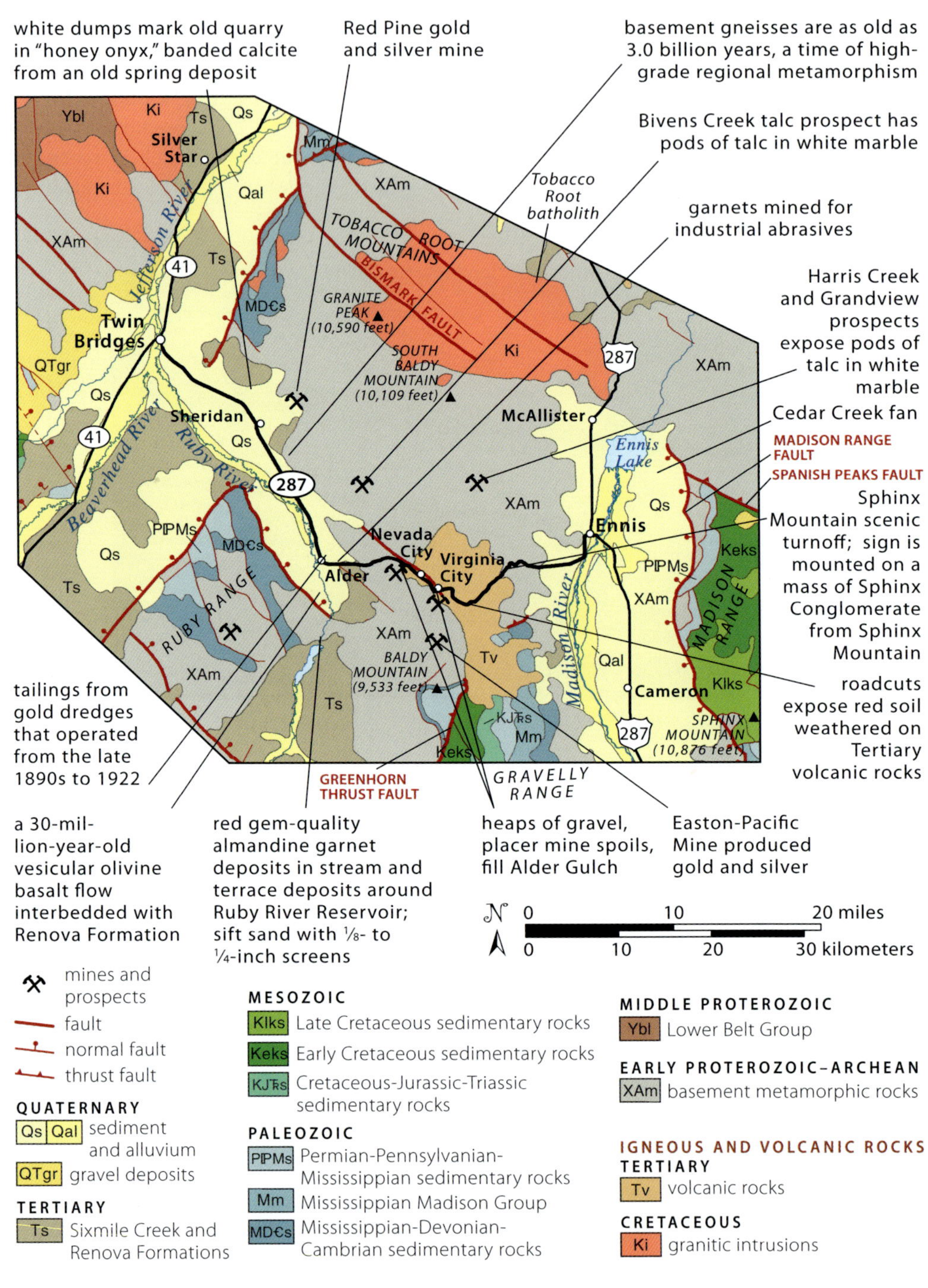

Geology along MT 287 between Twin Bridges and Ennis.

Garnets in Archean metamorphic rocks about 4 miles east of Alder.

From Twin Bridges to Alder, the highway follows the Ruby Valley. The Ruby Range, named for red garnets in its Archean rocks, is prominent to the southwest. It was a Laramide structure in late Cretaceous time, as shown by its core of basement metamorphic rocks flanked by folded Paleozoic and Mesozoic sedimentary rocks. Steeply dipping slabs of Mississippian Madison Group limestone are particularly conspicuous to the southwest en route to Alder.

The northeast end of the Ruby Range is abruptly cut off by an active northwest-trending normal fault that raises the range and drops the Ruby Valley. The fault and the northwest-trending topography formed by the fault are very young because they cut through gravels and tephra of the Sixmile Creek Formation, which were deposited by northeast-flowing streams of the ancestral Missouri River system from 16.5 to 4.5 million years ago. The Ruby Range and the Tobacco Root Mountains could not have been here, or they would have blocked the streams. As a result, the Ruby River has only been flowing in its present valley for the last 4.5 million years. White tephra (ash) of the Sixmile Creek Formation occurs on the northern flank of the Ruby Range and can be reached by taking most of the dirt roads south from Sheridan. The best place to see spectacular exposures of tephra and other deposits of the ancestral Missouri River is at Timber Hill, about 24 miles southwest from Alder on the largely unpaved Sweetwater Road.

Between Alder and Virginia City, furrowed rows of gravel, some separated by shallow ponds, occur along the entire length of the Alder Creek drainage. These odd piles are tailings left by dredge barges over one hundred years ago. Today, few people live in this drainage, and garnets—not gold—are mined for their value as industrial abrasives. However, this place was the site of a gold rush that drew thousands of people and produced one of the richest placer gold deposits ever discovered.

On May 26, 1863, a group of prospectors traveling back to the gold fields of Bannack discovered gold in Alder Creek. Agreeing to keep their discovery quiet, they

Cross section view of Timber Hill and the Sweetwater fault. The flat-topped mesas are capped by basalt that erupted from a vent in Idaho and flowed down a channel of the ancestral Missouri River 6 million years ago. Erosion of softer rocks around the basalt left the basalt-filled channel standing high. The deposits are offset by an active northwest-trending normal fault (Sweetwater fault), evidence that the topography didn't exist when the lava flowed down the valley. —Courtesy of Marli Miller

Tailings left from gold mining on Alder Creek over a century ago. The bouldery ridges across the photo were laid down by the floating dredge as its boom waved back and forth.

returned to Bannack for supplies, but word got out, and within three months ten thousand people were in the drainage. So many camps sprung up in Alder Gulch that it was referred to as the Fourteen Mile City. The largest of the camps was known as Verona. On June 16, 1863, a claim was filed on 320 acres to be used as a townsite named Varina, after Varina Davis, the First Lady of the Confederate States of America. When the charter was presented, a Unionist judge objected and changed the name to Virginia. By February 7, 1865, Virginia City became the capital of Montana Territory and one of the most prominent cities in the Rocky Mountains.

In the beginning, miners used hand tools and small sluice boxes to separate the gold from the gravels. A sluice box is a long, wooden trough with slats (riffles) across the bottom that trap the heavier gold particles as water washes the lighter materials out of the box. In 1867, hydraulic mining was employed as claims were consolidated and more money could be spent to remove the gold. An expensive system of dams and ditches was built to bring water from the mountains to create high-pressure jets that blasted the stream sediments, creating a slurry that was washed through even larger sluice boxes. By 1889, an estimated $90 million in gold had been extracted, a figure that is equivalent to no less than $40 billion today! It was not a metaphor when people referred to Alder Gulch as a "river of gold."

Mining activity and the area's population had greatly diminished as early as 1875, leaving Virginia City with fewer than eight hundred people, many of whom were Chinese who'd been successfully working abandoned claims for years. By the fall of 1898, new life was breathed into the placer mining with the arrival of the *Maggie A. Gibson*, a steam-powered dredge barge that was able to dig deeper and extract gold that hydraulic mining methods could not reach. The dredge barge had been in Bannack and was dismantled, moved, and rebuilt in Alder. Dredge barges make their own pond by digging the gravel in front and dumping the waste rock behind. A sluice box on the boat removes the gold, and a conveyer belt swings back and forth, dumping the dredge spoils into distinctive rows in the creek's channel. The *Maggie A. Gibson* consumed about twenty cords of wood each day to power its steam engine.

Hydraulic dredge barge on Alder Gulch 3 miles below Nevada City in 1900. The tailings boom on the left spews waste gravel out onto growing piles as it swings back and forth. —Courtesy of Montana Historical Society

By 1906, electrically powered dredge barges joined the fleet, including *Dredge No. 4*. Built in 1911 at a cost of $296,000, it was reported to be the largest dredge in the world. By 1922, the entire floodplain of Alder Creek had been consumed and an additional $10 million in gold had been removed from Alder Gulch. Virginia City was never dug up because it was not built on gold-bearing gravels, but the original town of Nevada City was removed by dredging in the early 1900s. Many of the cabins you see there today were moved there. The gold was processed into bars in the nearly vanished town of Ruby, near Alder, carried by "gold vest couriers" to the nearest bank, and shipped east. Harvard University shared in the profits through the estate of one of the owners of the primary dredge operator, the Conrey Placer Mining Company.

From the start of the gold rush, lode (bedrock) mining took place in the upper reaches of Alder Gulch, but it was never as productive as the placer mining. The placer gold in the gulch must have come from a source in the hard rock, but the exact location and the processes that produced it are not known. The gold in the hard-rock mining district occurs in quartz veins that intruded Archean metamorphic rocks, including some loaded with garnets. The source and age of the vein mineralization is uncertain, but it's likely related to the intrusion of the Tobacco Root batholith in Late Cretaceous time.

As you head over the Virginia City hill toward Ennis, Eocene and Oligocene andesite and basalt rest directly on Archean gneiss. The Paleozoic and Mesozoic sedimentary rocks that once covered the gneiss were eroded prior to the deposition of the volcanic rocks, probably when this area was raised during Laramide compression in Cretaceous time. Since the lava flowed in topographic low areas, it must have been raised after it erupted, probably by basin and range extension starting in Miocene time.

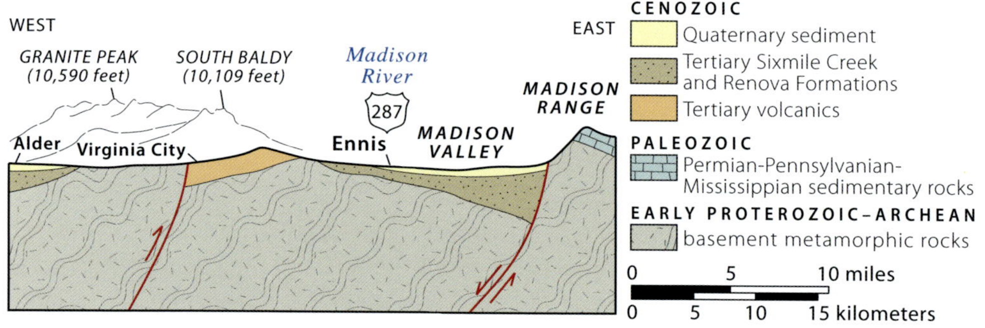

West-east cross section along the line of the highway between Alder and Ennis. Blocks of basement rock moved up and down along big normal faults to form the mountain ranges and valleys.

Before descending into the Madison Valley, stop at the striking exposure of dark Archean gneiss cut by white pegmatite dikes on the south side of the road. These former sedimentary rocks were metamorphosed by the collisions of continents several billion years ago. Some of the rocks were heated to melting, producing igneous intrusions that cut across the metamorphic layering (foliation) to cool as the white dikes. To be able to see rocks that formed so deep in the Earth is something special; a tremendous amount of uplift and erosion must have occurred to expose them at the surface. If you look closely, you'll notice that the banded gneiss is loaded with red garnets of many sizes. These minerals can be very useful in helping geologists determine the pressure and temperature history of the rocks in which they are found.

On the hill above the Madison Valley and Ennis, a scenic overlook on the south side of the road provides an extraordinary view of basin and range topography that began forming around 17 million years ago. Much of the Madison Range was

Pegmatite dike cutting across foliation in garnet-rich gneiss.

uplifted in the last 4.5 million years as the Yellowstone hot spot became an important factor in crustal extension in southwest Montana. Standing as a sentinel to the southeast over the Madison Valley is Sphinx Mountain, composed of Late Cretaceous conglomerate shed off mountains to the west during Late Cretaceous compressional mountain building, before either the Madison Valley or Madison Range had formed. There is also a great view of the immense Cedar Creek alluvial fan, a textbook example of a fan built mostly by streams flowing from glaciers during the Pinedale and Bull Lake glaciations. Modern streams have cut into the fan by as much as 31 feet because the Madison Range has been rising and thereby steepening the fan. (For more information on the geology of this area, see the US 287: Three Forks—West Yellowstone road guide.)

STRATIGRAPHIC COLUMN FOR CENTRAL AND EASTERN MONTANA

ERA	PERIOD	EPOCH	ROCKY MOUNTAIN FRONT / ADJACENT HIGH PLAINS / CENTRAL AND EASTERN MONTANA / NORTH DAKOTA (WEST ← → EAST)	DESCRIPTION OF MAJOR UNITS
CENOZOIC	QUATERNARY	HOLOCENE		
CENOZOIC	QUATERNARY	PLEISTOCENE	glacial deposits	glacial till, alluvium
CENOZOIC	QUATERNARY		clinker	baked and melted rock from burned-out coal seams; red, orange, and black; on and in Fort Union Formation
CENOZOIC	TERTIARY	PLIOCENE / MIOCENE	Sixmile Creek Formation in southwest MT; Flaxville Gravel	terrace gravels left behind as streams carved valleys sandstone and gravel on terraces
CENOZOIC	TERTIARY	OLIGOCENE / EOCENE	Renova Formation in SW MT (48–20 mya); includes Dillon Volcanics; Arikaree Fm. Wasatch Fm.	Renova Formation: sand, mud, and ash; deformed before Sixmile Creek Formation was deposited
CENOZOIC	TERTIARY	EOCENE	Lowland Creek and Absaroka Volcanics	
CENOZOIC	TERTIARY	PALEOCENE	Fort Union Formation — MAJOR MEMBERS: Tongue River; Lebo; Tullock K-T boundary	Tongue River Member: yellowish-gray to brownish, light to dark-gray thick beds of sandstone; some gray to black shale; has most of the region's coal Lebo Member: gray to brownish shale Tullock Member: yellowish-brownish-gray sandstone; thin coal beds
MESOZOIC	CRETACEOUS	LATE	Beaverhead Group (SW MT); Willow Creek and St. Mary River Fm.; Livingston Group (78-75 mya); Hell Creek Formation with ash of Livingston Group; Lance Formation (69–66 mya)	Hell Creek Formation: tan to brown and gray sandstone to light greenish shale, some bentonitic, a few thin coal beds; some dinosaur bones Lance Formation: brownish gray or buff, often cross-bedded sandstone
MESOZOIC	CRETACEOUS	LATE	Horsethief Formation; Livingston Group (78-75 mya); Fox Hills Sandstone	Fox Hills: light-colored gray to greenish-brown sandstone and shale Livingston Group: water-laid volcanic conglomerate, sandstone, shale; near Livingston; intertongues with Claggett through Fort Union
MESOZOIC	CRETACEOUS	LATE	Livingston Group (78-75 mya); Bearpaw Shale; Pierre Shale	Pierre Shale: blue-gray to dark-gray shale with bentonite Bearpaw Shale: dark-gray to brownish-gray, some flaky shale, some bentonite beds, some sand, many iron-rich concretions
MESOZOIC	CRETACEOUS	LATE	Two Medicine Formation (83–70 mya); Judith River (80–75 mya); Claggett Shale	Two Medicine Formation: sandstone rivers and deltas, some crossbeds; some bentonite Judith River Formation: yellowish to brownish-gray sandstone, cross-beds; some mudstone and coal Claggett Shale: brownish-gray shale, flaky, with concretions; some sand
MESOZOIC	CRETACEOUS	LATE	Gammon Formation	yellowish-brown to grayish shale to silt
MESOZOIC	CRETACEOUS	LATE	Virgelle Sandstone; Eagle Sandstone	off-white to slightly orange or buff; massive, cliff-forming
MESOZOIC	CRETACEOUS	LATE	Telegraph Creek Formation	yellowish-brown shale and some sandstone
MESOZOIC	CRETACEOUS	LATE	Frontier Fm. in SW Mont.; Marias River Shale; Colorado Group: Niobrara Formation	dark brownish-gray, flaky calcareous shale; some with concretions, some bentonite
MESOZOIC	CRETACEOUS	LATE	Colorado Group: Carlile Shale	black to gray flaky shale; some with concretions
MESOZOIC	CRETACEOUS	LATE	Colorado Group: Greenhorn Formation	dark-gray calcareous shale, weathers light gray; some with concretions
MESOZOIC	CRETACEOUS	LATE	Colorado Group: Belle Fourche Shale	dark-gray siliceous shale, flaky; iron-rich concretions; bentonite beds
MESOZOIC	CRETACEOUS	LATE	Colorado Group: Mowry Shale (98–94 mya)	Mowry Shale: light-to-medium-gray silicified shale and sandstone, fine grained; layers of dark shale
MESOZOIC	CRETACEOUS	LATE / EARLY	Bow Island Sandstone Member	Blackleaf Formation: shale, siltstone, sandstone
MESOZOIC	CRETACEOUS	EARLY	Blackleaf Formation; Muddy Sandstone; Thermopolis Shale; Fall River Sandstone	Fall River: tan-colored fine-grained sandstone, some with thin, dark shale partings
MESOZOIC	CRETACEOUS	EARLY	Kootenai Formation (125–112 mya)	Kootenai Formation: conglomerate, sandstone, shale; often purplish or green
MESOZOIC	JURASSIC		Morrison Formation (156–146 mya)	mudstone, sandstone, limestone; light to greenish-gray to red; river, floodplain deposits
MESOZOIC	JURASSIC		Ellis Group	brown to yellow sandstone
MESOZOIC	TRIASSIC		Chugwater Formation	

Major Cretaceous and Cenozoic sedimentary rocks in Central and Eastern Montana.

CENTRAL AND EASTERN MONTANA

VAST HORIZONS WITH BUTTES AND BADLANDS

Central and eastern Montana are dominated by Cretaceous and Tertiary sedimentary rocks laid down in shallow ocean water or on a gently sloping coastal plain. While usually hidden beneath the rolling sagebrush hills and prairie, they are spectacularly displayed in breaks, highlands, and badlands and along the Missouri River and its tributaries. Visit places such as Bighorn Canyon, Makoshika State Park, and Medicine Rocks State Park in spring, when the land is waking up from the long cold winter and all is green.

The Rocky Mountain front is one of the most striking topographic features in Montana. The mountains rise abruptly from the plains because great slabs of Proterozoic and Paleozoic rocks were shoved eastward up and over younger, softer, more easily eroded Cretaceous sedimentary rocks. These giant slabs of rock were moved up to 100 miles from where they were deposited due to plate collision to the west—when North America collided head-on with the Farallon plate, part of the Pacific Ocean floor. Sedimentary rocks that had accumulated on the continent since the deposition of the Belt Supergroup, nearly 1.47 billion years earlier, were delaminated and pushed eastward across Archean basement rocks. These thrust sheets of rock were folded and stacked in a thick pile called the Overthrust Belt. Immediately east of the Overthrust Belt is a region of more easily eroded folded and thrust faulted rock called the Disturbed Belt.

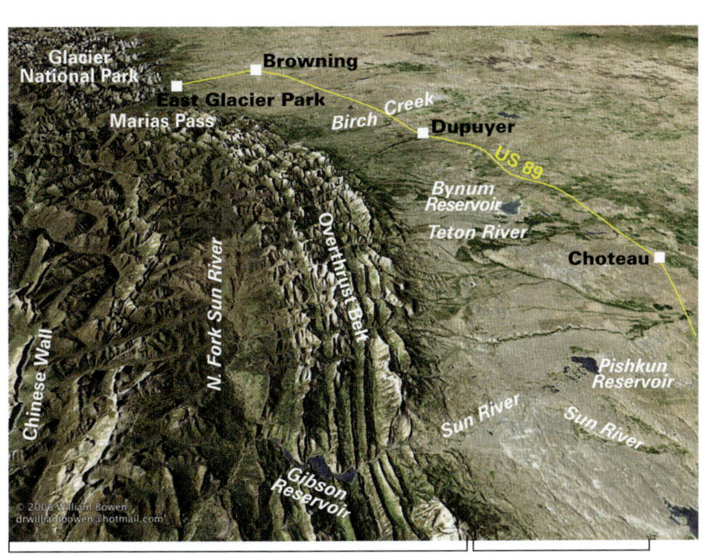

View north over the Overthrust Belt and US 89 from Choteau to Browning. —Courtesy of William Bowen

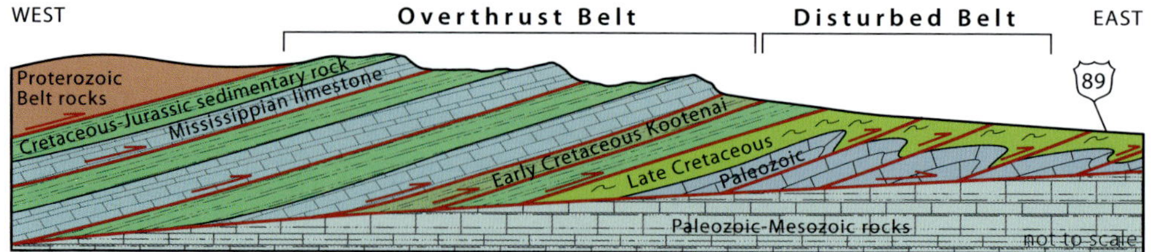

Overthrust Belt and Disturbed Belt along the Rocky Mountain front.

East of the Rocky Mountain front, the landscape changes dramatically to rolling plains dotted with isolated mountain ranges. Thrust faults, broad crustal folds, arches, and troughs occur in this region, but nothing like those found in the tightly folded and thrust-faulted rocks to the west, which look more like a crumpled carpet.

We know from a few outcrops, as well as from oil wells and seismic studies, that lying deep below the broad upland surface of Montana's plains are hard Archean and early Proterozoic crystalline basement rocks. The few exposures of basement rocks that do occur in central and eastern Montana are important to our understanding of the tectonic evolution of North America. Near Neihart in the Little Belt Mountains, for example, granite, diorite, and gneiss tell the story of continental growth that occurred around 1.8 billion years ago, when the Wyoming craton, the original core of the North American continent, collided with the Medicine Hat block, an event called the Big Sky orogeny. The suture zone, called the Great Falls tectonic zone (also called Trans-Montana fold-and-thrust belt), is buried elsewhere in central Montana, so this is a crucial window into this ancient continental collision. Middle Proterozoic Belt sedimentary rocks, discussed in detail in the Northwest Montana chapter, are also exposed in central Montana. Here they were deposited in the no-longer-active extensional basin called the Helena Embayment, which was part of the much larger Belt Basin.

Deeply eroded Archean and early Proterozoic basement rocks were buried by sediments from a tropical ocean that spread across North America at the beginning of Paleozoic time. As seas rose and fell, they deposited a thick stack of marine and nonmarine sedimentary rocks. Some of these formations, such as the Late Devonian to Early Mississippian Bakken Formation, were loaded with oil-creating organic matter. Paleozoic rocks are well exposed along the Rocky Mountain front and in large dome-like structures of the Little Belt, Big Snowy, and Little Snowy Mountains. The most conspicuous of all the Paleozoic sedimentary rocks is the Madison Group limestone, deposited in shallow tropical seas that covered most of the state during Mississippian time, some 350 million years ago. This massive limestone resists weathering in central and eastern Montana's dry climate, so wherever it appears it makes prominent cliffs of pale-gray rock. It also contains much of the oil and gas in central Montana.

Western Interior Seaway and Dinosaurs

Much of the Mesozoic rock exposed at the surface in central and eastern Montana is Late Cretaceous marine sedimentary rock deposited in the north-trending Western Interior Seaway and nonmarine sedimentary rocks deposited on a coastal plain between the seaway and the Rocky Mountain front to the west. The seaway was created

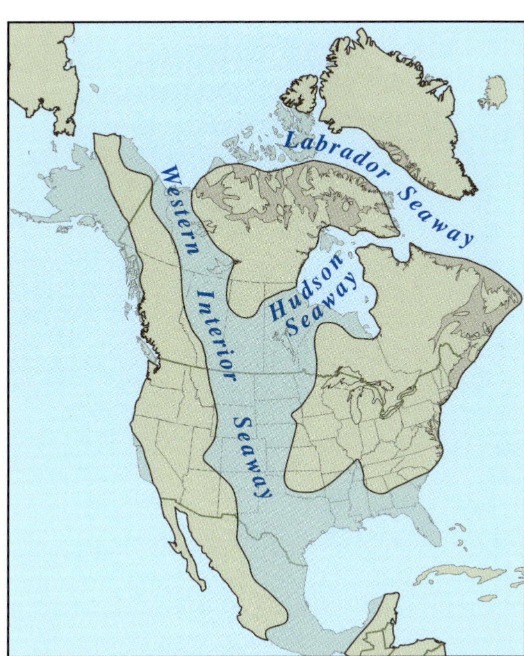

The shallow Western Interior Seaway about 83 million years ago, during Late Cretaceous time.

when the Farallon plate was subducted beneath the North American continent, causing thrust-faulted slabs of rock to pile up as the initial Rocky Mountains. Their weight depressed the continent east of the rising mountains, creating a foreland basin.

Throughout Cretaceous time, the western shoreline of the seaway shifted to the west and to the east, creating a stack of interbedded marine and nonmarine sedimentary deposits. Between about 83 and 70 million years ago, central and eastern Montana was a broad, flat coastal plain, with palms and cypress trees bordering the Western Interior Seaway. Conifers, hardwood trees, and brush bordered the rivers. The sediments deposited in these environments became the widespread Judith River and Two Medicine Formations: gray siltstone and sandy shale; gray to brown shale, siltstone, sandstone, and soft lignite coal; and swelling clays derived from volcanic ash blown in from the west. Duck-billed and horned dinosaurs, along with the meat-eating *Albertosaurus*, are preserved in them.

About 70 million years ago, the last significant sea to spread into Montana deposited dark-gray Bearpaw Shale over much of central and eastern Montana. This formation is loaded with fossils of clams, oysters, and squid-like ammonites, some of which are coiled, such as *Placenticeras*, whereas others have straight shells, such as *Baculites*. Twenty-foot-long swimming lizards called mosasaurs ate the ammonites. These creatures had strong swimming tails and limbs modified as flippers. Dinosaurs roamed the coastal lowlands, wary of *Deinosuchus*, a 40-foot-long crocodile that ambushed them at water's edge. The first remains of this monster were found in the Bearpaw Shale near the town of Mosby in the early 1900s.

The Western Interior Seaway retreated eastward into North Dakota in Late Cretaceous time, leaving a coastal plain—similar to today's Texas Gulf Coast—sloping eastward from the Rockies all the way across central and eastern Montana. Sandstone

Ammonite fossil from the Bearpaw Shale of Montana. Faint curved lines radiating out from the center are growth stages. Scale bar in centimeters.

and clayey shale of the Hell Creek Formation were deposited in stream channels and on floodplains, and sometimes in swamps and estuaries along the edge of the seaway. The climate was humid, and the swampy lowlands were cloaked in flowering trees, conifers, ferns, and ginkgoes. This was the habitat of dinosaurs, crocodiles, lizards, snakes, turtles, freshwater stingrays, and even opossums and rodent-like animals. The Hell Creek Formation is world-famous for dinosaur remains, including those of *Tyrannosaurus rex*, triceratops, and duck-billed dinosaurs.

Dinosaurs first appeared in the geological record about 243 million years ago. For more than 170 million years they evolved and lived in a variety of environments until they suddenly disappeared about 66 million years ago. Not all were the lumbering giants that we read about and are awed by in museums; some were no bigger than a chicken. Although it's the big guys, some up to 130 feet long, that draw most of our attention; about seven hundred species in a bewildering array of shapes and sizes have been named. Until the mid-twentieth century dinosaurs were thought to be cold-blooded. We now know that most were warm-blooded, built nests, laid eggs, and tended their young. Some dinosaurs were strictly plant eaters, others meat eaters, and some both.

In order for dinosaur bones to be preserved—an unlikely event—the animal must be buried rapidly during or immediately after death; otherwise scavengers quickly devour most of the body, and bacteria and insects destroy the rest. Only under extraordinary circumstances are soft parts preserved. In the mid-2000s, Mary Schweitzer, a paleontologist studying a fragment of a Montana T. rex under the microscope, found red blood vessels and blood cells in a scrap of soft tissue. Comparing them with cells of a modern ostrich, the two looked very similar. Then in 2011, a dinosaur collected from 110-million-year-old rock at a tar sands mine in northern Alberta included skin—even skin color. Caught in a flooding river, the animal was swept into shallow sea mud where it remained very well preserved. It was 18 feet long and weighed about 3,000 pounds. It confirmed that the reconstructions done by paleontologists in recent years have been remarkably good renditions of the real thing.

Montana's dinosaurs are curated and displayed in many museums, mostly in the central and eastern parts of the state. Most are free to the public, though often open only in the afternoon. Dinosaurs in Montana have been collected primarily from a few specific formations that also provide snapshots of the environments in which they lived. For example, red to green and gray mud, silt, sand, and limestone of the Late Jurassic Morrison Formation were deposited in floodplains, streams, lakes, coal swamps, desert sands, and sand dunes on the virtually flat, semiarid coastal plain from about 156 to 145 million years ago. Ferns, palmlike cycads, and ginkgo trees, among others, grew in these environments. Giant plant-eating stegosaurs and sauropods lived alongside meat-eating allosauruses that were of similar size.

Many of Montana's dinosaurs have been found in the famous Hell Creek Formation of latest Cretaceous time. It is overlain by the Tullock Member of the Fort Union Formation, a slightly more coal-rich coastal plain deposit of Paleocene time (earliest Tertiary). The Fort Union Formation sedimentary rocks are almost identical to the Hell Creek beds except that they contain no dinosaur remains. The primary difference between the two formations is the abrupt change in the animal and plant fossils they contain. It appears that the dinosaurs and many other species died after a major asteroid, 6 to 10 miles across, struck the Earth 66 million years ago. That impact left its imprint in the sedimentary rocks of Montana and elsewhere as a thin, dark layer in otherwise similar sediment. Known as the Cretaceous-Tertiary (or K-T) boundary, it separates the otherwise similar Hell Creek and Fort Union Formations. It's well exposed in the badlands of eastern Montana, such as at Bug Creek southeast of Fort Peck Lake on MT 24.

The similarity of the Hell Creek and Fort Union rocks indicates that not much changed in the general coastal plain, beach, and shallow sea environment of central and eastern Montana at that time. However, this dark layer everywhere has an unusually high iridium content, a trace element normally concentrated in Earth's mantle

The Cretaceous-Tertiary boundary exposed in the badlands of Makoshika State Park near Glendive. The boundary is approximately where the geologist's hand is, on the lower of the two dark-colored coal beds.

and in asteroids. It is virtually absent in most rocks in Earth's crust. The iridium at the K-T boundary was dispersed globally because of the impact. The impact layer is also marked by grains of shocked quartz, a distinctive deformation pattern thought to be produced only by extreme pressures imposed by hypervelocity impact. Recent research has also found shocked quartz in rocks within one foot of a lightning strike. Nonetheless, the asteroid theory holds up, because it is unlikely that scattered lightning would produce shocked quartz grains in the thin K-T boundary sediment and not in sediments of other ages.

It's generally thought that the asteroid struck the northern Yucatán Peninsula of eastern Mexico. Other localities have been suggested, and in fact it's quite possible that multiple impacts happened at nearly the same time.

Exactly how the impact doomed the dinosaurs has been a hot topic among researchers. Near ground zero, flash incineration likely killed most. The incredible energy generated during impact would have oxidized much of the nitrogen in our atmosphere; combined with moisture in the air, that would have produced enough nitric acid rain to cause mass extinction. That combined with dust and soot from wildfires would kill vegetation, either directly or by obscuring sunlight so that the temperature would quickly drop to freezing. The dinosaurs also could have been poisoned by nickel vapor released from the asteroid. Heavy volcanic ash in some dinosaur-rich beds is also suspicious, suggesting volcanic eruptions may have played a role in their demise.

Not all scientists agree that a killer asteroid was the culprit. Many paleontologists, including some experts on dinosaurs, agree that the dinosaurs, as well as many other life-forms, were exterminated in a very short time at the K-T boundary. They also agree that extreme levels of acid rain, wildfires, soot, and volcanic ash in the atmosphere contributed to the extinction event, but some see a quite different culprit. Methane clathrate, a methane-ice compound, occurs in large amounts at temperatures below the freezing point of water at modest depths, such as 1,500 feet below sea level on the continental shelf. (In fact, some companies and governments have been eyeing these deposits as a potential abundant source of natural gas.) When clathrate deposits become destabilized, they release methane, a greenhouse gas twenty times more potent at trapping heat in Earth's lower atmosphere than carbon dioxide.

Some paleontologists infer the impact released so much methane that runaway global warming killed the dinosaurs and many other life-forms. A warmer climate would have meant warmer ocean temperatures, ocean acidification, and the die-off of most ocean life, including plants that produce much of Earth's oxygen. Higher atmospheric temperatures would cause greater areas of drought, wildfire, and insect-decimated plants. The rapid depletion of land plants would have led to much greater competition for food and perhaps the rapid demise of the dinosaurs. Do these changes bring to mind things we're beginning to see in today's environment?

Cenozoic Badlands and Clinker

Cenozoic sedimentary rocks once covered all of Montana's Mesozoic rocks between the Rockies and North Dakota; today they mostly occur in the eastern third of the state. Most of the rocks belong to the Paleocene Fort Union Formation, which was deposited immediately after the extinction of the dinosaurs. Following their demise, mammals rapidly evolved. Fossils of many plant species, freshwater invertebrates, salamanders, frogs, turtles, lizards, snakes, crocodiles, and alligators have also been recovered from the Fort Union. Like the Cretaceous Hell Creek Formation before it, the Fort Union was deposited on a broad coastal plain extending from the Rockies to the rapidly diminishing Western Interior Seaway. Northeastward-flowing meandering streams developed vast floodplains with swamps and peat bogs where plant material accumulated and, with time and burial, was transformed into coal. Economic deposits of coal are mostly in the Tongue River Member of the Fort Union Formation in the Powder River Basin in Wyoming and Montana.

Striking aspects of the landscape in central and eastern Montana include colorful badlands, with their barren, striped hills and tangles of gullies and ravines. They may be "bad" because they are difficult to cross in contrast to the flat plains, but these

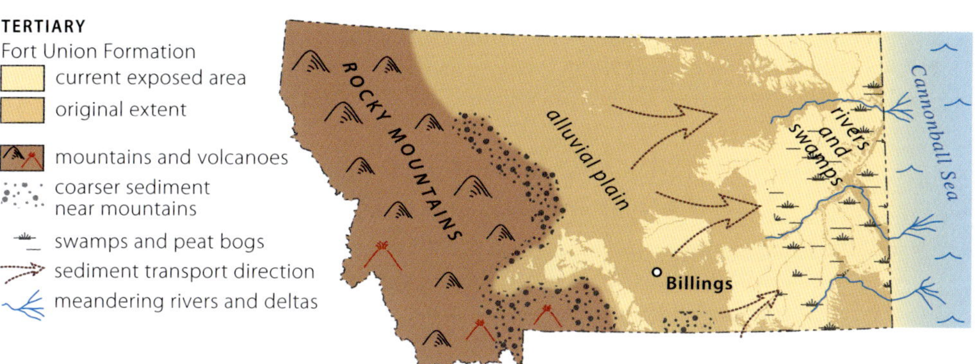

The depositional environments, sediment transport direction, and distribution of the Paleocene Fort Union Formation. The Cannonball Sea was a remnant of the Western Interior Seaway, which was largely retreating in Paleocene time.

A bluff at Terry Badlands, east of Miles City, capped by an erosion-resistant layer of clinker. Inset is a close-up of a clinker fragment.

are truly magical places to explore and get lost in—literally, so be careful! Badlands are a type of dry terrain where softer sedimentary rocks and clay-rich soils have been extensively eroded by wind and water. The recipe for making badlands starts with nearly flat-lying, soft sedimentary rocks with variable resistance to erosion. A dry climate with sparse vegetation is essential, because plant cover tends to reduce erosion. Water-resistant clays also help, because they shed rainwater off the slopes when it does rain, sometimes in turbulent flash floods, producing rills and deepening the gullies. Sediment collects at the bases of the slopes as coalescing fans.

Red-colored clinker caps many of the hills in eastern Montana. Clinker is so named because of the sound it makes when walked on or struck with a hammer. It forms when sedimentary rocks above and below burning coal layers are heated, baked, and even melted; melting temperatures can be around 1,800 to 2,500 degrees Fahrenheit. The brick-red color is from iron oxide. Coal fires can be ignited by grass fires or lightning, and once started they can burn for years, sometimes deep underground. A burning coal layer just northeast of Terry was extinguished as recently as the 1990s.

Island Ranges of the Central Montana Alkalic Province

Most of the isolated mountain ranges of central Montana are round, domal structures cored by distinctive 50-million-year-old igneous rocks. As the magma rose, it bulged up the overlying sediments, some of which now tilt steeply away from the intrusions. These mountains formed when blocks of basement rock were raised in early Tertiary time, during the Laramide orogeny. Most of the igneous rocks are rich in alkali elements, such as sodium and potassium, so these domes make up what's called the Central Montana Alkalic Province. Many of the intrusions are laccoliths, sheetlike masses that were injected between layers of sedimentary rock. Erosion of the overlying sedimentary rocks of many of the domes has exposed the alkalic igneous rocks in their cores.

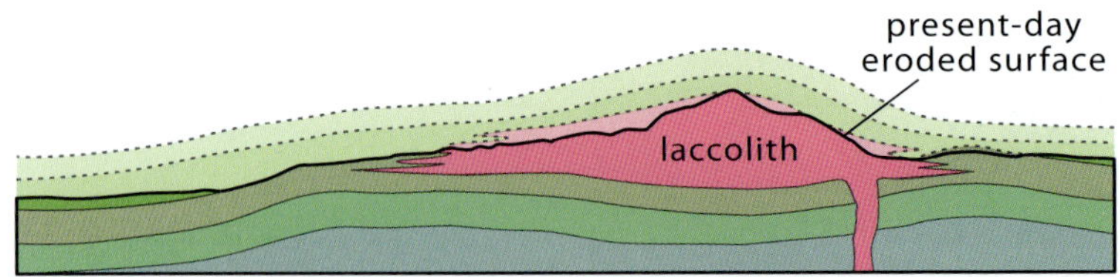

A laccolith is a sheetlike intrusion that domes the overlying sedimentary rocks; when this cover erodes away, the igneous rocks in the core of the dome are exposed.

The origin of the alkalic magma is unclear. Some geologists suggest that the slab of oceanic crust to the west—the Farallon plate—was sinking at an unusually shallow angle beneath western North America, thereby thickening the crust. As a result, the upper mantle was then pushed deeper into the Earth, where it melted at higher pressure. Lab experiments suggest that this would form small amounts of magma unusually high in potassium and sodium—just the unusual compositions that we see in the igneous rocks of the Central Montana Alkalic Province.

The oldest alkalic igneous rock in this province is Late Cretaceous shonkinite exposed in the Adel Mountains west of Cascade. This rare rock has a composition

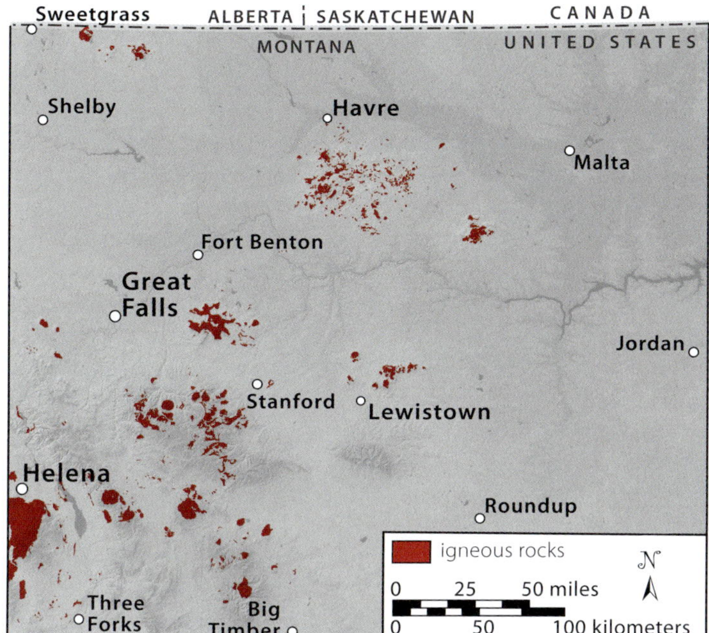

The distribution of igneous rocks in the Central Montana Alkalic Province.

Shonkinite that crystallized in one of the intrusions. A close look reveals shapely crystals of glossy black pyroxene (augite) in a dark-green matrix of tiny crystals of augite and potassium feldspar.

Shaped like a big pancake, the Shonkin Sag laccolith in the Highwood Mountains is made of dark-gray shonkinite that's about 200 feet thick and 1 mile wide. The magma was injected between horizontal layers of the off-white Late Cretaceous Eagle Sandstone, slivers of which are preserved at both the top and base of the laccolith. At its eastern edge (far right), dark fingers of magma were injected into the pale sandstone. Shonkin Sag, the low area in the foreground, was the bed of the ice age Missouri River.

similar to basalt but is greatly enriched in potassium. Large volumes of shonkinite exist in the Highwood and Bears Paw Mountains, but very little appears in the other mountains of central Montana. The rock was named for the Native American name for the Highwood Mountains and is something of a Montana specialty, even though it does exist in a few other places in the world. Shonkinite sills injected into sedimentary rocks in the Highwood Mountains make striking exposures.

Many of the isolated buttes and clusters of buttes that are widely scattered across central Montana are large intrusions of another breed of peculiar igneous rock, called alkali syenite. These rocks are abnormally rich in sodium and to some extent potassium; they are found in the Judith and Moccasin Mountains, the north end of the Little Belt Mountains, the Little Rocky Mountains, the Sweet Grass Hills, and several other places. Unlike shonkinite, which is much the same everywhere, the alkali syenites vary greatly. Most consist of white feldspars along with scattered dark minerals and little to no quartz.

The intrusions appear to have been large circular blobs of magma that formed stocks or laccoliths several cubic miles in diameter. Most are between 55 and 45 million years old. Had you been there to watch such an intrusion form, you might have seen a low hill slowly bulge at the surface over a period of a few months. Most of the intrusions crystallized within a few thousand feet of the surface, but a few did break the surface and erupt as volcanoes. Erosion over the last 50 million years has stripped away the cover of overlying sedimentary rocks. The more resistant igneous rocks stand high, in bold erosional relief, to make mountains and clusters of mountains. Some of the pronounced bulges that make hills in the sedimentary rocks of the plains probably mark places where large igneous intrusions are still buried.

Most of these intrusions contain concentrations of gold, silver, and lead. In a few cases, the ore minerals are in the igneous rocks, but more commonly they occur in veins around the margins of the intrusions. Nearly every large syenitic intrusion sponsored a gold rush, and some resulted in significant mine operations, such as the Zortman-Landusky Mine, a cyanide heap-leach operation in the Little Rocky Mountains. It's now closed, and its contaminated ground is being remediated.

The alkali-rich rocks of central Montana also include several dozen diatremes, very small igneous intrusions composed of hydrated peridotite that melted very deep within Earth's mantle. Most of the diatremes are a chaotic mass of rounded fragments of peridotite mixed with rounded pieces of the older rocks they intruded, which were blasted upward in a swirling carbon dioxide–rich magmatic fluid. Some of the chunks of older sedimentary rocks came up from below, while others sank from above. The escaping gas broke an upward path through the older rocks of the crust, shattering the crystallizing igneous rock and keeping all the pieces milling about in suspension long enough for their sharp edges to be rounded. Once at the surface, the magma likely shot a great column of volcanic ash and broken rocks high into the air. No such eruption has occurred in historic times, but it was likely a noisy, violent event.

Diatremes are less than a quarter mile across and easily erode; most remain as shallow, round depressions a few tens of feet across. They are an important window into the Earth's deep interior, and they can contain diamonds and sapphires. Diamonds in the diatremes of Montana are rare, but a jogger found a yellow diamond pebble near Craig in the late 1980s that sold, uncut, for around $80,000. Its source remains a mystery. More common and of more value to Montana are sapphires, which share honors with agates as the state's gemstone. Sapphires are a variety of corundum, an aluminum oxide mineral with a hardness of 9 out of 10; diamond is 10. Those from Yogo Gulch in central Montana are widely celebrated for their deep-blue color with no hint of the greenish tinge typical of most blue Montana sapphires.

High Plains Surface

About 15 million years ago, the climate became much drier, leaving mountain valleys drained by intermittent streams that fed an enormous desert plain east of the Rockies. As the valleys filled, the gravel deposits buried the hills between them and finally spread to make a nearly continuous blanket across a vast plain. Remnants of this plain are known as the High Plains Surface. That surface has probably survived due to the permeability of the gravel; it absorbs surface water, thus preventing erosion by surface runoff. The gravel extends east from the Rocky Mountain front at least as far as a north-south line drawn through central North Dakota down to central Texas. In places, the flat surface was eroded into Cretaceous formations. In northeastern Montana, the gravel deposit on the High Plains Surface is called the Flaxville Gravel. It is interlayered with lesser amounts of fine sediment. Southeast of Montana in the Dakotas and Nebraska, the same gravel is called the Ogallala Formation.

The gravel that covers broad, high-level surfaces that slope gently away from the Rocky Mountain front is not as well studied as the Flaxville Gravel. The surfaces certainly formed during a dry climate of latest Miocene-Pliocene time, long after Cretaceous sediments had been deposited and before Pleistocene glaciation.

The High Plains gravel provides excellent aquifers filled with large quantities of good water. The thickness of the High Plains gravel ranges from a few feet to several hundred feet, no doubt because it buries an older landscape. Occasional wells drilled in eastern Montana penetrate an uncommonly thick section of gravel and produce an extraordinary volume of water. Those wells were probably located, by sheer luck, in one of the buried valleys. If geologists someday piece together a map of the large streams that flowed through eastern Montana during the wet part of Miocene time, it would be possible to locate such wells intentionally.

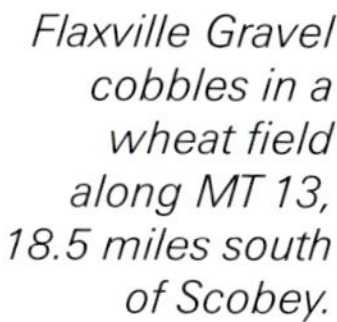

Flaxville Gravel cobbles in a wheat field along MT 13, 18.5 miles south of Scobey.

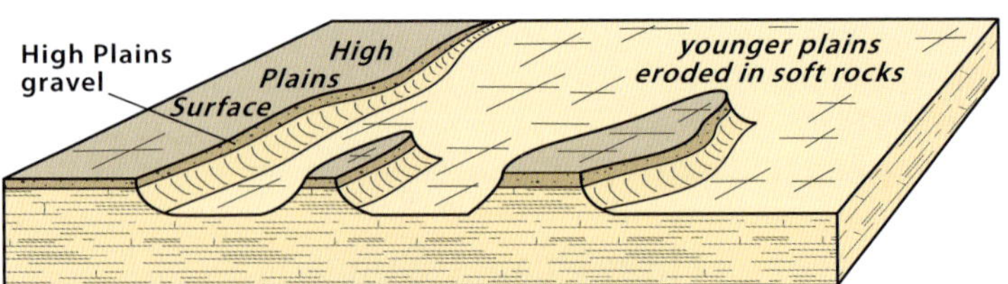

The High Plains Surface erodes in areas where a gravel cap is absent and the underlying rocks are easily eroded.

The High Plains gravel contains very little petrified wood and no fossil leaves because of the desertlike conditions that existed during its deposition. Here and there, the gravel contains bones and teeth of mammals of all sizes, from mice to elephants. The region's fossils include early versions of animals that thrive in dry regions today, such as horses and camels.

Coal and the Colstrip Power Plant

All those shorelines of the Western Interior Seaway that shifted back and forth across Montana during Cretaceous and earliest Tertiary time left a legacy of coal in central and eastern Montana, especially in the Powder River and Williston Basins, thick deposits that collected in long-lived depressions tens of miles across. Swamps and marshes thrived along the tropical shoreline of the seaway, which was likely similar to the west coast of Florida today. In them, thick deposits of peat, mucky sediment that consists mostly of partially decayed plant material, accumulated. Later sediments buried the peat to great enough depth to put it under enough pressure to turn it into coal.

Demand for central Montana coal blossomed in the late 1800s and early 1900s to power railroads and smelters that served mining districts. Several large coalfields were active, and production increased steadily until the 1920s. Almost all central Montana mines worked underground because the beds of higher-quality coal, found in older formations that had been deeply buried, were steeply tilted. Central Montana contains very little coal at shallow enough depth and in thick enough beds to make strip-mining feasible, so there is no alternative to underground mining. That's not the case in eastern Montana. With the advent of mechanization, new railroad spurs, and a growing demand for coal-fired electric power, strip-mining developed in eastern Montana; that spelled the end for expensive underground coal mining in the rest of the state.

Historic coal mining in central Montana spurred the development of many small communities, such as the town of Belt southeast of Great Falls, but it also left a legacy of contamination. Some of the old, abandoned underground coal mines discharge acidic water loaded with dissolved metals, such as iron, sulfur, aluminum, arsenic, cadmium, lead, and zinc, that severely pollute the streams. Much of the acidic drainage from coal mines is sulfuric acid formed by the oxidation of small amounts of pyrite, an iron sulfide found in the coal. In Belt Creek, for example, iron and sulfur concentrations are so high that the water can be orange, especially during low flows from late summer through early spring. State agencies plan on treating the water from some of these mines in the future.

Strip-mining in the Powder River Basin. The coal is at the bottom of the pit.
—Courtesy of Dave Mogk

The Powder River Basin contains mainly moderately low-grade subbituminous coal, including that mined near Colstrip. Farther northeast it's low-grade lignite, which retains more moisture and provides lower heat content per ton burned. In terms of coal's heat-generating capacity or quality, coal ranges from lignite to subbituminous, bituminous, and finally anthracite. The highest-quality coal, anthracite, has the lowest moisture content and provides the greatest heat when burned. Such coal requires deeper burial and heat from within the Earth to drive off most of its moisture. It is primarily found in the Appalachian Mountains of the eastern United States and is extracted from underground mines. Eastern Montana's lower-grade coal has two main advantages: it occurs in horizontal layers at shallow depths, so it's amenable to strip-mining, and it's low in sulfur, so when it's burned in power plants it generates less acid-rain-producing sulfur dioxide that must be scrubbed out of the stack gases.

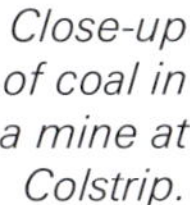

Close-up of coal in a mine at Colstrip.

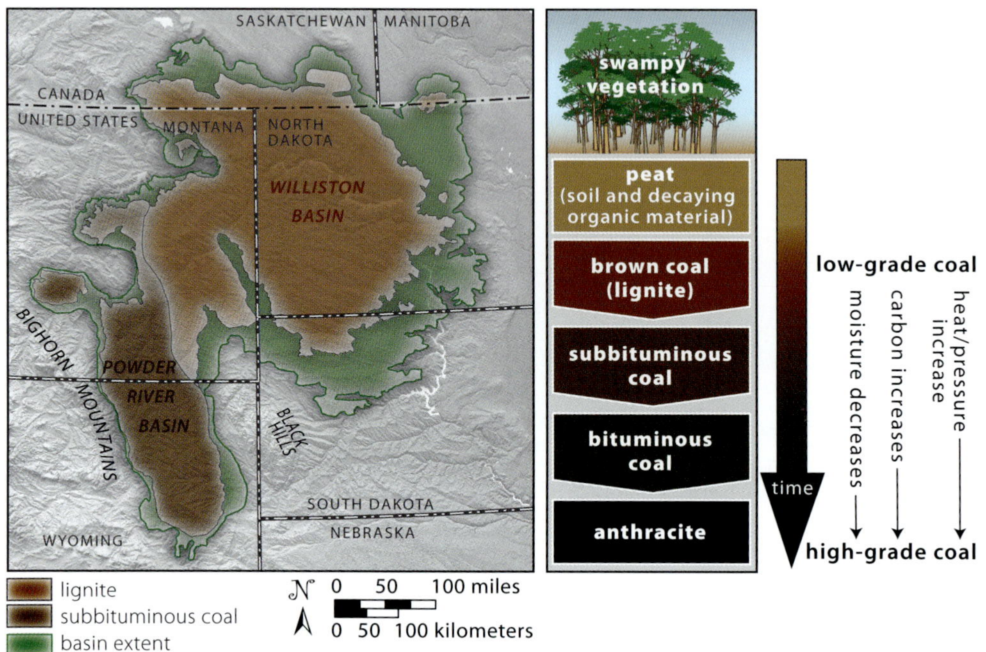

Coal in Montana and adjacent states. —Modified from Downey and Dinwiddie, 1988

Trees are scarce in most of eastern Montana, so the first European settlers were glad to find coal and quick to begin mining it for domestic use. Demand for Montana coal increased dramatically with the coming of the railroads during the 1880s, with their fuel-hungry steam engines. Then cheap oil intervened. Diesel engines, oil furnaces, gas furnaces, and gas-fired smelters all contributed to the downfall of Montana coal. Production went into a steep and steady decline that closed one mine after another until the end came during the 1950s. The Northern Pacific Railway, Montana's largest, switched to diesel engines in 1958, and that seemed to doom the Colstrip Mine.

Montana Power purchased the mine, however, and built two large coal-fired electric power plants between 1975 and 1976, then two more between 1984 and 1986. By 2006, Montana was the sixth-largest coal producer in the United States, with 41.8 million tons produced that year. With more than 100 billion tons of recoverable coal, it had the largest coal reserve of any state in the country. As of 2019, about 25 percent of Montana-mined coal was burned in the four Colstrip plants. Much more was shipped to other states and countries to generate electricity there.

Coal-fired plants produce air pollutants, and laws that restricted emissions of these pollutants, and the 2007 US Supreme Court decision designating carbon dioxide as a pollutant, spelled big trouble for coal. In 2015, Montana's seven coal-fired power plants produced 18.2 million tons of carbon dioxide per year, 55.6 percent of the state's total, along with 36,000 tons of nitrogen oxide and 18.2 million tons of sulfur dioxide. The sulfur dioxide forms from the oxidation of sulfur in pyrite in the coal.

Coal faced opposition from the public and ranchers over air and water pollution and the heavy use of water by power plants, and others were opposed to coal being shipped overseas from coastal ports, spelling more trouble. In the early 2000s coal was still the lowest-cost fuel for generating electric power, but by 2019 that was no longer the case. Costs increased with necessary new equipment to reduce pollutants and with repairs to aging power plants. Natural gas has become the go-to fuel because it produces far fewer emissions and is relatively easy to get thanks to the development of fracking technology, although it comes with its own environmental issues. In 2016, natural gas surpassed coal as the fuel that produced most of the electric power in the United States. Solar and wind were about on par with coal for cost. If pollution and cleanup costs are included, coal is much more expensive than these alternatives.

In 2020 only the largest two of Montana's coal-fired power plants—Colstrip units 3 and 4—remain in operation. The nearby Rosebud mine will continue to provide coal to the power plant complex at Colstrip. At the mine the overlying sandstone and shale waste rock is drilled and blasted so it can be stripped off with giant power shovels and clamshell cranes to be piled in a long ridge next to the long excavation. Power shovels then load the underlying coal into huge trucks, which transport it to a miles-long conveyor belt that feeds the coal to the power plant. Following mining, the mined-out pit is backfilled with waste rock, covered with stockpiled soil, and replanted with native vegetation, as dictated by Montana law.

Many coal seams serve as aquifers that hold large quantities of water that feed area wells, so the effect mining has on the availability and quality of groundwater is a serious issue for residents. Pumping the water out of a mine certainly does lower the water level in the immediate surrounding area, but that level will restore itself naturally within a few years after the mining and pumping cease. However, a return of the original water table doesn't necessarily restore the original water quality, because some of the aquifer has been replaced with waste rock. In most cases, the broken rubble used to backfill mine pits during reclamation should contain enough open space to be a reasonable aquifer, better in some cases than the original coal. But minerals dissolving in the freshly broken backfill tend to make the groundwater both salty and alkaline, so much so that in some cases it's unusable. The effect will certainly extend far in the direction of groundwater flow and last for a very long time.

Early miners left an ugly legacy. Enormous tracts of formerly beautiful and productive land were furrowed as though by a gargantuan plow. Long trenches from which the coal was stripped remain unfilled to this day, with parallel ridges of waste rock. The slow, natural processes of weathering and soil formation cannot return such devastated landscapes to any semblance of productivity for thousands of years. Modern mine reclamation is now so highly developed that recently worked strip mines are visible only to a trained eye.

Oil and Gas

Oil wells and pump jacks dot the landscape in parts of central and eastern Montana, especially along the High Line (the northern fringe of Montana, generally along US 2, and the route of the Burlington Northern Railroad) and on Montana's eastern border. Oil typically forms in organic-rich muds or shales at depths of 11,000 to 20,000 feet and at temperatures between 140 and 284 degrees Fahrenheit. The buried organic material must be nearly free of oxygen that would otherwise oxidize it. Because oil and especially natural gas are lighter than water, they percolate upward in water and float on top of it. In fact, oil and gas would rise to the Earth's surface and escape unless they were somehow trapped.

Finding oil and gas fields requires locating such traps. A structural trap may be an anticline or dome in which an impermeable layer such as shale prevents oil and gas from rising higher. Or oil and gas rising in a dipping layer may be trapped where a fault seals them in, preventing further rise. A stratigraphic trap typically forms where a layer of permeable rock such as sandstone grades into impermeable rock such as shale, in which the mineral grains are too small to permit oil or gas to continue to rise.

As an example, after having mapped the Cedar Creek anticline by 1910, the earliest geologists in eastern Montana immediately recognized its tilted beds as the perfect structure for trapping oil and gas. The straight northwest-trending crest of the fold stretches for more than 110 miles, from south of Baker near the North Dakota border to Glendive. A successful wildcat well found natural gas in the crest of the Cedar Creek anticline in 1912.

Oil tends to occur in places where the veneer of sedimentary rocks that covers the continental basement is abnormally thick. Low places in the continent, or basins, are places where large volumes of sediment accumulated because they were below sea level for longer than other parts of the continent. Because oil typically occurs in sedimentary formations laid down in shallow seawater, the former basins of these persistent inland seas are the best places to look for oil. Montana has exploited two large basins: the Williston and the Powder River.

The Williston Basin is a broad depression—an enormous shallow saucer filled with sedimentary rocks—centered in western North Dakota and also covering a large

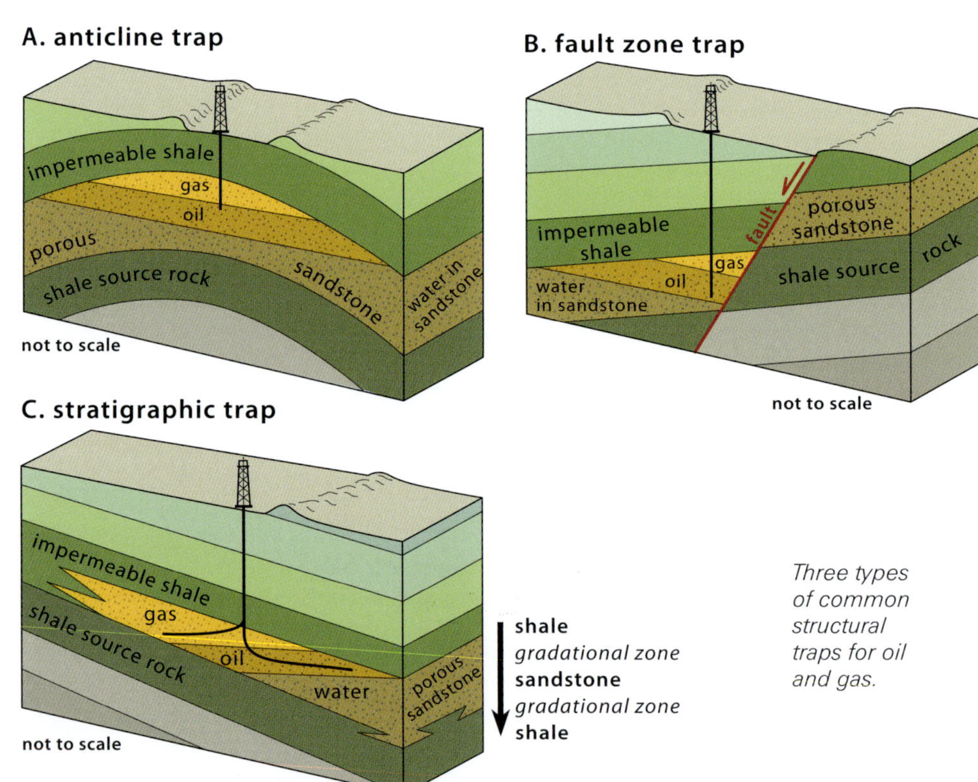

Three types of common structural traps for oil and gas.

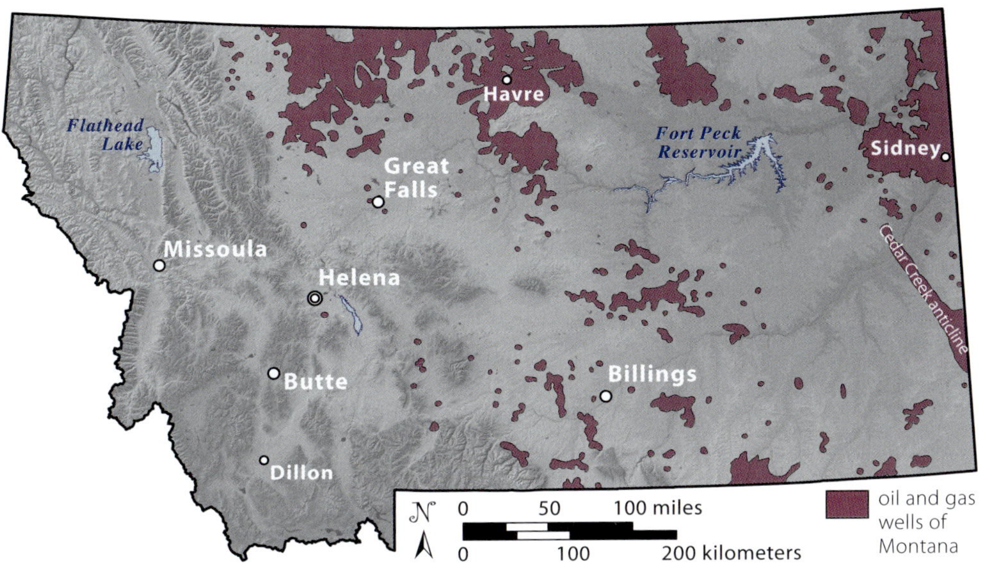

Oil and gas fields in Montana.

area of eastern Montana and southern Saskatchewan. The western part of the basin, in northeastern Montana, contains a large fraction of the state's oil and gas wealth. The deepest part of the basin, in North Dakota, is 16,000 feet thick. Throughout most of Paleozoic time, starting at least 500 million years ago and continuing well into Mesozoic time, the Williston Basin was a persistently low area. When shallow seas invaded the continent, they flooded the basin sooner, grew deeper, and remained longer than in other nearby areas. Many formations in the Williston Basin contain thick beds of salt, marking former isolated arms of inland seas where evaporating seawater dumped loads of salt.

The Williston Basin was among the last oil fields discovered in North America. Wildcat drillers began working the basin in the years after the Second World War; they brought in the first oil well in the Montana portion in 1951. Production climbed rapidly and peaked in 1966. Vigorous exploration drilling during the oil shortage years of the late 1970s and early 1980s discovered a number of new oil fields that greatly slowed, but did not quite reverse, the long downward trend in production. At the end of 1983, the Montana part of the Williston Basin contained a total of 1,446 active wells, which produced almost 21 million barrels of oil that year—more than twice the combined production of all other oil wells in Montana at the time.

The Bakken Formation of the Williston Basin in Montana and North Dakota is a distinctive black shale, siltstone, dolomite, and sandstone formation of Late Devonian to Early Mississippian age; it thickens from nothing to almost 100 feet in the far northeastern corner of Montana. The dolomite and brownish-gray siltstone are underlain and capped by the flaky black shale source rock. Oil was first discovered in the formation in 1951, but its fine grain size made it difficult for the oil to move, so production remained uneconomic. The use of horizontal drilling greatly extended the reach from individual wells, and hydraulic fracturing, or fracking, beginning in

2000 created millions of wider cracks through which the oil could seep. Between 2004 and 2008, the Bakken produced about 10 million barrels of oil per year.

By early 2016, oil had dropped to $30 per barrel, so few new wells were being drilled, especially in Montana, and production dropped sharply. By 2018, the price had recovered to $50 per barrel, but not enough to encourage new drilling in Montana. What is the future of oil extraction in the Bakken? As in the past, it will depend on the field's economic viability going forward, but since these are nonrenewable fossil fuels, the supply is not endless, and the oil will eventually run out.

In addition, insufficient pipeline capacity forced producers to ship crude oil to refineries by rail. Bakken oil is highly volatile and flammable, and during major derailments it has proved violently explosive. Adding to the environmental concerns and production costs is what to do with the wastewater and brine that are by-products of fracking. Fresh groundwater is often injected down a well with fracking fluids that help float the oil; that water mixes with salt and brine in the reservoir rocks, and some brine, fracking fluids, and lubricant come up with the oil. These waste products must be separated and properly disposed.

Unfortunately, much of the natural gas produced from the Bakken is wasted—burned off because of insufficient pipeline capacity to ship it to market. Oil is much more valuable than natural gas, so companies are unwilling to curtail oil production. Burning off the gas is their solution. Aside from wastefulness, this also contributes to global warming and climate change. Companies also vent gas directly into the atmosphere. Natural gas (methane) is twenty times more potent than carbon dioxide, the most-cited contributor to global warming.

Most of the Powder River Basin, the other large basin in Montana, is in Wyoming between the Bighorn Mountains and the Black Hills. The basin owes its existence more to deformation of the Earth's crust during the compressional development of the Rocky Mountains than to a long history of crustal subsidence, such as that which produced the Williston Basin. It formed due to the rise of the Bighorn Mountains in the west and the Black Hills of South Dakota in the east, during the Laramide deformation in early Tertiary time, about 50 million years ago. In its deepest part, sedimentary rocks above the basement rock are about 18,000 feet thick. Sedimentary rocks in the north end of the Powder River Basin, the portion in Montana, are less than 5,000 feet thick.

Oil actually seeps to the surface in the Powder River Basin, so drilling began early. Wildcatters discovered a shallow oil field in 1887 by drilling near a seep, and then another in 1889. Drilling in the crests of folds exposed at the surface led to the discovery of several important oil fields in the early twentieth century in Wyoming. All surface folds were drilled before 1925 because they were the obvious traps.

The only big oil discovery in the Montana part of the basin came in 1966 with the discovery of the giant Belle Creek field. Like many other oil fields in the Powder River Basin, the Belle Creek field produces from Early Cretaceous sandstone. Oil and gas were trapped in a single, highly porous and permeable, 20-to-30-foot-thick formation that permitted oil and gas to percolate upward along the bedding to where the formation became impermeable, trapping the petroleum. The nature and distribution of the oil and gas in the formation, largely at a depth of 4,500 feet, was clear from the beginning, so by 1967 production had been maximized.

As oil recovery dwindled in the field and it was recognized that increasing carbon dioxide in the atmosphere was a serious contributor to climate change, facilities were built to capture and use the carbon dioxide. Beginning in May 2013, gas was pumped back into the producing zone to enhance oil recovery (push more oil back up the wells) and to convert the former oil zone into a reservoir for future use and for possible use in other oil fields elsewhere. Carbon dioxide from power plants is captured and pumped back into the sandstone to be permanently sequestered (stored).

Glacial Lakes and the Missouri River during the Ice Age

As the glaciers of the Bull Lake and Pinedale glaciations reached their maximum extent in the mountains, continental ice of the massive Laurentide ice sheet was moving southwestward from the vicinity of Hudson Bay into central and eastern Montana. As in the Rockies, two distinct glacial advances of the Laurentide ice sheet are recognized in Montana: the Illinoian and Late Wisconsin. The older Illinoian is roughly correlative with the Bull Lake glaciation. The younger Late Wisconsin glaciation is roughly equivalent to the Pinedale glacial advance. The portion of the ice sheet that reached into Montana was relatively thin and moved very slowly. It probably only stretched into Montana as each ice advance approached its climax, then began to melt a few thousand years later. Ice lapped onto the northern edges of the Bears Paw, Highwood, and Little Rocky Mountains, as well as of the Sweet Grass Hills, but did not cover any of those ranges. The southern margin of the ice sheet appears to have been extremely irregular, but it closely paralleled the Missouri River.

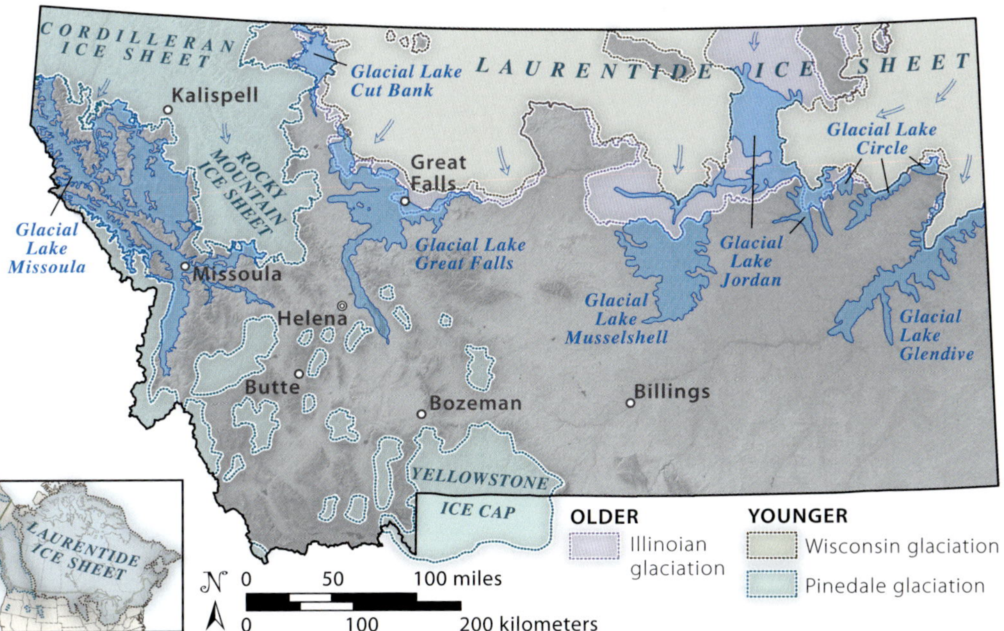

The approximate maximum extent of the Illinoian and Late Wisconsin continental ice sheets in central and eastern Montana. The diverted Missouri River was pushed to the south, and north-flowing streams backed up as lakes against the ice. —Modified from Fullerton and others, 2012

The effects of glaciation on the landscape of north-central and northeastern Montana are surprisingly minor. In most areas, the only obvious evidence is widely scattered erratic boulders. All are made of blocky pale granite or streaky basement gneisses, quite unlike the somber, platy, gray sandstone and shale that we see in Montana outcrops. The granite and gneiss came from basement rocks of central Canada well to the north. Here and there, tracts of hummocky moraines form long ridges that precisely record the ice margin; layered fine-grained sediments show the location of glacial lakes, and a few abandoned valleys show us where torrents of glacial meltwater once flowed.

Prior to glaciation, the Missouri and other rivers flowed generally northward and eastward, down the regional slope of the land to Hudson Bay. When the ice sheet advanced, it blocked those rivers. Because the land surface sloped down to the north, glacial meltwater was trapped along the southern edge of the ice. The meltwater grew into glacial lakes, which were connected by rivers along the edge of the ice sheet. Had you lived in Montana during the maximum of the Late Wisconsin glaciation 23,000 years ago, you could have paddled a canoe along the edge of the ice almost all the way from Cut Bank to Glendive. The largest of those lakes, Glacial Lake Great Falls, covered a vast expanse of the plains between Great Falls and Cut Bank and flooded the site of Great Falls to a depth of some 600 feet. If that lake still existed today, we would think it an inland sea comparable to the Great Lakes. Generally similar but larger and more continuous lakes existed along the edge of the Illinoian-age ice sheet when it reached its maximum about 140,000 years ago.

The continental ice sheet formed high cliffs of blue ice along the northern shores of the lakes. Icebergs that weighed thousands of tons broke off and smashed into the water with a thundering roar, then drifted across the lakes like floating islands, carrying boulders from the north, which they dropped when they melted. Great herds of horses, antelope, elk, giant bison, and camels roamed the hummocky grasslands around the lakes, while Columbian mammoths, mastodons, ground sloths, and musk oxen banded together into smaller groups. Predators such as the American lion, dire and gray wolves, 2,000-pound bears, and saber-toothed cats with 7-inch canines were an ever-present danger.

The old shoreline benches, scattered boulders dropped from melting icebergs, level plains floored with deep deposits of fine sediment laid down on the lake bottoms, and bones of Pleistocene animals are all that remain. Even where they are not conspicuous, old lake shorelines are fairly easy to recognize as horizontal benches that look exactly like modern lake shorelines, except that there's no lake. Wind-driven waves on the downwind sides of lakes eroded the benches. Larger lakes, stronger winds, and long-lasting lake levels on softer rocks form the most pronounced shorelines.

The Missouri River was rudely shoved south out of its regular channel during the Pleistocene ice advances. In 1953 an observant US Geological Survey geologist, W. C. Alden, pointed out that the broad valley of the lower Milk River is about the size of the Missouri River Valley downstream from Fort Peck Reservoir. He suggested that the Missouri may have occupied the broad valley before the ice sheet pushed it south; when the ice melted, the Missouri continued to flow in the narrow channel it had established along the edge of the glacier. The Milk River, however, established itself in the large, former Missouri River channel from near Havre to near Fort Peck. Big

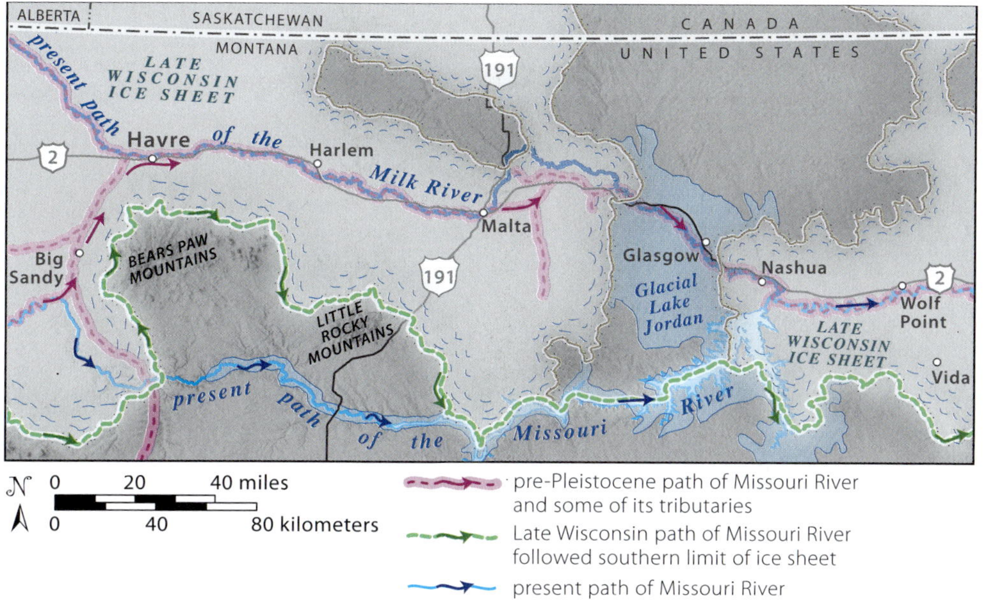

The Milk River now flows in the pre-Pleistocene channel of the Missouri River.

Sandy Creek follows another long segment of the former valley of the Missouri River between Big Sandy and Havre.

The continental ice sheet probably retreated from Montana sometime around 12,000 years ago and didn't return, not even during the Little Ice Age cold snap from AD 1300 to 1870. With the ice went the mammoths, mastodons, saber-toothed cats, and other unique ice-age animals, but humans survived the warming climate on a landscape shaped by ice.

Saline Seeps and Alkali Water

White salty crusts in low parts of fields are indicative of saline seeps, caused by salty groundwater rising to the surface and evaporating. The spreading development of saline seeps is one of the biggest agricultural problems in large areas of central and eastern Montana. Ironically, precipitation in excess of that used by plants, or even by excessive irrigation, can cause such seeps.

To a considerable extent, the problem is the result of farming practices, though seeps do emanate from natural springs that produce water heavily laden with dissolved salts. The saline water varies in composition but normally contains some combination of chlorides (sodium, calcium, and/or magnesium), carbonates, sulfates, and nitrates—an evil alkaline brine that tastes salty and bitter. The first sign of a developing saline seep is a patch of almost unnaturally dark-green vegetation that consists largely of sedges, reedy plants with tall stems that are triangular in cross section. Sedges like plenty of water, and they don't mind a bit of salt, so they thrive in saline seeps.

Most of the seeping brine evaporates as it reaches the surface, leaving its dissolved salts behind. Within a few years, those salts accumulate in the soil to a concentration

Saline seep south of Sweetgrass.

that kills the sedges, along with everything else. Then the seep becomes a patch of barren ground that turns into sticky mud during wet weather and acquires a glaring white crust of bitter salts when the weather turns dry.

The dry plains of Montana have always had a few saline seeps. Now there are a great many, and more appear every year. Most of the new ones are probably the result of dry-farming practices that became widespread during the 1920s. Dry farming entails planting a crop on alternating strips of land while keeping those between fallow and bare of all plants. With no plants to use it, an entire season's worth of water accumulates in the fallow strips. The next crop is planted in the fallow strips, while those just harvested are laid fallow. The effect is to use the moisture of two seasons to raise one crop. The problem is that the method uses a bit too much water.

Water accumulating in the fallow strips soaks down through the soil to the groundwater reservoir and raises the water table. Normally, without irrigation, little to no water would pass all the way through the dry soils, so the groundwater reservoir would be recharged mainly by seepage through the beds of streams. The new source of recharge would be good if not for the large amounts of salts in the soils of Montana's dry regions. As the excess water passes through them, it dissolves the salts, so the water is salty and alkaline by the time it reaches the water table. As the water table continues to rise, it eventually intersects the ground surface, forming new seeps. Wells that had produced sweet water become bitter. The local drinking water often contains high concentrations of dissolved alkalis, especially calcium sulfate, that can act as an effective laxative. Bring your own water when traveling through this region.

There is no quick solution for a problem that developed over a period of more than fifty years. In the long range, changes in dry-farming techniques can probably limit the future development of new saline seeps. The only cure for most of those that already exist is to let the rain leach the accumulated salts out of the soil. That will be an extremely slow process in a region as dry as central and eastern Montana.

Slippery Roads on Gumbo Clay

As you travel central and eastern Montana in search of geological treasures, be wary that the dirt roads are notoriously slippery when wet. Even the slightest drizzle will turn dirt tracks into a skating rink more reminiscent of black ice, even on a hot, midsummer day. Volcanic ash delivered from western volcanoes has gradually turned to bentonite (smectite clay), a notorious clay that swells when wet. The clay takes even the smallest amounts of water into its molecular structure, expanding its mineral grains by about 40 percent by volume. When a car or truck drives on such a road, its weight squeezes out the water, collapsing the structure, and the vehicle can skid out of control. With a little more moisture, the sticky "gumbo" clay can build up on tires and the tire tracks can turn to ruts. A ranch driveway graveled with red clinker from a burned coal layer is a sign of gumbo clay—and a rancher's solution to it. Popcorn weathering of the ground is another sign of swelling clay.

Clay that's weathered to a popcorn-like texture in the Late Cretaceous Hell Creek Formation in Makoshika State Park.

UPPER MISSOURI RIVER BREAKS

Coal Banks Landing—Judith Landing

88-mile float trip

About 10 miles southwest of Big Sandy, at Coal Banks Landing, the Missouri River takes an abrupt turn southeast, dropping into a much more pronounced, narrower valley. About 20 miles farther southeast, the river digs deeper below the High Plains Surface—almost 1,000 feet. The Missouri River has cut this path and canyon only since the retreat of the ice sheet of the Late Wisconsin glaciation, which pushed the river south out of its original channel. This spectacular canyon led to the river being designated a National Wild and Scenic River in 1976 and the creation of the Upper Missouri River Breaks National Monument in 2001. In spite of its remoteness, the national monument gets thousands of visitors each year, mostly floaters and paddlers.

Downstream of Fort Benton, the rocks in the riverbanks and along the valley are almost all Late Cretaceous sandstones and shales lying in nearly horizontal layers, all deposited about 90 to 70 million years ago along the coastal plain and in the shallow waters of the Western Interior Seaway. The harder sandstones, deposited by streams and on beaches or offshore sandbars, stand as cliffs; softer shales, deposited farther offshore, form gentler slopes. Cross beds, the inclined internal layers in the sandstone,

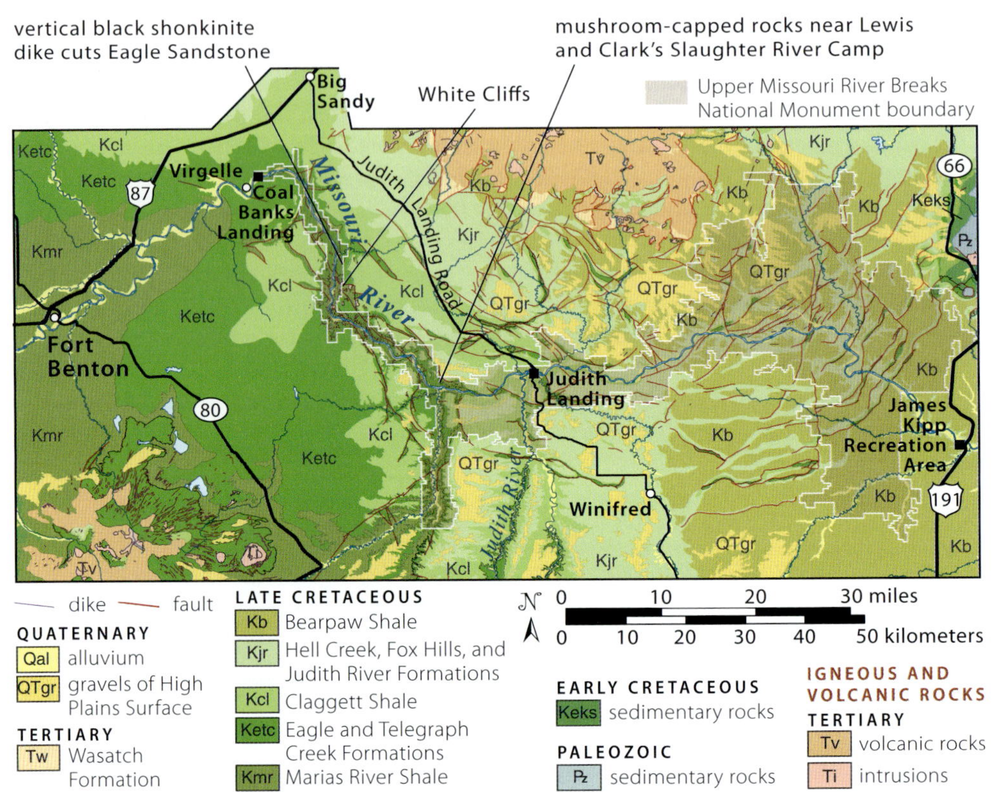

Geology of the Missouri River Breaks.

indicate the downwind layers in a sand dune or the down-current layers in flowing water that transported sediment generally eastward toward the seaway.

Between river miles 48 and 110 (the BLM uses Fort Benton as mile 0, Judith Landing as mile 88.5), dramatic cliffs that give this section of river the name White Cliffs are mostly quartz sand of the Eagle Sandstone (Virgelle Member). Ranging in thickness from 100 to 350 feet, the sandstone is nearly white to slightly yellowish or brownish. It was primarily deposited by migrating coastal stream channels draining eastward into the seaway from the Rocky Mountains. The dark-colored formation below, the Marias River Shale, is a clay-rich marine mud deposited in the quiet water of the seaway. The change from marine mud to nonmarine sand shows that the ocean retreated eastward, and the sand migrated over ocean floor mud with the retreating shoreline.

The cliff walls form due to vertical cracks in the rocks that widen over time as water freezes and expands in them. Erosion of the underlying soft shale undermines the cracked rock, which falls off, leaving near-vertical cliffs. In places, the cracks in the sandstone have eroded enough to form narrow slot canyons hardly wide enough to squeeze through. Pockmarks in the sandstone, several inches or more across, form where saline water dries and growing salt crystals pry individual sand grains off the rock surface. The process repeats itself when it rains, and wind blows the loose grains out of the hollows.

White cliffs of Eagle Sandstone (Virgelle Member) exposed in Upper Missouri River Breaks National Monument. —Courtesy of Bureau of Land Management

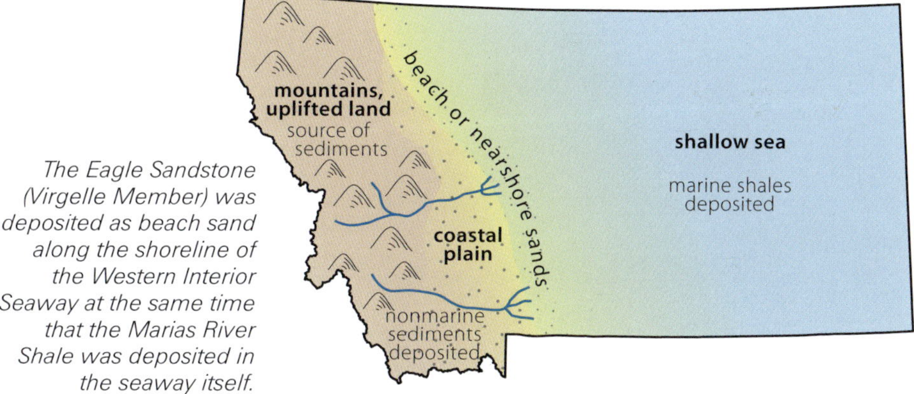

The Eagle Sandstone (Virgelle Member) was deposited as beach sand along the shoreline of the Western Interior Seaway at the same time that the Marias River Shale was deposited in the seaway itself.

As the vertical fractures widen, columns form. Pedestal rocks, looking like giant toadstools, form where a resistant layer of Eagle Sandstone protects the underlying softer sandstone (Virgelle Member) from erosion. You might also notice dark spherical concretions in the sandstone that range in size from baseballs to cannonballs or even larger. The concretions formed within the sediment as minerals precipitated around a nucleus, such as a shell, to form a hard mass that resists weathering better than the surrounding sandstone.

The nearly black rocks standing as thick vertical walls in the White Cliffs are dikes of shonkinite. The magma was injected into vertical fractures in the sandstone around 50 million years ago, during Eocene time. The igneous rock is much harder than the sandstone, so it weathers out in relief with time. These strange rocks, rich in the alkali elements sodium and potassium, are found elsewhere in the Central Montana Alkalic Province. The igneous rocks in the Bears Paw Mountains to the northeast are the likely source of the magma that formed these dikes. (See the US 87: Great Falls—Havre road guide for a discussion about the Bears Paw Mountains.)

Pedestal rock with a capstone of a hard, iron-oxide cemented unit of Eagle Sandstone protecting the softer underlying sandstone (Virgelle Member) from erosion. —Courtesy of Rod Benson, Bigskywalker.com

A conspicuous dike of hard black shonkinite juts out from the white Eagle Sandstone in the White Cliffs area. —Courtesy of Rod Benson, Bigskywalker.com

The white sandstone has yielded dinosaur fossils and petrified wood, and the underlying shale contains fossils of coiled squid-like animals called ammonites. Native Americans carved petroglyphs on the rocks, and the dreams of homesteaders were left behind in abandoned buildings.

Downstream to the east, the Missouri River cuts through higher and higher (younger and younger) layers of Late Cretaceous sedimentary rock formations. Light-gray, distinctly bedded shales and sandstones of the Judith River Formation form cliffs on the north bank of the Missouri River at the McClelland Ferry crossing.

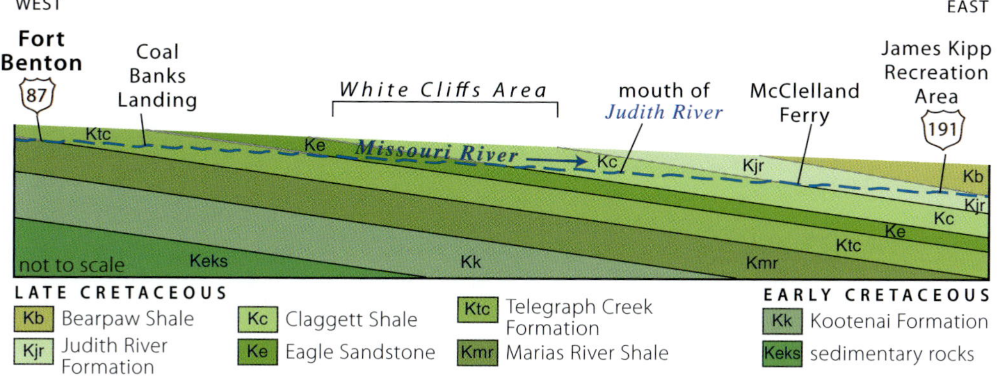

West-east cross section along the line of the Missouri River showing formations exposed in the cliffs on the north bank.

I-15
Helena—Great Falls
89 miles

Helena is near the south end of the Helena Valley, where Quaternary sands and gravels lie between mountains eroded in granite of the Boulder batholith and the Big Belt Mountains to the northeast. Between Helena and Wolf Creek, I-15 angles across a portion of the Overthrust Belt, big slabs of rock, mostly colorful Proterozoic Belt sedimentary formations, that were shoved eastward over Cretaceous sediments.

The Scratchgravel Hills rise west of the road 5 or 6 miles north of the US 12 junction in Helena. This isolated group of hills sparsely covered with trees is a bit east of the main mass of mountains. They contain a granitic intrusion that was injected into Belt sedimentary rocks. It crystallized about 85 million years ago, timing that probably means its magma is related to the earliest stages of the nearby Boulder batholith. A thin sheet of gravel that covers part of the surface on the north side of the hills contains gold nuggets that early settlers collected by ploughing and raking the land, hence the name. The Franklin Mine worked these meager bedrock deposits of gold from about 1870 until 1919, finally recovering about $500,000 in gold and some silver. In the 1970s cyanide leaching was used to recover more gold from old mine waste. Old mine dumps contain pyrite, galena, chalcopyrite, cerussite, and pyrolusite.

About 12 miles north of US 12 and 14 miles west of the interstate (access from exit 200) is the old mining town of Marysville. The first gold in nearby stream gravels was found in 1862, but richer deposits were found in 1864 at the top of the bedrock, which is about 15 to 20 feet below the surface. The main mine, the Drumlummon, mined silver and gold from both surface and underground workings discovered in 1876. The veins are in the Late Cretaceous Marysville granodiorite-to-quartz-diorite stock that intruded Belt rocks. Mines in the Marysville mining district were especially active from 1876 to the 1890s, reportedly recovering $28 million of primarily silver and gold, with lesser amounts of manganese and copper, through 1911. Old mine dumps nearby have pyrite, galena, sphalerite, chalcopyrite, fluorite, cerussite,

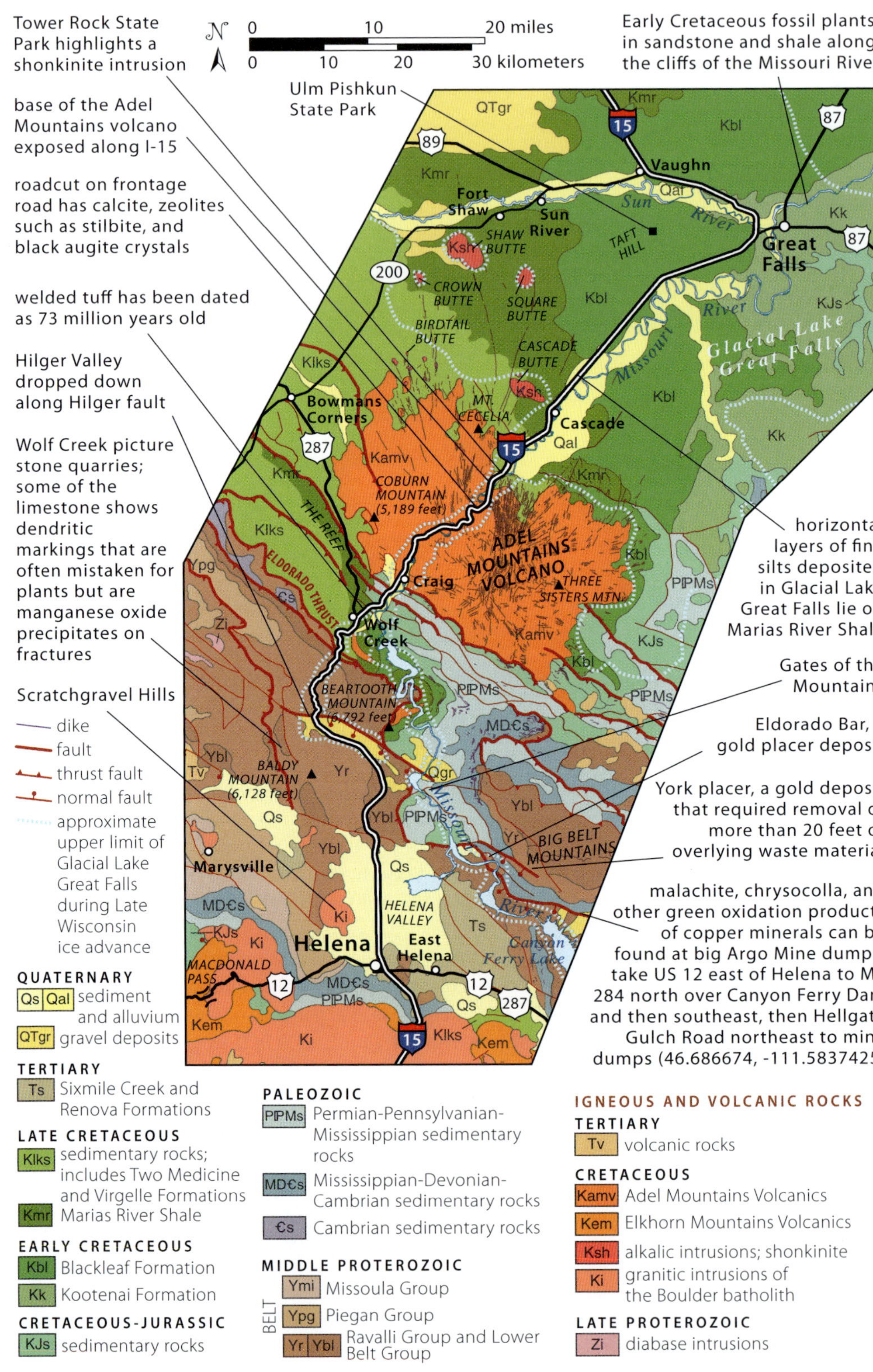

Geology along I-15 between Helena and Great Falls.

malachite, and some quartz crystals in cavities. Now essentially a ghost town, many of the original buildings in Marysville survive, and the few residents must commute to Helena for work.

About 10 miles north of Helena, I-15 crosses an east-west fault, north of which the highway climbs into hard Ravalli Group mudstones and sandstones of the Belt Supergroup. These beautiful rocks were deposited in the Belt Sea, a vast inland sea that covered northwestern Montana and parts of Idaho, Washington, British Columbia, and Alberta between 1.47 and 1.4 billion years ago. The rocks spread east of Helena well into central Montana in a west-to-east arm of the basin called the Helena Embayment. Look for the strikingly red exposures next to the road. These rocks are loaded with mud cracks, ripple marks, and other evidence of deposition on a mudflat that was episodically exposed and inundated with shallow water.

About 17 miles north of Helena a side road leads east to Gates of the Mountains, an impressive canyon cut through white Madison Group limestone of Mississippian age. Historians identify this as the canyon that Lewis and Clark named "Gates of the Mountains" because they considered it their point of entry into the Rocky Mountains. It's possible that the historical identification is wrong because Lewis and Clark

View looking upstream at the north end of what Lewis and Clark might have called the Gates of the Mountains, cut in dark shonkinite of the Adel Mountains volcano.

The south end of the possibly misnamed Gates of the Mountains, cut in pale-gray Madison Group limestone.

were traveling upstream (to the southwest) and described a canyon cut through "a black grannite" with a much lighter-colored rock above. The first deep canyon that Lewis and Clark encountered on the Missouri is about 10 miles southwest of Cascade on I-15. There the river cuts through dark-gray shonkinite of the Adel Mountains volcano, not the nearly white limestone seen by boaters at the south end of the canyon north of Helena. The "black grannite" would likely be the dark shonkinite, with pale limestone above.

The river probably began cutting the canyon across the Big Belt Mountains in Miocene time, when basin and range extension formed the Hilger fault, a northwest-trending range-bounding normal fault that raised the northern end of the Big Belt Mountains and dropped the Helena Valley. As the mountains went up, the Missouri River got stuck in hard rock, cutting its canyon in the Madison Group limestone. The fault is likely still active and shows the tell-tale signs of recent activity, such as linear fault scarps and faceted spurs (chopped-off ridge ends) near the base of the range. The distance between the upper canyon walls of Gates of the Mountains is much wider than that lower down, suggesting the range has risen faster over the last few million years, causing accelerated incision by the river.

During the summer months, excursion boats provide regular trips through the canyon past spectacular ledges of tightly folded Madison Group limestone. If you get off the tour boat at the Meriwether Picnic Area, be sure to take a close look at the limestone, which is loaded with seashells that collected on the bottom of a tropical ocean about 360 million years ago. Please do not collect the fossils, since they should remain for everybody to see.

South and west of the boat launch, there are exposures of light-colored lake sediment left by Glacial Lake Great Falls when it flooded the canyon during the Late Wisconsin glacial advance. Distinct light and dark annual layers called varves were deposited as the lake changed with the seasons. During the summer, abundant sediment-laden glacial meltwater transported light-colored coarse silt and fine sand to the lake; it settled to the bottom. When the lake froze in the winter, the water was calm enough for clay particles and bits of organics to sink, forming the finer-grained, darker layers. Exposures of varves occur downstream along the banks of lower Holter Lake.

Between about 18 and 26 miles north of Helena, I-15 passes through the Hilger Valley. It looks exactly like a stream valley, complete with tributaries, and it's big enough to hold a stream larger than the Missouri River, but the valley contains no stream worth mentioning. The valley likely started forming in Miocene time as it dropped along the Hilger fault, creating a short but broad valley. Little Prickly Pear Creek at the headwaters of the valley catches most of the water; from there it flows northeast into the Missouri River, leaving the Hilger Valley dusty and dry.

About 27 miles northeast of Helena, I-15 crosses the Hilger fault. It's easy to recognize because it marks where I-15 leaves the Hilger Valley and enters the mountains through the canyon of Little Prickly Pear Creek. The raised side of the normal fault exposes hard sedimentary rock of the Belt Supergroup in the narrow, marvelously picturesque canyon. The rock-bound canyon exposes huge roadcuts of colorful red and green mudstones and sandstones of the Spokane Formation, a Belt Supergroup deposit named for the Spokane Hills east of Helena. The bedding surfaces are covered with mud cracks and ripple marks made in sediment deposited in the shallow Belt Sea more than 1.4 billion years ago.

Belt Supergroup rocks exposed in the Little Prickly Pear Creek canyon along I-15 north of Helena. The inset shows a side view of mud cracks filled with gray sand. The dark red, broken pieces are mud chips.

About 2 miles south of exit 226 to Stearns-Augusta Road (MT 434) in the town of Wolf Creek, I-15 crosses a major northwest-trending thrust fault, north of which are much softer volcanic-ash-rich sedimentary rocks of Jurassic to Cretaceous age. The Proterozoic rocks were shoved northeastward along the southwest-dipping fault over rocks that are more than 1 billion years younger. Between Wolf Creek and Craig, I-15 crosses several more northwest-trending thrust faults, the southeastern, tapering continuation of major thrust faults in the Overthrust Belt south of Glacier National Park. Because some thrust faults carried Proterozoic Belt Supergroup rocks over the Late Cretaceous Adel Mountains volcanic rocks, the thrust faults must have moved at least as recently as Late Cretaceous time.

Northeast from Craig, both I-15 and old US 91 (parallel to it) slice right through the Adel Mountains volcano in a narrow, steep-walled canyon of the Missouri River that Lewis and Clark's party dragged their boats upstream through more than two hundred years ago. Lieutenant John Mullan avoided building his road through the canyon, and it wasn't until 1931 that builders ventured through it with US 91. You can still drive that spectacular road all the way from Wolf Creek to Cascade. It's a more relaxed alternate to I-15 that still uses two original steel truss bridges built with the original highway.

New radiometric age dates indicate that the Adel Mountains volcano, formerly thought to be about 50 million years old, was active from about 76 to 73 million

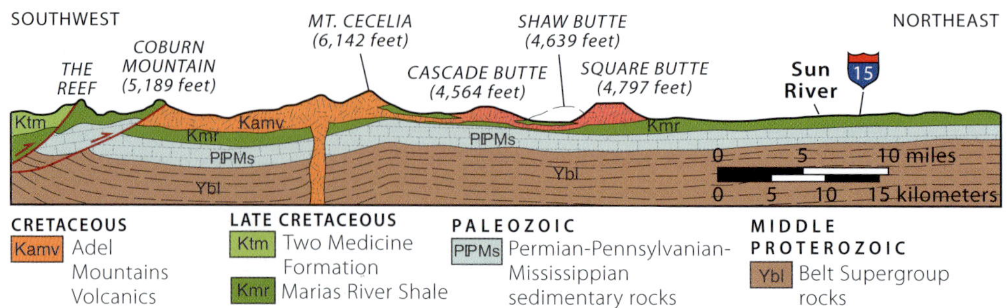

Southwest-northeast cross section west of I-15 between Wolf Creek and Great Falls. Thrust faults at the southwest end are the eastern edge of the Overthrust Belt.

years ago. The volcano erupted on top of the Cretaceous Blackleaf and Two Medicine Formations and spread subsurface magma tentacles far beyond its northern fringes. At roughly 24 to 27 miles long by 11 to 17 miles across at its narrowest part, where thrust faults have cut some of it off, it's not a big volcano. Most of the intrusive rocks in the Adel Mountains are shonkinite, a hard, dense, nearly black rock that's similar to basalt but with a lot more potassium. The rock has large crystals of shiny black augite set in a dark-gray matrix composed of very small crystals of augite and feldspar.

Many of the roadcuts reveal crude layering that formed as eruptions laid down new layers of ash and lava on the sloping flanks of the volcano. The weight of the growing volcano bowed the soft sediments below it. The base of the volcanic pile is well exposed, resting on top of the Cretaceous sediments on the volcano's north side, where the canyon narrows. We all understand that a volcano begins by erupting on top of rocks that were already there, but here we can actually see under the base of the volcano.

Shonkinite of the Adel Mountains volcano rests on top of Cretaceous shales about 8 miles north of Craig. The base of the volcano dips gently south, bowed down by the weight of the volcano. View northwest from I-15.

The Missouri River slices right through the middle of the Adel Mountains volcanic pile. Because the river could not have climbed up and over the volcano, the river's ancestor must have been here before the volcano erupted, and thus it maintained its path as the volcano grew. If a lava flow temporarily dammed the river, it didn't succeed in permanently diverting it.

Dikes of shonkinite magma squirted out intermittently in a radial pattern from the volcano's central feeder conduit; the magma rose from deep in Earth's mantle. Such radial dike swarms often form in response to the downward load of an overlying volcano. The magma at shallow depth spread outward in vertical sheets because the vertical stress of the volcano's load was greater than the horizontal stress in the Earth.

Some of these dikes spread for an amazing 22 miles or more from their source near Three Sisters Mountain, east of I-15 and 14 miles due south of Cascade. Erosion of the soft Cretaceous sediments over the last 70 million years has left the hard, dark shonkinite dikes standing like ruined walls from the highest part of the Adel Mountains. Roadcuts south of Cascade expose some of the big dikes that fed magma into Cascade Butte, 3 miles northwest of Cascade, and to other laccoliths farther northwest.

Shonkinite dikes look like ruined walls converging toward Three Sisters Mountain. View southeast from I-15 a few miles south of Cascade.

These laccoliths, big blisters of shonkinite injected between Late Cretaceous layers, formed right at the ends of several of the long radial dikes. There the magma pressure must have waned. The magma found it easier to flow between the soft sediment layers than to continue outward across the layers. The biggest laccoliths north of the volcano were injected at the top of the Virgelle Sandstone, perhaps between it and the overlying shale of the Two Medicine Formation. We can't know for sure, however, because the sediments originally overlying the laccoliths have been eroded off.

North of Cascade on the way to Great Falls, the prominent flat-topped butte to the north is Square Butte, another laccolith. Shaw Butte and Crown Butte, just west of that, are best seen south from MT 200, west of Great Falls. All were fed by the long shonkinite dikes radiating outward from the volcano.

Montana has more than one Square Butte. The Square Butte west of Great Falls that Lewis and Clark called "Fort Mountain" is dark gray, with a flat top and steep cliffs all around it. The Square Butte in the Highwoods east of Great Falls is broader

Square Butte west of Great Falls and its vertical feeder dike, as seen from the southeast.

and more gently sloping, with a ragged crest on its west side. It has a paler-gray-to-white (syenite) upper half and a very dark-gray (shonkinite) lower half. The next time you see a painting of Square Butte by the famous cowboy artist Charlie Russell, you'll be able to tell if it's the Square Butte west of Great Falls, in the northern Adel Mountains, or the one east of Great Falls, in the eastern Highwood Mountains (see the US 87: Great Falls—Lewistown road guide).

From Ulm (exit 270) it's 3.5 miles north to the First Peoples Buffalo Jump. An impressive visitor center has excellent exhibits of Native American artifacts and local animals. A good gravel road from there leads to the top of the buffalo jump, which offers good exposures of gray sandstone of the Cretaceous Blackleaf Formation and wonderful views in all directions.

During the Late Wisconsin glacial advance, the continental ice sheet northeast of Great Falls dammed the north-flowing Missouri River, forming a glacial lake that reached southwest through the Gates of the Mountains along I-15 all the way into the Helena Valley. Light-colored glacial lake sediments with varves are well exposed in several places along this route. Glacial Lake Great Falls left two prominent shorelines, beaches where the lake lapped against hills, one at an elevation of 3,860 feet, the other somewhat lower. It left other signs of a vanished glacial lake—a landscape littered with boulders that sank to the lake bottom as they fell from melting icebergs, and vast, level expanses of a former lake bed. We can reconstruct a map of Glacial Lake Great Falls in all its glory simply by tracing the elevations of the shorelines on a topographic map. At Great Falls, the water was about 600 feet deep.

See the US 87: Great Falls—Havre road guide for information about the waterfalls and dams along the Missouri River near Great Falls.

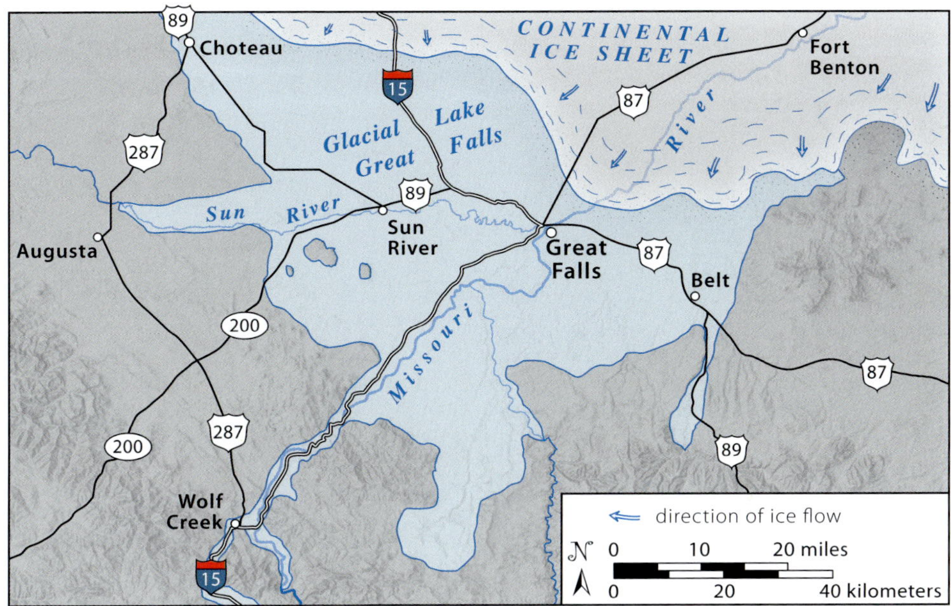

Glacial Lake Great Falls formed where the continental ice sheet blocked the north-flowing Missouri River. During the Late Wisconsin stage (shown here), its surface was 3,715 feet. Modern rivers and roads are shown beneath the lake and ice for location purposes.

I-15
Great Falls—Sweetgrass (Alberta Border)
119 miles

From Great Falls to Vaughn, 12 miles to the northwest, I-15 crosses fine-grained silts laid down by Glacial Lake Great Falls, which formed when the continental ice sheet was 7 or 8 miles north of the city, damming the Missouri River. The lake rose as high in elevation as 3,860 feet, putting Great Falls under as much as 600 feet of water. Widely scattered blocks of pink granite and streaky gray or pink gneiss, basement rock carried in from northern Manitoba, dot the fields in places along I-15. They tell us ice was once here, because nothing but a glacier could have carried big rocks so far. The low, hummocky hills just south of Dutton comprise the moraine that marks the farthest reach of the ice sheet during the most recent glacial advance, the Late Wisconsin glaciation, which reached its maximum around 23,000 years ago.

North of Great Falls, low hills and some roadcuts in soft shale and sandstone of the Early Cretaceous Kootenai and Blackleaf Formations protrude above the silts of Glacial Lake Great Falls. Farther north toward Power are lime-rich beds, then more shale and sandstone, all in the same formations. From 2 or 3 miles south of Dutton to Sweetgrass at the Canadian border, I-15 crosses more Cretaceous rocks that are almost completely covered by glacial till left behind by the continental ice sheet. About the only place you can easily see any of these rocks is where this upland surface drops into the Late Cretaceous Marias River Shale in the meandering valleys of the Teton River, 5 miles north of Dutton, and the Marias River, about 5 miles south of Shelby.

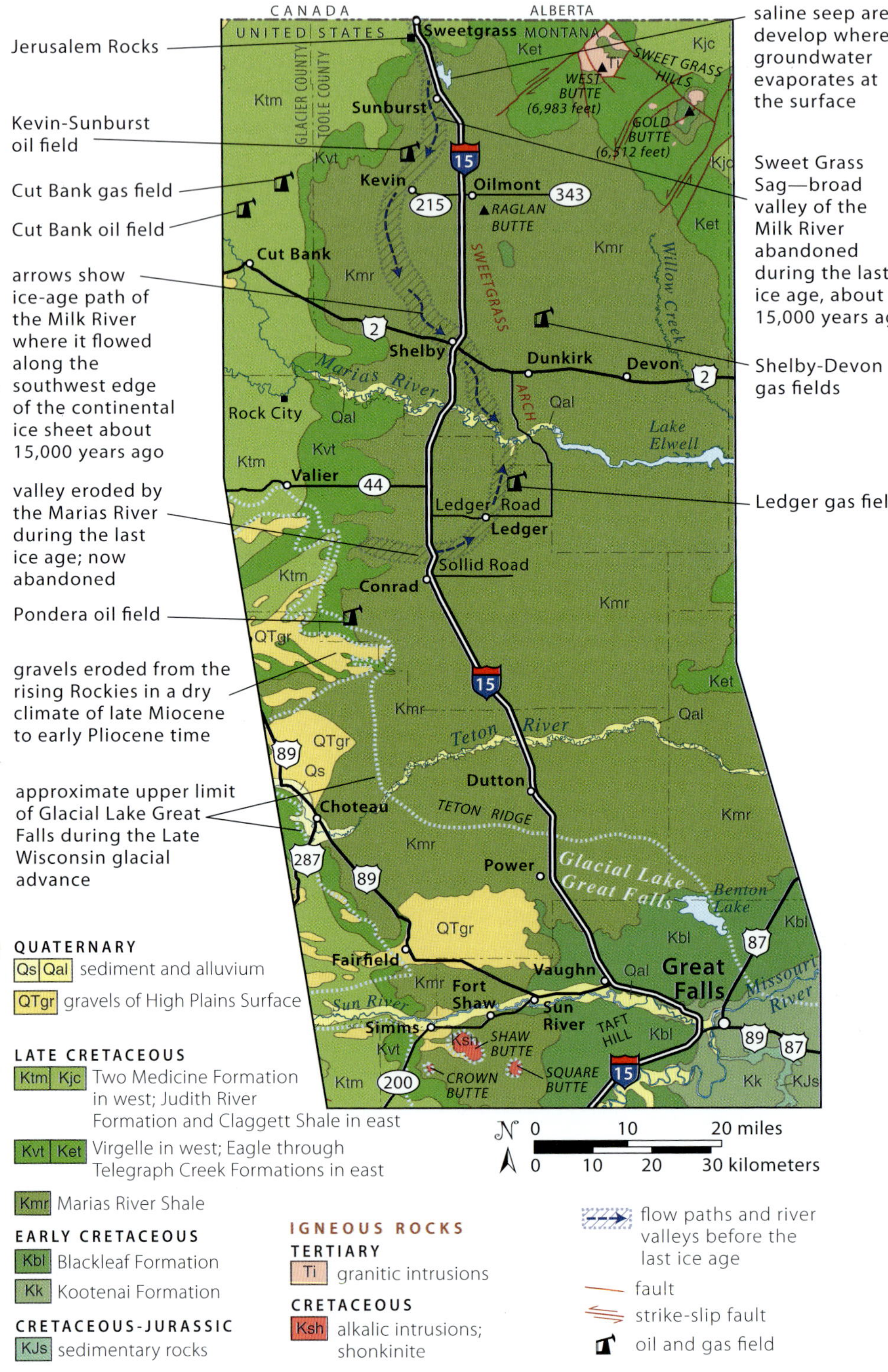

Geology along I-15 between Great Falls and Sweetgrass.

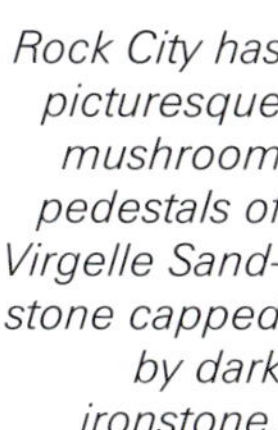

Rock City has picturesque mushroom pedestals of Virgelle Sandstone capped by dark ironstone.

Rock City, northwest of Conrad, is a fascinating place. At exit 348, take MT 44 west for 14 miles to Valier, then turn north (right) on the Cut Bank Highway and proceed straight north for about 7 miles, staying straight on the Rock City Road. The natural rock formations were eroded into the Late Cretaceous Virgelle Sandstone, beach sands deposited along the northern and western shore of the Western Interior Seaway between 80 and 70 million years ago. Along the edge of a terrace eroded by the Two Medicine River are "mushroom rocks," rock pillars with wider, flat ironstone caps that are more resistant to erosion. The dark ironstone is rich in iron oxide (iron carbonate) that precipitated in these sandstone layers from iron-rich water. Rain falling on the ironstone cap mostly flows off it, but some seeps underneath it to eat away at the more easily eroded pedestal rock.

From Conrad to the Canadian border, the highway travels along the length of the broad Sweetgrass arch. Petroleum migrating upward in the tilted rocks accumulated at and near the crest of the arch to form many oil and gas fields, including several with inspiring production records. Most of the wells in these fields have been producing for many years now and have been down to 1 or 2 barrels a day for years. They are known as "stripper wells."

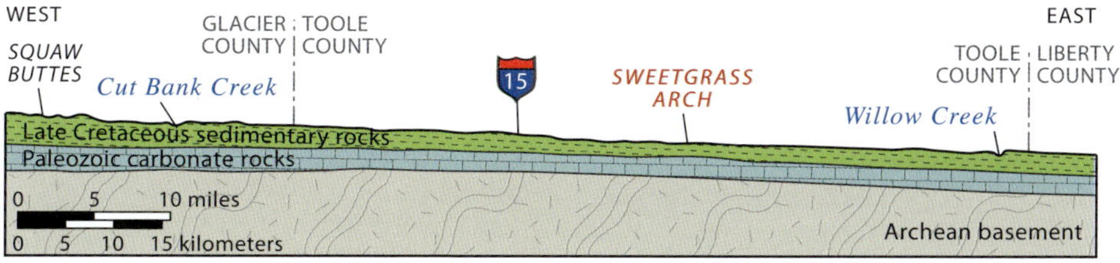

West-east cross section south of Shelby. The compressional crustal movements that created the initial Rocky Mountains left these rocks untouched, so they lie almost as flat as they were laid down. Even the big petroleum trap of the Sweetgrass arch hardly shows in this section drawn to true vertical scale.

Several hundred stripper wells steadily pump their small daily portion of oil from the Pondera field a few miles west of Conrad. Oil was discovered there in 1927 in Mississippian Madison Group limestone at depths between 1,000 and 2,000 feet. That is shallow by oil field standards. You can infer the shallowness of the wells by looking at the light pumps with their small counterweights.

Shelby is sheltered from the wind in the floor of a clearly defined river valley that contains no worthy stream. The dry valley, called the Sweet Grass Sag, continues north through Kevin and Sunburst to Sweetgrass, and then to Milk River, Alberta, where it connects with the present valley of the Milk River. South of Shelby, the dry valley loops south and east to join the present Milk River at Havre. Some geologists argue that the Milk River came this way before the continental ice sheet drowned its valley in Glacial Lake Great Falls. Others contend that the dry valley is an old glacial meltwater channel.

On March 14, 1922, near the town of Kevin, a wildcat driller had an exciting night when he tapped into a rich oil-bearing horizon that blew out through the top of his 20- or 25-foot oil derrick. The next day the rush for black gold was on. In short order hundreds of new wells were drilled within a few miles of Kevin. The wildcatters likely knew about the broad, gentle-sided, almost completely imperceptible Kevin-Sunburst dome, which stretches from south of Shelby to the Canadian border, because a US Geological Survey report six years earlier had described it, but up until that night there had been no big finds. Oil was finally discovered in quantity in porous sandstone of the Jurassic Ellis Group and at the top of Mississippian Madison Group limestone at depths of around 2,600 feet. The oil came from organic-rich shale below the Madison, rising along fractures within the limestone before soaking the Ellis Group sands, where it became trapped. Drillers "float" the oil by pumping in water from the underlying Madison Group limestone.

The Sweet Grass Hills east of Sunburst. Inset photo of syenite from East Butte with white crystals of feldspar that are several millimeters wide. —Courtesy of Rod Benson, Bigskywalker.com

The small group of mountains on the horizon about 16 miles east of Sunburst is the Sweet Grass Hills, an isolated group of igneous rocks that intruded Cretaceous sediment about 54 to 50 million years ago, in Eocene time. These much softer sediments eroded away, exposing the buttes by Pleistocene time. Their highest peaks reach almost 7,000 feet, approximately 3,000 feet above the High Plains Surface. The continental ice sheet was a little more than 1,000 feet thick during advances of Pleistocene time, so the three main buttes stood as islands more than 1,000 feet above the ice. Although they intruded Cretaceous sediments of the High Plains Surface, erosion removed the soft sediments, exposing the buttes by the time the ice advanced. Like most of the other igneous rocks in the Central Montana Alkalic Province, these are rich in alkalis, in this case mostly pale syenite and related rocks composed mainly of feldspars rich in sodium and potassium.

Picturesque rock sculptures were eroded into the rim rocks of the Sweet Grass Sag just east of Sunburst and at Jerusalem Rocks, a somewhat smaller version of Rock City in the same formations. To get there, drive northwest from Sunburst about 7 miles on Loop Road, then zigzag north and northeast on Little Jerusalem Road for 6 miles. The rocks occur where the road drops into Buckley Coulee. The more-resistant caprock of Virgelle Sandstone overlies the more easily eroded sandy shale of the Telegraph Creek Formation.

I-90
Billings—Wyoming Border
101 miles

The high and often snowcapped Beartooth Mountains on the distant skyline southwest of Billings make up the eastern end of the relatively flat Beartooth Plateau, an immense block of basement rock that was initially shoved up to the north along thrust faults about 50 million years ago, during the Laramide orogeny. More recently it has been raised again by Tertiary extension and thermal crustal swelling related to the nearby Yellowstone hot spot to the southwest. The very broad Pryor Mountains rise along the skyline almost directly south of town. This broad Laramide arch brings large areas of Madison Group limestone to the surface, along with younger sedimentary formations exposed along its flanks. Farther southeast, that low bulge rises to become the towering Bighorn Mountains, the snowcapped range hovering on the skyline southwest of the highway between Hardin and the Wyoming line.

The dramatic Rimrocks, composed of Eagle Sandstone, stand 500 feet above the north side of Billings. Locally cross bedded, the sandstone was laid down as sandy beach deposits, perhaps on a barrier island, about 80 million years ago, in Late Cretaceous time. Being resistant to erosion, the sandstone forms high pale-brownish-gray cliffs, but big blocks occasionally break off to destroy buildings below the "Rims." (See the I-94: Billings—Miles City road guide for more about the Rimrocks.)

Pictograph Cave State Park is located in the same thick Eagle Sandstone as the Billings Rimrocks. The buff-colored cliffs are pockmarked with small to very large holes and caves, including two very large open-air alcoves, or rock shelters: Pictograph Cave and Ghost Cave. The alcoves formed as water slowly dissolved the calcite cement that binds the sand grains together. The pictographs are faint, but informative signs illustrate and point them out well. Archeologists have discovered more than one hundred

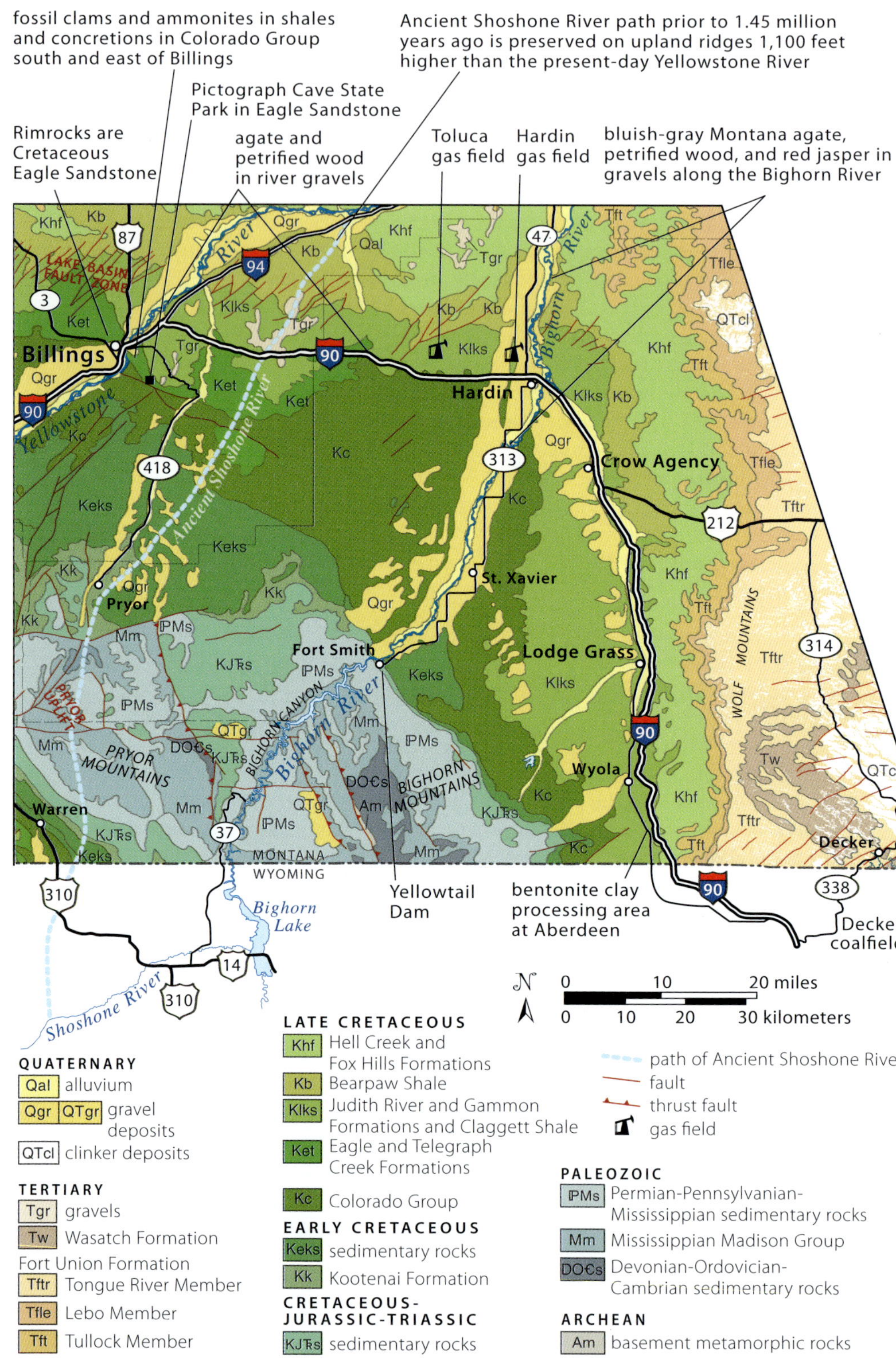

Geology along I-90 between Billings and the Wyoming border.

Pictograph Cave in the Eagle Sandstone near Billings.

pictographs, painted between 2,145 and 200 years ago. Pictographs, which are drawn or painted, are sometimes confused with petroglyphs, which are made by scraping or chipping part of a rock surface. To reach the park, take exit 452 on I-90 for old US 87E and drive about 5 miles south on Coburn Road.

Northeast of Billings for 4 or 5 miles, I-90 follows the south side of the Yellowstone River, passing hills eroded in the brownish-gray Late Cretaceous Claggett Shale overlain by sandy light-brownish-gray shale of the Judith River Formation. On the north, the low gravel terrace next to the river, 50 or so feet high, was deposited in the ice-age channel of the Yellowstone River; the river has cut downward since then, leaving its old channel deposits stranded.

I-90 leaves the Yellowstone River valley and turns abruptly east, immediately climbing for 5 miles through the Claggett Shale across a ridge of nearly horizontal Judith River Formation sandstone that somewhat resembles the Eagle Sandstone of the Rimrocks, but here only the thicker parts of the sandstone make prominent outcrops. Remnants of gravels that were likely deposited by an old river in Pliocene time cap the ridge a half mile south of the interstate. Dropping down the east side of the ridge you go through the reverse sequence down to Pryor Creek. East of Pryor Creek, I-90 crosses the same rock sequence in a wider second ridge that's 14 miles across.

Stream gravels cap the crest of that second high ridge for about 3 miles along the north side of I-90. At an elevation of about 1,100 feet above the nearest section of the Yellowstone River, this is a decidedly inappropriate place for stream gravel to be. Its pebbles consist mainly of granitic gneiss, schist, and quartzite that apparently came from ancient basement rocks of the Beartooth Plateau southwest of Red Lodge, 50 miles to the southwest. The gravel also contains Absaroka Volcanics that also hail from the southwest, along the east side of Yellowstone National Park. By tracing this distinctive gravel upstream on high ridges that gradually increase in elevation to the southwest, geologists have pieced together an ancient river system. Called the Ancient Shoshone River, it drained northeast across this area at an elevation about 1,100 feet higher than the present Yellowstone River. Volcanic ash in its sediments has

BIGHORN CANYON

Hardin is on the west bank of the broad, north-flowing Bighorn River. Yellowtail Dam, 41 miles southwest of Hardin on MT 313, impounds the river to flood its canyon in the north end of the Bighorn Mountains. The dam is anchored in massive Mississippian Madison Group limestone at a sharp flexure where the Paleozoic rock dips steeply to the northeast. Fossils are widespread in the Madison and include well-preserved *Spirifer* brachiopods, clam-like creatures that lived in tropical ocean water that covered this area more than 350 million years ago.

Bighorn Canyon to the northeast from WY 37 near the Montana-Wyoming border. The inset shows Spirifer *brachiopod fossils in the Madison Group limestone near Barrys Landing boat launch.*

The Bighorn River cuts a spectacular notch through the mountains, big folds that formed over blocks of basement as they were first raised during the Laramide orogeny. Vertical canyon walls in the middle of the canyon are up to 1,500 feet high, with another 500 feet submerged below the reservoir level. These high cliffs are eroded mostly in Madison Group limestone, but some of the canyon bites through the Madison into older formations, including the conspicuous Ordovician-age Bighorn Dolomite. The deepest, oldest exposed rocks are Cambrian in age.

At the south end of Bighorn Lake, far south of Yellowtail Dam, these older rocks lie in the core of the Big Bull Ridge anticline. This fold formed during the Laramide orogeny but was later slowly raised in the path of the Bighorn River by regional crustal thinning and swelling, maybe as recently as 1 or 2 million years ago. Because the entrenched river has the sinuous shape of a low-gradient meandering stream, such as the Mississippi River, previously flat land must have risen slowly over a broad area, allowing the river to cut down into the hard limestone and maintain its former channel shape. The river no longer meanders back and forth because it's stuck in the hard rocks.

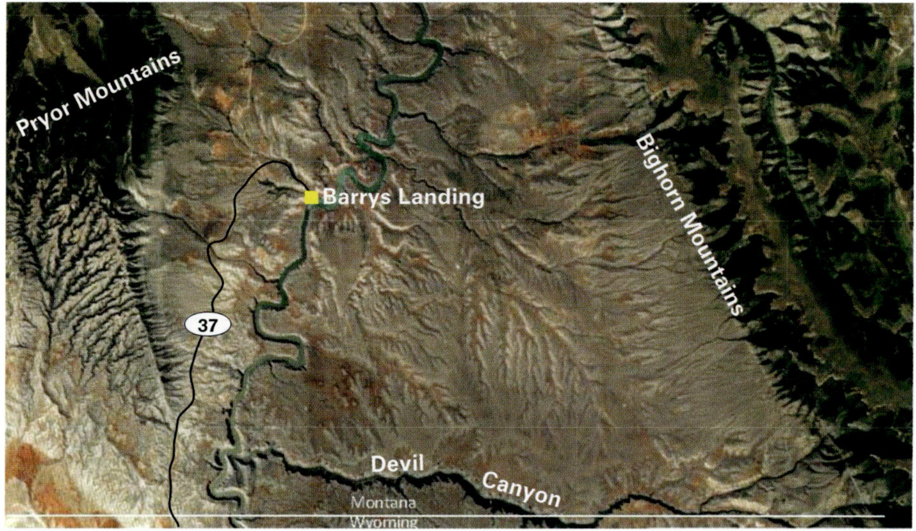

Landsat satellite view of Bighorn Canyon showing the river's meandering path.
—Courtesy of National Aeronautics and Space Administration

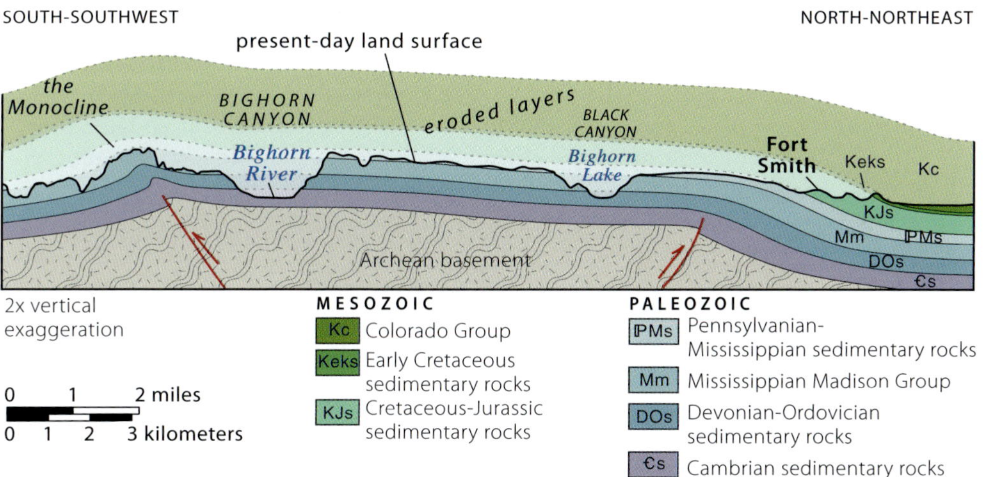

Cross section of Bighorn Canyon showing what the stratigraphy and topography looked like prior to erosion. **—Courtesy of National Park Service**

The easiest access to the spectacular south end of Bighorn Canyon is to head south from Laurel on US 310 through Warren to Lovell, Wyoming, and then north on WY 37. Rocks along the Montana section of this road in the Bighorn Canyon National Recreation Area are mostly pale-gray limestone of the Mississippian Madison Group and limestones and shales of the overlying reddish-toned Amsden Formation. The red color in the Amsden might have formed during many thousands or millions of years of weathering, which concentrated and left behind red oxidized clays. It's also possible that the red comes from oxidation of the iron-bearing sediment grains in the Amsden that occurred shortly after they were deposited. In either case, runoff from the Amsden trickles down the light-gray cliffs of the Madison Group limestone and stains them.

been dated to late Pliocene time, so the river was active at that time. Research shows that the upper part of that channel was captured by the present Shoshone River about 1.45 million years ago.

For 6 miles east of the second ridge, I-90 drops through Judith River Formation, Claggett Shale, Eagle Sandstone, yellowish-brown shale of the Telegraph Creek Formation, and dark-brownish-gray shale of the Niobrara Formation to Fly Creek (near exit 478). From milepost 478 to Hardin (near milepost 495), I-90 alternates up and down through horizontally layered yellowish-brown silts of the Gammon Formation and slightly older, dark-brownish-gray flaky shales of the Niobrara. The Niobrara is noted for beds rich in smectite (bentonite), the notorious slippery swelling clay.

South from Hardin, I-90 remains in Late Cretaceous sedimentary rocks all the way to the Wyoming border. At the bridge across the Bighorn River, at the east edge of Hardin, and for 8 miles southeast, yellowish-brown marine silts and shales of the Gammon Formation appear both east and west of the floodplain of the Little Bighorn River. Brownish Claggett Shale, along the west edge of the floodplain, and yellowish-gray to greenish sandstone of the Judith River Formation, on the east, appear for the remaining 6 miles to the US 212 junction at Crow Agency. South of the junction, muddy sandstones of the Judith River occur both east and west of the river. All of these rocks were deposited either in or along the western margin of the Western Interior Seaway.

About 7 miles south of the US 212 junction, I-90 jogs abruptly to the east about 1 mile, presumably to provide a more stable foundation on solid rock instead of the Little Bighorn River bottom. Distant hills 5 or 6 miles to the east with brilliant caps of red clinker, the usual sign of coal country, become conspicuous near Lodge Grass, north of the Wyoming border; these are in the Paleocene Fort Union Formation.

About 4 or so miles north of Wyola, the Little Bighorn River has eroded down into the soft Claggett Shale, which the river continues in almost to the Wyoming border.

About 2 miles north of the Lodge Grass exit, sandstone of the Late Cretaceous Lance Formation forms a prominent cliff a half mile east of I-90. Slightly older sandy shale underlies the sandstone in the foreground.

About 1 mile south of Aberdeen, I-90 again veers eastward onto Lance Formation sandstone and follows it for the remaining 3 miles to the Wyoming border.

West of the interstate, Judith River rocks are a few hundred feet upslope. The thin Fox Hills Sandstone overlies the Judith River, and the higher, more-ragged hills farther east are made of Lance Formation sandstone. About 20 miles east of Wyola, the town of Decker, just north of the Wyoming border, is located on the upper Tongue River sediments of the Paleocene Fort Union Formation. There, large open-pit coal mines extract low-sulfur subbituminous coal, which is shipped by train from Sheridan, Wyoming, to power plants around the United States.

I-94
Billings—Miles City
142 miles

High, buff-colored cliffs of Eagle Sandstone, the Rimrocks, rise 500 feet above Billings. The massive sandstone appears to be the remains of a barrier island, perhaps similar to Galveston Island in Texas, that stood between a coastal lagoon west of Billings and the shallow inland sea that still flooded much of North America during Late Cretaceous time. In some places the sandstone is cross bedded; currents and waves formed these internal angled layers. The Eagle Sandstone is known for very large, hard spherical masses called concretions, some up to 15 feet in diameter. They formed when minerals in groundwater precipitated around a nucleus, such as a shell or a bit of organic matter, while the deposit was still sand. Concretions can form in concentric layers around the nucleus, or as a single mineralized mass. The orange color of many suggests that iron oxide minerals were involved in their formation.

Giant slabs separate from the cliff face along vertical cracks in the hard sandstone; that allows moisture to seep into the cracks and pry them apart when it freezes and

Big cracks are spreading through the Rimrocks above Billings.

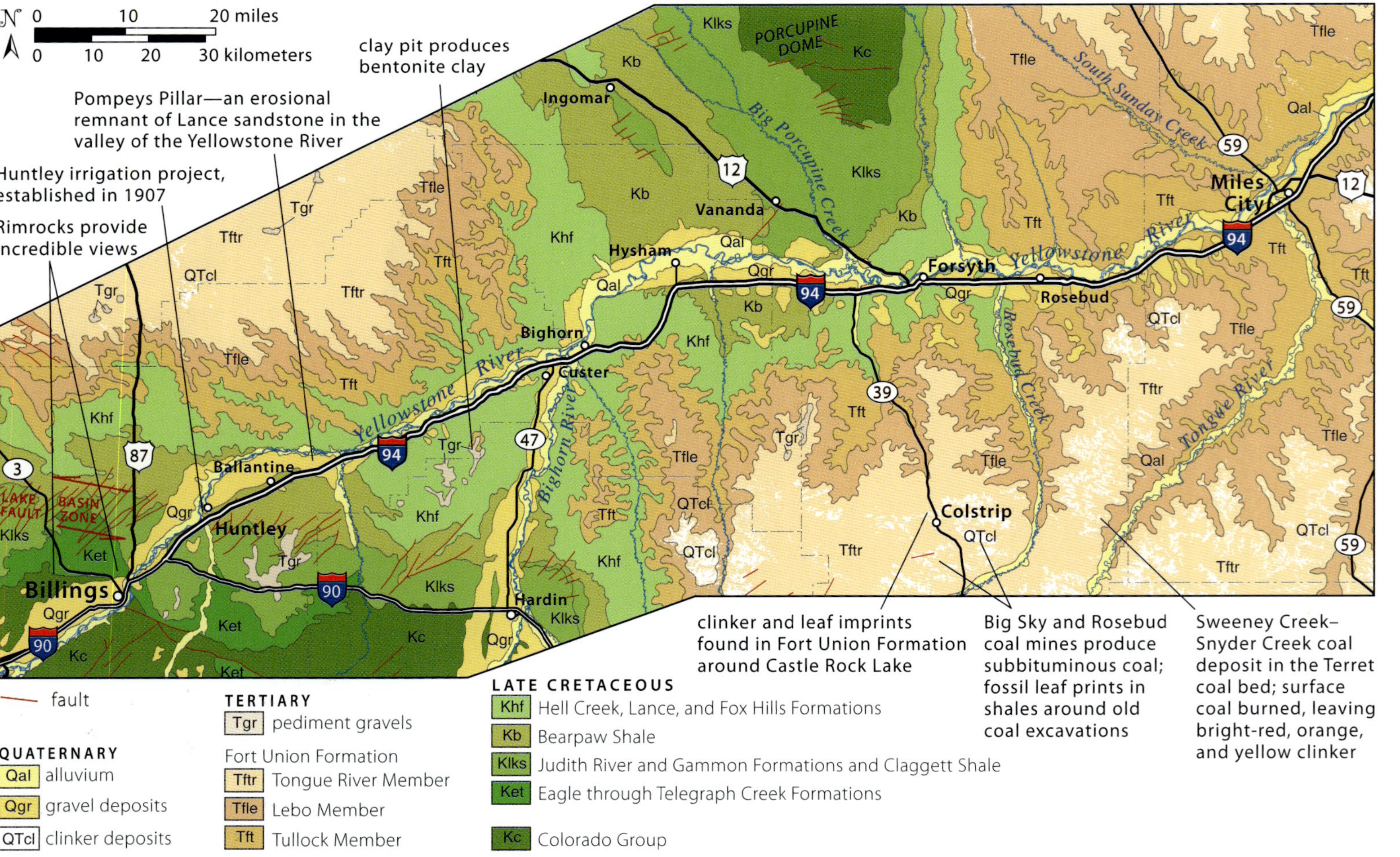

Geology along I-94 between Billings and Miles City.

expands. Some slabs have dropped only a few feet and remain a threatening presence that may drop catastrophically onto homes and businesses below. On October 9, 2010, a giant boulder 25 feet across destroyed a house below the cliffs, narrowly missing the occupant. Because the underlying shale is more easily eroded, it leaves the sandstone cliff overhanging and unsupported.

Between Billings and Miles City, I-94 follows the south side of the Yellowstone River valley. The road alternates between short stretches on the valley floor and longer routes that cross the nearly level high ground slightly to the south, remnants of the Pleistocene channel of the Yellowstone River. Watch for pebbles strewn through the fields and for occasional roadcuts that expose beds of gravel. From the east edge of Billings to several miles east of Huntley, the same low terrace is visible north of the river, a few tens of feet above the present river. The upslope edge of the terrace is variably Late Cretaceous Claggett Shale, sandy shale of the Judith River Formation, or the Bearpaw Shale. The higher hills north of the terrace are light-brownish-gray Judith River Formation almost to Huntley, then the Bearpaw Shale to where the low terrace peters out.

West of Ballantine, the Bearpaw Shale becomes confined to low elevations, and the ragged cliff-forming Lance Formation sandstone of the higher hills comes down close to river level. South of the Yellowstone River, the hills near Huntley are Claggett Shale, but 1 mile northeast of town it's all flaky, dark-gray Bearpaw Shale except near Pompeys Pillar, where the Bearpaw is confined to a few feet above the low terrace gravels.

Between 7 and 10 miles east of Billings, near the community of Huntley, I-94 crosses an east-west band of faults called the Lake Basin fault zone, a major structure in central Montana. Watch for roadcuts and outcrops that expose tilted beds of tan Cretaceous sandstone; these are unusual in a place where most sedimentary rocks lie flat. The slightly southeast-trending fault zone consists of a long row of small faults, most 3 to 4 miles long, that are oriented northeast. The continental crust south of the fault zone probably moved southeast relative to that to the north, tearing the sedimentary rocks above into the pattern we see on the map. The faults are likely old, having been formed in the basement rocks during Archean time, and then reactivated with changes in plate stresses over time. Previous breaks in the crust tend to get used many times if plate stresses can take advantage of the weaknesses.

Pompeys Pillar (exit 23) is a stump of sandstone of the Lance Formation (equivalent in time to the Hell Creek Formation). It resembles the pale-gray to beige Eagle Sandstone of the Rimrocks at Billings but is a little younger. The pillar stands isolated from the main line of bluffs on the south side of the Yellowstone River because the river gradually cut through a meander and left it stranded. It is one of Montana's most widely celebrated rocks, both because of its visual prominence and because Captain Clark of the Lewis and Clark Expedition carved his name in the soft sandstone in 1806. The Lance Formation was deposited by streams along the western shore of the Western Interior Seaway coastal plain. The plant fossils it contains indicate the climate was subtropical, and the rocks have yielded the bones and teeth of a variety of dinosaurs, pterosaurs, crocodiles, lizards, snakes, turtles, frogs, salamanders, fish, and mammals, as well as birds from groups that exist today.

From Pompeys Pillar east to the MT 47 junction, I-94 passes more pale-gray beds of Lance Formation sandstone and softer shale. Some of the sandstone has weathered

Pompeys Pillar, a sandstone stump of the Lance Formation. Captain William Clark named it after the son of Sacagawea and Toussaint Charbonneau in 1806.

into rounded outcrops with striking holes. Closer to MT 47, the roadcuts are in horizontal layers of soft, pale-gray shale. The prominent thick beds west of Custer, near milepost 45, are sandstone.

At a rest stop near milepost 41, west of Custer, big red boulders and gravel of clinker were used for landscaping. The clinker that caps most of the higher hills north and south of I-94 formed when underground coal fires melted the enclosing sediments of the Fort Union Formation.

I-94 crosses the Bighorn River 2 miles east of the MT 47 junction. This river starts in the Wind River Range of Wyoming, 460 miles south of here, and flows through Bighorn Canyon National Recreation Area on the Montana-Wyoming border before joining the Yellowstone River here. East of the Bighorn River, I-94 climbs 700 to 800 feet through the Lance Formation as the Yellowstone River swings to the north. At the crest of the ridge the road passes through a half mile of the Fort Union Formation and then back down through the Lance Formation again. Before reaching the old ice-age terrace of the Yellowstone River near Hysham, the road passes down into the underlying Late Cretaceous Bearpaw Shale, which continues intermittently east to the MT 39 junction.

In some places the gravels along the Yellowstone River contain bluish-gray agates, red jasper, and brown petrified wood, especially between Pompeys Pillar and Glendive. All of these semiprecious stones are composed almost entirely of silica, having precipitated slowly from water rich in dissolved silica. Differences in color and texture are related to small amounts of iron or manganese oxides and the style of deposition. Such rocks form in areas with silica-rich volcanic rocks, such as volcanic ash, fine shreds of silica-rich volcanic glass. Water in the pore spaces of the volcanic rock slowly dissolves the fine particles of glass. The silica-rich water then precipitates silica in larger open cavities to form agate and jasper, or it replaces the organic molecules in wood with silica to petrify it. In Montana the volcanic rocks erupted millions of years ago in and around Yellowstone National Park. The 50-million-year-old Absaroka Volcanics around the park are the most likely source of the agates and petrified

Unpolished agates (left), about 2 inches across, collected from terrace gravels near Forsyth. Polishing with a rock tumbler makes them beautiful (right).

wood brought here by the Yellowstone River. Agates are also found in the older gravel terraces within a few miles of the river. The best time to find them is when they wash out after heavy rainstorms, such as in the spring.

East from the MT 39 junction, about 6 miles west of Forsyth, I-94 cuts through light-brownish-gray sandstone of the Lance Formation overlying flaky dark-gray Bearpaw Shale. The high sandstone cliffs on the south side of Forsyth are also Lance Formation. The Bearpaw Shale was deposited during a major incursion of the Western Interior Seaway into Montana, during which its shoreline extended a bit farther west than Livingston. The overlying Lance Formation records a major regression (retreat) of that ocean water, which didn't return to Montana in any significant way.

Immediately east of Forsyth, I-94 climbs a couple hundred feet out of the shale onto gravel terraces of the ice-age Yellowstone River; they overlie the Lance Formation that the river eroded into when depositing the gravel of the terraces. Because the sedimentary layers dip very slightly to the east, 2 miles farther east the Paleocene Fort Union Formation is under the terrace gravels. I-94 then alternates between flat areas of ice-age Yellowstone River gravels and the Fort Union Formation for almost 30 miles, through Hathaway and all the way to Miles City. The Fort Union sediment east of Rosebud is the Tullock Member, yellowish-gray sandstone with smaller amounts of shale and coal; it's mostly exposed in gullies and tributary river valleys and elsewhere capped by ice-age gravels. High bluffs along the road include thick, light-brownish sandstone and thinner, darker shales of the Fort Union Formation. Rounded pebbles of red jasper, granite, quartz, and other rocks along the road are from those higher-level gravels.

From elevated areas of the interstate east of Rosebud, you can see to the south what appear to be groups of small, steep-sided volcanoes. If curiosity gets the best of you and you drive south from Rosebud (for example, on MT 447), you will soon see that those sharply pointed hills display horizontal layering. They are, in fact, remnants of the Fort Union Formation and not volcanoes at all. Hard caps of clinker protect the underlying rock from erosion.

Miles City is at the mouth of the Tongue River where it joins the Yellowstone River. The Yellowstone has eroded down several hundred feet below the broad, gently

What look to be a row of volcanic cinder cones south of Rosebud are in fact irregular hills protected by erosion-resistant clinker in the Fort Union Formation.

rolling uplands of the Fort Union Formation, so moderately steep slopes flank the river a couple of miles south and north of town. The highest elevations south of the river, about 7 miles east of Miles City, have large exposures of ragged red clinker.

I-94
Miles City—Glendive—Wibaux
104 miles

I-94 northeast from Miles City follows the south side of the Yellowstone River as far as Fallon, then the north side from there to Glendive. Along most of the route the road stays on a Pleistocene terrace of the Yellowstone, overlooking the valley the river eroded into it. Except near Glendive, bedrock along the entire route is the Paleocene Fort Union Formation, mostly Tullock sandstone and Lebo shale, which tend to erode into level plains.

The town of Terry is built on a lower terrace of Yellowstone River gravels, almost on the floodplain of the river, 6 or 7 miles northeast of the mouth of the Powder River.

Brownish shale over a black, coal-rich layer in a roadcut of the Fort Union Formation about 8 miles northeast of Miles City.

Agates, which are usually dull when found and sometimes coated with dull precipitates from groundwater, can be collected from those river gravels.

The Terry Badlands is a hidden gem less than 3 miles north of Terry. Take MT 253 to Scenic View Road and drive west (left) to the overlook above the badlands. Three members of the Fort Union Formation are exposed here, starting with the Tullock at river level, overlain by dark shales of the Lebo, and topped by lighter-colored sandstone, shale, and red clinker of the Tongue River Member. These layered sediments were deposited during Paleocene time on the coastal plain of a river system that extended from the Rocky Mountains to the Western Interior Seaway.

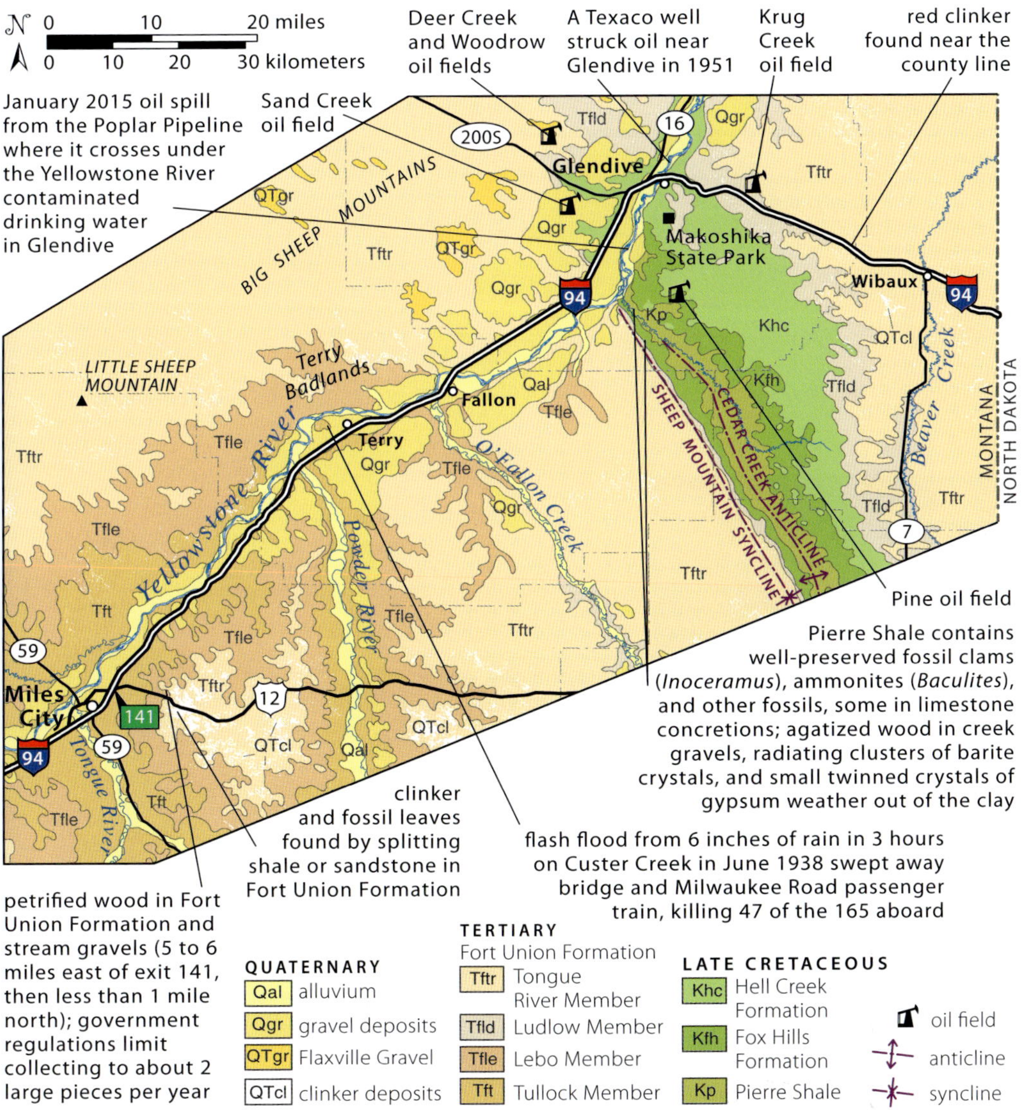

Geology along I-94 between Miles City and Wibaux.

As Captain William Clark headed downriver to the east on July 31, 1806, he noted the striking red clinker found in the bluffs along the Yellowstone River near Terry: "I observe Several Conical [mounds] which appear to have been burnt." Clinker forms when sedimentary rocks above and below burning coal layers are heated, baked, burned, melted, and even fused together. Essentially, the rock has been metamorphosed by fire. A coal fire just northeast of Terry was extinguished in the 1990s, and one day the clinker formed by that fire just might get exposed at the surface by erosion.

East of the rest stop at exit 192, the big bluffs on the south side of the highway are in a thick layer of light-colored shale. Midslope, this shale caps thick sandstone beds

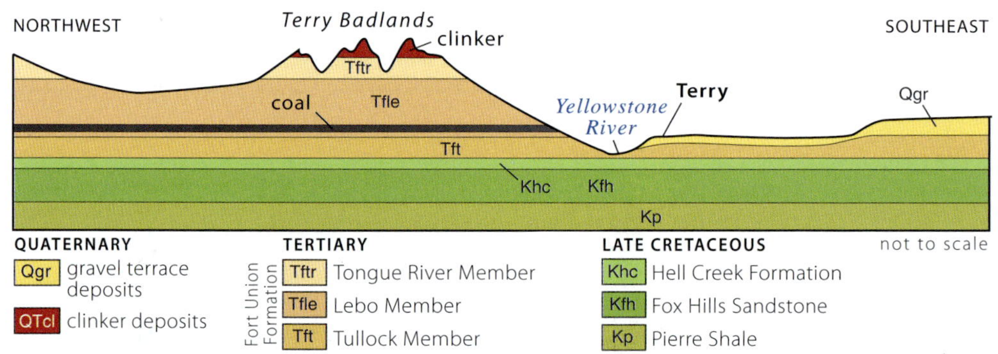

Cross section showing the members of the Fort Union Formation at the Terry Badlands, as well as clinker capping the buttes. —Modified from Vuke and Colton, 2003

A hard cap of red clinker in the Terry Badlands prevents the underlying, softer sandstone and shale of the Tongue River Member from eroding away. The heat broke the clinker into fragments, melting some of it to a red, foamy mush.

that are full of cavities. These cavities form as the cement between the sand grains dissolves and wind blows the loosened grains away. This boundary may mark the Cretaceous-Tertiary boundary, where the Cretaceous Lance Formation is overlain by the Tullock Member of the Fort Union Formation.

East of the Yellowstone River crossing, it seems like the world is as flat as it can be as far as the eye can see. It's almost as flat as the shallow sea bottom and coastal plain that these Cretaceous sediments were deposited on. I-94 parallels the northwest side of the Yellowstone River from Fallon to Glendive. The heavily gullied Fort Union Formation makes up the lower slopes, and broad, flat, gravelly surfaces of the Flaxville Gravel make up the grassy uplands.

About 17 miles past the river crossing (10 miles southwest of Glendive), the familiar horizontal sedimentary rocks southeast of the Yellowstone River abruptly steepen to 10 to 15 degrees, even reaching 30 degrees to the southwest on the west flank of the Cedar Creek anticline. Glendive is on the northern end of the anticline, the crest of which extends southeast for more than 100 miles into South Dakota. Although it is the most striking geologic structure in eastern Montana, the rocks visible from the road give little hint that the anticline exists. The Hell Creek and related sedimentary rocks extend for 15 miles across the crest of the fold. The fold is steep on the west side, gentler on the east side. It warps the younger Fort Union Formation, so its deformation must have continued until sometime after the last of those sediments accumulated about 55 million years ago. (See the MT 7: Wibaux—Ekalaka road guide for more information about the fold.)

I-94 leaves the Yellowstone River valley at Glendive to head almost due east. East of Glendive, many patches of badlands north and south of the road provide good exposures of the Tullock, the lowest member of the Fort Union Formation. The rock is pale-gray mudstone that erodes into rugged little buttes covered with rills between ledges of brown sandstone that jut from the slopes. A few scraps of red clinker are on both sides of the highway, especially near the line between Dawson and Wibaux Counties. Farther east, the badlands become less numerous and then disappear as the road passes onto the overlying shale of the Lebo Member; from there the country flattens into a grassy prairie that stretches out to the eastern horizon. Near Wibaux, the Tongue River sandstone in the upper part of the Fort Union Formation contains beds of coal.

MAKOSHIKA BADLANDS

At Makoshika State Park, located at the south end of Snyder Street at the south edge of Glendive, roads and miles of trails delve deep into surrealistic natural features sculpted by wind and rain. The museum at the state park focuses on dinosaurs and includes a triceratops skull. This ungainly looking 9-foot-tall plant-eating dinosaur lived 68 or 66 million years ago, right before the asteroid impact that exterminated all dinosaurs.

Rain splash and surface runoff have carved soft silt and sandstone of the Hell Creek Formation into fantastic shapes suggestive of hundreds of fairyland castles fallen into ruins. Evidently something, perhaps an intense range fire, destroyed the plant cover in this area centuries ago, permitting splashing raindrops to attack the rocks. The "toadstool caps" are a grainy cross-bedded sandstone rich in iron, probably grains of magnetite, an iron oxide. The caps were cemented to ironstone,

perhaps during the oxidizing process, leaving them more resistant to erosion. Water dribbling down from these ironstone caps stains the underlying white silty sandstone a rusty color as well.

Ironstone caps of cross-bedded sandstone protect the underlying softer sandstone from rain-splash erosion to create these pillars in the Makoshika Badlands.

A thin bed of coal, possibly deposited in a swamp on a floodplain of the Western Interior Seaway, overlies floodplain mud at the base of the sequence. The plants that made the coal were buried by sand deposited by streams flowing roughly from left to right, as shown by the dip direction of the tilted layers. Once exposed at the surface, runoff eroded narrow gullies called rills into the soft sandstone. Yellow notebook is 6 inches high.

Most of the rocks exposed in the Makoshika Badlands are the uppermost siltstone and shale of the Cretaceous Hell Creek Formation. The lower part of the Paleocene Fort Union Formation caps some of the eastern hills. The boundary between the formations marks the great mass extinction at the end of Cretaceous time and the beginning of Tertiary time (Paleocene), referred to as the K-T boundary. The boundary is preserved on the highest ridges of the park, but there's nothing of note about how the rocks look that would allow one to identify the boundary. The two formations look alike except that the Hell Creek contains a few widely scattered fragments of dinosaur bone but the Fort Union doesn't.

A clay layer at the K-T boundary contains unusually high levels of the trace element iridium and shocked quartz, good indicators that an extraterrestrial object struck the planet. A spike in fossil fern spores, recorded in the iridium layer, suggests that vegetation was wiped out and ferns rapidly colonized; spores are more easily distributed by wind than are seeds. A similar fern spike was seen in the blast zone around Mount St. Helens shortly after it erupted in 1980.

US 2
Browning—Havre
160 miles

The mountains of Glacier National Park rise from the plains west of Browning. To the east, halfway between Browning and Cut Bank, US 2 passes from the area covered by glaciers spreading out from the mountain valleys in and near Glacier National Park to that covered by the continental ice sheet. The ice coalesced on the plains to form piedmont glaciers. One of these big piedmont glaciers, the Two Medicine glacier, stretched almost 10 miles east of Browning; as it melted, it left a vast expanse of ground moraine, a chaotic hummocky landscape with small ponds snuggled amidst lumpy little hills. The most obvious distinction between the two glaciated areas is the kind of erratic boulders that litter the surfaces of the moraines. Piedmont glaciers coming from Glacier National Park and the Rocky Mountain front carried mostly Proterozoic Belt Supergroup sedimentary rocks, often dark colored; those left by the continental ice sheet are older, Archean basement rock—light-colored granite or light-gray gneiss—from far to the north and northeast in Canada. Watch for these telltale "out-of-place" boulders scattered along most of the route.

Between 20 miles west of Browning and Cut Bank, US 2 rides on the flat, broad surface of terrace gravels eroded from the mountains to the west at the end of Miocene time. Known as the High Plains Surface, it hosts giant farms with fields of hay and grain. Cut Bank sits on the High Plains Surface next to the eroded banks of meandering, aptly named Cut Bank Creek. The name the Blackfeet Indians gave it translates to "the river that cuts into the white clay bank." Originally the town was built on the west bank when the railroad came through in the 1890s; it was moved to the east side when it was discovered that it was on reservation land. The creek's meandering channel is entrenched in Cretaceous Two Medicine Formation sandstone and shale, deposited in braided streams and floodplain coal swamps.

Cut Bank Creek has sliced a narrow, meandering channel below the High Plains Surface, nicely exposing sandstone and shale of the Two Medicine Formation.

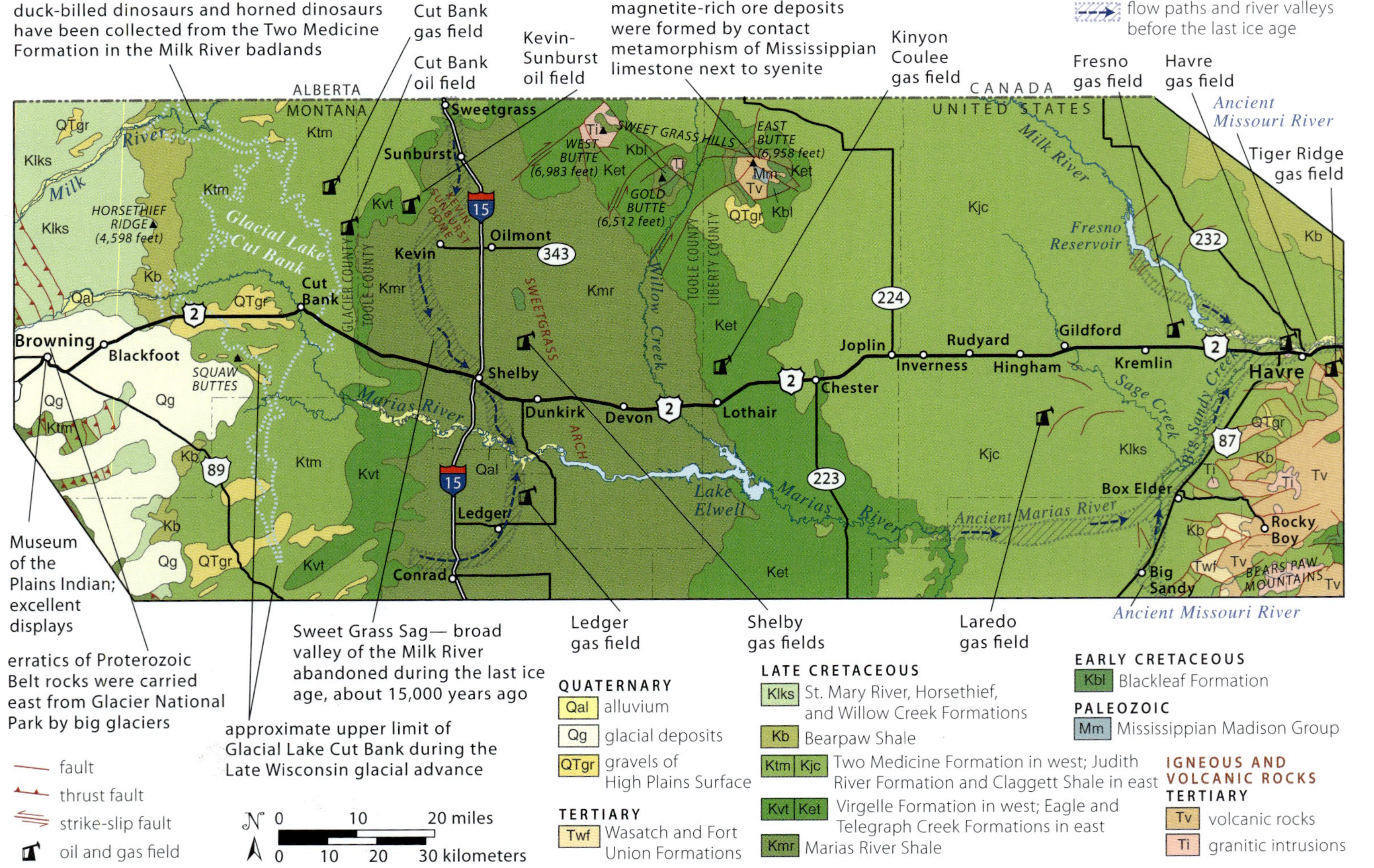

Geology along US 2 between Browning and Havre.

Around Cut Bank, US 2 passes old wells, separators, and storage tanks of the Cut Bank oil and gas field. A well drilled in 1926 found natural gas but was abandoned for lack of a market. Production finally began in 1931 after another hole struck both oil and gas. Most of the wells produce from sandstones in the Kootenai Formation at a depth of about 3,000 feet. Several dozen deeper wells produce from Mississippian Madison Group limestone.

Glacial Lake Cut Bank shallowly flooded the town's area during the Late Wisconsin ice advance. Water was impounded by ice and ponded at an elevation of 3,900 feet before spilling south into Glacial Lake Great Falls.

A big wind farm caps moraine hills on the south side of US 2 about 8 miles east of Cut Bank (19 miles west of Shelby). The westerly winds at this latitude rise and expand over the Rockies before descending to the east, picking up speed as downslope winds. The jet stream, traveling at about 100 miles per hour, also travels west to east over this general area. The best places in Montana for wind power are on the plains at the base of the Rockies, such as here, and on the east side of small mountain ranges, such as the Sweet Grass Hills, Bears Paw Mountains, and Beartooth Mountains.

Landowners receive royalties by leasing space on their land to companies that erect the giant wind turbines. The towers are 250 to 300 feet tall, sporting blades 100 to 130 feet long from rotation axis to blade tip. They only spin when wind speeds are at least 7 miles per hour. If the wind is too strong, more than 50 miles per hour, they stop spinning to protect the turbine from damage. The big blades sometimes kill birds, but new turbine designs use underground lines between turbines to reduce

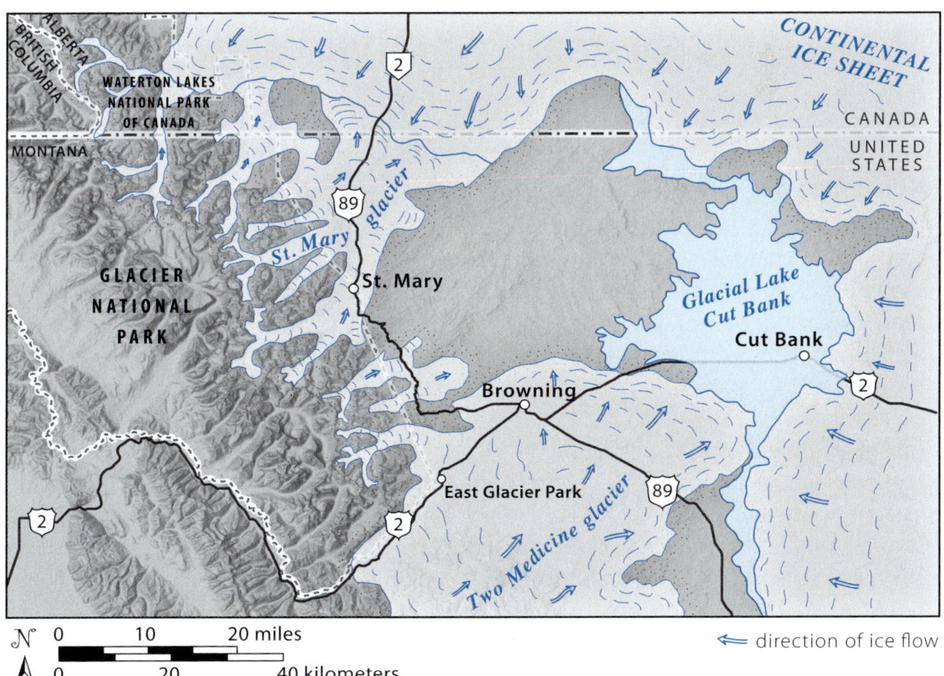

Glaciers and Glacial Lake Cut Bank when the Late Wisconsin glaciation was near its maximum, around 23,000 years ago. —Modified from Fullerton and others, 2013

places for birds to perch. For perspective, wind turbines are estimated to kill 100,000 to 400,000 birds per year, flying into windows about 100 million to 1 billion, and cats about 500 million to 3 billion.

The prominent rounded hills about 4 miles west of Shelby are glacial moraines. They're marked by scattered big white boulders of granite and metamorphic basement rocks that were carried south in the great continental ice sheet. The pre–ice-age path of the Milk River trends north-northwest through Shelby in a valley called the Sweet Grass Sag. The river came through the town of Sweetgrass, to the north, passed through Shelby, and drained southeast of Chester to ultimately join the Marias River. White salt pans 1 mile or so west of Shelby are saline seeps. (See the I-15: Great Falls—Sweetgrass road guide for information on the oil fields north of Shelby.)

Between Shelby and Chester, the three big buttes of the Sweet Grass Hills punctuate the skyline to the north. Good gravel roads head north to the hills from Chester and several small towns to the west. From this distance the hills look small, but they rise more than 3,000 feet above the plains. The highest peak, West Butte, stands at 6,983 feet; the highest peak in East Butte is only 25 feet lower.

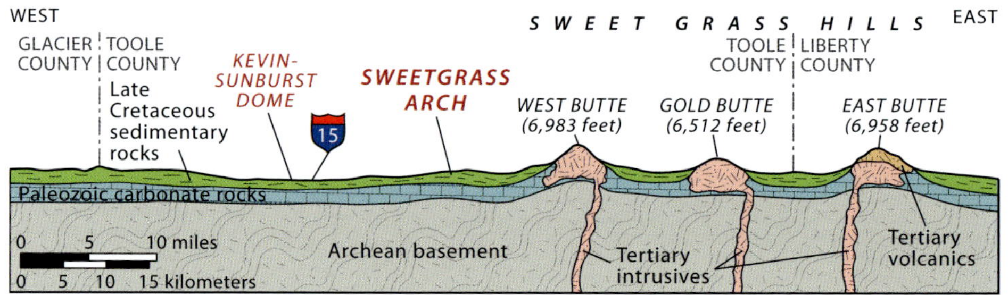

West-east cross section north of and parallel to US 2 across the Sweet Grass Hills.

Each butte is a miniature mountain range composed mainly of syenite, rock composed mostly of potassium and sodium feldspars, as well as other related igneous rocks. Grassy and Haystack Buttes, much smaller hills south of the main buttes and hardly visible from US 2, were built from single igneous intrusions, whereas the larger buttes were built by multiple intrusions. About 50 million years ago, the magmas domed and poked through Paleozoic and Mesozoic sedimentary rocks, which now occur in roughly concentric patterns around the buttes. The sedimentary rocks were cooked and metamorphosed to varying degrees during the intrusion, resulting in mineralization that produced some low-grade gold on all three main buttes.

About 20 miles east of Shelby on the south side of US 2 (between Devon and Galata) is a small patch of a very lumpy glacial moraine. This patch is near the southern limit of the ice sheet during the Illinoian glaciation. During both the Late Wisconsin and Illinoian glaciations, the continental ice sheet flowed south around the three big buttes of the Sweet Grass Hills, leaving them standing like islands above a sea of ice.

North of Rudyard, many dinosaurs have been excavated from Kennedy Coulee, a gash carved out of the plains by badlands erosion of the Judith River Formation. For decades the owners of Redding Farms have opened their private land to professional

Badlands in the Cretaceous Judith River Formation along Kennedy Coulee north of Rudyard. —Courtesy of Jack Horner

paleontologists and students to conduct research, resulting in the discovery of dozens of dinosaur skeletons and the education of countless students. The site is known for its armored and duck-billed dinosaurs and is considered one of the richest dinosaur localities in the region.

US 2 vividly illustrates how flat the layers of sedimentary rock in central and eastern Montana remain, having been beyond the reach of crustal movements that crumpled and shoved the rocks exposed in the Rocky Mountains. All the bedrock under US 2 from Chester to Havre is the Judith River Formation. This sandy shale was deposited in coastal environments adjacent to the shallow Western Interior Seaway between 80 and 75 million years ago. It was a coastal plain deposit, with the Claggett Shale deposited at the same time as mud immediately offshore.

A few miles west of Kremlin, scattered boulders, a foot or so in diameter, poke out of roadcut exposures of ground moraine. They are composed of granite and pale, grainy, highly metamorphosed rocks that aren't from around here; they must have been piggybacked here on the continental ice sheet from source rocks in Manitoba or northern Saskatchewan.

The big roadcut in black, unlayered, rubbly-looking rock 1 mile west of the US 87 junction at Havre is shonkinite, a potassium-rich basalt-like rock. It's the most abundant rock of the Bears Paw Mountains to the south. Visible grains of black pyroxene and white potassium feldspar are the main minerals making up the rock. The structure suggests it may have been rubble on the flank of the Bears Paw volcano from which it erupted. Why it lies so far north from the main mass of the volcano is not clear, but the volcano is about 50 million years old, and erosion would have lowered the landscape by 1,000 feet or more since then; perhaps its flanks stretched this far. Alternatively, the major landslides documented on the south flank of the volcano may have happened here as well, possibly carrying this big mass of rock far out from the base of the volcano. Springwater oozes from a steeply oriented crack in the roadcut; the crack may be a fault surface.

Shonkinite rubble in a roadcut on US 2, 1 mile west of US 87, is related to igneous rocks in the Bears Paw Mountains.

At Havre, steep banks up to 100 feet high mark the pre–ice age valley of the Missouri River, now occupied by the much smaller Milk River. Intermittently from at least 1,000 to about 355 years ago, Native Americans herded buffalo off a steep terrace of the Milk River at the west end of Havre, and then killed the injured animals and butchered them for food and skins. The Wahkpa Chu'gn Archaeological Site behind the mall at the west end of Havre is open to visitors from June 1 to Labor Day. The site, an ongoing archeological excavation, preserves layers of buffalo bones where the animals were prepared and cooked for consumption.

US 2
Havre—Malta
88 miles

Between Havre and Malta, US 2 follows the pre–ice age valley of the Missouri River, now occupied by the relatively small Milk River. During the ice age, the continental ice sheet pushed the Missouri River south out of the channel in which the Milk River now wanders aimlessly. It's obvious that the Milk could not have eroded the spacious floor of its broad valley.

The few exposures of bedrock along US 2 between Havre and Malta are of the Judith River Formation: siltstone, sandstone, and swelling clay, with scattered coal seams. The first exposures are of a soft brownish-gray shale about 5 miles east of Havre, where US 2 climbs out of the Milk River valley. The sediments were deposited in the channels and floodplains of eastward-flowing rivers in Late Cretaceous time, between 80 and 75 million years ago. An amazing variety of fossils have been found in the Judith River Formation, including plant-eating dinosaurs such as horned ceratopsians and duck-billed hadrosaurs, bony needlefish such as gars, stingrays, amphibians, alligators, lizards, and turtles. Along with them, of course, were meat-eating predators, such as tyrannosaurs and feathered theropods, or raptors. A few miles north and south of US 2 the slightly younger Bearpaw Shale overlies the Judith River Formation.

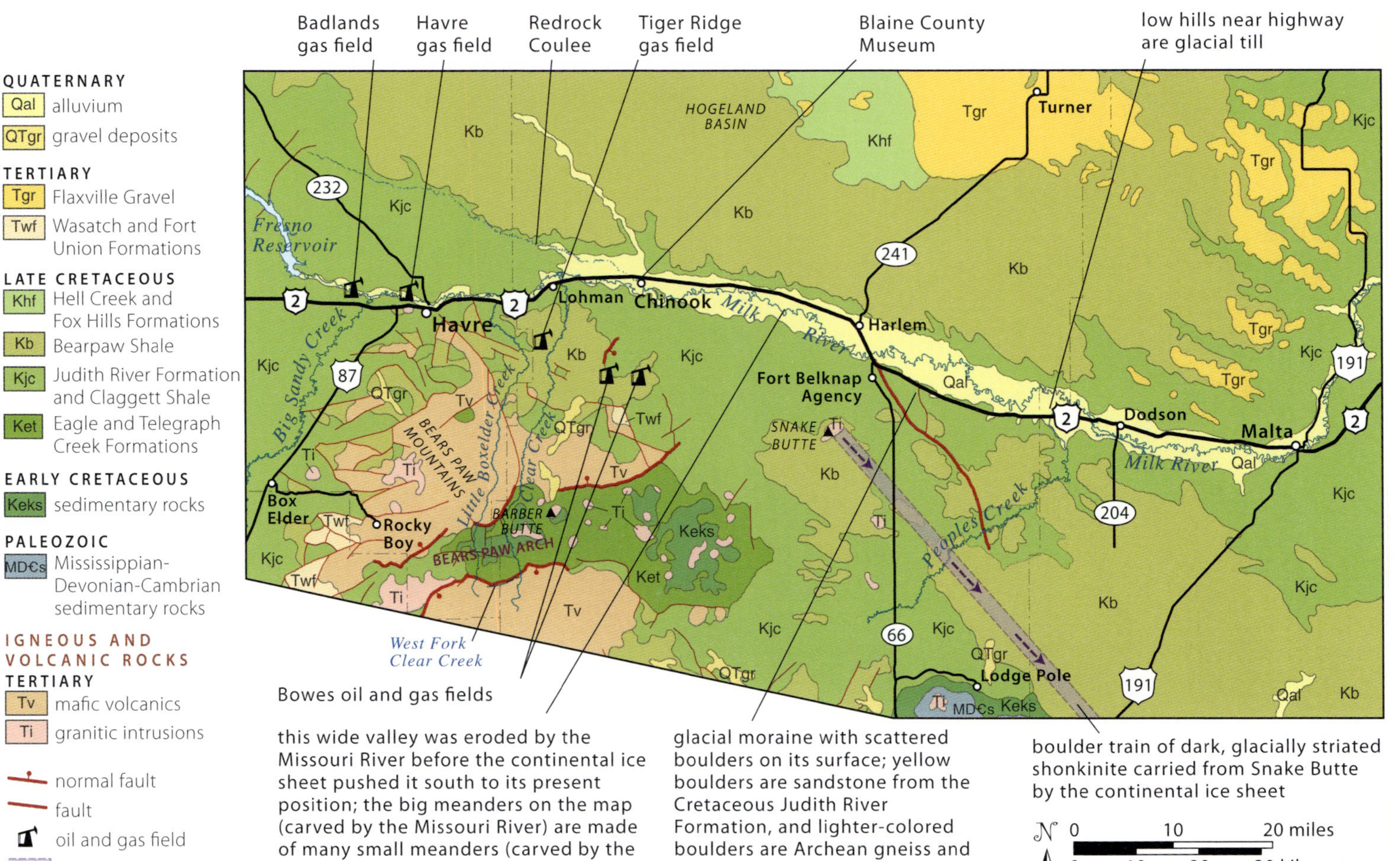

Geology along US 2 between Havre and Malta.

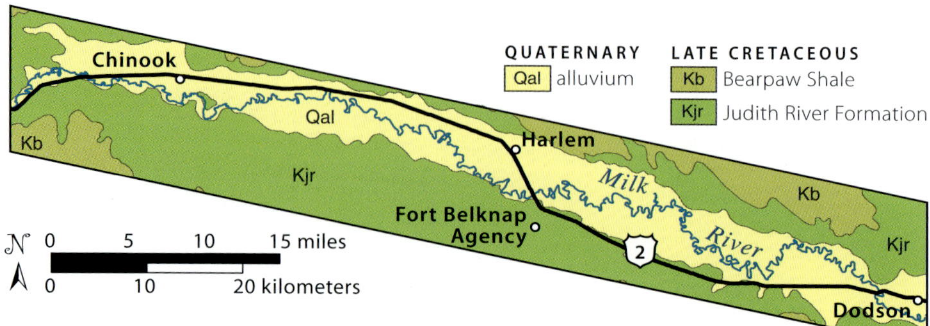

The thin squiggly blue lines are tiny meanders of the Milk River. On a larger scale, those meanders are part of one big northward bend from Chinook to Harlem, and another two between Harlem to Dodson. These big bends from one side of the big overall valley (yellow on map) to the other trace big meanders of the much larger pre–ice age Missouri River.

The undersized, meandering Milk River at the west edge of Havre occupies the pre–ice age valley of the Missouri River.

This shale was deposited during an expansion of the Western Interior Seaway and contains abundant fossils of a number of marine animals, including ammonites, a type of squid with a chambered shell.

The Blaine County Museum in Chinook, one-quarter mile south of US 2 on 5th Street, displays dinosaur fossils from northern Montana, including a huge *Gorgosaurus* skull. These toothy meat eaters were among those that roamed lowland coastal areas in search of duck-billed hadrosaurs.

The Bears Paw Mountains sprawl along the distant horizon directly south of Havre and Chinook. This volcanic pile erupted about 50 million years ago, in middle Eocene time, along with the Highwood Mountains volcano south of the Missouri River. The Bears Paw is the largest volcano in the Central Montana Alkali Province. The magma of the volcano was intruded as a laccolith, pushing the Cretaceous sedimentary rocks up into a broad arch that trapped enormous volumes of natural gas, along with some

Excavation of a very complete Brachylophosaurus *fossil from the Judith River Formation near Malta. Dark-colored bones are exposed across the bottom of photo.* —Courtesy of Jack Horner

oil. The Bowes oil and gas field produces from a peculiar reservoir on the north flank of the Bears Paw Mountains.

The Eagle Sandstone there rests on layers of smectite (bentonite), extremely slippery clay that forms from weathered volcanic ash. During Cretaceous time, the Elkhorn Mountains volcano was erupting between Helena and Butte, spewing volcanic ash that probably accumulated in the area of the present-day Bears Paw uplift, as well as elsewhere. While the laccolith was bulging the Cretaceous rocks, the Eagle Sandstone broke into large blocks that slumped downslope into the Bowes graben on the slippery smectite. The Claggett Shale above it dropped as well, blocking the broken edges of the Eagle Sandstone with impermeable smectite in the shale, creating a barrier that trapped the gas at the faults, preventing it from escaping to the surface through the sandstone.

The Tiger Ridge gas field, the largest in Montana, also produces from the Eagle Sandstone in the Bears Paw uplift. The especially productive sandstone was deposited as beaches and probably barrier islands along the coast of the Western Interior Seaway during Late Cretaceous time. It's a rich source of gas, oil, and water because the open pore spaces between its sand grains hold fluids.

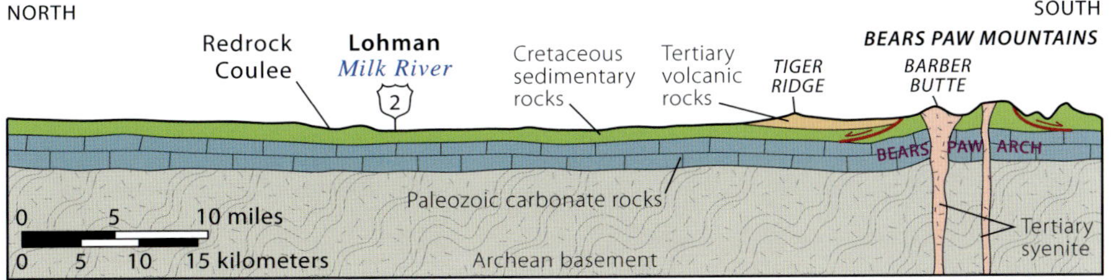

North-south cross section across US 2 through the Bears Paw Mountains just east of Havre. Barber Butte stands high because the hard igneous intrusion it's composed of resists erosion.

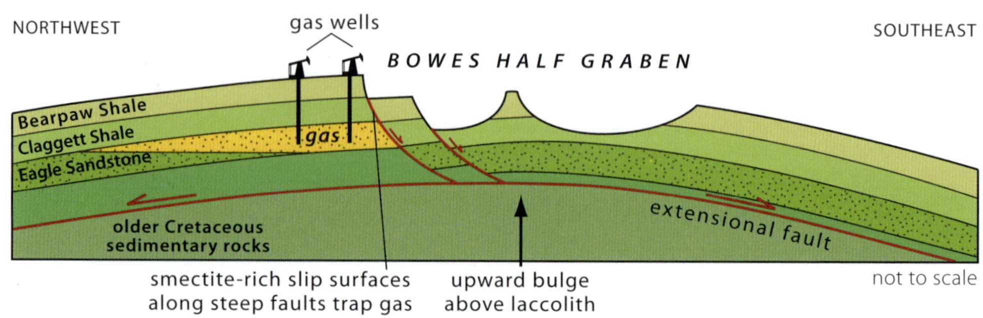

Smectite in the impervious Claggett Shale smeared along steep faults as they dropped past the Eagle Sandstone of the fault blocks; that sealed the ends of the porous sandstone beds, trapping the gas in the Bowes field south of Chinook.

Snake Butte, 10 miles south of Harlem, is one of the unusual potassium-rich intrusions scattered around central Montana. Its composition is so unusual that when a geologist found a huge, black angular boulder on the plains, he was able to easily match it to its source—Snake Butte. One of these big black boulders is on the west side of MT 66 (right at milepost 36), 14 miles south of Fort Belknap Agency. It rests near the fence line, dropped in place by the continental ice sheet. Winter ice has split it into several parts since.

This boulder is clear evidence that in this area the continental ice sheet moved in a southeasterly direction, having picked the boulder up at Snake Butte and then dropped it here. Other shonkinite boulders from Snake Butte form what is known as a boulder train, a line of boulders that stretches toward the community of Lodge Pole. All of the other abundant ice-dropped boulders in the area are pale, rounded granites from far north in Canada. These are especially obvious along MT 66 closer to Fort Belknap Agency. Additional evidence of continental glaciation comes from spectacular glacial striations on Snake Butte. These scratches and grooves were gouged out

A big shonkinite boulder the continental ice sheet transported here from Snake Butte (in background).

Spectacular glacial striations and grooves on Snake Butte were gouged out by rocks embedded in the base of the continental ice sheet during the Late Wisconsin glacial advance around 18,000 years ago. The glacier flowed in the direction the observer is looking (southeast). —Courtesy of Rod Benson, Bigskywalker.com

of the shonkinite as the rock-studded base of the glacier slowly ground its way southeastward. Shonkinite was quarried at Snake Butte for the construction of Fort Peck Dam because hard rock is rare in this part of Montana.

The isolated cluster of low peaks on the distant skyline directly south of Harlem is the Little Rocky Mountains. The core of the range contains the only exposures of Paleozoic rock (and a little early Proterozoic basement rock) in this part of Montana, as well as large masses of igneous rock, such as syenite porphyry. This magma intruded about 50 million years ago, doming the older rocks. The massive, thick, nearly vertical layers of Madison Group limestone stand up in erosional relief to form a wall that surrounds the range. (See the US 191: Lewistown—Malta road guide for information about the long history of mining in these mountains.)

Between Dodson and Malta, the lumpy hills near the road are till of glacial moraines. They are sprinkled with off-white to pink granite boulders several inches to many feet across that the continental ice sheet carried here from ancient basement rocks in central Canada. A greater number of boulders occurs in the swales on the flanks of hills because erosion in the gullies stripped away the dirt, leaving the heavy boulders behind. Although in some places the continental ice sheet reached to south of US 2, elsewhere it didn't and boulders are not present. A dense line of trees to the south punctuates the open rangeland, marking the path of the Milk River.

In 2000, a juvenile duck-billed dinosaur (*Brachylophosaurus canadensis*) was found in the Judith River Formation in Phillips County near Malta. The creature died around 77 million years ago and was so well preserved that 90 percent of its body was still covered in fossilized soft tissue. The male hadrosaur, dubbed Leonardo, was 23 feet long, weighed nearly 2 tons, and was about 4 years old when he took his last breath. Though he has been described as a "mummified specimen," he consists of imprints and minerally replaced soft tissue, such as skin, muscles, foot pads, and internal organs—he's all stone! The remains of his keratin beak—the duck-billed part of the animal—are visible. The stomach contents were so well preserved that researchers determined that he ate ferns, conifers, and magnolias for his last supper and that he had parasitic worms.

US 2

Malta—Wolf Point—North Dakota Border

283 miles

The broad, flat uplands about 10 miles north of US 2 at Malta are the gravel-covered High Plains Surface, the Flaxville Gravel. Rivers spread this gravel over much of eastern Montana 5 to 2.5 million years ago, during Pliocene time, when this region was a desert. We don't actually cross this surface along US 2. The well-drained gravel provides perfect land for growing wheat. With the arrival of a wetter climate during Pleistocene time, year-round streams gradually eroded valleys in the High Plains Surface, which now lies 850 feet above the Milk and Missouri River valleys along this route. Malta has two great museums worth checking out: the Great Plains Dinosaur Museum and the Phillips County Museum.

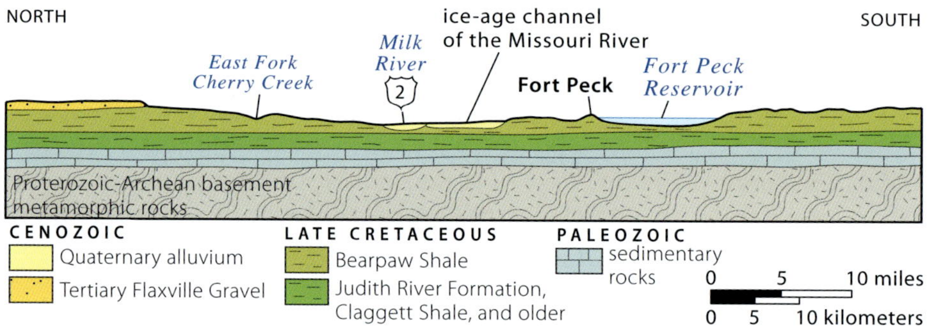

Cross section across US 2 between Glasgow and Nashua.

During both the Late Wisconsin and Illinoian glaciations, the continental ice sheet stretched from Canada to south of the line of US 2. By the time the ice reached northeastern Montana, it was near its southernmost extent, moving slowly and thinning rapidly. So far from its source, the ice sheet wasn't capable of eroding the landscape in this area, so the chief evidence of its passage is in deposits of glacial sediments; erratic boulders are the most obvious sign. Watch for the rounded rocks that litter the fields nearly everywhere. Farmers have picked many of them out of their fields and piled them along fences. Although most erratic boulders are smaller than a washing machine, some must weigh several tons. Yellow and brown Cretaceous sandstone boulders are from local sources; lighter-colored gray granite and streaky red, gray, and black gneiss boulders (for example, near milepost 531) are likely from the basement rocks of northern Manitoba, the nearest area where the ice could have picked them up on its way to Montana.

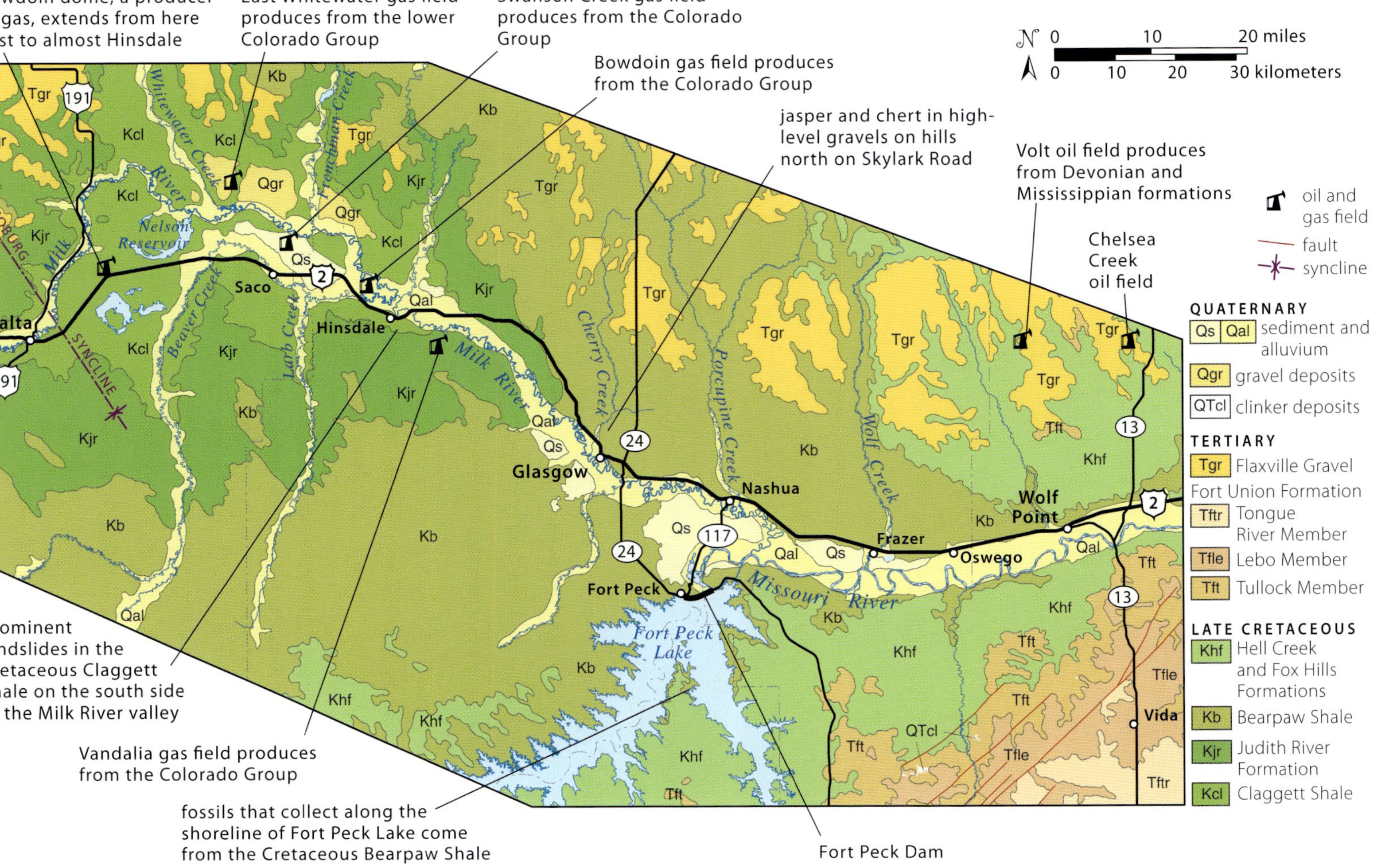

Geology along US 2 between Malta and Wolf Point.

Between Malta and Saco, US 2 crosses a very gentle bulge in the Cretaceous rocks, a giant dome likely formed during the Laramide orogeny of Late Cretaceous and early Tertiary time. The Bowdoin gas field was discovered in 1913 when a water well on a local ranch began to produce natural gas from a depth of 640 feet. Deliberate drilling for gas for local use began in 1916, and the area developed into a large gas field after a pipeline was laid to Malta and Glasgow in 1929. The field is now connected to a regional pipeline network. This dome traps gas primarily in two layers; one is a little less than 1,000 feet below the surface, the other a few hundred feet deeper. Most of the gas comes from Late Cretaceous sedimentary rocks of the Niobrara Formation and Belle Fourche Shale and related rocks.

Between Saco and Hinsdale, US 2 travels through the broad valley of Beaver Creek, which joins the Milk River at Hinsdale. Neither of those streams is big enough to have eroded the big valley; they both wander aimlessly back and forth as if looking for the downstream direction. That's because they're flowing in the former valley of the pre–ice age Missouri River. The Milk River follows this valley from Hinsdale to Fort Peck.

The Milk River is marked by erratic meanders that migrate gradually with every flood season, sometimes coming so close to an adjacent meander that they cut through the intervening sediment, taking a shortcut downstream. That leaves the old meander loop isolated as a curved, stagnating lake called an oxbow lake. A meander loop's size depends on the size of the river eroding it. A small river such as the Milk can make tight loops. A big river such as the Missouri is too wide, has too much water volume, and flows too fast to flow in such tight bends. (The map view of the two rivers near Nashua and Fort Peck Dam, as shown on the geology map at the beginning of this road guide, shows this contrast well.)

Just west of Hinsdale, US 2 passes a moraine, the tract of low, hummocky hills heavily littered with rocks in the valley floor south of the highway. The edge of the continental ice sheet lay here for a while late in the last ice advance (Late Wisconsin), and the melting ice dumped large quantities of debris to build the moraine. Meanwhile, glacial meltwater pouring out of the ice flowed between the edge of the ice and the south wall of the Milk River valley. Rocks in that valley wall consist of soft Claggett Shale overlain by a thick layer of hard brown sandstone of the Judith River Formation, both of Late Cretaceous age. The torrential rush of meltwater eroded the shale, thus undercutting the sandstone and causing the landslides that created the rough topography.

Between Saco and Hinsdale, and for miles farther east, the ground is unstable because smectite (a swelling clay) is widespread in the lower part of the Judith River Formation. Smectite is a clay mineral formed by the alteration of volcanic ash over millions of years. Water penetrates the clay layers, causing the clay to swell. Smectite is very slippery, making sediments susceptible to landslides and causing vehicles to slide off wet dirt roads.

East of Hinsdale, US 2 crosses to the north side of the Milk River valley and onto low slopes in sandstone and shale of the Judith River Formation, which it crosses for about 5 miles before passing onto the dark-gray Bearpaw Shale. Ammonite fossils are widely distributed in the marine Bearpaw Shale, and fish and various marine reptiles have also been found. The highway continues in the partially till-covered Bearpaw Shale between Glasgow and Wolf Point.

During the ice age, the continental ice sheet spread south beyond US 2 everywhere around here except between Hinsdale and Glasgow. To the north and into Canada, a bedrock topographic high called the Wood Mountain Upland diverted the ice east and west around it. The Missouri River Valley backed up in the area of Fort Peck Reservoir to an elevation of 2,600 feet, flooding the subdued topography between Hinsdale and Glasgow and forming the northern part of Glacial Lake Jordan.

At Wolf Point the Bearpaw Shale underlies the northern part of town; a half mile or so upslope to the north, the Late Cretaceous Fox Hills Sandstone holds up a slightly higher river terrace. Across the Missouri River to the south, the Fox Hills shows up low next to the river, with the latest Cretaceous Hell Creek Formation upslope. The Fox Hills was deposited as barrier island sands over the Bearpaw Shale as the Western Interior Seaway was retreating. The seaway continued to retreat from Montana through the end of Cretaceous time, coinciding with the deposition of the largely nonmarine Hell Creek Formation, famous throughout the region for its incredible

FORT PECK DAM

A short side trip to Fort Peck Dam, south of Glasgow on MT 24, is fascinating both for the story of the dam and for the magnificent displays and explanations of dinosaurs at the Fort Peck Interpretive Center. One of the most prolific and important sites of dinosaur finds in the state was at Bug Creek, farther south on MT 24.

Tyrannosaurus rex *skull on display at the Fort Peck Interpretive Center.*

Fort Peck Dam is a massive sediment-and-rock-fill blockage of the Missouri River. The dam has a clay core between two permeable shells of coarser material. Gravel toes were built at the base of each shell to anchor each side of the dam, but there was not enough hard rock in the area to build them, so shonkinite was brought in by rail from Snake Butte, a laccolithic intrusion 115 miles to the west, southwest of Harlem. Shonkinite is a massive, dense, tough basalt-like rock with tightly interlocking grains—just the stable characteristics engineers were looking for in fill material. The dam is sited on river deposits resting on 1,000 feet of Bearpaw Shale. To prevent water from getting under the dam, slabs of sheet steel were driven more than 150 feet into the Bearpaw Shale to create a waterproof barrier the length of the dam. The fill for the dam was hydraulically dredged upstream and sent to the construction site as a slurry through pipelines. Construction began in 1933 and the dam opened in 1940.

Not all went smoothly with the construction. On September 22, 1938, cracks appeared on the upstream side of the dam and the east end failed, allowing 5 million cubic yards of material to spill into the lake. The landslide moved at about 10 miles per hour, and thirty-four workers were carried with the sliding material. Eight died, and six bodies remain permanently entombed at the base of the dam. Analysis of the failure showed that the weight of the dam material probably caused slippage on clay (smectite) seams in the Bearpaw Shale. A board of consulting engineers conceded there was risk to completing the dam, but ultimately they recommended to move forward with construction; the vote was not unanimous.

Supposedly built for flood control, it was primarily for the benefit of President Franklin Roosevelt's work-relief program during the Great Depression of the early 1930s. Over six years it employed some fifty thousand men. The final dam, 250 feet high and more than 21,000 feet long (almost 4 miles), created a reservoir 135 miles long. The dam provides water for irrigation, generates 185 megawatts per hour of electric power, and controls floods.

The landslide on the east side of Fort Peck Dam that took the lives of eight men on September 22, 1938. **—Photo by Robert A. Midthun, US Army Corps of Engineers**

variety of fossils, including those of dinosaurs. Although Wolf Point doesn't have any dinosaurs that we know of yet, its museum displays a great collection of arrowheads, as well as antiques from the settlers who homesteaded along the Missouri River, especially after the Great Northern Railway came through in 1887.

A few miles east of Wolf Point, Hell Creek shale and sandstone are near the river, with Paleocene Fort Union shale and sandstone above. These sedimentary rocks look similar, but the Hell Creek was deposited while the dinosaurs were still alive, and the Fort Union after they all had died 66 million years ago following an asteroid impact.

Culbertson, which became a town in 1887 when the Great Northern Railway came through, is a regional center for cattle ranching, grain farming, and, when the price of oil is high enough, oil production. It lies in the Williston Basin, where oil is pumped from the Bakken Formation. The large Elm Coulee field extends from Sidney northwest toward the Missouri River south of Brockton. Southeast of Culbertson on MT 16 for 2 or 3 miles, on and near the cut banks of the Missouri River, thin layers of red clinker mark the beige Fort Union Formation. Clinker forms where sedimentary rocks are heated by burning coal beds ignited by lightning or wildfires.

The Medicine Lake area, 25 miles north of Culbertson on MT 16, has been designated a National Natural Landmark because of its great examples of continental glacial features: glacial till, erratics, outwash, wetland potholes, and eskers (sinuous ridges of gravel deposited by meltwater streams under the ice).

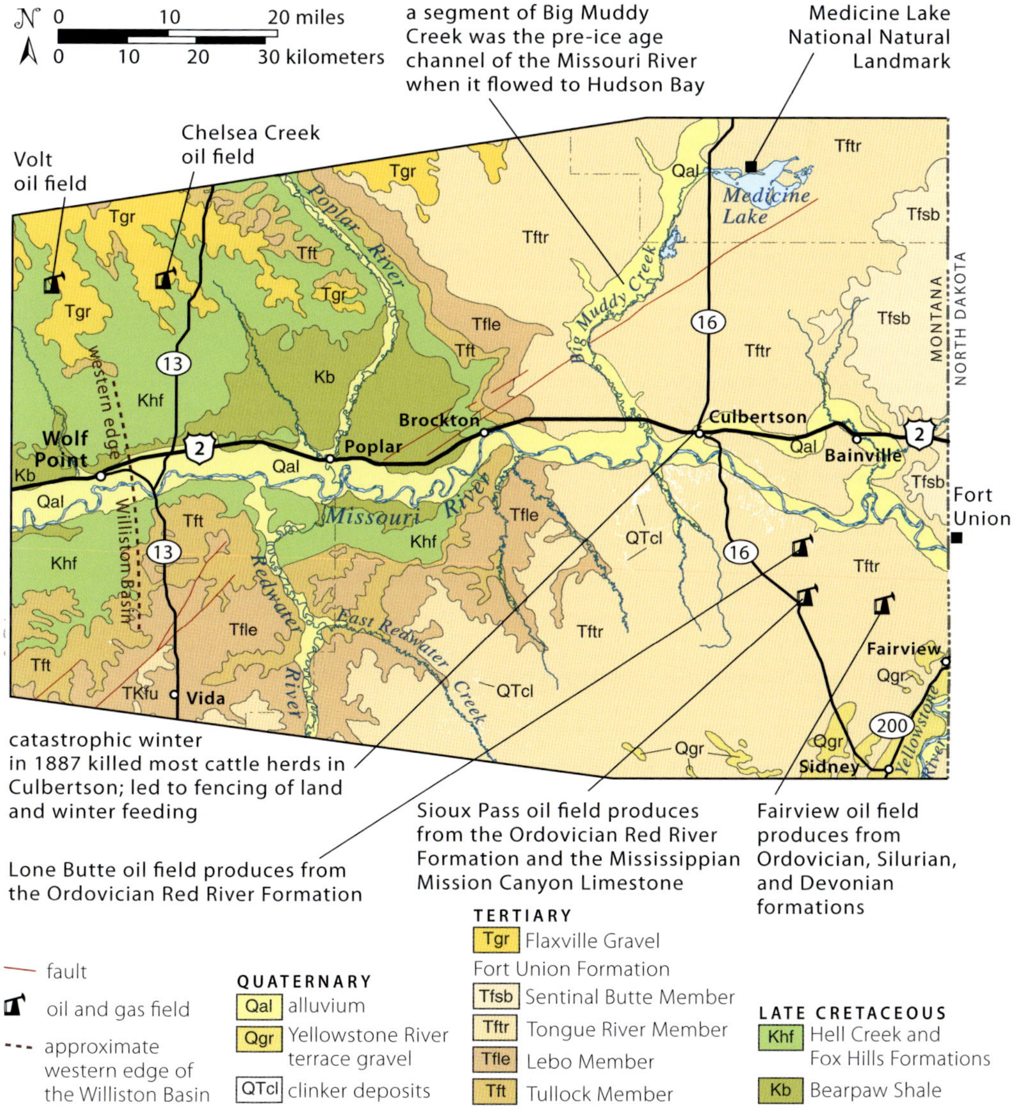

N
0 10 20 miles
0 10 20 30 kilometers
a segment of Big Muddy Creek was the pre-ice age channel of the Missouri River when it flowed to Hudson Bay
Medicine Lake National Natural Landmark
Chelsea Creek oil field
Volt oil field
Tgr
Tft
Tftr
Tfle
Tfsb
Kb
Khf
Qal
Qgr
QTcl
TKfu
Poplar River
Big Muddy Creek
Medicine Lake
Missouri River
Redwater River
East Redwater Creek
Yellowstone River
western edge
Williston Basin
MONTANA
NORTH DAKOTA
13
16
2
200
Wolf Point
Poplar
Brockton
Culbertson
Bainville
Fort Union
Vida
Fairview
Sidney
catastrophic winter in 1887 killed most cattle herds in Culbertson; led to fencing of land and winter feeding
Lone Butte oil field produces from the Ordovician Red River Formation
Sioux Pass oil field produces from the Ordovician Red River Formation and the Mississippian Mission Canyon Limestone
Fairview oil field produces from Ordovician, Silurian, and Devonian formations
fault
oil and gas field
approximate western edge of the Williston Basin
QUATERNARY
Qal alluvium
Qgr Yellowstone River terrace gravel
QTcl clinker deposits
TERTIARY
Tgr Flaxville Gravel
Fort Union Formation
Tfsb Sentinal Butte Member
Tftr Tongue River Member
Tfle Lebo Member
Tft Tullock Member
LATE CRETACEOUS
Khf Hell Creek and Fox Hills Formations
Kb Bearpaw Shale

Geology along US 2 between Wolf Point and the North Dakota border.

The striations on this glacial erratic along MT 16 north of Culbertson were scratched into it by rocks entrained in the bottom of glacial ice.

The study of preglacial sand and gravel deposits under glacial till just north of Poplar, and along a path northeastward through Homestead and Dagmar, appears to document the preglacial path of the Missouri River. From there it merged with the preglacial Yellowstone River before continuing north to Hudson Bay in central Canada.

The Fort Union Formation, which extends from Brockton to the North Dakota border, gets its name from Fort Union. This early trading post was on the Missouri River about 5 miles upstream from the confluence of the Missouri and Yellowstone Rivers, on the Montana–North Dakota border. It was built in 1828 at the request of the native Assiniboine people to facilitate the trade of buffalo robes and furs for beads, knives, guns, blankets, and cookware. Fewer than ten years later, an arriving steamboat brought smallpox that killed three-quarters of the natives.

US 12
Roundup—White Sulphur Springs
126 miles

Coal mining began between Roundup and Klein, 4 miles to the south, in 1906 when officials from the Milwaukee Road railroad were sent ahead to find coal to power their locomotives. Coal seams less than 10 feet thick were in horizontal layers only 200 to 300 feet below the surface. Underground mining left "rooms" with broad roofs that often collapsed, killing miners. The Roundup mines finally closed in 1956 when the Milwaukee Road switched to diesel locomotives.

Most of the route between Roundup and White Sulphur Springs follows the Musselshell River, the headwaters of which are in the Little Belt Mountains, north of White Sulphur Springs, and in the Castle Mountains south of town. The river drops 2,300 feet from its headwaters to Roundup, before turning north to join the Missouri River upstream of Fort Peck Lake.

Roundup sits on Paleocene shale of the Fort Union Formation, the layers of sedimentary rocks deposited after the worldwide catastrophic event that killed off the dinosaurs

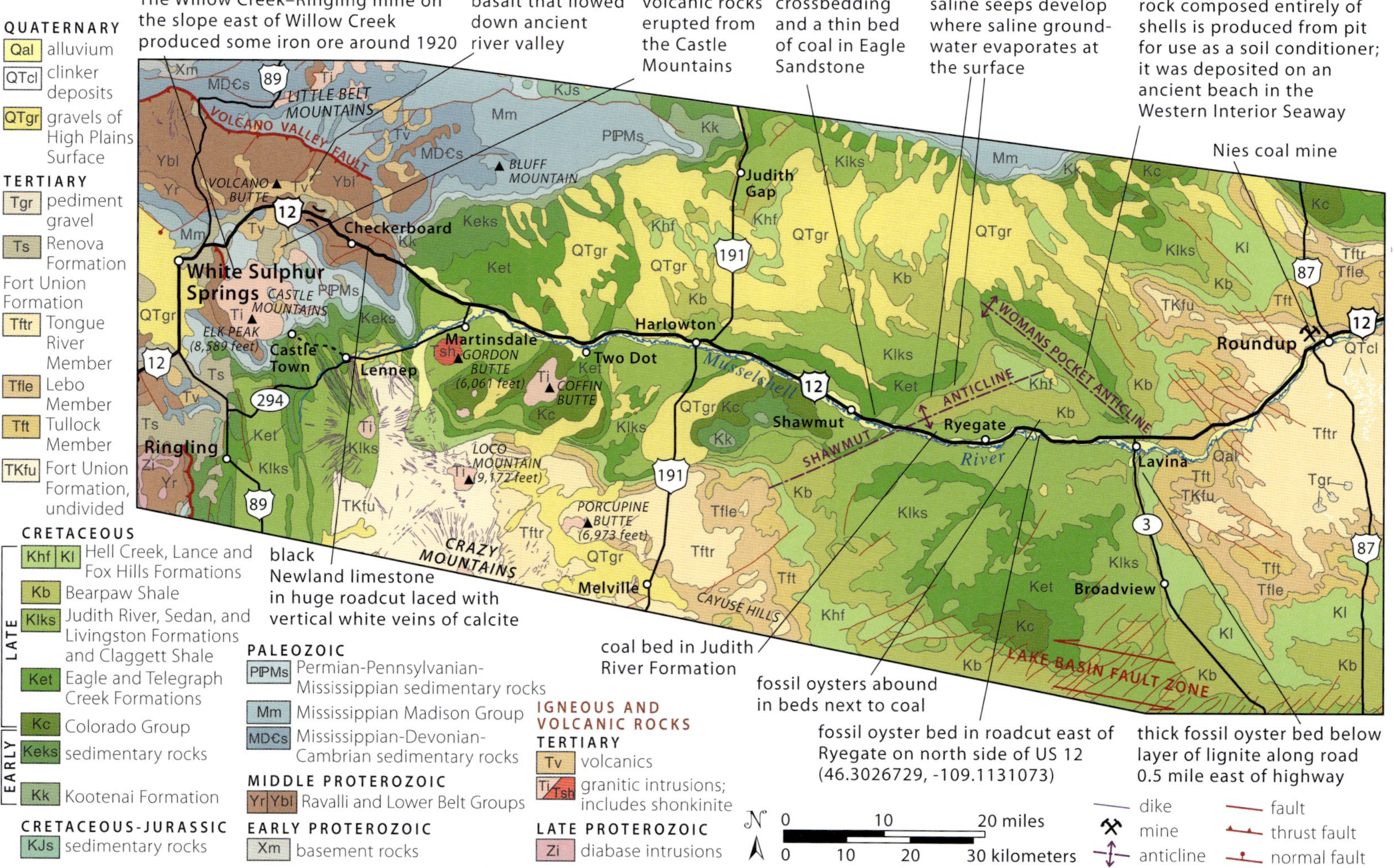

Geology along US 12 between Roundup and White Sulphur Springs.

Tongue River Member sandstone and shale on US 12 west of Roundup.

and many other species. For 12 or so miles southwest of Roundup, US 12 follows Fort Union shale at river level, but the overlying thick sandstone and shale of the nonmarine Tongue River Member forms the benches immediately above the river.

About 13 miles west of Roundup (11 miles east of Lavina), the highway climbs onto that bench of the Tongue River Member. About 22 miles west of Roundup (2 or 3 miles east of Lavina), the layers begin to dip down to the east as much as 20 degrees, so US 12 drops down into older rocks—from the Paloecene Fort Union to the Cretaceous Lance Formation sandstone and shale, Fox Hills Sandstone, Bearpaw Shale, Judith River Formation shale and siltstone, and finally Claggett Shale at Lavina. These rocks look very similar, but they have different physical characteristics and fossil content that allow geologists to tell them apart. These layers were deposited along or in the Western Interior Seaway, which waxed and waned across eastern and central Montana during Cretaceous time.

For about 3 miles west of Lavina, US 12 follows the Musselshell River bottom. The dark-gray Claggett Shale appears in patches along the river bottom, but the Judith River sandstone is more prominent around Ryegate. Look for high cliffs of the sandstone on the south side of the river along the 5-mile stretch east of Ryegate. The sandstone continues for 9 miles west of Ryegate.

The community of Shawmut lies near the core of the Shawmut anticline. Because the older layers are in the core of the anticline, you go deeper into the sedimentary pile as you head west, through Eagle Sandstone, sandy buff shale of the Telegraph Creek Formation, and finally Marias River Shale in the core, about 4 miles southeast of Shawmut. The highway then descends into the valley bottom of the Musselshell River again, following it to Harlowton. The Late Cretaceous Judith River Formation yielded an excellent specimen of the distinctive plant-eating, horned dinosaur *Avaceratops* near Harlowton in 1981. A full-sized replica is in the Upper Musselshell Museum in Harlowton.

Fort Union Formation sedimentary rocks southwest of Harlowton have provided abundant remains of land animals that lived in earliest Cenozoic time, shortly after

Late Cretaceous Judith River sandstone and shale overlying the Claggett Shale (not well exposed here) on the north side of US 12 just west of milepost 141 and Lavina.

the demise of the dinosaurs. Fossils were discovered in 1902 by Albert Silberling, a local homesteader and self-taught paleontologist, and Earl Douglass, the founder of the Dinosaur National Monument site in Utah and the first person to earn a master's degree (1899) from the University of Montana. Their discoveries include *Ptilodus*, a small tree-climbing, plant-eating, squirrel-like mammal considered one of the most successful of all mammals. It first appeared around 160 million years ago, in Jurassic time, survived the environmental changes that caused the demise of dinosaurs, and persisted until about 35 million years ago. Collected from the same rocks were remains of creodonts, carnivorous doglike mammals ranging in size from that of a small cat to twice the size of an adult grizzly bear.

About 9 miles west of Harlowton, US 12 crosses gravel of the High Plains Surface. About 3 miles east of Two Dot, nearly horizontal bedrock formations begin tilting up to the west, and the highway heads into older formations: the Claggett Shale and then the Eagle Sandstone.

Broad, flat-topped Coffin Butte stands on the horizon southwest of the Two Dot junction. Magma of this Tertiary laccolith intruded the surrounding rocks and gently domed them. Gordon Butte, just southwest of Martinsdale and about 5 miles south of the highway, is a perfectly circular, steep-sided laccolith. Both buttes stand about 1,000 feet above the surrounding plain. Gordon Butte is shonkinite, a fifty-fifty mix of augite and potassium feldspar and the most common rock of the Central Montana Alkalic Province. The butte's shonkinite is higher in sodium than most. The magma intruded between layers of the Late Cretaceous Judith River Formation, forming an almost perfectly circular shape when viewed from above. Although the intrusion bulged overlying rocks, its heavy weight pushed down the rocks under it, so the surrounding sedimentary rocks dip 10 to 30 degrees in toward the laccolith; those that were once above it have eroded away.

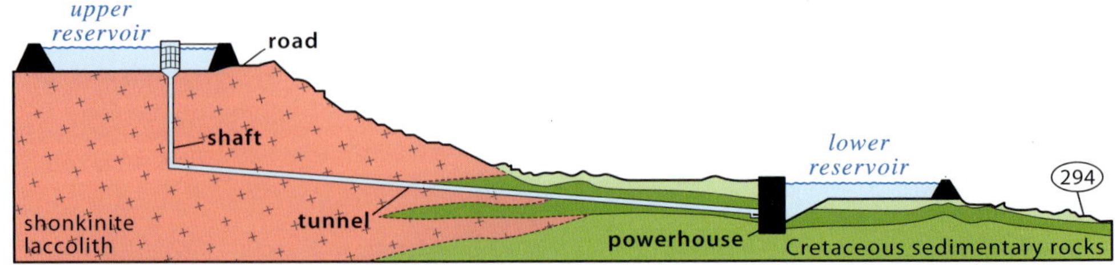

Excess electric power from a nearby wind farm will pump water from a reservoir at the base to one at the top of the butte. When the wind doesn't blow, water from the upper reservoir will flow down to electric turbines at the base to generate electricity to be fed into the electric power lines. —Courtesy of Gordon Butte Energy Park, Absaroka Energy

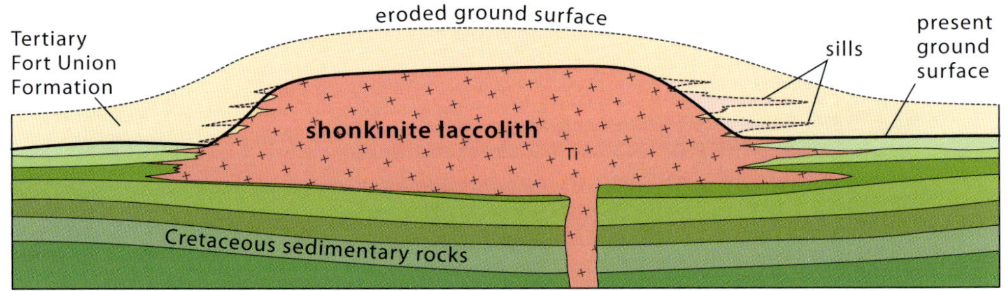

Magma intruded Cretaceous and Tertiary sedimentary rocks, forming a laccolith that bulged the layers above it. The layers below were pressed downward.

Wind turbines are a highly efficient source of renewable electric power that cost very little to run once power has been brought to them. But what happens when the wind doesn't blow? Absaroka Energy plans to pump water up to a reservoir atop Gordon Butte; the stored water will be released during high-demand periods to drive electric turbines at the base of the butte when the wind isn't blowing. In essence, the reservoir will act as a high-tech 400-megawatt battery for storing electricity.

Between the Martinsdale turnoff and White Sulphur Springs, US 12 skirts around the north side of the Castle Mountains, a large mass of granite 4 to 8 miles across. The magma rose close to the surface about 50 million years ago, and some of it erupted to form lava flows and volcanic ash, part of which still remain north of the mountains. (See the US 89: Livingston—White Sulphur Springs road guide for information about the Castle Mountains and the ghost town of Castle.)

About 7 miles west of the Martinsdale turnoff, Early Cretaceous Kootenai Formation rocks gradually steepen in the span of 1 mile; the road then passes through steeply dipping sedimentary rocks of Cretaceous, Jurassic, Permian, Mississippian, Cambrian, and Proterozoic (Belt Supergroup) age. Large white outcrops in the canyon are Madison Group limestone, a sedimentary formation deposited in shallow, tropical ocean water between 359 and 326 million years ago. These are on the east flank of a big, steep-sided anticline with a core of Proterozoic Belt sedimentary rocks. The Cretaceous Kootenai Formation, dipping up to 25 degrees away from the core of the fold, is the youngest. Sediment of these rocks was deposited between about 125 and 112 million years ago, indicating that the deformation that formed this anticline

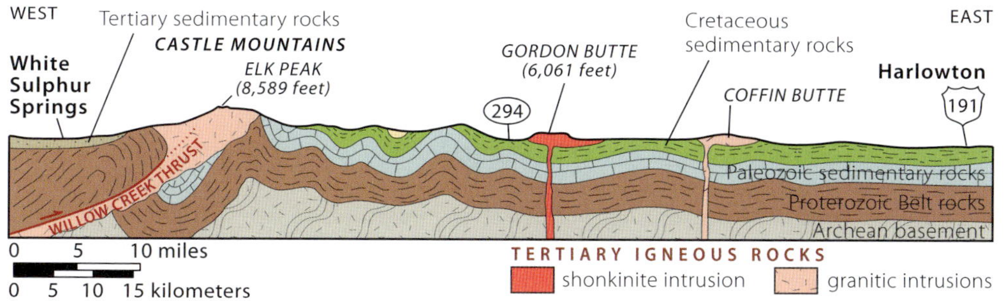

West-east cross section slightly south of US 12 between White Sulphur Springs and Harlowton.

happened sometime afterward. The Kootenai is nonmarine, deposited mainly in streams and lakes at the western edge of the shallow Western Interior Seaway, likely as the Rockies were being lifted for the first time during the Sevier orogeny.

Starting about 8 miles west of the Martinsdale turnoff, US 12 crosses onto the Greyson and Newland Formations of the early Belt Supergroup. About 2 miles east of Checkerboard are roadcuts in the nearly black Newland Formation limestone cut by big white calcite veins deposited in cracks that formed when the rock was folded and stretched. Calcite is soft enough to scratch with a pocketknife, whereas similar-looking quartz is too hard. A handy rule of thumb is that soft calcite veins form in limestone, whereas hard quartz veins form in quartz-rich rocks such as granite. Water deep underground dissolved calcite from limestone and then precipitated it here in the cracks.

Big white veins of calcite fill cracks in black Newland Formation limestone 2 miles east of Checkerboard. The veins are up to 6 inches wide.

One mile west of Checkerboard, big cuts in light-gray to buff quartzite and mudstone are the Greyson Formation. The highway crosses a pass 5 miles farther west at almost 5,800 feet. The pass is on the divide between the North Fork of the Musselshell River and the North Fork of the Smith River, which flows west through White Sulphur Springs on its way to Great Falls.

Immediately west of Lake Sutherlin for about 1 mile the road passes Oligocene-age basalt flows that lie on top of Proterozoic rocks on the north side of the highway. This long, 0.5-to-1.5-mile-wide belt of basalt flows trends east-west and appears to have flowed down an Oligocene-age river valley sometime between 34 and 23 million years ago. The large river drained eastward about parallel to the Musselshell River and US 12.

When James Brewer heard about hot springs that local Indians used as "medicine waters," he built a bathhouse, general store, and stables and began charging local settlers and prospectors for whisky and a bath—the start of commercial enterprises at White Sulphur Springs. The hot water at the modern-day town comes from a geothermal well that was drilled at the location of the original spring. The springs occur where water penetrates deep into the crust, is heated by hot rocks at depth, and migrates to the surface through porous and permeable rocks broken up by movement on recently active faults.

US 12
Roundup—Forsyth
102 miles

Roundup was a busy coal-mining town throughout the first half of this century. In those days, US 12 passed big tipples, platforms with ramps to load ore onto railcars, at sprawling railway yards near the west side of town. Then the coming of diesel and electric locomotives, cheap oil, and natural gas eroded the market for coal. Mining ended during the 1950s. The tipples are gone now, as are the railway yards and most of the miners. All the mines in the Roundup area were underground, so the scars they left on the landscape were relatively minor. The Bull Mountain Mine, 20 miles south of Roundup, it reopened in 2008, still underground; it made money for a few years, then struggled with reduced coal exports, decrease in demand for coal because of cheap oil and gas, and increased environmental concerns. In 2016 the mine, now the Signal Peak Mine, scaled back production by more than 30 percent and laid off 20 percent of its workforce.

At the north edge of Roundup, sandstone of the Late Cretaceous Lance Formation caps an east-west hill. To the east, US 12 follows the meandering Musselshell River past bluffs of Tongue River Member sandstone of the Paleocene Fort Union Formation between Roundup and a couple of miles west of the town of Musselshell. The overlying sandstone easily breaks off, and fallen blocks of it cover the underlying Lebo shale, also of the Fort Union Formation; the Lebo consists of gray shale that erodes easily and forms flat or gentle slopes. The shale was derived from older, Cretaceous marine shales that were raised during the Laramide orogeny and then eroded. It contains the fewest coal beds of the three main members of the Fort Union Formation. Across the river to the south, the more rugged and higher slopes are all in Tongue River sandstone, the coal-rich member of the Fort Union Formation. The flat upland surface 3 or 4 miles south of the river is held up by hard, resistant clinker left by a thick, naturally burned coal bed. Close to the town of Musselshell, a low ridge on the north is capped by thick-bedded light-brownish-gray sandstone of the Tongue River Member.

About 1 mile or so east of Musselshell, US 12 swings north out of the Musselshell River valley, and then east again. Gently dipping layers of the Tongue River sandstone

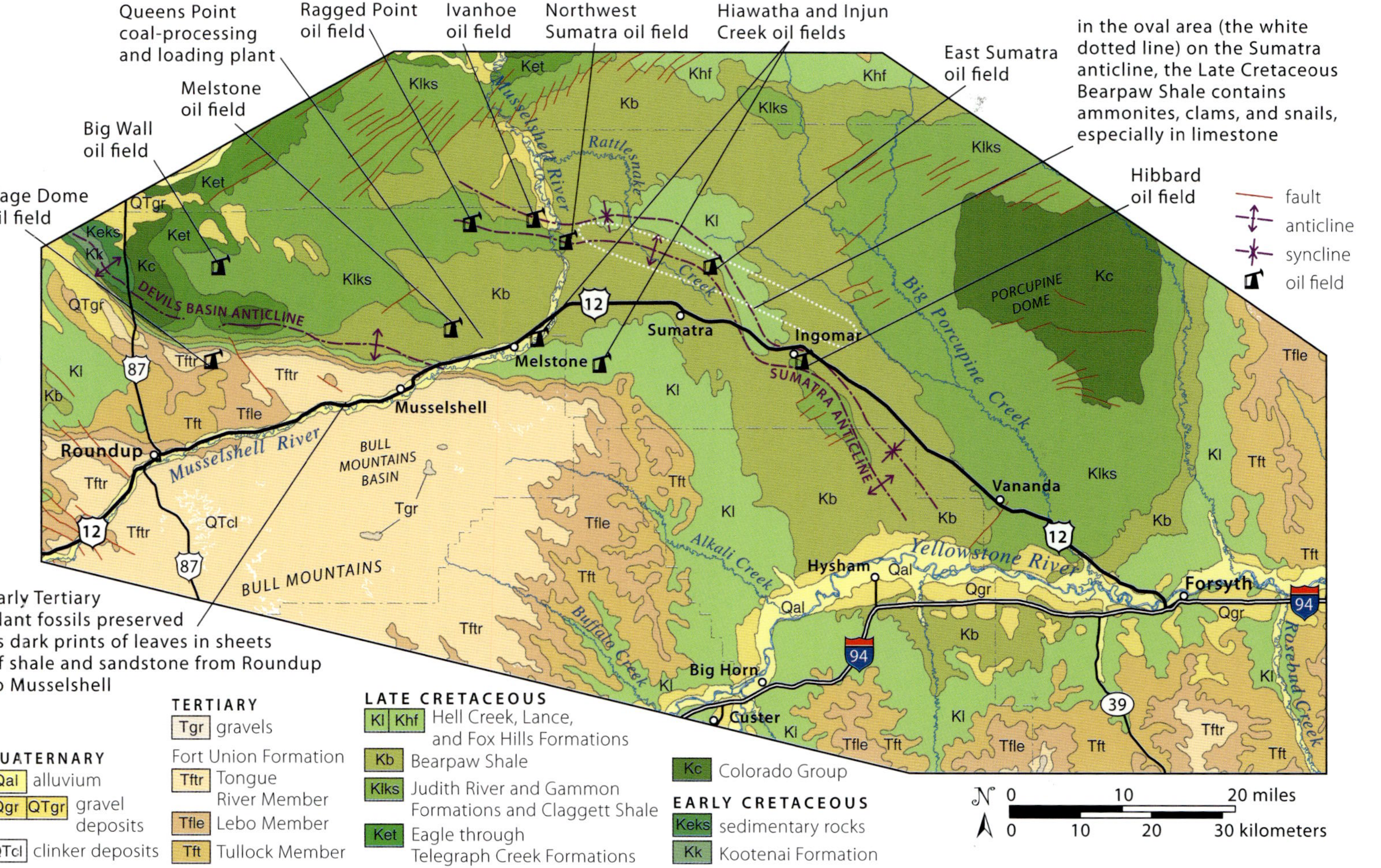

Geology along US 12 between Roundup and Forsyth.

Thick-bedded, light-brownish-gray sandstone of the Tongue River Member over softer light-gray shale of the Lebo Member near milepost 190, 1 to 2 miles west of Musselshell.

abruptly turn up to 20 to 30 degrees, with dips to the southwest, and then return to gentle dips again, all within 1 or 2 miles. This flexure is the Devils Basin anticline. Because the layers bend up ahead, the highway dives deeper into the sedimentary rock section, passing from widespread light-colored cross-bedded sandstone of the Tongue River, often covered with thick growths of pine trees, down through gray Lebo shale, yellowish to brownish-gray Tullock sandstone, cliff-forming brownish-gray sandstone of the Late Cretaceous Lance Formation, and finally into widespread medium-gray Bearpaw Shale. Oil is produced from Mississippian and Pennsylvanian formations under the anticline at depths of about 3,000 feet.

US 12 crosses the southeast-trending Sumatra anticline between Sumatra and Ingomar. The regional folding happened during the Laramide orogeny, creating these structures that trap oil and gas. The adjacent Sumatra syncline, in which the exposed

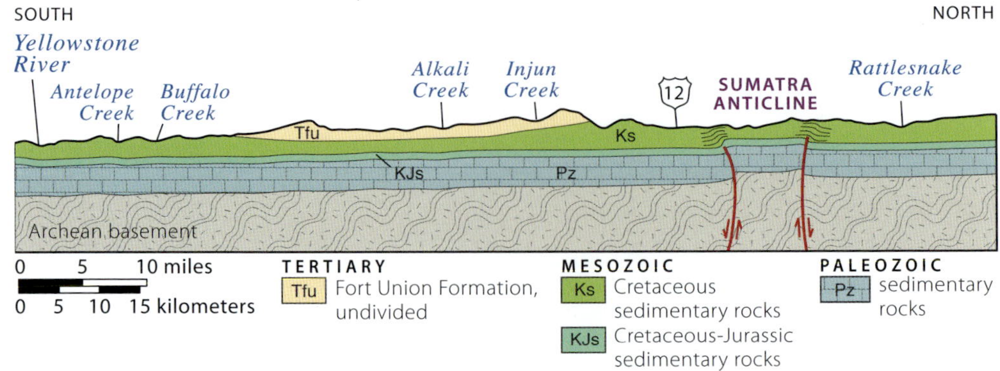

South-north cross section across US 12 between Melstone and Sumatra. Like many folds in eastern Montana, the Sumatra anticline drapes over faults in the continental basement. The folds formed from compression during the Laramide orogeny.

rocks of the Cretaceous Fox Hills Sandstone are folded downward, is 2 miles northeast of the crest of the anticline. The highway briefly follows the axis of its trough near Ingomar. The whole northwest-southeast anticline-syncline pair has been traced for 90 or so miles.

East of Melstone an old oil well lingers 100 yards north of the highway, but no pump jack. The area north of Melstone and Sumatra contains a number of small oil fields and one large one, all along the crest of the Sumatra anticline. The oil comes from several sandstone layers within the Tyler Formation, which is exposed in the Big Snowy Mountains north of US 12, to the west.

The oil in the Tyler Formation migrated into it from underlying Mississippian rocks. During the Mississippian Period, from 359 to 323 million years ago, shallow seawater flooded most of Montana. Several of the sedimentary formations deposited then in central and eastern Montana contained large amounts of dark organic matter, the raw material of oil. Geologists call such formations source rocks. During Pennsylvanian time, most of Montana rose above sea level, and streams carved an erosional landscape into the Mississippian rocks. The Tyler Formation consists largely of sand deposited in those stream channels. Sandstone makes a good reservoir rock for oil because the pore spaces between the grains are numerous enough to hold quite a bit of oil and large enough to let it seep through the formation. It's fortunate that the Tyler sandstone is where it is, because the pore spaces in the underlying Mississippian source rocks are too small to permit oil to flow through it and into a well. It can only be pumped out of the Tyler.

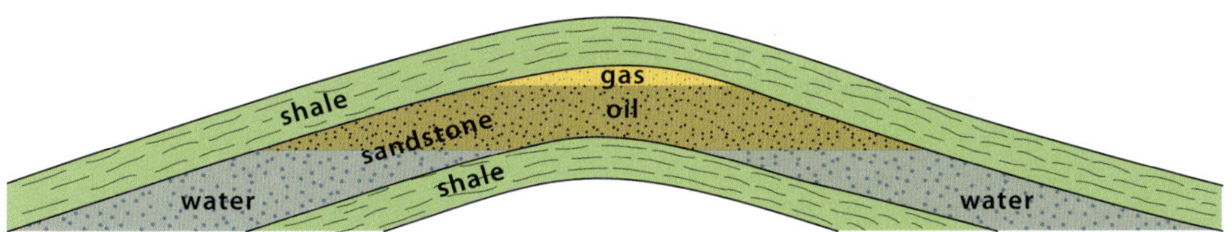

Cross section of the Sumatra anticline showing how oil and gas accumulates in the crest of the fold. The dips on the flanks of the anticline are exaggerated to better show the main features.

The Sumatra field, the largest in the area, was discovered in 1948 and had produced some 40 million barrels from the Tyler sandstone by the end of 1983. The Hiawatha field, about 1 mile northeast of Melstone, is fairly typical of the small oil fields that produce from the Tyler sandstone along the crest of the Sumatra anticline. It was discovered in 1967 at a depth of about 5,000 feet.

The dark Bearpaw Shale, which the road passes over to about 4 or 5 miles northwest of Vananda, was deposited during an expansion of the Western Interior Seaway during Cretaceous time. Fossils of oysters, clams, snails, and squid-like animals called ammonites are fairly abundant, especially in hardened concretions. If you find some loose concretions, break them open and see what's inside.

Just east of Vananda, the highway skirts the southwestern edge of the Porcupine dome, a broad uplift in the sedimentary rocks that doesn't show in the landscape. The dome lies directly southeast of the Cat Creek anticline and appears to be part of the same structural trend. Rocks exposed in the central part of the dome normally lie more than 2,000 feet beneath the surface in this part of Montana. The broad dome has drawn considerable interest from wildcat drillers and petroleum companies for more than a century but without success; its main product has been salty water. From the dome's top, its flanks dip several degrees outward in all directions, providing a closed cap that could trap rising fluids, but a couple of steep faults that cut up through the rocks near the crest of the dome could have bled off petroleum, if any was there in the first place.

Judith River Formation sandstone alternates with Claggett Shale along the northeast side of the highway between Vananda and the north edge of the Yellowstone River valley. For much of the remaining distance to the Yellowstone River at Forsyth, the same kind of brownish-gray Judith River sandstone and some Bearpaw Shale are exposed in low bluffs along the road. Sandstone of the overlying Lance Formation forms prominent, massive cliffs. Streams deposited it between 69 and 66 million years ago on the coastal plain of the Western Interior Seaway. Similar in age to the Hell Creek Formation and just as rich in fossils, the Lance has yielded tens of thousands of fossils of Late Cretaceous dinosaurs, fish, frogs, salamanders, and even birds.

US 87
Billings—Roundup—Lewistown
125 miles

In the broad valley of the Yellowstone River, Billings seems guarded on all sides by bold, vertical, beige to pale-orange cliffs 200 to 300 feet high. Known as the Rimrocks, the cliffs are Eagle Sandstone, deposited as beaches and barrier island sandbars along the western margin of the shallow Western Interior Seaway about 83 to 81 million years ago, during Late Cretaceous time. The more easily eroded shales of the Telegraph Creek Formation underlie the Eagle Sandstone. (See the I-94: Billings—Miles City road guide for more information about the Rimrocks.)

The spectacular Rimrocks cliffs are at the west edge of US 87 as it heads north upslope from the city. The Billings airport rests on the eroded flat surface atop the Rimrocks. About 2 miles north of the airport turnoff is brownish-gray Claggett Shale, the soft rock that eroded from the top of the Eagle Sandstone. About 3 miles north of that the highway crosses about 4 miles of Judith River Formation, which is mostly light-brown sandstone and sandy shale.

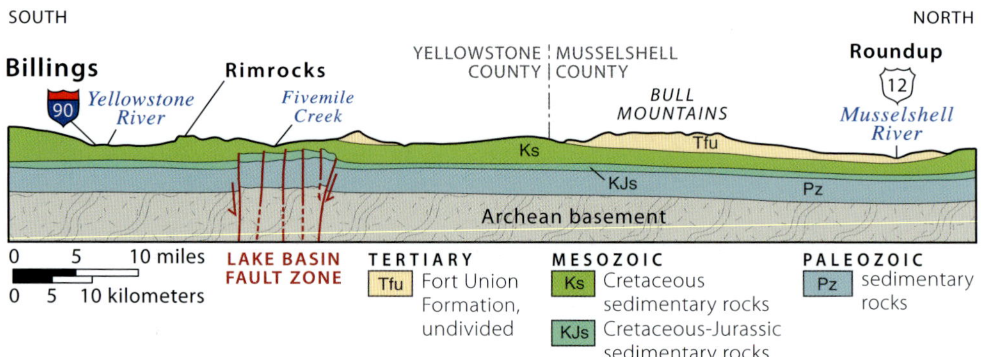

South-north cross section along the line of US 87 between Billings and Roundup.

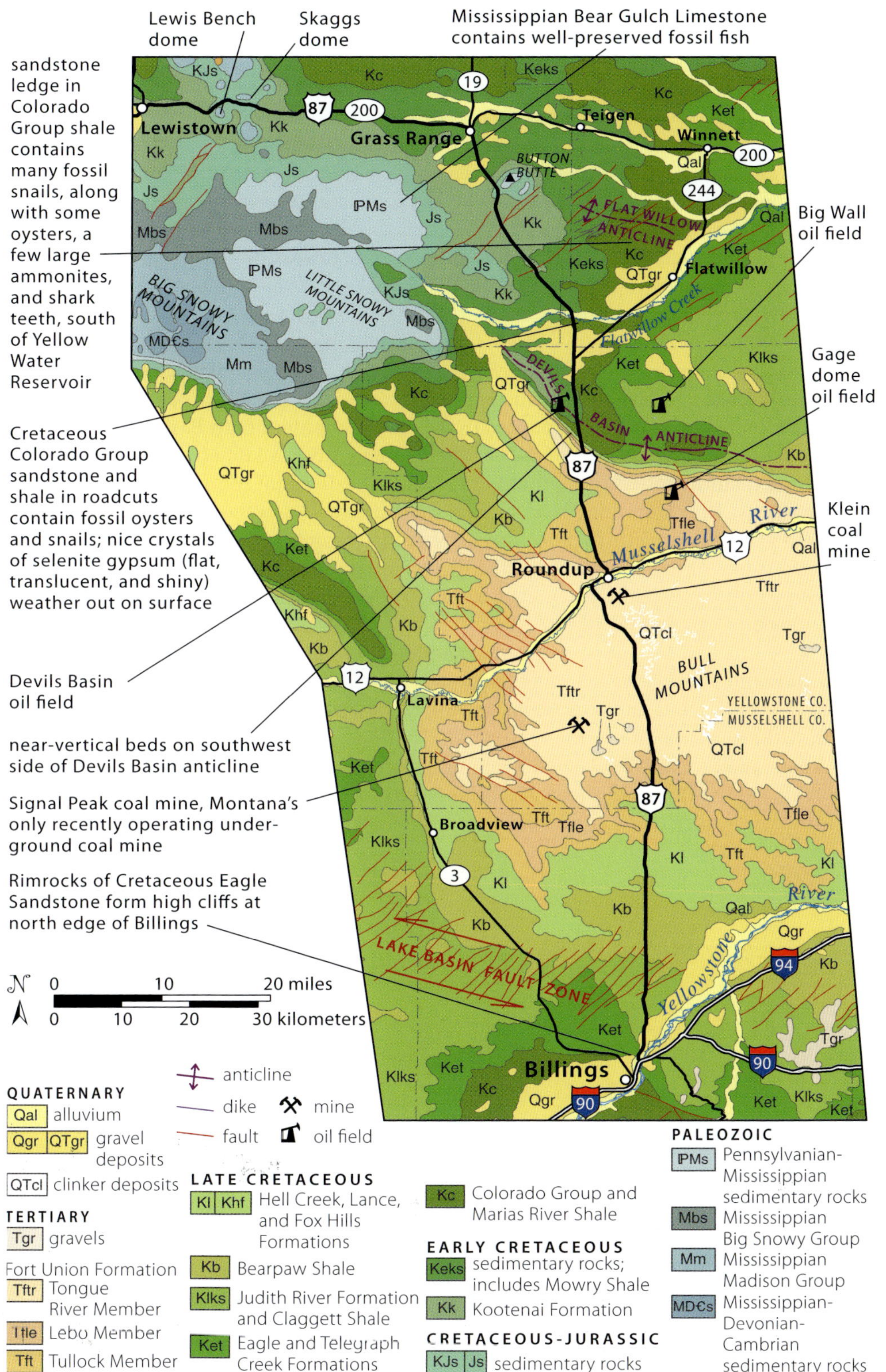

Geology along US 87 between Billings and Lewistown.

Between 5 and 10 miles north of the Rimrocks, the highway crosses a roughly northwest-southeast belt of northeast-trending faults, the Lake Basin fault zone. The whole zone extends more than 50 miles west of US 87 and 30 miles to the east. The faults cut through all of the Late Cretaceous formations here, and the more recent faults moved 50 to 40 million years ago. However, information from subsurface drill holes suggests that the basement rocks in the zone were first deformed in Proterozoic time, and then again in Paleozoic time, and that weak zones in the basement controlled the overlying faults. Crustal weaknesses such as this tend to get used repeatedly by changing tectonic forces over time.

From 17 to 20 miles north of I-90, US 87 crosses flat-topped ridge crests held up by Cretaceous Fox Hills Sandstone. From this area north to Roundup, the highway crosses the Bull Mountains, a northeast-trending arc of rugged hills. Sediments of the Fort Union Formation were deposited in the broad Bull Mountains Basin in Tertiary time. Thick sandstone ledges of buff-colored Tongue River Member, the youngest of its three main members, are prominent throughout the Bull Mountains. Scattered pine forests, especially along ridges where moisture resides in the cracks in the hard rock, make for very picturesque country. Between horizontal, buff-colored sandstone-capped ridges of the Tongue River, the highway crosses rolling grassland with some sagebrush and hayfields; black coal-bearing beds appear locally, along with buff-gray shale.

The Tongue River sandstone in this area was deposited in meandering and braided streams with floodplains rich in vegetation that eventually accumulated to form peat and then coal. Shale and coal seams are widespread but not well exposed. The main coal bed, from 2 to 17 feet thick, is marked by thin layers of volcanic ash, an impurity that makes it less desirable for mining. Many of the coal seams near the surface burned, baking the rocks above and below them into red clinker. Watch for red caps of clinker on ridge crests and broad bands of red on their flanks.

Local ranchers were the first to use these surface exposures of coal in the upper part of the Fort Union Formation in the 1880s. The first commercial mine opened

Tongue River sandstone holds up bold ridges in the Bull Mountains; this one is 18 miles south of Roundup.

in 1907, mostly to supply steam engines of local railroads. The first coal miners had some difficulty finding places to live, and a few of them built tiny stone houses against overhanging cliffs of sandstone. One of these is preserved near the community of Klein, just south of Roundup. Demand for the coal peaked in the 1940s and 1950s. The Signal Peak Mine (formerly the Bull Mountain), south of Roundup, mines high-temperature/high-grade coal from the Paleocene Fort Union Formation. It is Montana's only recently operating underground coal mine.

About 6 miles north of Roundup, US 87 crosses a syncline in buff-colored Fort Union Formation sandstone; rocks to the south dip northward, and those to the north dip south. A little more than 1 mile to the north, at Willow Creek, the dips of the beds of the Lance Formation reverse direction. About 11 miles north of Roundup, the sedimentary beds quickly steepen so they dip about 40 or so degrees southwest, some almost vertical. With all the flat-lying sedimentary rocks for miles around, it's startling to see rocks standing on end! A couple of miles farther north the steeply tilted beds quickly roll back to horizontal at the crest of the Devils Basin anticline, the axis of which trends northwest-southeast. This fold leans over sharply to the southwest. The Lake Basin fault zone, northwest of Billings, is also oriented northwest-southeast. Both the Devils Basin anticline and the Lake Basin fault zone may have formed in early Tertiary time due to deep crustal stress related to the Laramide orogeny. Within a couple of miles, US 87

View facing northwest of the upturned beds on the southwest flank of the Devils Basin anticline, 11 miles north of Roundup.

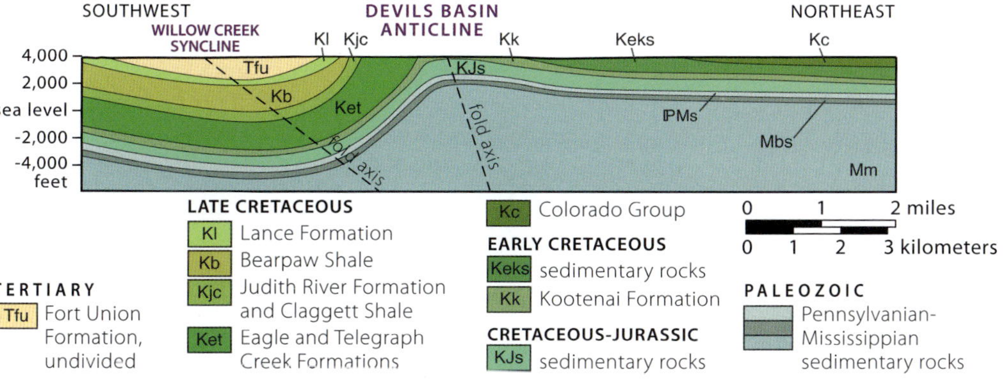

Southwest-northeast cross section across the Devils Basin anticline north of Roundup.
—Modified from Porter and others, 1996

passes northward through older, deeper beds, from the Paleocene Fort Union Formation to Early Cretaceous Kootenai Formation sandstone, and back.

These prominently tilted layers caught the attention of oil geologists a century ago. A wildcat well drilled on the anticline struck oil in 1919—not much, but that first discovery in central Montana sparked a rash of drilling elsewhere that led to the discovery of a number of good oil fields.

Three miles north of the MT 244 junction, as you descend into the Flatwillow Creek valley, you may see bright reflections that look like fingernail-sized mica or chips of glass in shale roadcuts. They're actually crystals of soft gypsum (calcium sulfate) that grew in the sulfur-bearing shale. To the north, US 87 crosses flat-to-rolling country, intermittently cutting through low hills that expose beige Cretaceous sandstones. About 15 miles north of MT 244 (8 miles south of Grass Range) are low roadcuts in beige sandstone and in dark-red Jurassic-Cretaceous shale. Tree-covered Button Butte, east of the highway, probably obscures an igneous intrusion—not yet exposed by erosion—related to those we'll pass closer to Lewistown.

On private land about halfway between Grass Range and Lewistown, a lengthy outcrop along Bear Gulch has two claims to fame. The extraordinary Bear Gulch Limestone of Late Mississippian age exhibits rarely preserved life-forms. The deposit is the source of one of the world's most diverse and well-preserved fossil fish collections, including bony fish, sharks, and enigmatic swimming animals, including a soft-bodied animal once thought to be the origin of mineralized toothlike fossils called conodonts. The preservation is so complete that some specimens show veins and skin pigments. The sediments accumulated in a very quiet-water embayment in a tropical sea, where fish and other animals were buried rapidly enough that they weren't scavenged or decayed by bacteria. The other claim to fame is the Bear Gulch pictograph site, accessible only by a guided private tour. It has an equally amazing display of Native American rock art.

About 21 miles west of the MT 200 junction, US 87 crosses the south edge of the Skaggs dome. About 3 miles farther west it skirts the north edge of the Lewistown Bench dome. Paleozoic rocks in the cores of both domes conceal buried intrusions that are part of a larger, fascinating group of intrusions, including the Judith

A 10-inch-long male ratfish, with a prominent dorsal spine, from the Bear Gulch Limestone west of Grass Range. —Courtesy of H. Zell, Creative Commons SA 3.0 license

Mountains to the north, as well as the North and South Moccasin Mountains to the west (discussed in the US 191: Lewistown—Malta road guide).

US 87
Great Falls—Lewistown
106 miles

East of Great Falls for about 11 miles, US 87 crosses the broad High Plains Surface eroded on the Early Cretaceous Kootenai Formation. It has a thin coating of glacial till deposited by the stagnant edge of the continental ice sheet, and Glacial Lake Great Falls silt. For 4 miles the highway crosses the same surface on the Kootenai Formation, albeit slightly more eroded, and then it rapidly drops through light-gray to very red Kootenai sandstones into the canyon of Belt Creek, near the town of Belt.

Cliffs at the north edge of Belt expose remnants of the coal-rich rocks that provided the main impetus for the development of the town. Coal in the Great Falls–Lewistown coalfield is in the Jurassic Morrison Formation. The coal developed from tropical peat marshes that formed on a floodplain along the northern margin of a large lake around 150 million years ago. Mountains raised during the Sevier orogeny during Cretaceous time shed sand and gravel eastward, burying the swamp deposits. The red color of the overlying sandstones, the lowest part of the Kootenai Formation, may be from deep tropical weathering that leached out soluble minerals and concentrated red iron oxide. Leaf impressions are abundant in black shale of the Morrison Formation, 15 feet below the Kootenai Formation.

John Castner discovered an outcrop of coal near Belt in 1877. He hauled it in freight wagons to Fort Benton for use as steamboat fuel and for heating homes. Marcus Daly and the Anaconda Company bought the operation in 1893 and expanded it a few years later, eventually employing 1,200 workers. Daly constructed one hundred coke ovens and shipped coal to his smelters in Great Falls and Anaconda. The coal also found a use powering Great Northern Railway locomotives.

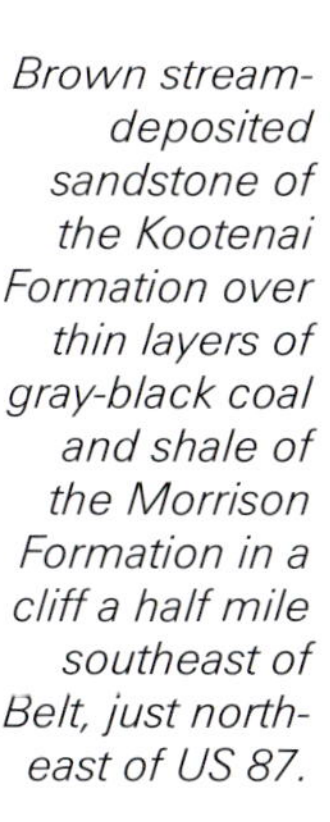

Brown stream-deposited sandstone of the Kootenai Formation over thin layers of gray-black coal and shale of the Morrison Formation in a cliff a half mile southeast of Belt, just northeast of US 87.

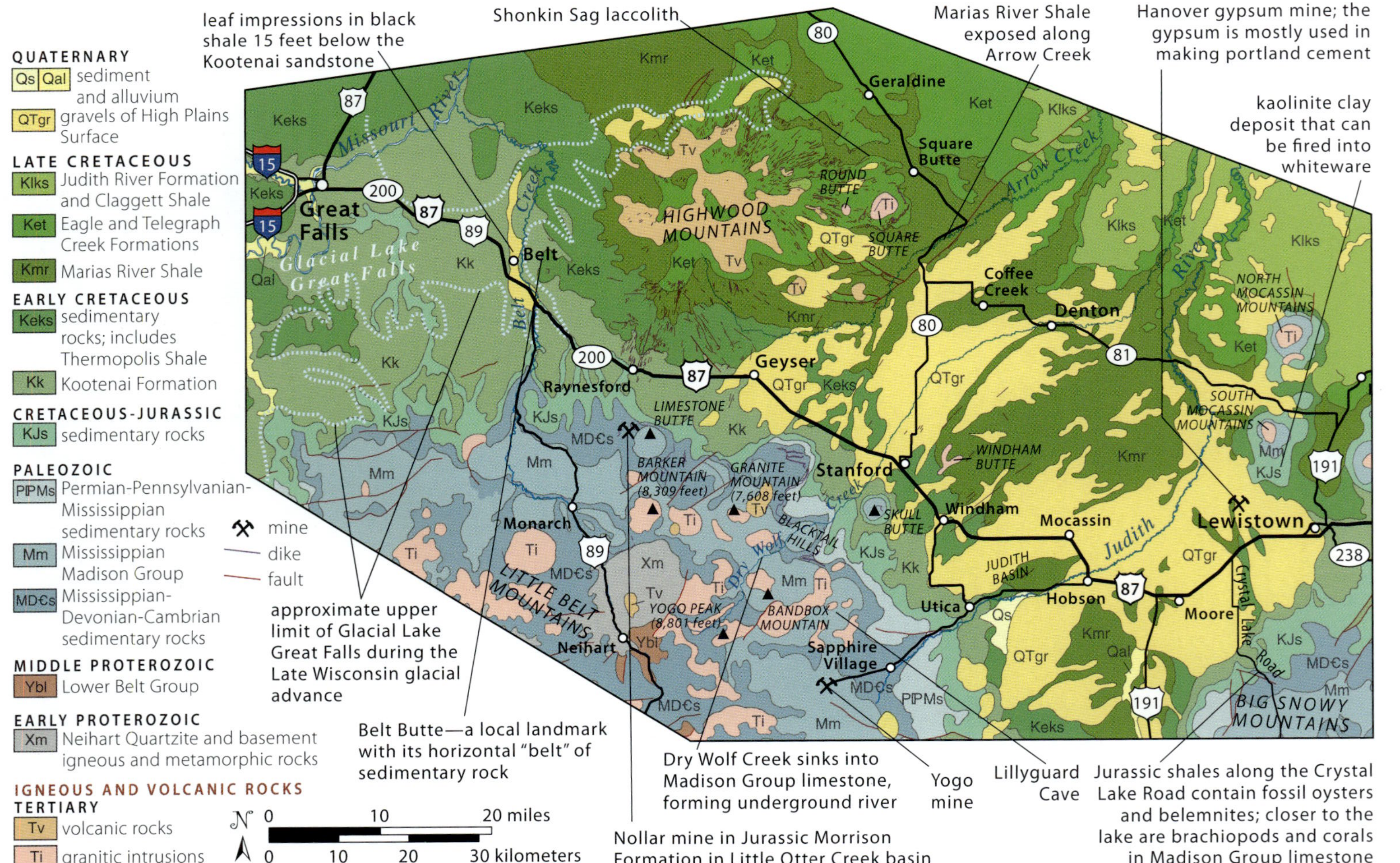

Geology along US 87 between Great Falls and Lewistown.

Mining slowed after the end of the century as railroads converted to diesel and the Anaconda smelter converted to natural gas. The Belt mine closed in 1924. Smaller mines continued until the early 1960s. Acidic rusty water continued to drain from abandoned mines for decades after, especially at times of low water. Iron, aluminum, sulfur, and other toxic metals hinder the use of the water. The iron and sulfur probably originate from the oxidation of iron sulfide (pyrite) in the coal. Environmental cleanup was slated to begin in 2019.

A distinctive volcano-shaped butte of Cretaceous sediments a couple of miles directly east of Belt was named Belt Butte because of a very prominent, dark, cliff-forming "belt" around it about halfway up. The butte gave its name to the town, the stream that flows through the town, the Little Belt Mountains to the south where the stream originates, and the Big Belt Mountains farther south. These mountains, in turn, gave the name to the Belt Supergroup, the Proterozoic-age rocks that dominate northwest Montana but were first studied in these mountains.

The rest area just east of Belt, at the junction of US 87, MT 200, and US 89, borders a quarry in rusty pale-gray Kootenai Formation sandstone that breaks into large blocks. Thin dark-gray layers, with small black grains of chert to color them, accentuate thin cross beds that formed from water currents bouncing sand grains downstream from highlands to the west in Cretaceous time. Kootenai Formation sandstones continue from the rest area all the way to Geyser.

Rusty light-gray Kootenai Formation sandstone with thin, dark, chert-rich layers and small-scale cross beds at the rest area east of Belt. The slope of the cross beds in the inset indicates that the current that transported the sand moved from left to right.

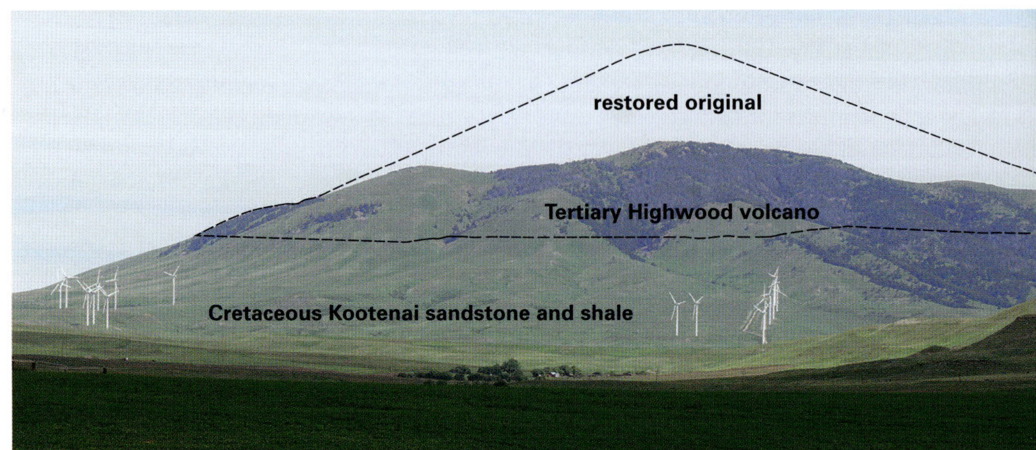

The Highwood Mountains volcano, as seen to the north about 4 miles west of Geyser. Trees (the dark areas) mostly grow on the volcanic rocks, whereas grass mostly grows on the Kootenai Formation sandstone and shale. Note the wind turbines taking advantage of the wind energy on the south base of the range.

The Highwood Mountains, north of the highway between Belt and Geyser, beautifully illustrate prominent features of the small mountain ranges of the Central Montana Alkalic Province. Like many of the others, this Eocene volcano's dark basalt-like igneous rocks are rich in potassium, an element generally almost absent in such dark rocks. Much of the central volcano is mafic phonolite, a volcanic rock much like shonkinite but with less pyroxene. These rocks rest on top of latite, an igneous rock much like rhyolite. These volcanic rocks rest, in turn, on top of Cretaceous sedimentary rocks, including the Kootenai Formation and Eagle Sandstone. The feeder conduit for the volcanic eruptions came up through the Cretaceous sedimentary rocks and spread out on top of them.

Radial dikes extend out from the center of the Highwood Mountains volcano, especially south toward Raynesford, which is about 34 miles southeast of Great Falls. If you watch carefully 1 to 1.5 miles east of Raynesford, you can see dikes just north of the highway. Radial dikes are a curious spectacle around some volcanoes. The load of a large volcano on soft, recently deposited horizontal sedimentary rocks can bow them down and thereby stretch and crack them, causing weaknesses that allow magma to move laterally, forming dikes.

High Plains Surface gravels eroded from the Highwood Mountains and other nearby ranges cover older rocks from west of Geyser to Stanford. These were deposited during Pliocene time in a desert environment, probably similar to that of present-day Nevada. Intermittent storms carried the sand and gravel downslope, where it filled valleys. A lack of through-going streams prevented the gravel from moving farther downslope, so it spread out to form broad sheets. Geyser, a stage stop between Lewistown and Great Falls, was named for some bubbling mud springs nearby that spouted mud and water into the air when a local rancher pushed a long pole into the mud. The town was moved to its present site between 1907 and 1908 when the Great Northern Railway came through.

The Little Belt Mountains, described in more detail in the US 89: Great Falls—White Sulphur Springs road guide, form a broad crustal arch south of the highway between Belt and Hobson. Once a significant Eocene volcano and now deeply eroded, the Little Belt Mountains feature a myriad of small buttes, granitic intrusions that rose into Paleozoic Madison Group limestone and older sedimentary rocks. The intrusions are laccoliths, 2 or 3 miles across and circular in shape when viewed from above. Their magma squirted out from a central intrusion and spread between the sedimentary beds, bulging them to form igneous-rock blisters. These intrusions look like bull's-eyes on geological maps, with the oldest sedimentary rocks in the center of a bull's-eye transitioning to younger formations at the edges. Most of the intrusions are granitelike quartz latite, a rock consisting mostly of feldspar with a little quartz.

SQUARE BUTTE AND ARROW CREEK

About 20 miles north of Geyser and Stanford are flat-topped Square Butte, to the east, and the smaller Round Butte, to the west. Both are big laccoliths that were fed by radial dikes that spread east from the Highwood Mountains volcano. Square Butte has a conspicuous white syenite cap over a dark shonkinite base. Shonkinite is an unusual igneous rock that's similar to basalt except it's potassium rich. (Note that the term *shonkinite* is used for fine-grained volcanic rocks and fine-grained and coarser-grained intrusive rocks.) It is about half black pyroxene and half white potassium feldspar. Syenite is mostly white potassium-rich feldspar. Many geologists studying this and similar intrusions concluded that the lower-density potassium feldspar crystals separated and rose to the top of the magma as it crystallized. Others, pointing to round blobs of syenite the size of marbles to golf balls or even small oranges, suggest that the syenite separated as immiscible liquids that floated to the top of the magma like oil on water. Almost all agree that the intrusion formed as a single magma that separated to form the white caprock.

The Square Butte laccolith, as seen northwest from Arrow Creek, has a white syenite cap over dark shonkinite.

If you drive north on MT 80 to get a closer view of Square Butte, you'll enter the deep valley of Arrow Creek about 20 miles north of Stanford. The Missouri River flowed through the valley during the Late Wisconsin glacial advance (see the US 87: Great Falls—Havre road guide for more information) and exposed a thick section of the Marias River Shale. The marine shale contains many spherical to oval concretions, hard, compact masses of sedimentary rock. They often form as a mineral cement precipitates around a nucleus, such as a shell or bit of organic matter, after the sediment is deposited. Many are cracked (septarian concretions) from shrinkage and filled with minerals or sediment. Since concretions are usually harder than the surrounding sedimentary rock, they weather out of the host rock. They are often confused with dinosaur eggs but can be distinguished by the absence of any eggshell mineral coating.

Cretaceous-age Marias River Shale in the canyon of Arrow Creek north of Stanford. Inset shows a septarian concretion, with scale in centimeters.

The closest of these buttes are Limestone Butte, 3 to 4 miles south of Raynesford, and Skull Butte, 4 miles southwest of Stanford. These perfect domes have Mississippian-age sediments on top, with younger rock layers on the lower flanks. The laccoliths themselves remain buried. Other laccoliths farther south are fully exposed.

For several miles around Windham, erosional windows through the Pliocene gravel expose dark-gray Late Cretaceous shales with slatelike cleavage. Windham Butte, east of Stanford and north of Windham, is an igneous intrusion with part of its original cover of Cretaceous sedimentary rocks still intact.

Flakes of gold in gravels in Yogo Creek, 20 miles southwest of Hobson, sparked a stampede in the spring of 1879, but there was more hype than substance, and it soon died. Five years later more talk led to the building of a big ditch to bring water

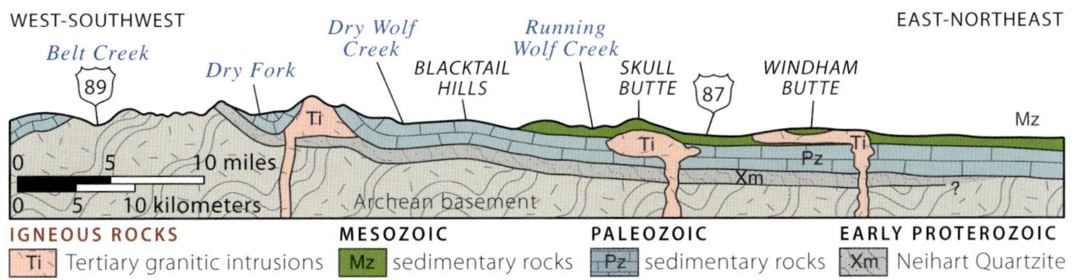

Cross section across the northern edge of the Little Belt Mountains near Stanford.

to sluice boxes, but still almost no gold was recovered. Glassy blue pebbles in the riffles were ignored until 1896, when they were recognized as sapphires. Sapphire mines opened in 1900 and produced large quantities, including from underground mines. A combination of legal and financial difficulties, partly from the development of synthetic sapphires, forced the mines to close in 1929. The mine, with its spectacular blue Yogo sapphires, was sold to a British company, and for many years most of the stones were sold in London jewelry stores. The property has gone through many hands over the years, and Yogo sapphires are still available in Montana stores. The intermittently active claims at Sapphire Village are still worked by a few locals, but the mines are closed to the public.

These rare, beautiful, deep-blue crystals occur in very solid rock that when mined must be broken into pieces and weathered for many years until the fine-grained rock softens up enough for the larger hard sapphire crystals to be released undamaged. Sapphire is an aluminum oxide mineral, a variety of the mineral corundum that is colored blue by trace amounts of iron and titanium. At the Yogo mine, the sapphires occur in an Eocene-age dike of biotite lamprophyre, an alkaline igneous rock similar to basalt. The sapphires may have originated in an older rock, such as corundum-bearing basement gneiss, and later were assimilated in the lamprophyre magma at depth. High-temperature experiments suggest that sapphires generally form under high-temperature and high-pressure conditions at great depth.

Madison Group limestone is well exposed along the northeastern edge of the Little Belt Mountains. You can find fossils in loose rocks at the base of cliffs near

Sapphire embedded in an alkaline igneous rock from Yogo Creek. —Courtesy of Dick Berg, Montana Bureau of Mines and Geology

Forest Road 266, west of Sapphire Village. A highly fossiliferous reef in the limestone crops out on the south side of Bandbox Mountain, northwest of the Sapphire mine. Because of the soluble nature of limestone, caves and sinkholes are common here. For example, Lillyguard Cave, northwest of the village, has 1,400 feet of passageways and a few stalactites. Wolf Creek, which US 87 crosses just west of Stanford, has two tributaries, Running Wolf Creek and Dry Wolf Creek. The latter sinks into a hole in Madison Group limestone, forming an underground river. Its surface channel remains dry except during spring runoff.

From a few miles east of Windham to a few miles west of Lewistown, flat desert gravels, likely of the High Plains Surface, overlie Late Cretaceous Telegraph Creek Formation sandstone and Marias River Shale, both of which are exposed in many of the larger gullies. Early Cretaceous Kootenai Formation and Thermopolis Shale appear in valleys farther east. The sedimentary layers are nearly horizontal most of the way.

The gravels exposed in the gullies in this area present an interesting problem. The flat surface looks very much like the High Plains Surface, and the underlying gravel looks like Pliocene Flaxville Gravel, deposited in a desert environment during Pliocene time. The very dark layer doesn't quite fit, however, because it's an organic-rich buried soil, something not typical of desert environments. Could the deposits be glacial or recent alluvial gravels? How do we test these various explanations?

It's unlikely that the deposits are glacial, because any local mountain glaciers were small and not close to this location. The continental ice sheet was located far to the north, and any gravel-transporting meltwater would have had to flow a long distance to get here, as well as up the regional slope of the land, something that is highly improbable. Could this gravel be a recent stream deposit? Yes, but there are no major rivers in the area, and significant flows are required to move pebbles and cobbles of this size over a broad area, so that also seems unlikely. The most probable explanation is that this deposit is Flaxville Gravel, and the flat surface is the High Plains Surface.

Possible Flaxville Gravel exposed beneath the High Plains Surface under wheat field stubble east of Moore. A nearby gravel quarry is mining the gravel for use in construction.

The problematic soil layer between the two gravel layers might have been a low, wet, and vegetated area on the alluvial plain where fine-grained sediment accumulated and plants provided organics for soil. In order to test this idea, we would need to find fossils or possibly a datable deposit, such as a layer of volcanic ash. Carbon dating could be helpful, but only if the buried soil is less than 40,000 years old.

The Big Snowy Mountains, 20 miles south of Lewistown, are an eastward extension of the same broad crustal arch the Little Belt Mountains are part of. Paleozoic rocks are exposed at the center of the range, with Jurassic and Cretaceous rocks around the edges, but no igneous rocks of the range's core are exposed yet. As is the case elsewhere in Montana, Madison Group limestone is prominently exposed and loaded with fossils of corals, crinoids, bryozoans, and brachiopods that lived in shallow, tropical marine water that covered Montana more than 300 million years ago. The flat limestone summit of the Big Snowy Mountains, accessed via a strenuous hike from Crystal Lake, has numerous sinkholes and caves, including a cave containing ice year-round. (See the US 191: Lewistown—Malta road guide for information about the Judith Mountains and North and South Moccasin Mountains on the north side of Lewistown.)

US 87
Great Falls—Havre
113 miles

Lewis and Clark, working their way up the Missouri River toward present-day Great Falls on June 13, 1805, encountered the first of several closely spaced waterfalls they had heard about. Lewis was mesmerized by the falls, but they were all concerned about how they would portage the 18 miles around the falls with their heavy canoes and supplies. Ultimately it took them a month, battling rough ground, prickly pear cactus spines, marauding grizzly bears, and vicious wind, rain, and hail. The Lewis and Clark Interpretive Center in Great Falls, on the south side of the river, has outstanding interactive displays and dioramas of their adventures.

The Great Falls in the summer of 1880, long before Ryan Dam was built here in 1915. —Courtesy of Haynes Foundation Collection, Montana Historical Society

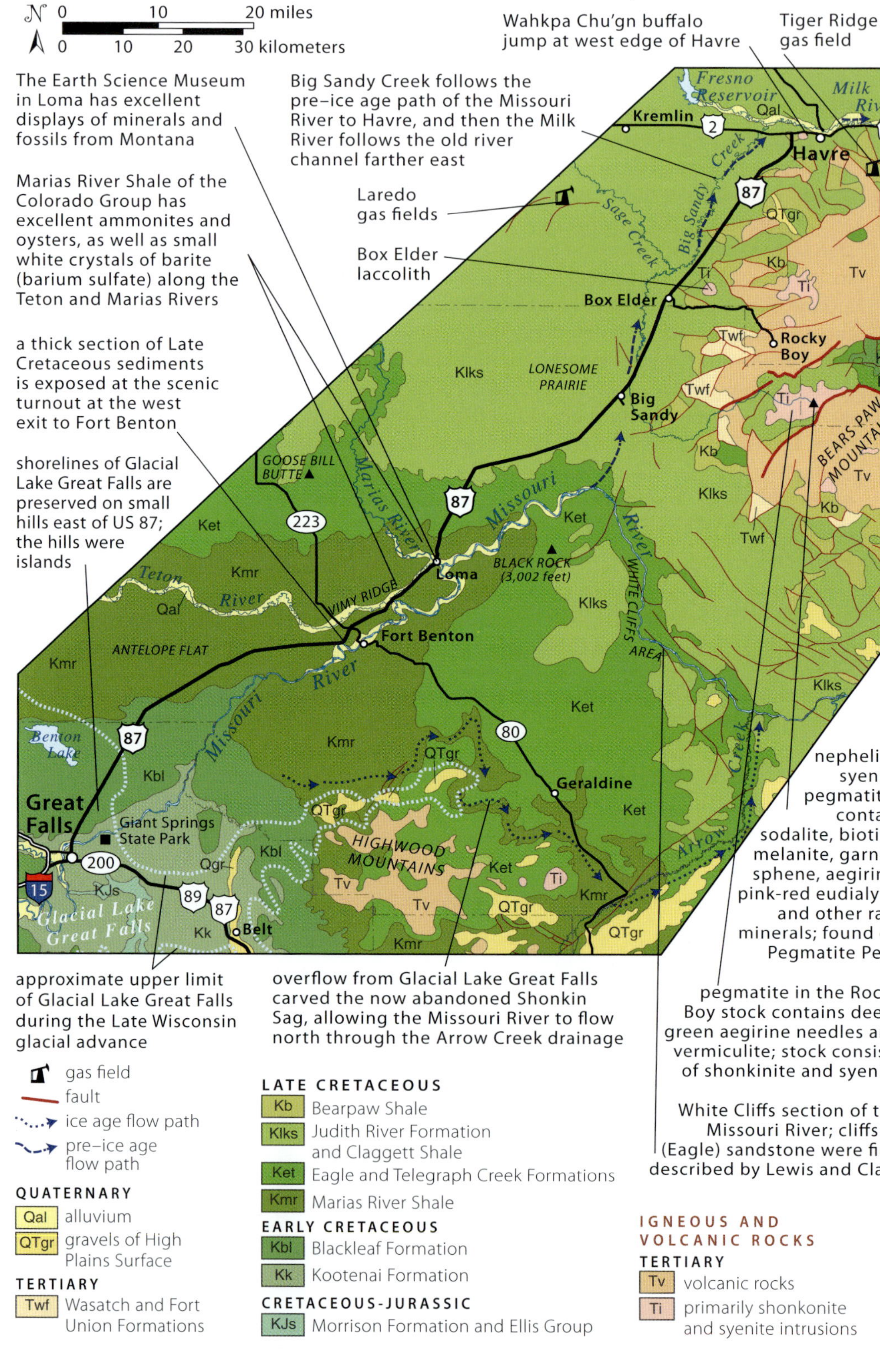

Geology along US 87 between Great Falls and Havre.

At 87 feet, the Great Falls at Ryan Dam, also called Big Falls, is the highest of the five waterfalls that make up the Great Falls of the Missouri. To reach Ryan Dam, head north for 3.5 miles on US 87 from the highway bridge over the Missouri River, then 5 miles east on Morony Dam Road to Ryan Dam Road, then down to the river. There are excellent interpretive signs about the geology of the falls on Ryan Island. Black Eagle Falls, 26 feet high, is about 1 mile downstream from the south side of the US 87 bridge. Rainbow Falls is best seen from the south side of the river about 4 miles downstream from the US 87 bridge (1 mile downstream from Giant Springs State Park). It's 47 feet high but dries to a trickle in summer because most of the water is drawn off for hydroelectric power.

The falls have formed on resistant ledges of sandstone in the Cretaceous Kootenai Formation. Typically, the thicker the layer of sandstone, the higher the waterfall. Big Falls cascades over a particularly thick and hard layer of marine sandstone deposited around 120 million years ago in the Western Interior Seaway. The other falls, all upstream of Big Falls, cascade over thinner layers of sandstone interbedded with mudstone deposited by streams on the coastal plain. The falls migrate upstream over time as falling water erodes the soft rocks underneath the sandstone holding up the lip of the falls, undercutting the sandstone, which collapses along fractures. Each dam, however, forms a barrier to that migration, at least during the relatively short duration that humans will be around to maintain the dams. Four of the five different falls are now dammed for hydroelectric power, but enough water still pours across them to provide an impressive sight.

View of Ryan Dam and Big Falls (also known as Great Falls) during low flow on the Missouri River. The Early Cretaceous Kootenai Formation here consists of a lower marine sandstone that forms the falls, and overlying stream-deposited sandstone and mudstone that forms smaller waterfalls upstream. The curve-shaped layers in the marine sandstone are tidal channels.

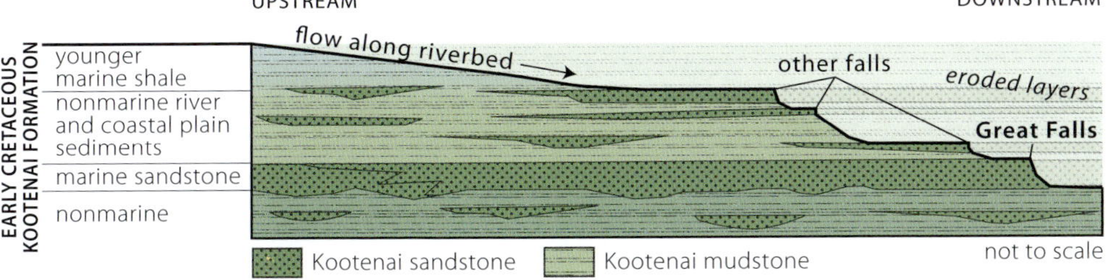

Simplified cross section through the Missouri River gorge at the Great Falls, showing that the falls have developed over resistant layers of marine and nonmarine sandstone interbedded with softer mudstone. —Modified from Thomas and Roberts, 2000

Giant Springs is the centerpiece of a little state park on the south bank of the Missouri River near the eastern edge of the town of Great Falls. Enormous volumes of water, about 650,000 gallons per hour (or 10,830 gallons per minute), well up through fractures in Kootenai Formation sandstone and pour into the river. The spring serves as the headwaters of the Roe River. At 200 feet long, it was once considered the shortest river in the world! Giant Springs is 3 miles east of the US 87 bridge over the Missouri River via River Drive North and Giant Springs Road.

It seems likely that the water discharging from Giant Springs is rising from Paleozoic Madison Group limestone, which lies several hundred feet below the surface. Limestones are the usual source of such large springs because they commonly contain caverns capable of conducting enormous flows. If Madison Group limestone is indeed the source, then the water probably comes from the Little Belt Mountains about 35 miles southeast of Great Falls, where the Madison surfaces.

Look for plant fossils nicely preserved as black impressions on the sandstone bedding surfaces of the Kootenai Formation around Giant Springs. The abundance of plant fossils and the complete absence of marine animal fossils show that the sandstone was deposited on land.

Except in the Great Falls area, the entire region along the line of US 87 to Havre lay beneath glacial ice when the Late Wisconsin continental ice sheet was at its maximum, around 23,000 years ago. Great Falls was under water at this time because the ice sheet

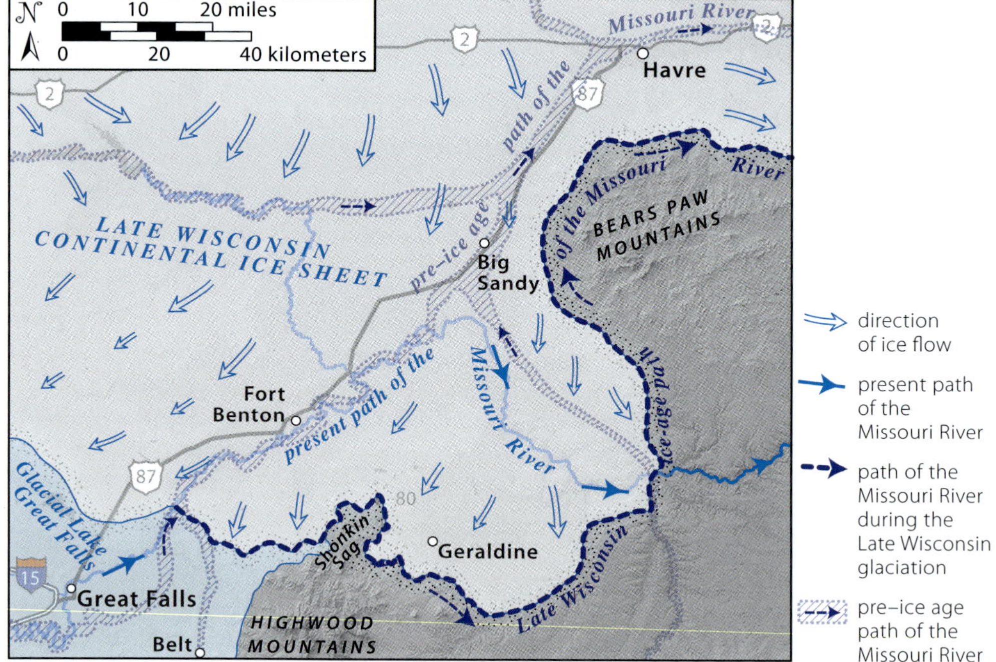

The continental ice sheet blocked the path of the Missouri River many times during Pleistocene time, diverting it east along the ice margin. During the last (Late Wisconsin) ice advance, Glacial Lake Great Falls spilled over the ice at its eastern end to scour a new channel, the Shonkin Sag. —Modified from Fullerton and others, 2012

flowed against the north edge of the Highwood Mountains and blocked the northeast-flowing Missouri River, impounding Glacial Lake Great Falls. The lake filled to an elevation as high as about 3,860 feet before overflowing along the southern end of the ice sheet, skirting the north edge of the Highwood Mountains through what are today the tiny communities of Shonkin and Highwood. The big channel, now dry, is called the Shonkin Sag.

Did the ice dam burst as a sudden event, or did the flow of glacial meltwater gradually increase? How many times did it happen? Did excavation of the sag start during an earlier glacial advance (Illinoian), only to be widened and deepened during later glacial advances? The depth of research on the Shonkin Sag is remarkably thin, so we have plenty to learn. After each flood, the ice would readvance and form a new dam and lake, resulting in another outburst flood. What is certain is that it all came to an end with the retreat of the Late Wisconsin ice around 11,700 years ago, leaving the Shonkin Sag as the record of the event. (See the US 87: Great Falls—Lewistown road guide for information about the Highwood Mountains.)

While the overflow from Glacial Lake Great Falls was rushing through the Shonkin Sag, the old course of the Missouri River from the Fort Benton area to near Havre lay under ice. When the ice sheet melted back to the north, the river reclaimed its original valley from Great Falls to halfway between Fort Benton and Big Sandy. From there, deprived of its old channel by deep glacial deposits, it established a new course to near Fort Peck.

A string of shallow lakes now exists along the floor of Shonkin Sag, including Lost Lake, which sported a half-mile-wide waterfall that plunged 300 feet during releases from Glacial Lake Great Falls. Today, the Dry Falls of Lost Lake is all that remains of an ice-age waterfall twice the size of the largest of today's Niagara Falls! You can easily reach the sag by heading south on MT 80 from Fort Benton and taking a secondary road that heads south near Harwood Lake. Drive about 3 miles south to White Lake and the old community of Montague. There, the now-abandoned tracks of the Central Montana Railroad follow the Shonkin Sag. Secondary roads west of

The ice-age Missouri River carved giant meanders on the north flank of the Highwood Mountains. Arrows indicate flow direction. —Aerial view courtesy of Hal James, Montana Bureau of Mines and Geology

Cliff of Dry Falls and Lost Lake east of Shonkin in Shonkin Sag. View looking upstream at what was a waterfall when Glacial Lake Great Falls drained. —Courtesy of Rod Benson, Bigskywalker.com

Geraldine follow the sag through the tiny communities of Shonkin and Highwood. It is one of the most spectacular glacial meltwater channels found anywhere, on par with those of the Channeled Scablands of eastern Washington that were scoured out by the Glacial Lake Missoula floods.

For the 8 or so miles northeast of Great Falls, US 87 crosses the floor of Glacial Lake Great Falls. Shorelines, formed by waves lapping against and eroding soft sedimentary rocks on what were then small islands in the lake, are well preserved near US 87 about 5 miles northeast of Great Falls. Farther northeast, the highway crosses uplands with a thin coating of glacial till from the most recent (Late Wisconsin) glaciation. The area under the ice sheet was not deeply scoured because the ice was not thick near this southern limit; erratic boulders that sparsely litter a few of the fields beside the road are the most obvious evidence of glaciation. Many are granite and gneiss, hard rocks full of glittering crystals that do not even slightly resemble the local bedrock, having been brought here from Canada. These glacial deposits thicken northeastward to 200 feet or more.

Just west of the southwest exit to Fort Benton, a turnout is perched on a cliff overlooking a tight meander bend of the Missouri River. Exposed in the cut bank below

Glacial Lake Great Falls shorelines viewed east from US 87, 4 or 5 miles northeast of Great Falls on small hills of Early Cretaceous sandstone and shale, former islands in the lake. Lower shorelines formed as the glacial ice dam melted back to the north.

the parking area is a high cliff of Marias River Shale. The shale was deposited as marine muds in the Western Interior Seaway when it flooded this part of Montana 90 to 85 million years ago, in Late Cretaceous time. Fossils of the ammonite *Clioscaphites*, an extinct squid-like animal, have been found in the shale. These swimming animals lived in open water, possibly feeding on plankton and using gas stored in the chambers of their shells for buoyancy. About 7 miles east of Fort Benton, another big meander of the Missouri River exposes high cliffs in the same sedimentary rocks.

A big sweeping meander of Missouri River at the scenic turnout near the southwest exit to Fort Benton exposes black to gray Marias River Shale with a lighter-colored sandstone layer on top. The flat High Plains Surface is visible in the distance.

Fort Benton was established in 1850 as a trading post for the Blackfeet tribes, on the north side of the Missouri River just below the rapids that thwarted further river navigation. Shallow-draft paddle wheelers managed to get upstream this far. Fort Benton was a major shipping point for buffalo robes and furs. After Lieutenant John Mullan completed the military road from Fort Benton to Walla Walla, Washington, in 1860, Fort Benton became a bustling, noisy commercial center. Completion of the Northern Pacific (1883) and the Canadian Pacific (1885) Railways quickly deflated the Missouri River steamboat stern-wheelers, which couldn't compete on speed or cost.

About 5 miles north of Fort Benton, the old valley of the Missouri River, now containing the much smaller Teton River, lies on the west side of US 87 for 6 miles. The Teton River joins the Marias River at Loma, where the combined flow enters the Missouri River. Long stretches of the highway from south of Big Sandy to Havre follow Big Sandy Creek, which is also too small to have eroded its wide valley. The Missouri River flowed through this valley, too, before the continental ice sheet pushed it south into its present course. The outlines of the old valley are hard to see from the ground because the valley contains deep deposits of glacial sediment.

Square Butte (also called Box Elder Butte), east of US 87, 4 miles east of Box Elder, is one of at least three buttes with this name within 100 miles of Great Falls. Like the others, this one also is a laccolith that formed as a blister of shonkinite magma pooled between layers of sedimentary rock around 50 million years ago. The upper part of that laccolith is sparkling white syenite composed almost entirely of potassium feldspar. As with several other such laccoliths in the Central Montana Alkalic Province, syenite magma may have separated from the molten shonkinite, as oil separates from water, and floated to the top of the intrusion to form the white cap.

East of Big Sandy and south of Havre, the Bears Paw Mountains (Bearpaws to the locals) rise to 6,900 feet, 4,200 feet above the plains. The range grew as a significant volcanic cluster about 50 million years ago, with a diameter between 28 and 30 miles. These Central Montana Alkalic Province rocks are high in potassium and sodium, mostly dark shonkinite and pale latite. The Rocky Boy stock, forming the highest, western part of the range, is shonkinite; in a few places it's associated with a little carbonatite, a very unusual calcite-rich rock erupted as magma. This peculiar rock is rich in rare minerals that are high in uranium, thorium, columbium, tantalum, and rare earth elements. The stock was explored in the 1970s but determined to be uneconomic. With increasing demand for rare earth minerals, which are used in everything from electronics, extremely strong magnets, lasers, LED lightbulbs, and camera lenses to hybrid cars, that economic assessment may someday change.

Like several other igneous centers in central Montana, the Bears Paw Mountains stand on the crest of a broad arch in the Earth's crust. A big laccolith magma chamber, 5 to 6 miles across, was injected below Early Cretaceous rocks, bulging them into a broad east-west arch, and erupted volcanic lavas and ash above. Many smaller intrusions, at least some of them laccoliths, intruded the flanks and fringes of the arch.

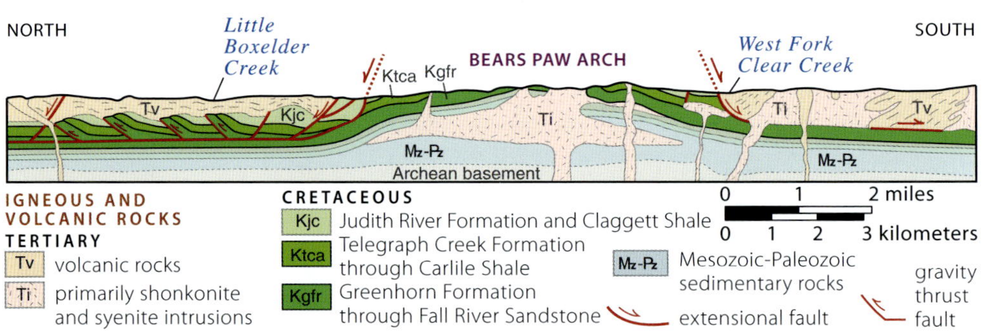

North-south cross section across the Bears Paw arch. Small thrusts formed on slippery surfaces smeared with smectite clay. —Modified from Hearn, 1976

A broad view of the mountains reveals a wide east-west gap through the volcano, which creates a depression across the crest of the arch. The Cretaceous sedimentary rocks under the volcano are exposed in the gap. It appears that the volcano split along that line, and the northern and southern halves were pulled apart by about 4 to 7 miles, slowly sliding down the flanks of the gentle arch. Late Cretaceous rocks, mostly the Niobrara Formation of the Colorado Group, slope gently south of the volcano toward the Missouri River.

Sliding on such a gentle slope must have involved special conditions. If you have driven on dirt roads in central Montana in wet weather, you have probably experienced easy sliding on almost no slope. That sliding is facilitated by the expanding clay mineral smectite (bentonite), which exists in some shale layers of the Late Cretaceous Telegraph Creek and Greenhorn Formations. When that mineral gets wet, it soaks up water and expands, making for extremely weak and slippery layers.

But there was more sliding than that which pulled the volcano apart. Sliding also occurred on the most slippery surfaces between layers. Individual layers slid on other layers, and as these sliding slabs ran into others, they rode up to the surface. The final result is a series of small thrust faults starting at a particular slippery layer and curving up to the surface.

US 89
Livingston—White Sulphur Springs
71 miles

Established in 1882, Livingston was a major junction and repair center for the Northern Pacific, then the Burlington Northern, and now Montana Rail Link. Seven miles east of the junction of US 89 and I-90 in Livingston, US 89 heads north along the Shields River across the western part of the Crazy Mountains Basin, in which sediment accumulated during latest Cretaceous and early Tertiary time. The southern margin of the basin is sharply defined at I-90 and the Yellowstone River, where its rocks ramp up against the Absaroka Range to the south. The bowl-shaped basin is dominated by the Paleocene Fort Union Formation, rim to rim, but is underlain by thinner sections of volcanic-rich sandstone and mudstone of the uppermost Cretaceous Livingston Group, which is the same age as the Hell Creek Formation in Eastern Montana.

The Crazy Mountains east of US 89 mark the center of the basin. The Bridger Range well west of the highway defines the western margin of the basin. (The Crazy Mountains are discussed in the US 191: Big Timber—Lewistown road guide.)

The Crazy Mountains Basin is a deep trough that buckled down in the crust. A number of smaller folds within it are obvious from the air but hard to see from

The east side of the Bridger Range as seen from near Wilsall. Glacial sculpting of Mississippian Madison Group limestone, rock that resists weathering in Montana's arid climate, makes the rugged crest of this range.

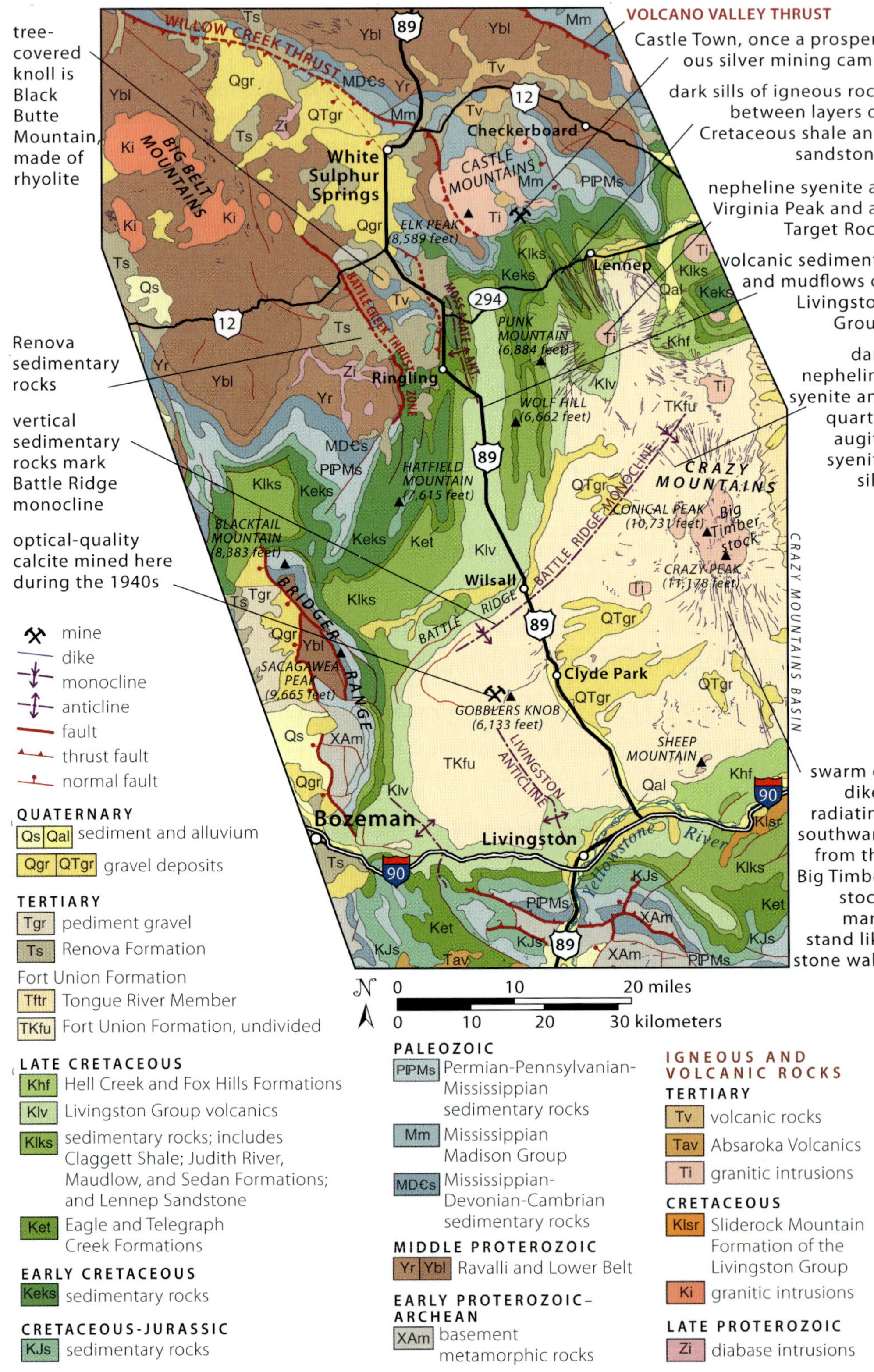

Geology along US 89 between Livingston and White Sulphur Springs.

the road. The originally near-horizontal beds, including the Fort Union Formation and all of the older rocks in this western edge of the basin, were deformed by Sevier folding and thrusting in late Paleocene to early Eocene time. Laramide uplift of the Bridger Range to the west overlapped and deformed the Sevier structures, continuing the deformation in Eocene time.

The basin and the folds within it have greatly disappointed those who hoped to find oil. An uncommonly thick sedimentary basin, deposited over a long period of time, is usually a good place to find oil. The Crazy Mountains Basin is about 3 miles thick, with its oldest rocks (Cretaceous to Paleocene) resting on Archean metamorphic rocks that never contain oil and gas. The basin is thick enough to cook its organic material, the remains of billions of microscopic plants and animals, to mature it over millions of years to form oil and gas. The younger igneous intrusions in the center of the Crazy Mountains, which strongly cooked the central core of the sedimentary pile, likely overdid it there, so the best prospects would be away from those intrusions.

On US 89 about 4 miles north of I-90, a large sandstone knob on the west side of the highway is one of the few decent outcrops of the Fort Union Formation along the Shields River. Shale of the Fort Union, too soft to maintain outcrops, underlies the road for the 18 miles north to Wilsall. There are a few outcrops of Fort Union shale a few miles south of Wilsall. The near-vertical beds at Wilsall are at the deformed western edge of the Crazy Mountains Basin.

Brownish-gray to dark-red shale of the Livingston Group extends north for about 17 miles from Wilsall. The Livingston Group consists of volcanic andesite-rich mudflows and fragments erupted from the Elkhorn Mountains Volcanics near the Boulder batholith, to the west, and locally from the smaller Sliderock Mountain volcano, about 50 miles to the southeast (south of Big Timber). The volcanoes were active between about 78 and 75 million years ago, during Late Cretaceous time. About 4 miles southeast of Ringling, US 89 takes a sharp bend to the northwest and passes through a small anticline and syncline and into progressively older Cretaceous

A large knob of Fort Union Formation sandstone as seen from US 89, 4 miles north of I-90. Some layers in the sandstone have leaf impressions.

formations, including volcanic-rich sandstone and mudstone; these belong to the Sedan Formation of the Livingston Group.

Ringling is named for John Ringling of the famous Ringling Brothers circus, who purchased thousands of acres in the area in 1903 to raise cattle and sheep. North from Ringling for 3 or 4 miles, US 89 follows the axis of a tight syncline, then it passes into older rocks buried under Tertiary-age sedimentary rocks. North from MT 294 all the way to White Sulphur Springs, the post-deformation, nearly flat-lying Tertiary rocks include mudflow deposits eroded from nearby ranges and airfall volcanic ash, along with broken fragmental rocks that occur in isolated channels. Most were deposited in an arid, desertlike environment; mapped with various different formation names, including the Fort Logan Formation, they fall broadly under what we call the Renova Formation in southwest Montana. The source of the volcanic ash is not clear, but large volumes may have come from volcanoes in central Oregon. Bones of animals that lived in Oligocene to early Miocene time, including mammals and small reptiles, are preserved in the sedimentary rocks.

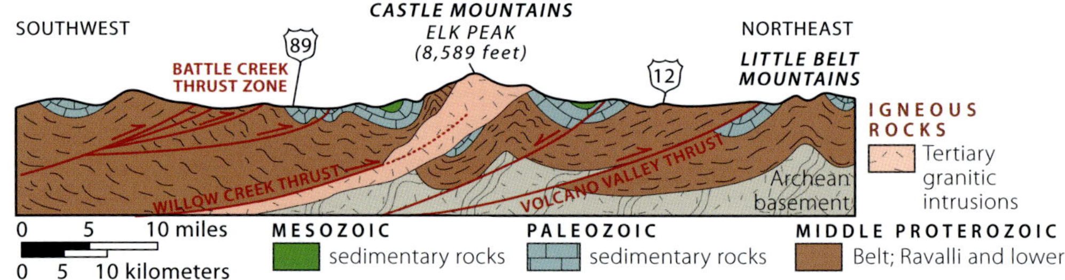

Southwest-northeast cross section across US 89 near the junction with US 12.

The Castle Mountains southeast of White Sulphur Springs have a large granite intrusion in the highest, central part of the range. On the geologic map it looks like a granite bull's-eye surrounded by crudely concentric rings of sedimentary rocks that become younger outward. The granite magma rose in Eocene time along the Willow Creek thrust fault to bulge the already folded sedimentary formations into a dome, now largely planed off by erosion.

The magma crystallized at extremely shallow depth, so we should probably view the Castle Mountains as a small, deeply eroded volcano, its erupted deposits having eroded away long ago. Like many granites that crystallized near the surface, that in the Castle Mountains contains blocky crystals of feldspar set in a matrix composed of tiny grains. These small grains crystallized rapidly in their cool, near-surface environment. The Castle Mountains granite has weathered into a softly rounded landscape punctuated by tall pinnacles of rock that rise through the deep mantle of soil. In dim light, and with the help of an active imagination, it's possible to see those rocky towers as ruined battlements of a castle left from some fabled age. The granite bulged the whole sequence of surrounding sedimentary rocks and redeformed already crumpled formations into tight, northwest-trending, steep-sided folds. Steep walls of weather-resistant Mississippian Madison Group limestone in the Castles are popular with rock climbers.

Legend has it that a prospector found a pebble of ore in the Musselshell River to the east and then located its source 50 miles upstream in the Castle Mountains by following similar pebbles. Mines began to produce silver, lead, and zinc in 1885 and reached their peak in 1892, when three smelters were in operation in the area. The ore formed in the Madison Group limestone, which reacted with hot fluids near its contact with the Castle Mountain granite. The main mine, the Cumberland, opened in 1886, but due to its remoteness, transporting ore and supplies by wagon 75 miles to the rail line in Lewistown was a major expense. By the time of Castle Town's incorporation in 1891, its population had reached 1,500. Famously, Calamity Jane came to open a restaurant, but before it opened, she was found in possession of a team of horses that she had not paid for.

A drop in the price of silver led to the "silver panic" of 1893, and the town went into a rapid decline. Most mines closed, but the Cumberland limped along off and on until 1950. Magnetite, pyrite, chalcopyrite, sphalerite, galena, and siderite are present in mine dumps. Castle Town can be reached from MT 294 at Lennep by heading 6 or 7 miles west to northwest on a public dirt road. Many of the town's buildings are still standing and in good shape, but they are on private land and must not be entered without permission.

The bank and Spa Hot Springs Motel and pools in White Sulphur Springs are heated by the natural springs from which the town got its name. The springwater contains dissolved bicarbonate, sulfate, and sodium and potassium chloride, and smell, as you might expect, faintly of sulfur. Water at about 130 degrees Fahrenheit is pumped from

Madison Group limestone exposed in Green Canyon in the Castle Mountains.
—Courtesy of Craig Zaspel

Fins of granite in Stone Temple Park in the Castle Mountains. —Courtesy of Bryan Gartland

a shallow well to feed the hotel's indoor pool, which is about 105 degrees Fahrenheit, and its outdoor pools, which are 98 to 103 degrees. It seems likely that the hot springs water has circulated deep below the surface along fractures, perhaps into the Willow Creek thrust fault or a more recently active extensional normal fault. Rocks at a depth of several thousand feet are hot enough to heat water to the temperature of the hot springs. The hydrogen sulfide probably comes from the dissolution of pyrite (iron sulfide), a common mineral in many kinds of rocks.

US 89
Great Falls—White Sulphur Springs
98 miles

The section of US 89 between Great Falls and Belt is described in the US 87: Great Falls—Lewistown road guide. Just east of Belt, US 89 heads south along Belt Creek and climbs over the Little Belt Mountains, the largest of the crustal arches in central Montana that formed during the initial rise of the Rockies. This big, elongate dome of Paleozoic sedimentary rocks is cored with old Proterozoic Belt rocks. The Little Belt Mountains, peppered with 50-million-year-old igneous intrusions, is one of the main intrusive centers of the Central Montana Alkalic Province.

At milepost 70, just south of the junction of US 87 and US 89, the gray cliffs on the horizon to the east are Cretaceous sandstone of the Kootenai Formation. A few miles farther south, as US 89 begins to climb the east edge of the Belt Creek canyon, a good gravel road to Riceville descends to Sluice Boxes State Park. A picnic area lies at the north end of a spectacular canyon cut in Mississippian-age Madison Group limestone. An old Great Northern Railway line, built in 1890 and abandoned in 1945, runs through the canyon. The canyon is accessible by boat or a 7-mile trail, which is usable at low water.

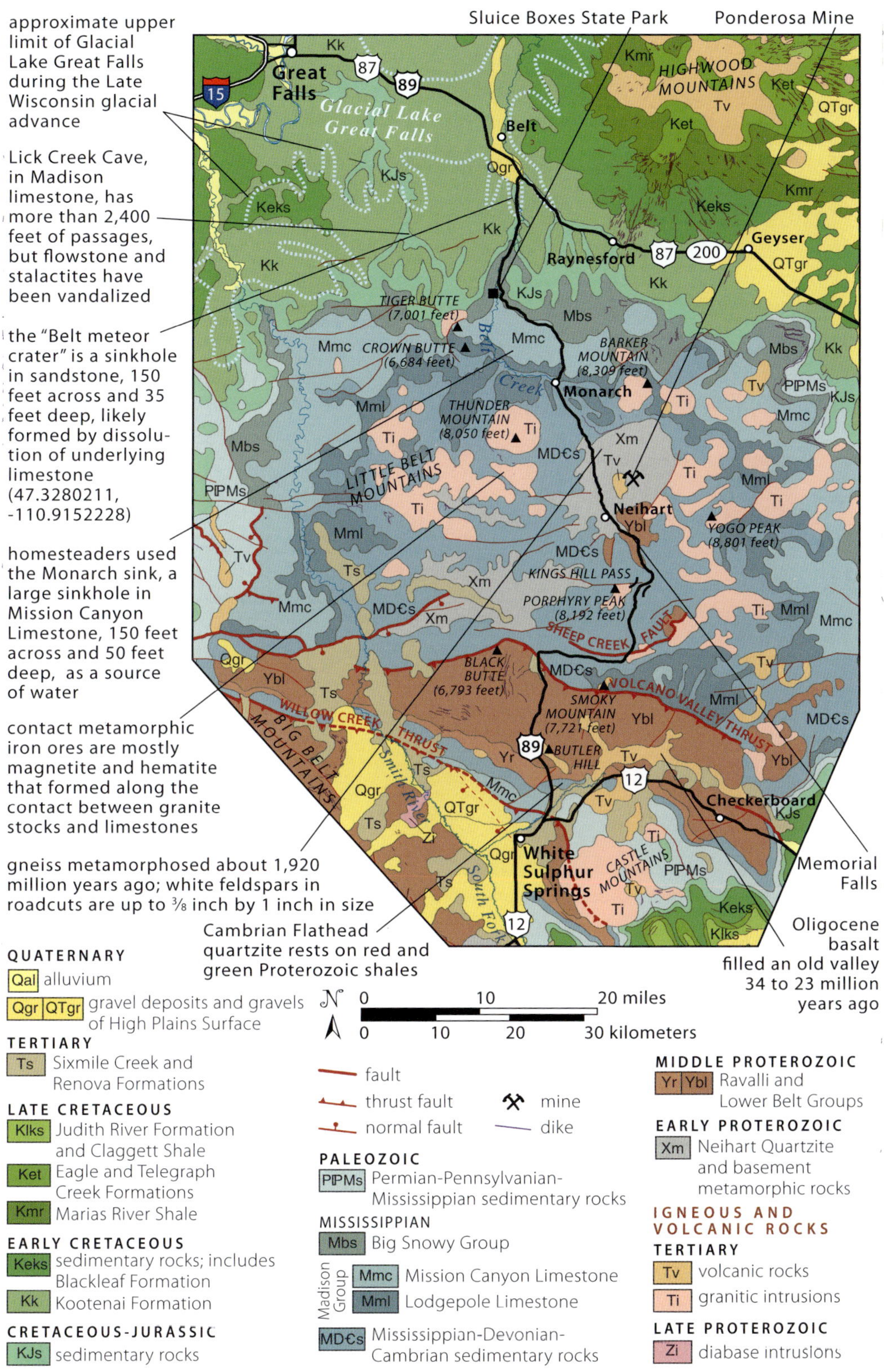

Geology along US 89 between Great Falls and White Sulphur Springs.

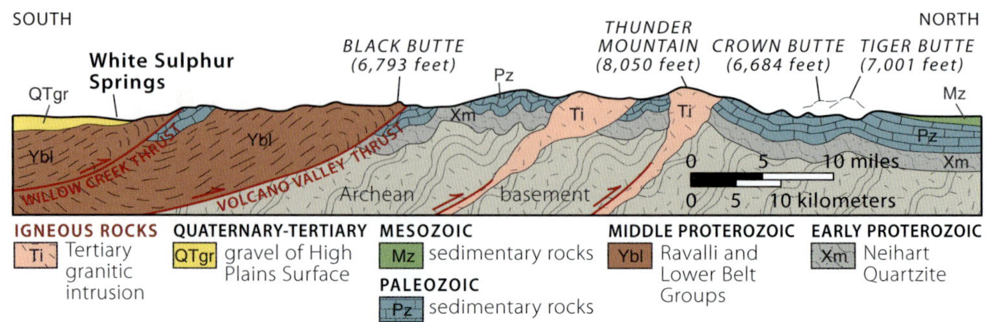

South-north cross section just west of US 89. Basement rocks were brought to the surface in the core of the Little Belt Mountains in a faulted crustal arch.

Cliffs of Madison Group limestone loom over the south end of the Belt Creek canyon near Logging Creek Bridge, about 12 miles from the main entrance to Sluice Boxes State Park. The cliffs to the east (left) are composed of thinly bedded Lodgepole Limestone, and the overlying, more massive Mission Canyon Limestone forms the cliffs in the background. Note the large cave in the highest cliff band.

A scenic overlook 10 miles south of the US 87 junction provides a nice view down into the mouth of the Belt Creek canyon. A wonderful roadcut across from the overlook exposes reddish sedimentary rock of the Big Snowy Group offset by a steep, clearly defined normal fault. Deep tropical weathering concentrated small amounts of red clay to give the rock its color. Red staining in the underlying Madison Group may have come from iron oxides that percolated down from this younger formation.

Southeast from the overlook, US 89 stays in the same Big Snowy Group for the next 14 miles. A couple of miles south of the overlook, the highway follows close to the upper contact of the Mission Canyon Limestone. In places the road passes up into pale-green shale of the Otter Formation of the Big Snowy Group, with its subtle horizontal layers. Farther south, the highway continues across a broad ridge on massive light-gray Mission Canyon Limestone, a formation within the Madison

A normal fault cuts the reddish sedimentary rock at the scenic overlook on US 89. The layers left (north) of the fault were dropped with respect to the layers on the right (south) side, which is toward the raised core of the Little Belt Mountains.

Group. It's typically much more massive and lighter gray than the younger Mississippian formations. The rock contains the fossils of horn-shaped corals, colonial corals, brachiopods, byozoans, crinoids, and other tropical marine animals that filtered clear, warm ocean water to make a living.

The highway follows a narrow canyon down to Monarch, on Belt Creek, well upstream from where it left the creek near US 87. Monarch originated in 1889 as a supply town for the silver mines in Neihart and nearby areas. The Lodgepole Limestone of the Madison Group forms the massive cliffs at the north edge of town.

The Belt Creek valley south of Monarch is one of the few easily accessible places in Montana where we can see most of the sedimentary rocks deposited in central Montana during early Paleozoic time. The sedimentary rocks here are almost horizontal, dipping less than 10 degrees north, so south of Monarch along the narrow valley of Belt Creek US 89 goes down the Paleozoic section into progressively older formations. In sequence, from Devonian to Cambrian age, they are dolomitic limestone of the Jefferson Formation, siltstone and shale of the Maywood Formation, Pilgrim Limestone, Park Shale, Meagher Limestone, Wolsey Shale, and finally Flathead Sandstone. More than 4,000 feet of Paleozoic rocks are preserved in the Little Belt Mountains. Pennsylvanian and Permian rocks were also deposited in this area, but they eroded away due to uplift and exposure of some land above sea level prior to the deposition of the Ellis Group in Jurassic time.

About 3 miles south of Monarch, a 1-mile stretch of US 89 cuts through a laccolith of Tertiary syenite. Injected between these Paleozoic formations and the underlying Proterozoic Belt rocks, the magma bulged the Paleozoic rocks. This is one of a swarm of laccoliths of granite and related rocks, including their finer-grained equivalents. In some cases, the magma was injected between layers in the Paleozoic sedimentary rocks.

South of the laccolith, the basement rocks in the core of the Little Belt Mountains are older than the Proterozoic Belt rocks. Earliest Proterozoic in age, the dark, metamorphosed gabbro and gneiss are both full of black grains of biotite and hornblende. Foliation in the metamorphic rocks dips moderately to steeply south. Ancient rocks such as these in the Little Belt Mountains provide a rare surface exposure of the Great Falls tectonic zone, a northeast-trending region of tectonic activity caused, in part, by the collision of the Archean Medicine Hat block to the north and the Wyoming craton to the south during the assembly of North America (Laurentia). This event is called the Big Sky orogeny and dates to around 1.8 billion years ago. Earliest Proterozoic rocks from the Little Belt Mountains are interpreted to have formed in a subduction zone that consumed oceanic crust just prior to the collision of continents.

In 1881, prospectors discovered silver and lead in basement rocks just north of Neihart, near the center of the Little Belt Mountains. Mining began at the Queen of the Hills Mine the next year and expanded rapidly, as did the towns of Neihart and Monarch. It was tough going for almost a decade because the ore had to be carried by horseback or oxcart to a small smelter about 12 miles east of Monarch. When it closed a couple of years later, the ore had to go all the way to Fort Benton, then by steamboat down the Missouri River, and eventually across the Atlantic to a smelter in Wales. Only the highest-grade ore was sufficiently valuable to warrant such travel.

Things improved when a spur line of the Montana Central Railway connected Neihart with a new smelter in Great Falls, which began operating in 1892. The financial panic of 1893 dropped the bottom out of the silver market, effectively stopping mining in the Little Belt Mountains until 1917, when the price of silver rose. In 1927, with the next big drop in the price of silver, mining again ceased. Interest resurged for zinc mining for a while in 1935 but didn't last.

The high ridge east of Neihart is the middle Proterozoic Neihart Quartzite, the oldest of the great stack of rocks that make up the Belt Supergroup. The quartzite was derived from a sheet of windblown quartz sand that covered vast areas of the North American continent prior to the existence of the Belt Basin—at least 1.5 billion years ago. As the Belt Basin gradually opened and subsided deep enough for sediment to accumulate, braided streams transported the sand into the basin.

Queen of the Hills Mine at the north edge of Neihart around milepost 38.

The high cliffs above the highway department pullout at the south side of Neihart are composed of vertically jointed Neihart Quartzite, the oldest deposit of the Belt Supergroup.

US 89 climbs south to reach those rocks and more Belt Supergroup mudstones and sandstones 2 or 3 miles south of Neihart. The Neihart Quartzite is well exposed at the Memorial Falls parking area, where the rock is pale, rusty metamorphosed quartzite, all sparkling with the sun on its quartz crystal faces. A nice trail leads from the parking area a few hundred yards up to the falls, which spill over the Neihart Quartzite. These earliest Belt rocks were deposited on top of ancient, much-eroded and deformed metamorphosed gabbro and gneiss.

Six miles south of Neihart, a thin rhyolite laccolith was injected between the Proterozoic Belt rocks and the overlying Flathead Sandstone of Cambrian age. Its scattered dark, rounded inclusions may be remnants of the basalt magma that invaded the crust deep below, melting it to form the rhyolite (granite) magma. A

Tilted layers of the Neihart Quartzite exposed in a roadcut about halfway between mileposts 32 and 33 are part of a limb of a big fold that's overturned downslope to the west. Notice the small-scale anticline and syncline within the layers.

couple of miles farther south, Neihart Quartzite beds hang over the highway in big folds. Geological engineers who specialize in landslides refer to such beds that overhang roads as "daylighted" because their broken edges "see daylight."

A dense swarm of Eocene-age granite sills invaded the Cambrian sedimentary rocks in this area. About 1 mile north of Kings Hill Pass a roadcut exposes one of these sills; several more roadcuts several miles south of the pass expose more sills of this swarm. The road passes through these and other Cambrian sandstones and shales across the pass and down (south) another 13 miles to the east-west Volcano Valley fault, at a big 90 degree turn to the south (18 miles north of White Sulphur Springs).

SMITH RIVER CANYON

The Smith River is a tributary of the Missouri River that flows northwest across the western edge of the Little Belt Mountains, cutting a spectacular canyon through the range. The river is noted for its scenery and blue-ribbon trout fishery and is so popular that to float it one must have a permit from Montana Fish, Wildlife and Parks. The lottery for permits is very competitive. The canyon exposes miles of Mississippian Mission Canyon Limestone with many caverns and Native American pictographs. The limestone formed in a shallow ocean that covered most of North America between 359 and 326 million years ago.

The pattern of the river is very sinuous where it cuts the canyon across the limestone. This shape shows that it once meandered across a low-gradient floodplain of soft sediment. It is now entrenched and stuck as a deep slot in the hard rock because the river continued to erode as the land rose. The land is likely rising due to mantle heat from basin and range extension and crustal thinning that started 17 million years ago and continues to this day.

Cliff walls of Mission Canyon Limestone of the Madison Group exposed along the Smith River northwest of White Sulphur Springs. **—Courtesy of Rod Benson, Bigskywalker.com**

About 5 miles south of Kings Hill Pass, US 89 crosses the east-west Sheep Creek fault and then the Volcano Valley and Willow Creek faults a few miles farther south. Faults are weak zones in the crust that tend to get reused over time as tectonic stresses change. During Proterozoic time, extension caused the north side of the Volcano Valley fault to go up. Belt rocks were likely eroded away, and Cambrian rocks were deposited on crystalline basement rocks. During Laramide compression in Eocene time, the south sides of these faults went up, shoving older Proterozoic rocks over younger Paleozoic rocks.

Extensional movement during Oligocene time formed several west-to-northwest-trending grabens that filled with sediment and basalt, an igneous rock type that often forms during crustal extension and decompression melting of the mantle as the crust thins. Oligocene rocks are preserved along US 89 in the Sheep Creek valley, which US 89 parallels westward for about 6 miles before abruptly turning south. It's also possible that these Oligocene deposits may have been preserved in grabens that dropped along these faults much later, during basin and range extension 17 million years ago.

At the abrupt south turn, US 89 crosses the Volcano Valley fault. For 11 or so miles south from Sheep Creek the highway passes through the Newland Formation and other Lower Belt rocks. Then it crosses a left-lateral strike-slip fault and passes into Mississippian Madison Group limestone, which was crumpled into tight folds under the Willow Creek thrust fault. Butler Hill, a prominent hill along the east side of US 89, is held up by hard Oligocene-age basalt that filled an ancestral stream valley trending eastward from here for 20 miles.

US 89
Great Falls—Browning
124 miles

Much of US 89 between Great Falls and Choteau crosses the now-eroded lake bottom of Glacial Lake Great Falls, an ice-age lake that formed when the continental ice sheet impounded the Missouri River. Between Great Falls and Sun River, US 89 follows the low vegetated swale of the Sun River. The prominent stump of Square Butte stands a few miles to the south-southwest. Farther west but still south of MT 200 is somewhat larger Shaw Butte. These and other laccoliths in this swarm are described in the MT 200: Bowmans Corners (US 287 Junction)—Great Falls road guide.

Between Sun River and Choteau, the road crosses the Cretaceous Blackleaf Formation and Marias River Shale, laid down in the middle of Cretaceous time in shallow marine to nonmarine environments. The Marias River Shale is mostly dark-gray mudstone with thin sandstone beds and some light-brownish concretions. Shells of clams, oysters, and ammonites are moderately common.

West of US 89 between Choteau and Glacier National Park, the high Sawtooth Range, part of the front range of the Rocky Mountains, exposes a giant stack of sedimentary rocks that was pushed east from western Montana about 70 million years ago. The ages of the rocks range from Proterozoic to Cretaceous, the same rocks that are buried deep under the High Plains to the east. Shoved up on thrust faults, slabs of the Sawtooths stand at steep angles, with older rocks resting on younger rocks. (For more on this story, see the US 287: Wolf Creek—Choteau road guide.)

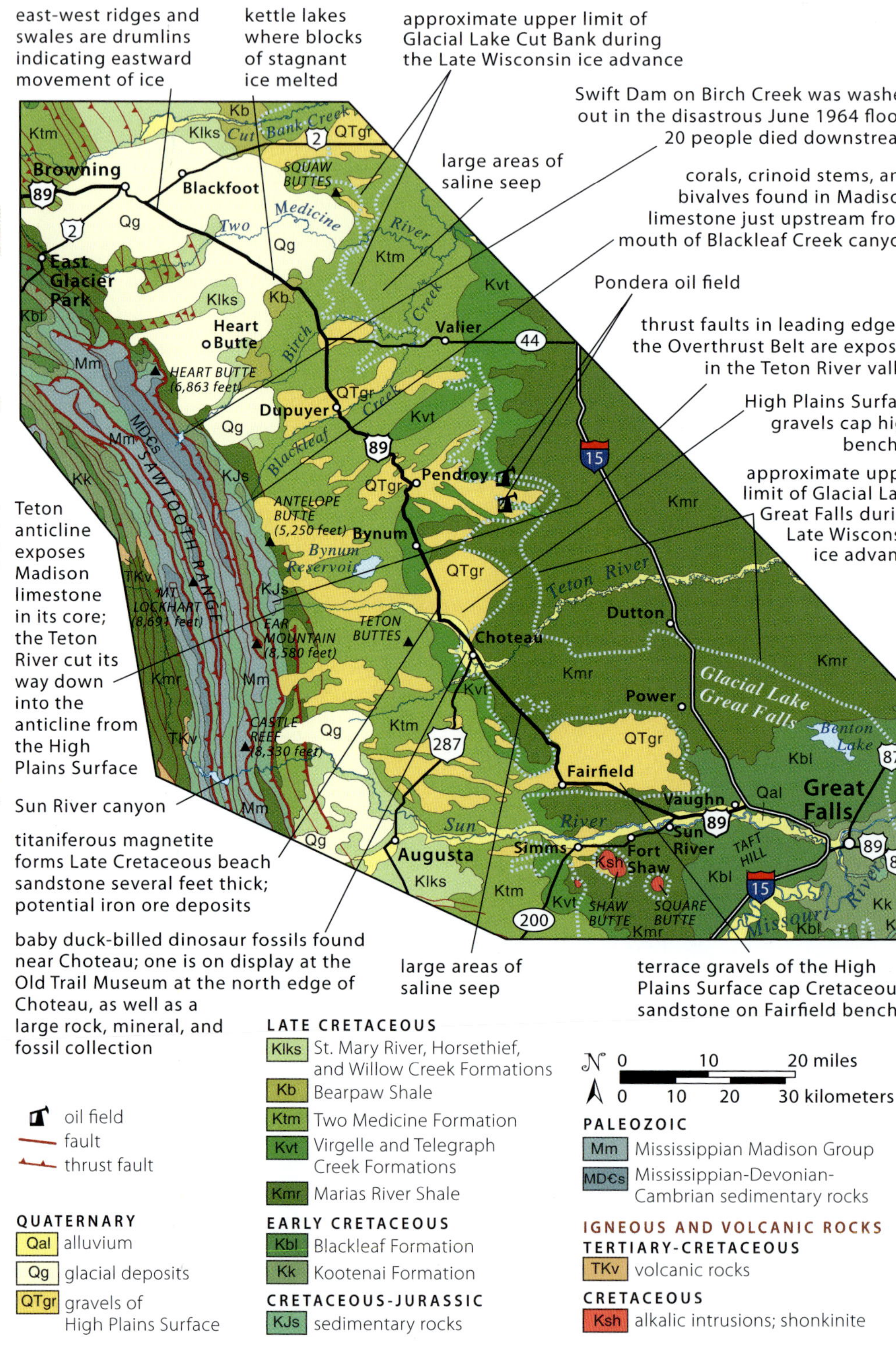

Geology along US 89 between Great Falls and Browning.

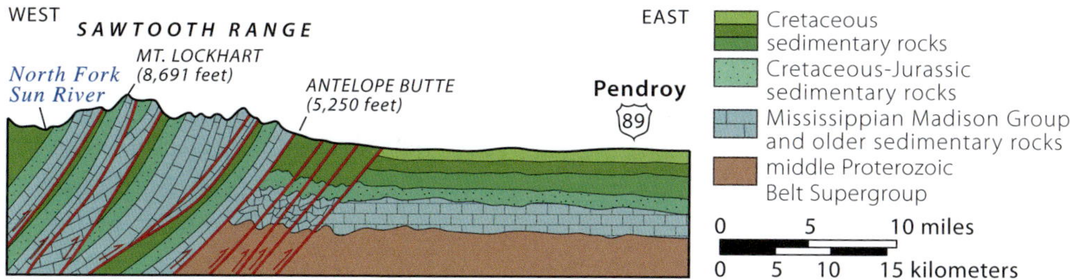

West-east cross section across the Sawtooth Range and US 89 at Pendroy. The strongly deformed, scrambled rocks of the Rocky Mountains are the same formations that lie flat beneath the plains.

Within 2 to 3 miles of Bynum, you can see scattered exposures of the Virgelle Sandstone, a rimrock-forming beach deposit, and sandstone and shale of the Telegraph Creek Formation. North of that the road crosses the Two Medicine Formation, gray-green sandstones and mudstones deposited in rivers and on deltas on the coastal plain of the Western Interior Seaway in Cretaceous time. Petrified wood is common in the lower, older part of the formation, and the remains of vertebrate bones are widespread in the upper part.

Dinosaurs dominated the coastal plain adjacent to the seaway, and dinosaur fossils are especially well preserved in the Two Medicine Formation. In 1977, Marion Brandvold, owner of a rock shop in Bynum, discovered the first baby dinosaur bones found in North America. A year later, dinosaur paleontologist Jack Horner walked into her shop; their cooperation led to the excavation of a colonial nesting site, now called Egg Mountain, south of Choteau. Duck-billed dinosaurs of what came to be called the *Maiasaura* genus, meaning "good mother lizard," cared for their young here. Prior to this discovery scientists hadn't demonstrated that some dinosaurs nurtured their young. Since the initial discovery, hundreds of specimens in all stages of life have been discovered at Egg Mountain, providing an unparalleled understanding of the behavior of these dinosaurs. They lived in herds in upper coastal plain settings, migrated seasonally for food, nested as a group, and raised their young until the entire herd could again move on. Another extensively studied dinosaur locality is about

Dinosaur egg cast from the Two Medicine Formation, displayed at the Two Medicine Dinosaur Center in Bynum. Specimen is 3.75 inches long.

10 miles south of Browning in the Two Medicine through Telegraph Creek Formations. At this site, called Shell Hill, abundant hadrosaurid eggshell fragments, along with dinosaur nests and teeth, have been found.

The Two Medicine Dinosaur Center is at the south edge of Bynum on the west side of US 89. The museum has a range of fascinating dinosaur fossils, including *Struthiomimus*, an ostrich-like dinosaur with a toothless beak; a 137-foot-long life-size model of *Seismosaurus* (now considered a species of *Diplodocus*), one of the world's largest and longest dinosaurs; and the first baby dinosaur bones collected in North America. In spite of its small-town location, this is an outstanding dinosaur-research museum. Paleontologists collect dinosaur remains from a site a few miles away and carefully clean and study the fossils and prepare them for further study. Amateur volunteers are trained and can participate in the unearthing and preparation of fossils in the field.

Cast of a juvenile Maiasaura *fossil from the Two Medicine Formation at Egg Mountain near Bynum, on display at the Two Medicine Dinosaur Center in Bynum. Specimen is 31 inches across.*

The Late Cretaceous muds of the Two Medicine Formation contain beds of volcanic ash that have been partly altered to smectite (bentonite), the clay mineral that expands and becomes extremely slippery when wet. At least some of the volcanic ash appears to have come from the Late Cretaceous Elkhorn Mountains Volcanics that erupted from the Boulder batholith in the Butte-Helena area. A fast-moving pyroclastic flow mowed down and buried a forest about 10 miles south of Choteau. The age of the clay indicates it was deposited 80 million years ago, when the Elkhorn Mountains volcano was erupting, and the consistent orientation of the tree trunks within the clay confirms that the flow came from the southeast—from the Butte-Helena region. Ash layers right on top of groups of dinosaur bones suggest that such volcanic flows, or even heavy ashfalls, might have killed many of the dinosaurs found in the Two Medicine Formation.

Swift Dam, about 19 miles west of Dupuyer, is a good access point for the Rocky Mountain front. Turn west on Swift Dam Road about a half mile north of Dupuyer. The road passes through numerous kettle lakes in glacial till dumped by a mountain glacier to the west. Kettle lakes form in the depressions left after chunks of glacial ice

Folded Mississippian limestone was thrust eastward over the crumpled Cretaceous shale at Swift Dam, west of Dupuyer.

buried in till melt. Just below the spillway and adjacent to the road is a spectacular exposure of folded Mississippian Madison Group limestone thrust faulted over crumpled Cretaceous shale.

Swift Dam failed in June 1964 due to a combination of unusual factors. June is typically the wettest month in Montana, but with annual precipitations of only 10 to 15 inches, that's not saying much. Cut Bank's annual precipitation is 11 inches, and 4 inches of that falls in June. Precipitation in the winter of 1964 was not unusual, and the total water content of the snowpack at Marias Pass on US 2, west of Browning, was not unusual. However, by the end of May, the snowpack had melted to 15 inches of water that completely saturated soils. Then, on June 6 and 7, warm moist air from the Gulf of Mexico collided with a huge low-pressure area blocked by a cold front just north of the Canadian border. Easterly winds (headed west) at the margin of the low-pressure area were pushed up against the front of the Rockies, where they rapidly cooled and unloaded their water.

More than 14 inches of rain fell between June 7 and 8 just east of the Continental Divide, all the way from north of Cut Bank to north of Augusta; the mountains west of Cut Bank received more than 16 inches. Thunderstorms were not reported, just steady, very heavy rain lasting about 30 hours. Falling on top of saturated ground, it all flushed downstream to form catastrophic floods. Thirty people died and $55 million in damages were reported in Montana. Heavy flooding inundated towns along the mountain front all the way east to Great Falls. Bridges failed, as did Swift Dam and Lower Two Medicine Dam, north of East Glacier Park.

Heart Butte stands boldly east of the main front of the Rocky Mountains west of Dupuyer and about 5 miles southwest of the community of Heart Butte. This mass of Mississippian and Devonian sedimentary rock, deposited 390 to 330 million years ago, was thrust over the top of Cretaceous sedimentary rock deposited about 90 to 80 million years ago. The Mississippian-Devonian rock of Heart Butte is part

Heart Butte consists of resistant Mississippian and Devonian sedimentary rocks thrust faulted over easily eroded Cretaceous sedimentary rocks (lower slopes). —Courtesy of Jeff Albrecht

of the Overthrust Belt preserved a bit farther east than the rest of the rock north and south of it.

From several miles north of the junction with MT 44, US 89 crosses the Bearpaw Shale and the St. Mary River Formation, but they are mostly covered by glacial deposits left by glaciers originating in the Rocky Mountain front and southern Glacier National Park. During the Pleistocene ice advances, ice spread eastward out of the mountains to form the Two Medicine lobe. It covered Babb, Browning, and the area west of Dupuyer and Choteau. The glaciers left souvenirs in the form of boulders scattered across the landscape for about 20 miles southeast of Browning. The erratic boulders of Belt sedimentary rocks around Browning must have come from Glacier National Park. Paleozoic and Mesozoic sedimentary boulders must have come from the Rocky Mountain front, including the few off-white limestone boulders that probably came from the Sawtooth Range west of the highway. Any boulders of granite and granite gneiss came from central Canada, but those should only occur well east of US 89.

US 191
Big Timber—Lewistown
101 miles

Bedrock beneath the plains crossed by US 191 is almost entirely sandstone and mudstone, sedimentary rocks deposited in and along the shallow sea that covered much of Montana in Cretaceous time. These formations weather so easily that outcrops are rare along much of the route. Between 80 and 70 million years ago, the Elkhorn Mountains volcano erupted between Butte and Helena, as did the smaller Sliderock Mountain volcano, centered 16 miles south of Big Timber. Enormous volumes of volcanic debris went eastward into the Crazy Mountains Basin to become the Livingston Group (included with Hell Creek Formation on the road guide map) in the southern third of the route.

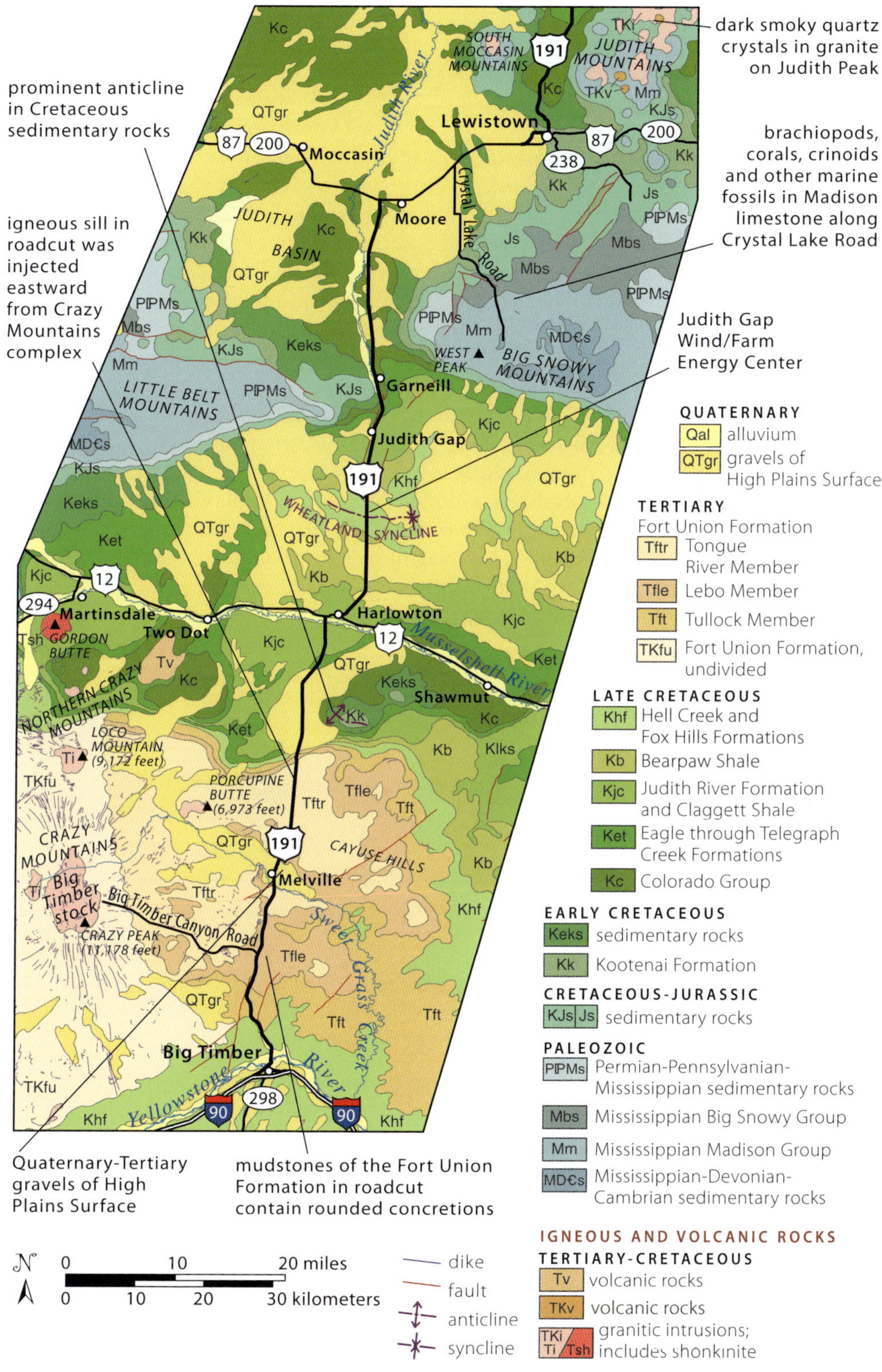

Geology along US 191 between Big Timber and Lewistown.

For the first 3 or 4 miles north from the Yellowstone River bridge at Big Timber, US 191 crosses shale of the Late Cretaceous Hell Creek Formation, the time of the last stand of the dinosaurs, but these soft rocks are poorly exposed. Farther north, scattered sandstone cliffs of the overlying Paleocene (Tertiary) Fort Union Formation appear above the road; these are interlayered with poorly exposed shales. The two formations look almost alike, but between them is the infamous Cretaceous-Tertiary boundary that marks the demise of the dinosaurs and many other animals.

Gravels of the High Plains Surface blanket much of the Crazy Mountains Basin. This basin is shaped like a plate, with a flat bottom and upturned edges. It developed in Late Cretaceous to Paleocene time in response to the Sevier orogeny and filled mostly with Paleocene Fort Union Formation sediments eroded from mountain ranges to the west and southwest.

The high, deeply glaciated southern part of the Crazy Mountains forms an alpine backdrop on the horizon northwest of Big Timber. This part of the range contains the Big Timber stock, a large, nearly circular intrusion of diorite to gabbro composition. The Fort Union Formation was sand and mud, still soft and full of water, when that big mass of hot magma invaded it about 50 million years ago. Heat and chemical-laden hot water escaping from the rising magma baked and altered the soft sediments as far as 1 mile from the contact, converting them into extremely hard rocks.

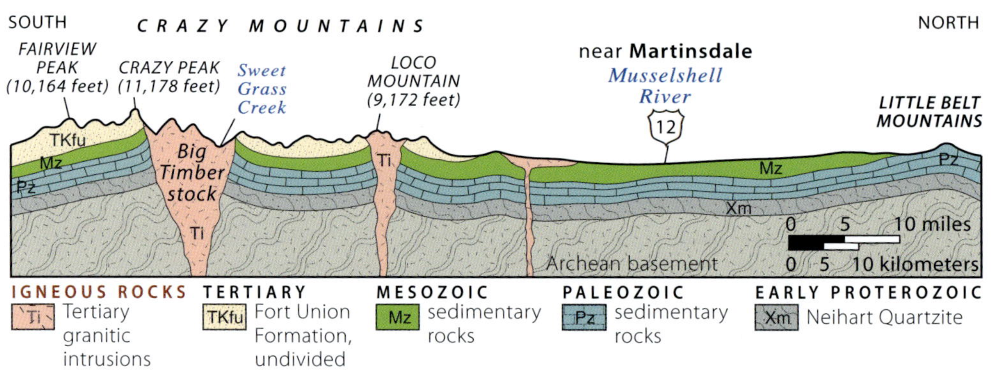

South-north cross section west of US 191 between Big Timber and north of Harlowton showing the Big Timber stock of the Crazy Mountains and the arch of the Little Belt Mountains.

When the magma arrived, the soft sediments of the Fort Union Formation must have extended across this entire countryside to an elevation about as high as the Crazy Mountains. Erosion has since removed most of that thick accumulation, leaving the resistant igneous rock and the hard-baked sedimentary rocks around its contacts standing in high relief as the Crazy Mountains.

Spectacular dikes radiate outward to the north and south from the central stock, and dikes spread similarly from two smaller stocks to the north. The weight of the volcanoes here (largely eroded away over the last 50 million years) loaded the sedimentary layers of the underlying Crazy Mountains Basin, stretching them and directing the underlying magma outward as dikes. The prominent orientation of the dikes is north to south, less so east to west. Geologists have noted this pattern for years but until recently did not come up with an answer for it. Modeling of dike intrusion

Crazy Peak and glaciated terrain in the Crazy Mountains. All of the rocks in this view are diorite of the Big Timber stock. —Courtesy of Rod Benson, Bigskywalker.com

in the Adel Mountains between Helena and Great Falls showed that a volcano that's more elongate in one direction will stretch the underlying sediments more in that direction, leading to more dikes being oriented in that direction. It seems likely that the same mechanisms would have operated here. Note that the overall distribution of intrusions (former magma chambers) in the Crazy Mountains is oriented in the same northerly direction as the dikes.

The only significant road into the high country of the Crazy Mountains heads west from US 191, up Big Timber Canyon, about 9 miles north of Big Timber. The bedrock along the road is almost entirely the Fort Union Formation, but it's obscured by grass and sagebrush. Some places are liberally littered with angular boulders of the Big Timber stock carried down-valley by big valley glaciers of the Pinedale (Late Wisconsin) ice advance. The valley walls become steeper as you go up-valley to the west. This stream valley is typical of those that were deeply gouged by mountain glaciers studded with abrasive rocks. The abrasive ice changed them from a V shape to a U shape. The road ends 1 mile north of Big Timber Peak, at its base.

The distinctly lower north end of the Crazy Mountains contains a swarm of small intrusions, most of them the peculiar Crazy Mountains shonkinite and sodium-rich nepheline syenite. The assortment of intrusions includes sills injected between sedimentary rock layers, vertical dikes formed where magma was injected into fractures, and laccoliths formed where magma pooled like a blood blister under a strong but thin layer, or "skin," of sedimentary rock. Shields River Road heads northeast from Wilsall, on US 89, into this part of the range.

East and north of Melville, an area still within the Crazy Mountains Basin, massive beds of sandstone in the Fort Union Formation form distinctively picturesque bouldery outcrops in the rugged Cayuse Hills. The pore spaces between its sand grains hold enough water to support a good growth of pine trees. About 3 miles north of

Big Timber Canyon, a big glacial valley. The big boulders are glacial erratics carried down-valley from high in the Crazy Mountains. Crazy Peak is 11,178 feet.

Melville and 1 mile west of US 191, a steep-sided, flat-topped butte is capped by an igneous-rock sill, an outlier of the alkaline rocks of the Crazy Mountains. Porcupine Butte, 6 miles farther west-northwest, is a related but larger intrusion.

At the north edge of the Crazy Mountains Basin, 10 miles north of Melville, the highway crosses a swale through a narrow ridge and drops, in short order, from the Paleocene Fort Union Formation down into Late Cretaceous rocks, starting with the Hell Creek Formation and continuing through the Fox Hills Sandstone, Bearpaw Shale, Judith River sandstone and shale, and finally Claggett Shale. For the next 5 miles US 191 crosses a broad plain of gravel eroded from the Crazy Mountains and sloping gently northeast away from the range. This is part of the High Plains Surface.

The Little Belt Mountains lie along the distant horizon northwest of Harlowton. They are part of a broad arch with early Proterozoic basement rocks exposed in its core. The low profiles of the Big Snowy Mountains northeast of the highway and the Little Snowy Mountains still farther northeast make them look exactly like what they are: relatively broad east-west bulges in the plains. Cambrian rocks were brought to the surface along the southern margin of the Big Snowy Mountains in this bulge. With a little more erosion the basement will be exposed, and possibly alkalic igneous rocks that some geologists think rose in the bulge. The Little Belt, Big Snowy, Little Snowy, and Judith Mountains all escaped glaciation during the ice age because they were too low to catch the enormous quantities of snow that big glaciers need for their nourishment. Contrast their relatively rounded profiles to the jagged crest of the ice-carved southern Crazy Mountains.

Judith Gap is a swale between the Little Belt Mountains and the Big Snowy Mountains. A forest of more than ninety wind turbines 5 to 9 miles south of the town of the same name takes advantage of the steady wind that sweeps through here, necessitating the endless snow fences that keep blowing snow from drifting onto the highway. The poles holding these wind turbines are 250 feet tall; each blade is 126 feet long, making the total height almost 400 feet. The best locations for wind power have relatively

These things are really big! A wind turbine blade rounds the corner between US 12 and US 191 at Harlowton, en route to Judith Gap.

consistent wind speeds. Winds become usable at 6 to 7 miles per hour, and the blades stop spinning when wind speeds are above 50 miles per hour to protect the turbines from damage.

Bedrock along US 191 is easily eroded and sparse. Soft, beige Late Cretaceous shale is exposed in low roadcuts 1 or 2 miles north of Judith Gap. About 3 miles north of Judith Gap, US 191 passes the end of the easternmost nose of the anticlinal arch of the Little Belt Mountains. In this area, the highway crosses Mowry Shale (Colorado Group), shale of the Telegraph Creek Formation, and Eagle Sandstone. Farther north it passes intermittent exposures of Telegraph Creek and Colorado Group shales all the way to the junction with US 87 near Moore.

About 10 miles north of Judith Gap is a large group of oil storage tanks that seem entirely out of place—no town, no factory, only wide-open grasslands. The site lies along the path of a buried oil pipeline between the Athabasca tar sands in northern Alberta and oil refineries in Casper, Wyoming. Tar sands oil is low-grade and only marginally profitable. It's stored here until the price of oil rises enough to make it profitable to refine it. (For the stretch of road between Moore and Lewistown, see the US 87: Great Falls—Lewistown road guide.)

US 191
Lewistown—Malta
134 miles

For several miles north of Lewistown, US 191 climbs through a series of Late Cretaceous shales and sandstones along the trough of a broad syncline between the Judith Mountains to the east and the North and South Moccasin Mountains to the west. All three mountain ranges are part of a tight group of igneous intrusions that includes Black Butte south of Roy. They formed as large masses of magma rose through the Earth's crust about 50 million years ago, pushing up older rocks buried under the plains. They are most commonly described as stocks, but given their similarity (including age) to the intrusions in the Little Belt and Highwood Mountains, and the prominent sills of the South Moccasins, they are more likely mushroom-shaped laccoliths.

All of the igneous rocks in the Judith Mountains crystallized below the surface. Some are granite, such as at Black Butte; heat and steam from its magma created a spectacular

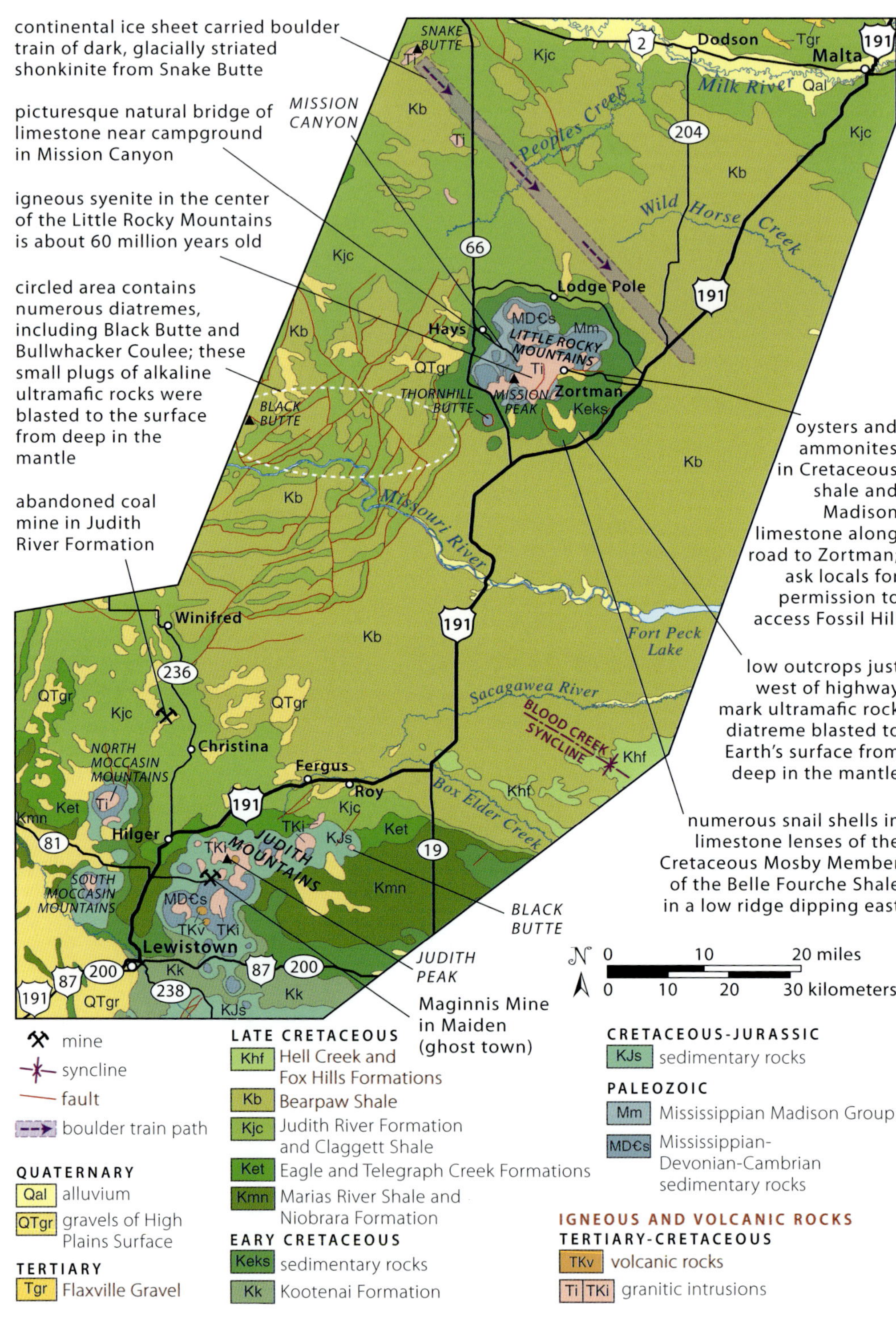

Geology along US 191 between Lewistown and Malta.

zone of metamorphism in the older sedimentary rocks within a few hundred feet of the contact. Others are peculiar rocks rich in the alkali elements potassium and sodium. The intrusions in the North and South Moccasin Mountains are mostly syenite, a rock consisting mostly of potassium feldspar, or orthoclase. Most of the igneous rocks are either fine-grained, meaning they crystallized fairly close to Earth's surface, or porphyritic, meaning that larger grains crystallized in the magma at deeper levels before the magma rose to a shallower level in the crust, where the remainder of the magma formed small crystals. The intrusions dragged the adjacent sedimentary rocks up as they rose, so Paleozoic sedimentary rocks, including large exposures of gray Madison Group limestone, stand steeply on end next to some of the intrusions. Those older rocks normally lie buried several thousand feet below the plains.

A side road east from Brooks, 9 miles north of Lewistown, leads to the ghost town of Maiden, in the Judith Mountains. Placer gold was discovered in the Warm Spring Creek canyon in 1880, leading to the usual rush of everyone within earshot. More profitable ores were soon discovered in the bedrock nearby, especially at what became the Spotted Horse, Maginnis, and Kentucky Favorite Mines. Three years later, Maiden had more than 150 buildings, more than one thousand people, and, of course, 13 saloons. Like most gold-quartz vein deposits, those around Maiden contain scattered pockets of bonanza-inspiring ore isolated within vast barrens of profitless quartz. More than a dozen buildings and a ten-stamp mill have been preserved. A couple of mines reopened in the 1970s. The Maginnis finally closed in 2016.

A road east and north from Maiden leads to an old radar site on Judith Peak and a distinctive and fascinating granite full of crystals of dark smoky quartz and smaller rectangles of white orthoclase. Although they are well-formed, the quartz crystals are so tightly bound within the rock that they are impossible to separate.

Looking southwest, toward Lewistown at granite along the flanks of Judith Peak just below the old radar station. Inset shows well-formed smoky quartz crystals that you can see along the road to the top of Judith Peak.

Placer gold was first discovered in the North Moccasin Mountains in the 1880s in shallow deposits of rocks altered to clay by the hydrothermal fluids of nearby magma. The gold was difficult to separate, even in underground lode mines, so mining efforts didn't last. The development of the process of extracting gold by dissolving it in cyanide solutions caused a sudden boom in mining. The lode mines, including the Kendall, North Moccasin, and Barnes-King Mines, were very profitable up through the 1920s, some even into the 1930s. Early cyanide processing lacked any environmental constraints. The crushed ore was leached in cyanide tanks, and the spent waste was dumped into nearby drainages. A large low-grade deposit discovered in 1987 was worked until 1996 using more-modern cyanide-leaching techniques; 300,000 ounces of gold were recovered. The old mining town of Kendall is north of Lewistown; head about 14 miles north on US 191, and then 5.5 miles west on North Kendall Road. Several crumbling foundations related to early-1900s mining remain. Back on US 191, yellow waste piles 11 miles northeast of Hilger were left from early gold mining in the North Moccasin Mountains to the west.

Northeast of Hilger, US 191 skirts the Judith Mountains on the north, crossing over Late Cretaceous Eagle Sandstone, Claggett Shale, Judith River Formation sandstone, and Bearpaw Shale, gradually going up the rock section, from older through younger formations, through gently rolling, grassy hills to the junction with MT 19. Just over 1 mile north of the MT 19 junction, US 191 cuts through a small hill that exposes the overlying Late Cretaceous Fox Hills and Hell Creek Formations. North of the hill, the highway crosses a flat, featureless plateau of grass and sage developed on poorly exposed Bearpaw Shale. Small gullies show white alkali evaporite (saline seeps).

North and south of the Missouri River, US 191 crosses the former floor of Glacial Lake Musselshell, which formed when the ice sheet moving southwest out of central Canada dammed the Missouri and Musselshell Rivers about 20 miles upstream from Fort Peck. It raised their water level to 2,600 feet to form Glacial Lake Jordan and its downstream extension, Glacial Lake Musselshell. At that level, the combined glacial lakes stood 300 or 400 feet higher than today's Fort Peck Lake, and therefore they covered a somewhat larger area in the very hilly terrain near the present lake. The Missouri River is at an elevation of 2,300 feet where US 191 crosses it here, so the lake was about 300 feet deep in this place but only about 1 mile wide.

As US 191 approaches the Missouri River, it drops rapidly through deep gullies in shale for about 1 mile, finally passing into the underlying Judith River Formation at Armells Creek. This small creek meandering in a big flat-bottomed valley is an underfit stream, meaning it's too small for its valley. The valley was carved by a much larger stream, perhaps larger flows of the last two ice advances. The Judith River Formation is exposed in the steep slopes along the bottom of this valley and for the 2 to 3 miles of road across the Missouri River Valley. The valley at this point is just downstream of the spectacular gorge of the Upper Missouri River Breaks but shows little hint of it here. The secondary road leading west to Winifred crosses a number of high vantage points that provide marvelous views of the colorful canyon walls.

The Lewis and Clark Expedition passed here in late May of 1805, employing their sails due to a "fine breeze" from the southeast. Lewis recognized different rock layers in the canyon walls, writing "the country high and broken, a considerable portion of black rock and brown sandy rock appear in the faces of the hills . . . the whole producing but little grass." He clearly noticed the dark Bearpaw Shale and underlying sandstone of the Judith River Formation. Little did he know that the Judith River

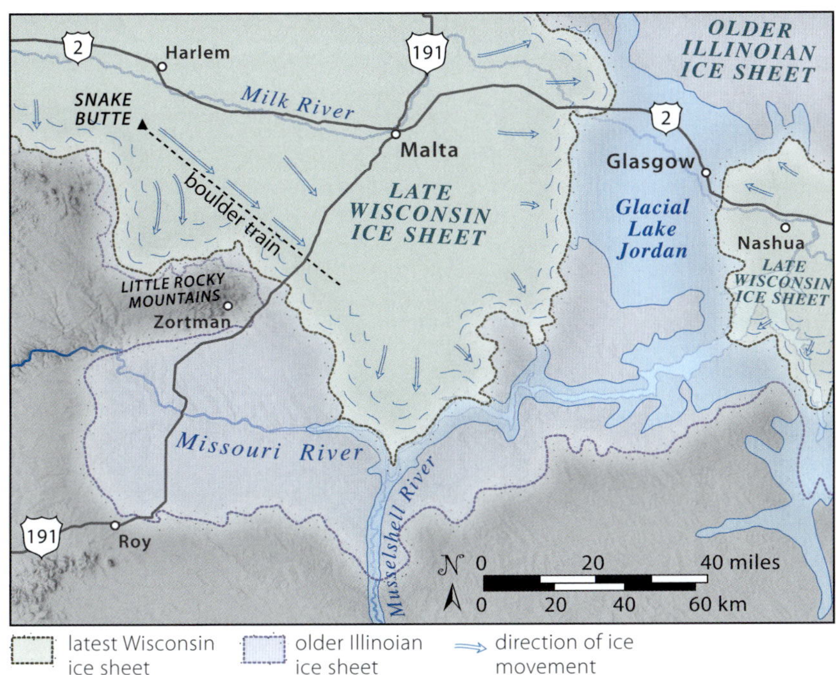

The continental ice sheet spread south in two lobes, one east of the Little Rocky Mountains, the other over Wolf Point and Fort Peck. The latter blocked the Missouri River during the Late Wisconsin glacial advance, raising its level to 2,600 feet and forming Glacial Lake Jordan, a larger version of today's Fort Peck Lake. Earlier, between 190,000 and 130,000 years ago, ice of the Illinoian glaciation covered a similar but somewhat larger area of northern Montana. Along US 191, it extended south of the Little Rockies almost to the town of Roy. —Modified from Fullerton and others, 2012

Formation was harboring at least forty different species of dinosaurs, including the now-famous *Tyrannosaurus rex*. He also noted that the small tributary streams were unfit for drinking because they were salty: "I have tryed it by way of experiment & find it moderately pergative, but painful to the intestens in it's opperation."

Northeast of the MT 66 junction, the Little Rocky Mountains rise west of US 191. Like several other isolated ranges of central Montana, the Little Rockies form a dome cored by Paleozoic sedimentary rocks that are cut by 60-million-year-old syenite, which punched up through the High Plains.

Thick Madison Group limestone, which resists weathering in today's dry climate, stands steeply tilted, wrapped around the Little Rocky Mountains. From a good vantage point and in the raking light of early morning or late evening, it's easy to imagine this limestone as the wall around the ruined fortification of a medieval city. Like elsewhere, caves occur in the Madison here. Azure Cave, also called Zortman Cave, has 1,580 feet of passages with stalactites, stalagmites, and flowstone. Spelunking trips require special permission from the Bureau of Land Management.

Gold- and silver-bearing veins were mined in shallowly emplaced syenite and related rock intrusions. Although Native Americans apparently knew of gold in the Little Rocky Mountains and there are accounts of a discovery in 1868, mining didn't

MISSION CANYON

Mission Canyon, the namesake of the Mission Canyon Limestone of the Madison Group, east of Hays, is a narrow, cliff-bound gorge with an amazing natural bridge. From US 191 head north about 16 miles on MT 66 to the community of Hays; go east and south through Hays for a few miles to where the road enters a narrow slot canyon. A short distance into the rock-walled canyon, the natural bridge arches over a short side canyon on the right. Limestone is very susceptible to solution in rainwater. Water dissolves carbon dioxide in the air and soil to create carbonic acid, a weak acid that slowly dissolves the limestone along cracks. The same process creates cavities and caverns underground; the arch may have started that way, eventually leading to this remnant natural bridge.

A natural bridge in Mission Canyon Limestone a few miles south of Hays.

For the next mile or so up the canyon, the road follows the narrow gorge of Mission Canyon in the same limestone. There are two possible causes of the local red coloration of the limestone: extensive solution of the limestone over millions of years could have led to the concentration of minute amounts of red iron-rich clay within it, or perhaps the red clay was derived from overlying formations and seeped into the underlying limestone.

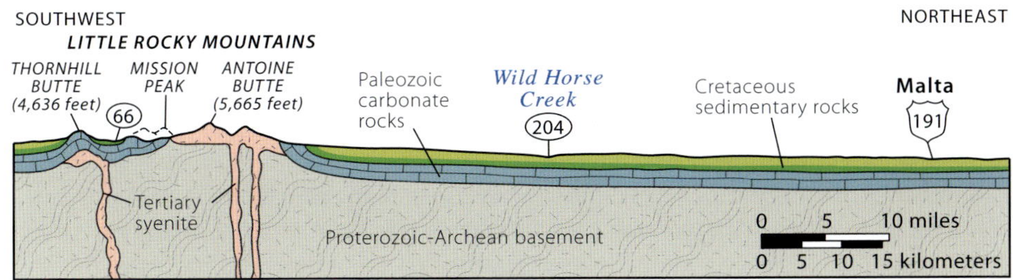

Southwest-northeast cross section between the Little Rocky Mountains and Malta.

begin until after prospectors found placer gold there in 1884. Several thousand of the free-floating miners of those days rushed to the Little Rockies, quickly cleaned out the easy placer deposits, and left within a few months. The discovery of extremely rich gold-quartz ore in 1890 triggered a new gold rush in the district. Landusky sprang up in 1894 near the southern margin of the intrusion, not far from MT 66.

The old mining town of Zortman can be reached from US 191, either on 7 Mile Road, 7 miles northeast of the junction of US 191 and MT 66, or on Bear Gulch Road, about 8 miles farther northeast. Today it's a hodgepodge of old miner's cabins and more recent buildings with a few dozen residents. Acres of spoil heaps and tailings in Ruby Gulch north of Zortman testify to the enormous amount of rock that was mined and fed through cyanide mills, most of it between 1894 and 1939. Mining began dying out in the 1950s. Interest in the district rekindled in the late 1960s with the development of cyanide heap leaching. In this process, crushed rock was spread out on the ground on an impervious pad and cyanide solution sprinkled on it to soak through and dissolve the fine particles of gold. The gold-bearing cyanide solution was collected and the gold recovered. Pegasus Gold began mining in 1977 and remained active while gold prices were high. The company declared bankruptcy in 1998. The bonds it had posted to cover reclamation were insufficient to clean up the site. The Montana Department of Environmental Quality and the US Bureau of Land Management now handle the reclamation.

At the entrance to the giant Veseth Ranch on the east side of US 191, about 20 miles north of Bear Gulch Road, are huge, 10-foot-diameter black boulders that look completely out of place. The isolated boulders scattered on the plains of northern Montana are almost always light-colored metamorphic rocks and granite, brought here from Canada by the continental ice sheet. But these are larger and composed of black shonkinite. Stranger yet, there aren't any such black rocks nearby. The most prominent exposure of black shonkinite is Snake Butte, southwest of Harlem. This butte provided the rocks used to build Fort Peck Dam. The continental ice sheet carried these boulders southeast almost 40 miles (the path is shown on the road guide map). Geologists use this kind of evidence to determine the directions that ice moved during the ice age. (Other large black boulders along this same boulder line from Snake Butte are pointed out in the US 2: Havre—Malta road guide.)

From the north edge of the Little Rocky Mountains almost all the way to Malta, the flat High Plains of eastern Montana rest on top of equally flat layers of easily eroded Bearpaw Shale. Only within 11 or 12 miles of Malta does this shale give way to Judith River Formation sandstone, and these rocks are exposed in only a few places along the highway. In Malta, check out the Great Plains Dinosaur Museum.

US 212
Crow Agency—Broadus—Alzada
166 miles

US 212 crosses the full width of the Powder River Basin, an especially thick accumulation of sedimentary rocks deposited in and along a shallow inland seaway in Late Cretaceous time. The thickest part (17,000 feet) of the 80-mile-wide basin is 10 to 13 miles east of its western margin. The rocks at the surface of the basin today are mostly Tongue River Member sandstones of the Paleocene Fort Union Formation. Thick beds of the Tongue River are the only rocks resistant enough to weather into hills and make prominent outcrops; this member also contains most of the coal. Forty-two percent of the coal used in the United States comes from Montana and Wyoming, much of it from the Powder River Basin.

Geology along US 212 between Crow Agency and Alzada.

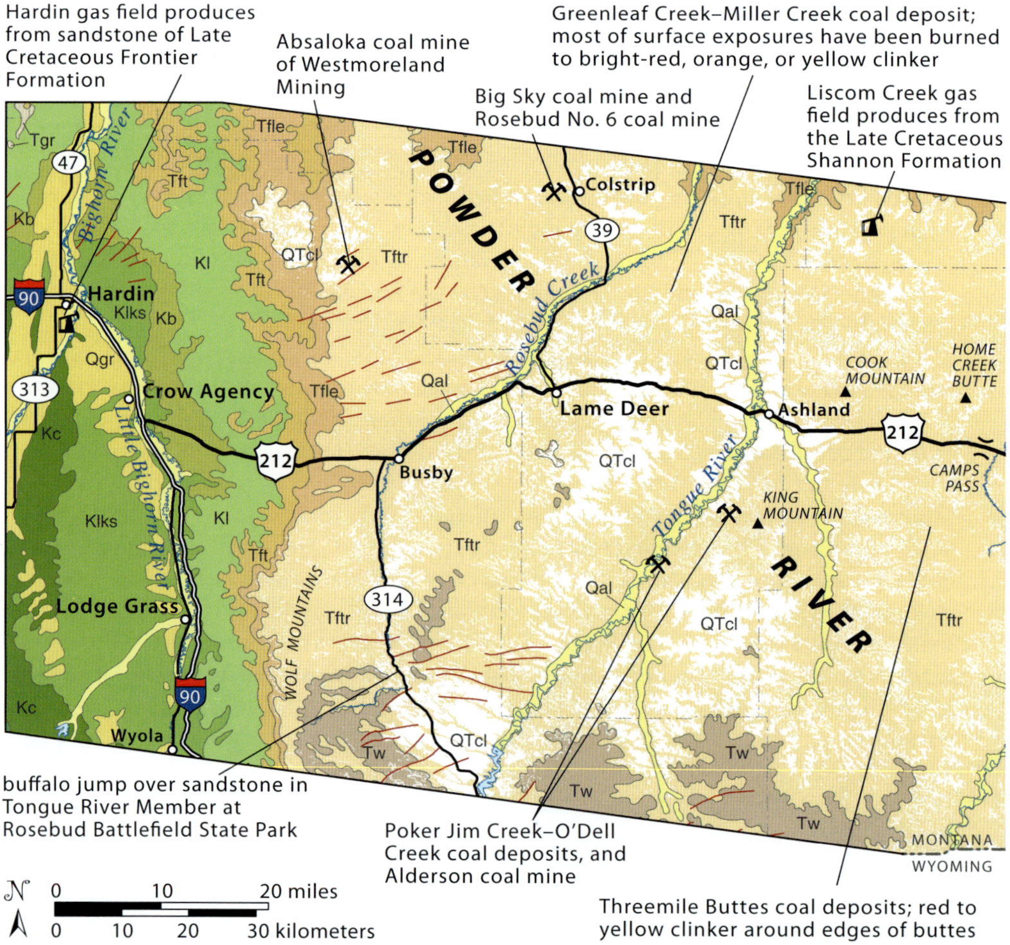

East of the I-90 junction near Crow Agency, US 212 passes through the Crow Inidan Reservation, mostly across open, rolling grassland and sagebrush with no trees and a few farms. It crosses Late Cretaceous, yellowish-gray Judith River Formation sandstone for the first mile, then 7 or 8 miles of younger, dark-gray Bearpaw Shale, then about 5 miles of brownish-gray Fox Hills Sandstone and Lance Formation. Though the sedimentary layers are nearly horizontal, they do dip very gently to the east, so the road climbs progressively up through the rock section, from older to younger rocks.

About 13 miles east of I-90, US 212 reaches yellowish-gray sandstone of the Fort Union Formation. Exposures of beige shale with some black coal at the base occur near milepost 20. The highway crosses Rosebud Creek at Busby, where low hills expose beige shale without the characteristic red clinker caps that mark former coal beds in the region. For 6 miles northeast of Busby, US 212 continues along the south bank of Rosebud Creek in the Fort Union Formation; scattered exposures of red clinker appear in some roadcuts and cap hills with prominent horizontal layering.

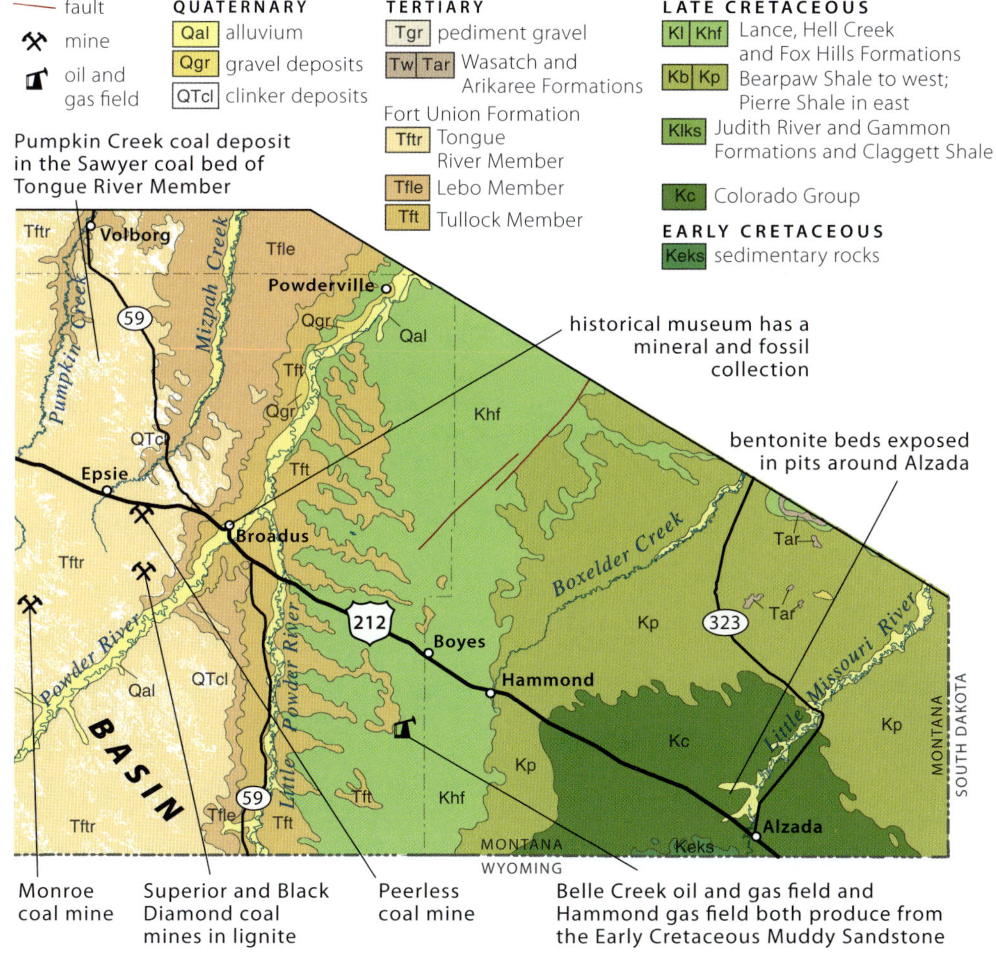

Red clinker caps hills of Fort Union sandstone about 9 miles west of Lame Deer, near milepost 33. The inset photo shows a drip of clinker, where it got hot enough to melt and drip over the sediment. —Inset courtesy of Sheila Roberts

The clinker formed when coal beds burned, baking the overlying sediments. Lightning strikes and wildfires have been igniting coal fires in central and eastern Montana for thousands of years; they burn and smolder underground for long distances. The reason we see so little coal exposed at the surface is because that which was exposed has burned. In some places the red clinker cap drips down over beige sediments; the rock was hot enough to melt and drip down the outcrop. Clinker typically caps hills because it's much harder than the underlying shale and sandstone and thus resists erosion. West of Lame Deer we see more clinker-capped buttes to the north, in some cases with pale-gray Fort Union shale and sandstone below.

At Lame Deer, MT 39 heads north to Colstrip, where Powder River Basin coal is strip-mined. (See the introduction to this chapter for more information about this area.) Between Lame Deer and Broadus, US 212 passes through areas of rugged hills that support a scattering of trees, typical landscape of eastern Montana's coal country. The highway crosses over the giant Pumpkin Creek coalfield, which contains about 2.5 billion tons of unmined coal in several major seams. It's the North American equivalent of the Persian Gulf oil fields. Whether this coal is ever mined depends on the economics of natural gas and renewable energies such as wind and solar, which are now cheaper than coal, while coal struggles more and more with its pollution and waste disposal.

The Tongue River sandstone, which contains most of the coal, weathers into pinnacles and picturesque outcrops of yellowish boulders. US 212 crosses the unit's namesake, the Tongue River, at Ashland.

Pale-gray shale is well exposed near Camps Pass, about 20 miles east of Ashland. Red-and-black clinker, some with melted surfaces and full of gas holes, perhaps caused by rising steam, is preserved near the tops of nearby roadcuts. Most of this

A coal bed a few feet thick peeks out from under pale-gray and beige clay at the base of a roadcut about 1 mile east of Ashland, around milepost 64.

pine-forested terrain is very hilly and rocky, composed of similar steep-sided gray hills with red-and-black clinker caps.

Closer to Broadus, the landscape becomes more gently rolling grassland with sagebrush and glimmers of gray shale. The lack of coal beds and their accompanying clinker to hold up the hills may explain the subdued topography. Broadus sits on the west bank of the Powder River. In 2016, a Columbian mammoth was discovered along the banks of the river. The site yielded ribs, leg bones, and a mostly intact skull with both curved tusks. The animal was about 7 feet long and died here between 20,000 and 10,000 years ago. The bones now reside at the Carter County Museum in Ekalaka.

Broadus has extraordinarily wide steets, especially for a town of only about five hundred people. In the early 1900s, the streets were platted to be wide enough for a horse-drawn carriage to make a U-turn. The rest area on the east side of the Powder

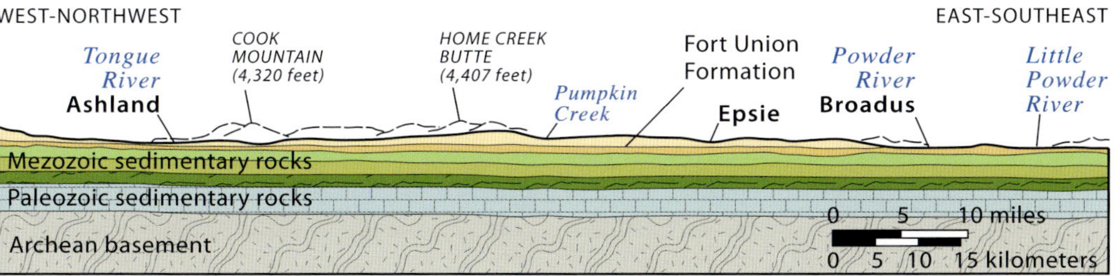

Cross section along US 212 between Ashland and Broadus. The sedimentary formations lie almost flat in this part of Montana, deeply covering the complex igneous and metamorphic rocks of the basement.

Soft beige shale with a few hard beds of sandstone, both part of the Fort Union Formation, 11 miles east of Ashland.

River has big, decorative boulders of red-and-black clinker—fractured, broken, some swirly-looking and deformed. When burning coal beds collapsed, broke up, and locally melted the overlying beds, they created these chaotic, recemented rocks.

For 8 miles east from Broadus, the landscape is more open and subdued, becoming mostly flat with a few low rises. East of the Little Powder River, 3 miles east of where MT 59 heads south, the sedimentary rocks on and near the road are the Fort Union Formation; farther east from the river the rocks look about the same but belong to the slightly older, latest Cretaceous Hell Creek Formation. These continue to the tiny community of Boyes, where clinker from burned-out coal beds caps nearby small buttes with pine forests.

The Belle Creek oil and gas field, in the southeast corner of Powder River County a few miles south of Boyes, was the ninth-largest oil producer in Montana in 2012, with about 350,000 barrels of oil, almost all of which came from the Early Cretaceous Muddy Sandstone. The sands were deposited as beach and barrier islands at the margin of the Western Interior Seaway. Until 2010, most of the oil was produced by pumping water into the sandstone to float the oil out. Since then carbon dioxide has been injected to enhance oil recovery.

Five or six miles east of Boyes, the Hell Creek gives way to a short interval of sandstone and then nearly horizontal, poorly exposed Cretaceous Pierre Shale for another 7 miles. For the remaining 20 or so miles to Alzada, the highway rises and falls very slightly, for a total drop of only 150 or so feet. The sedimentary layers are not well exposed but dip about 3 degrees northwest—that is, the layers gradually rise toward Alzada. Here we are in the northwestern extremity of the Black Hills uplift of South Dakota. The highway passes gradually into older layers of that uplift.

US 287
Wolf Creek—Choteau
65 miles

The entire route along US 287 between Wolf Creek and Choteau lies 10 to 20 miles east of the Overthrust Belt that forms the Rocky Mountain front. The highway is within the Disturbed Belt, a zone in which the rocks are almost all easily eroded Cretaceous-age shales and sandstones. They are deformed like those in the Overthrust Belt to the west but too soft to withstand weathering and erosion. In many places between Wolf Creek and Choteau, conspicuous ledges of resistant sandstone mark the trend of the sedimentary layers, most of which are too weak to form good outcrops. In places these ledges wind across the countryside in sinuous hairpin curves, evidence that the rocks are tightly folded.

Almost all of US 287 north of the I-15 junction crosses the Late Cretaceous Two Medicine Formation and Virgelle Sandstone. Nice roadcuts a little over 2 miles north of I-15 are steeply dipping Two Medicine mudstone and sandstone, part of the Disturbed Belt.

The high, mostly forested hills east of the road between Wolf Creek and Bowmans Corners are the Adel Mountains. They are the eroded remains of a volcano that erupted about 75 million years ago in Late Cretaceous time and are part of the Central Montana Alkalic Province. Most of the rocks are shonkinite, a very fine-grained dark rock made of black augite pyroxene and potassium-rich feldspar. High, ragged ridge crests well above the highway 10 to 30 miles north of I-15 are held up by hard dikes from the volcano. These black to brownish rocks are similar to basalt except they have unusually large amounts of potassium and sodium.

North of MT 200, US 287 rises high enough for the ragged peaks of the deeply glaciated Sawtooth Range, part of the Overthrust Belt, to appear on the western skyline. Most of the range that's visible from the highway is composed of westward-tilted slabs of pale limestone that were pushed up to the east along large thrust faults

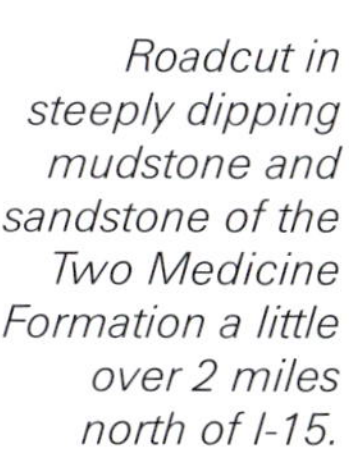

Roadcut in steeply dipping mudstone and sandstone of the Two Medicine Formation a little over 2 miles north of I-15.

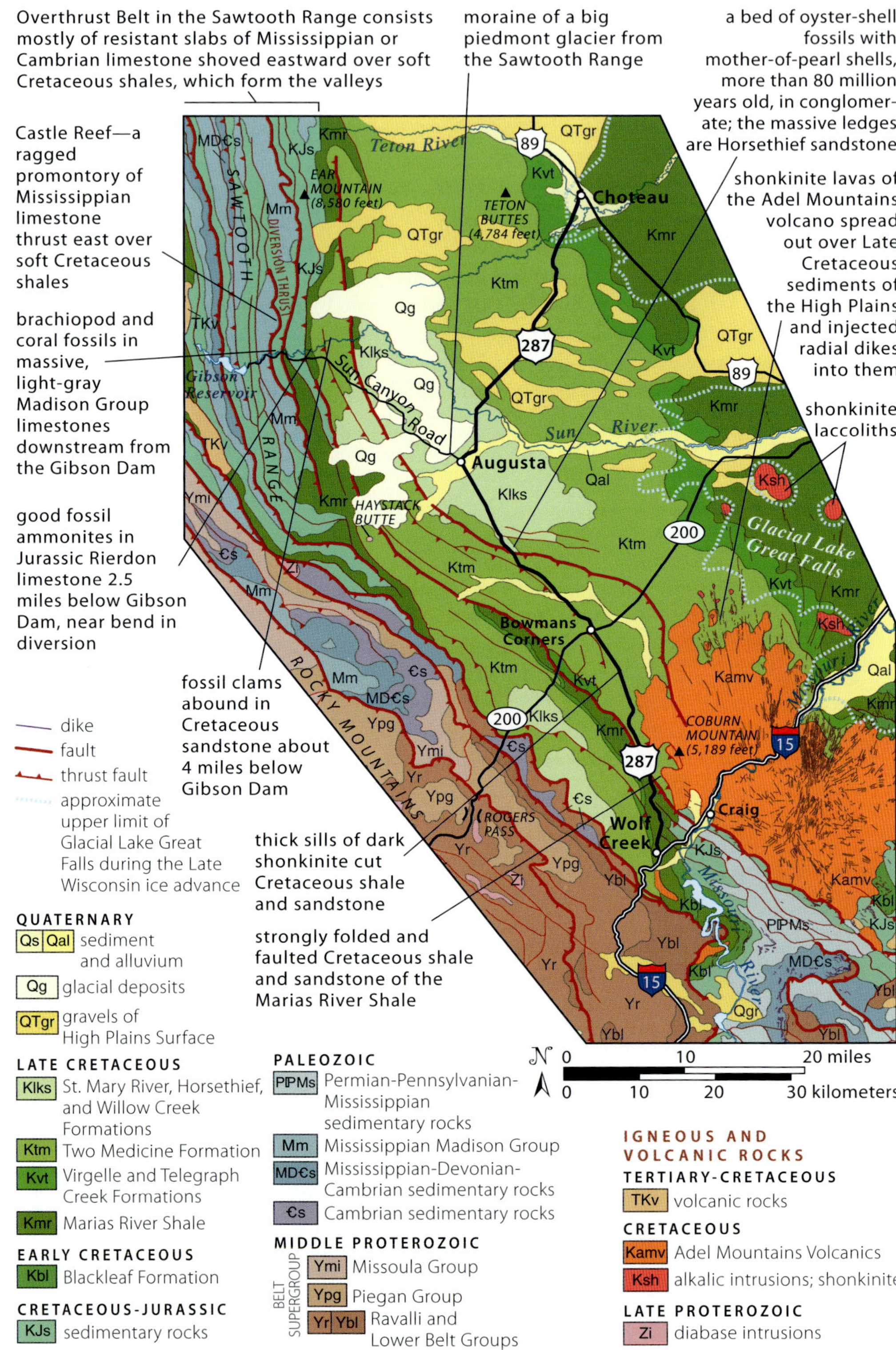

Geology along US 287 between Wolf Creek and Choteau.

The high, ragged ridge east of US 287, 13 to 14 miles north of the I-15 junction, is capped by dikes from the Adel Mountains volcano. The softer hills in the foreground are composed of the Two Medicine Formation and Virgelle Sandstone.

Sawtooth Mountain at the front of the Sawtooth Range west of Augusta. Giant sheets of limestone were thrust up to the east (right) over thin shale layers (at the base of the cliffs in photo). —Courtesy of Rick and Susie Graetz

during the Sevier orogeny in Cretaceous time. The prominent cliffs are made of either Mississippian-age Madison Group limestone or pale Cambrian limestones, which overlie much younger formations, most commonly shales of Cretaceous age.

About 10 miles north of MT 200 the road crosses the gray to gray-brown cross-bedded Horsethief Formation that was deposited in beach and lagoon environments of the Western Interior Seaway of Cretaceous time. North of there, the road crosses grayish-green siltstone and sandstone of the St. Mary River Formation, laid down in stream and floodplain environments on the western margin of the seaway.

Haystack Butte, an isolated, pyramid-shaped, 6,500-foot-tall mountain west of US 287 about 10 miles southwest of Augusta, is an alkaline igneous intrusion. Most of the Central Montana Alkalic Province is about 50 million years old, but recent radiometric ages of the Adel Mountains volcano show it to be closer to 75 million years old. That leaves the age of Haystack Butte in limbo—either 50 or 75 million years old.

One conspicuous anticlinal arch in the Cretaceous rocks of the Disturbed Belt, the Robertson dome, lies east of the highway between MT 200 and Augusta. Virtually all such anticlines exposed at the surface in Montana had wildcat wells drilled into them during the early years of oil exploration, before 1925. If you don't see producing wells and storage tanks on them now, you can safely assume that early wells were dry.

A few miles north of Augusta, the road crosses remnants of the High Plains Surface. Roadcuts expose the gravel at the edges of this conspicuously flat-topped surface in the broad valleys that have cut through it. Stream-rounded pebbles litter

SUN RIVER CANYON

The best place in Montana to get a close roadside view of geologic structures in the Overthrust Belt is along the road that heads northwest from Augusta to Gibson Dam. Just outside Augusta, the road crosses a large moraine, its hummocky surface littered with erratic boulders. The moraine marks the edge of a glacier that poured out of the Sawtooth Range, spreading sluggishly across the plains east of the mountain front as a piedmont glacier.

The road enters the Sun River canyon through a narrow gorge eroded through the slab of Madison Group limestone that forms the steep front of the Sawtooth Range. It is the easternmost overthrust slab and lies on top of the Cretaceous sedimentary

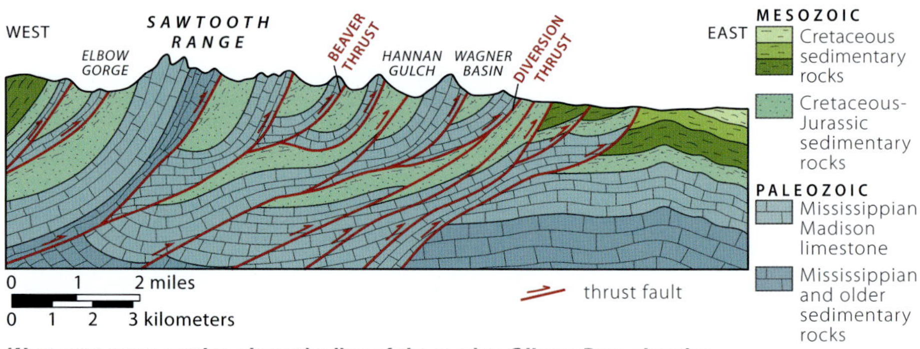

West-east cross section along the line of the road to Gibson Dam showing multiple slabs of Madison Group limestone stacked on each other, with slices of younger Mesozoic formations sandwiched between them. Only thrust faulting on many surfaces can create such a bizarre arrangement of sedimentary layers.

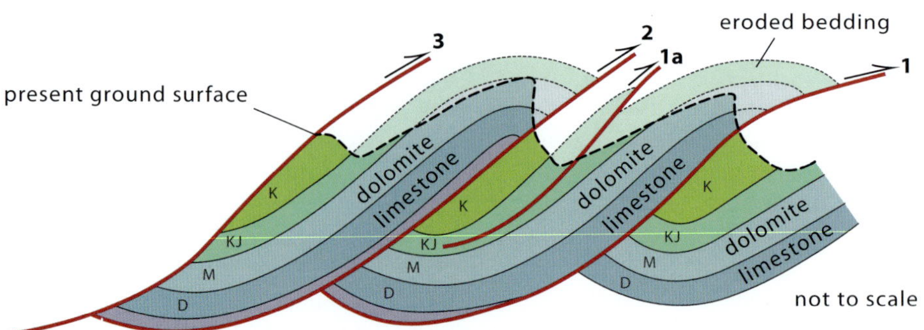

Style of folds and thrust faults in Sun River canyon, west of Augusta. The numbers 1, 2, and 3 refer to the sequence of fault movements. D *stands for Devonian;* M, *Mississippian;* KJ, *Cretaceous-Jurassic;* K, *Cretaceous.* —Modified from Mudge, 1972

rocks exposed at the mouth of the canyon. The valley widens in the Wagner Basin west of the first slab of limestone, where less-resistant, younger rocks are exposed. Then the valley narrows again to pass through a second slab of Madison Group limestone that slid up and over the younger formations on another thrust fault. Another wide place in the valley, Hannan Gulch, marks the second outcrop of the same set of less-resistant, younger formations exposed in Wagner Basin. The road then passes through a third narrow gorge in the third overthrust slab of Madison Group limestone, which here anchors Gibson Dam. Dating of clays that formed from the heat generated during the thrust faulting indicates the faults were active 67 million years ago. Throughout the northern Montana Overthrust Belt, the ridges trend generally from north to south along the upturned edges of resistant limestone slabs, and the long valleys follow the less-resistant rocks between them.

Madison Group limestone (right) thrust eastward over dark Cretaceous shale (left) in the Sun River canyon northwest of Augusta.

View looking southwest over the entrance to Sun River canyon. Multiple slabs of cliff-forming Madison Group limestone were shoved northeastward over valley-forming Cretaceous shale during Sevier compression in Late Cretaceous time. —Courtesy of Rod Benson, Bigskywalker.com

Badlands in Two Medicine Formation shale about 8 miles south of Choteau between mileposts 57 and 58. The view is downslope to the west behind the highway information sign describing Egg Mountain, discussed in the US 89: Great Falls—Browning road guide.

the wheat fields on the flat top. The High Plains Surface is a remnant of a desert plain that stretched continuously from the Rocky Mountains to the central Dakotas during late Miocene and Pliocene time, as recently as 3 million years ago, and was dissected during the wetter climates of Pleistocene time.

Nothing protects a land surface against erosion quite as effectively as a cap of gravel, which absorbs surface water that would otherwise flow across the surface and erode gullies. Perhaps an old alluvial fan made the gravel thicker here than elsewhere in the area, and so preserved this remnant of the High Plains Surface by preventing any streams from starting to flow across it.

These younger gravels, several feet thick here, were laid down on top of a flat surface that was eroded on Two Medicine Formation shale that had been deposited tens of millions of years earlier, during Late Cretaceous time. The shale is still visible under the gravels in some of the roadcuts. The Two Medicine shale continues to about 5 miles south of Choteau, where the highway passes into the underlying formations as it gently descends to Choteau in the valley of the Teton River.

MT 7
Wibaux—Ekalaka
80 miles

MT 7, in the southeastern fringe of the state, is one of the loneliest, least-traveled roads in Montana. For 25 miles south from Wibaux, MT 7 follows the west edge of gently meandering Beaver Creek through fields of wheat, corn, and hay. Cut banks in the creek and some low roadcuts locally expose soft shales of the Fort Union Formation. Red clinker remains on isolated hilltops farther south, a reminder left from burned-out coal seams that baked adjacent rock layers. Savvy ranchers spread clinker over dirt roads to avoid problems with gumbo clay when it rains.

About 33 miles south of Wibaux (11 miles north of Baker), MT 7 cuts into the latest Cretaceous Hell Creek Formation as the first hints of the Cedar Creek anticline

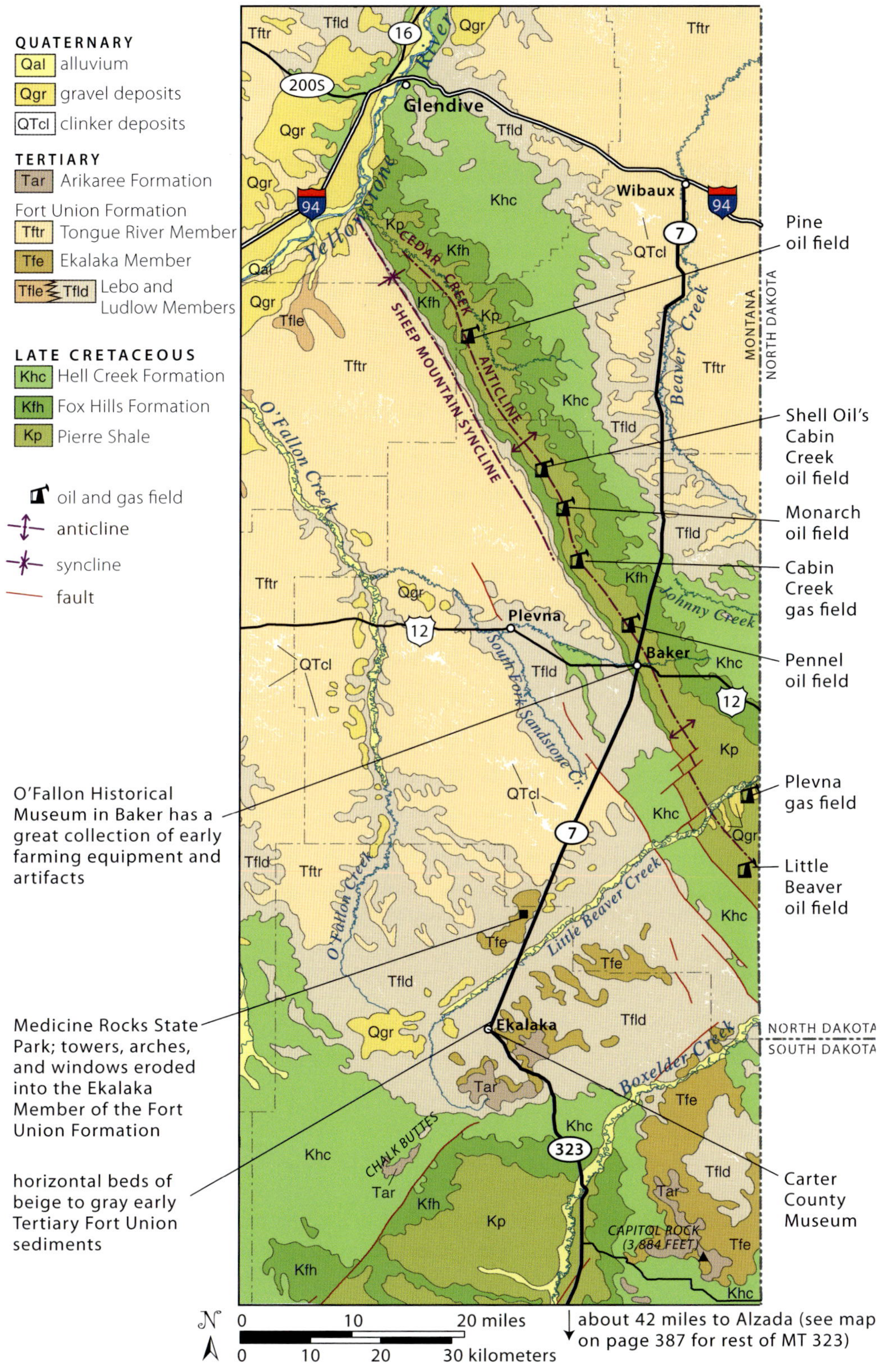

Geology along MT 7 between Wibaux and Ekalaka.

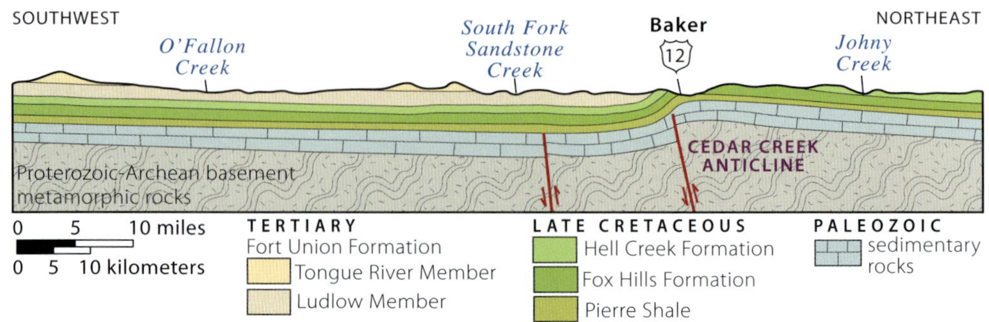

Southwest-northeast cross section through the Cedar Creek anticline at Baker.

rise across its path. The northwest-trending anticline has a very gentle northeast flank and a steep southwest flank and extends for 100 miles southeast from Glendive. Its layers resemble a thick carpet that's been shoved to the southwest. This sharp-crested fold is isolated in the otherwise flat Cretaceous-Tertiary sedimentary layers in the southwestern part of the Williston Basin. The Late Cretaceous Fox Hills Sandstone appears a couple of miles deeper into the fold, followed by beds of the Pierre Shale that continue across the fold crest into Baker itself.

Detailed study has shown that the same compressional forces that crumpled the Rockies to the west probably formed the Cedar Creek anticline. The fold formed during the Laramide orogeny, when a high-angle compressional fault (reverse fault) raised a block of Archean basement, and the sedimentary cover ended up draped over the block. The anticline may have started forming during Cretaceous time, because the Hell Creek Formation lies on progressively older parts of the underlying Fox Hills Sandstone toward the crest of the anticline, showing the fold was rising and eroding at that time. Similar evidence shows that it grew during the deposition of the Fort Union Formation in Paleocene time. Folding and uplift of the structure likely ended shortly after the Fort Union Formation had been deposited.

Oil and gas, slowly formed deep in Paleozoic sediments below, rose into the arch and collected under impervious layers. Gas was discovered accidently by a homesteader drilling a water well in 1913, and then oil was discovered in 1951. Major oil production has come from the Mississippian Mission Canyon Limestone and from the Ordovician Red River Formation, much of it from depths of about 7,000 feet. The cable that you see dropping into the ground below the head of a pump jack goes down to that depth. The Cedar Creek field natural gas production is mostly from the Cretaceous Eagle Sandstone. Natural gas is also actively pumped into the Cedar Creek anticline, making it the largest underground gas storage site in the United States.

Continuing southwest down the steep west flank of the fold from the middle of Baker, MT 7 drops through about a half mile of the Fox Hills and Hell Creek Formations to again reach the Fort Union Formation. Dips on that steeper flank are about 25 degrees just south of Baker. Farther south the landscape again quickly reverts to gently rolling hills with scattered low roadcuts in soft beige Fort Union Formation shales. South for the next 6 or 7 miles, the road stays on the Fort Union Formation but dips into the Hell Creek Formation in three small streambeds.

Cretaceous sedimentary rocks on the west flank of the Cedar Creek anticline just north of Baker.

About 24 miles south of Baker, between mileposts 11 and 12, the landscape suddenly changes, with the appearance of big, blocky vertical-sided sandstone formations protruding from the rolling grassy hills; full of big holes, they may look like nothing you've seen before. Head into Medicine Rocks State Park for a close look at these bizarre outcrops of thick, 61-million-year-old sandstone of the Ekalaka Member of the Fort Union Formation.

MEDICINE ROCKS STATE PARK

Medicine Rocks is a fairyland of broken towers of the Ekalaka Member of the Fort Union Formation. Pale-beige outcrops, 30 to 80 feet high, cluster over a few acres on a low hill. Each outcrop has steep sides and a surface etched into weird pockmarked hollows that leave the rock looking like Swiss cheese. Cross bedding, a faint layering at gentle angles to the overall nearly horizontal layering of thicker beds, is visible in the vertical faces of the monuments. The cross beds here look like those that form in sandy rivers. Around 61 million years ago, a freshwater river flowed to the southeast into the Western Interior Seaway, which was located in northwestern South Dakota at the time. The river carried fine-grained quartz sand in channels that migrated back and forth, forming trough-shaped cross-beds as the grains bounced down the river.

The river was influenced by tidal flow in the seaway to the east, and vertical burrows from animals that lived in a tidally dominated estuary occur in harder sandstones that cap the river deposits, showing that the sea expanded to inundate the river. This environment was rich in plants and animals, as shown by fossils of leaves, wood, snakes, amphibians, crocodiles, mammals, primate-like mammals, and distinctive scales from alligator gar, a fish that could breathe both air and water and could live in both fresh and salty water. They are still around today, considered a "living fossil" because they remain almost unchanged.

Because there are no small stream channels in the area today, the Medicine Rocks towers appear to be remnants left by erosional processes not related to running water. A good clue to one process of erosion is the tendency of the towers to stand in the middle of closed depressions, like miniature castles within their moats. It would be

impossible for any process of water erosion that operates at the ground surface to transport material away from the rocks without first filling those depressions to create a downhill path. However, it's easy to imagine the strong High Plains winds howling around these rocks, whipping sand away from their bases to create the depressions.

Broad, flat surfaces on the sides of some rock structures may represent former joints or fractures in the sandstone. Fractures widen to form fins of rock, which develop holes that widen into windows and eventually arches. As surrounding rock erodes away, fractured rock can evolve into hoodoos, pillars of rock protected by a resistant caprock. Large blocks of freshly fallen rock litter the ground and show that this process of joint failure has played a very important role in the formation and separation of the Medicine Rocks structures over time.

The holes in the towers themselves are caused by alkali-rich groundwater seeping up into the porous sandstone and drying out, allowing the dissolved salts that we see in some of the cavities to crystallize and pop out individual grains to form tiny cavities. Rainwater seeping in and freezing on cold winter nights would help enlarge the cavities, as would wind blowing out loose sand grains.

It's easy to detach sand grains from the rock simply by rubbing it with your hand, and there is clear evidence that sand grains are coming off the rock surface naturally at a fast rate. Many visitors to the area deface the rocks by carving their names and the date, but doing so is prohibited. The rock sheds its surface so rapidly that those inscriptions become unreadable within a few years. There are no crusty growths of lichens on the Medicine Rocks, apparently because their surfaces disintegrate too rapidly for such slow-growing plants to survive.

Medicine Rocks was considered a sacred place by the Plains Indians who used it for religious gatherings and camped here as far back as 11,000 years ago. They left pictographs and petroglyphs illustrating their experiences and beliefs. Please be respectful of this special place and leave it as you found it.

A relatively large window that formed as erosion widened a hole in a fin of sandstone over time. The inset photo is the armor plating (scale) from an alligator gar, a torpedo-shaped fish that lived in both fresh and salty water, and still thrives today in rivers, lakes, and estuaries along the Gulf of Mexico. The specimen is about 1.25 inches long.

The sandstone used for building stone in Ekalaka was quarried from rock east of town. Chalk Buttes, an area of massive limestone towers, is 15 miles southwest of Ekalaka on Chalk Buttes Road in Custer National Forest. Thick limestone layers that form cliffs have eroded into these flat-topped towers. The Carter County Museum in Ekalaka exhibits dinosaur and other fossils and minerals.

MT 323 TO ALZADA

MT 323 is a little-traveled, lonely road with mostly flat terrain, few houses, and no services. Do not assume that the gas station in Alzada will be open for business. The first 7 miles follows little Russell Creek through poorly exposed siltstone of the Fort Union Formation. The buttes above the road to the northeast, cloaked in pine forests with patches of grassland, are made of the same sandstone found in Medicine Rocks State Park. The high point in the road, about 7 miles south of Ekalaka, passes through gray Arikaree Formation sandstone, deposited during Miocene time. It lies above the Fort Union Formation.

Eighteen miles south of Ekalaka, MT 323 crosses Boxelder Creek, which it parallels for another 15 miles. A line of trees about 1 mile west of the road marks its course. Capitol Rock, a National Natural Landmark, is a white tower of volcanic-rich sandstone near the Montana–South Dakota border to the east. Reached on side roads from the Boxelder Creek crossing, it resembles the US Capitol.

Approximately 45 miles south of Ekalaka, the castle-like buttes 2 miles east of the highway are composed of erosion-resistant Arikaree Formation sandstone overlying Late Cretaceous Pierre Shale. The broad, lumpy slopes above the flat foreground are slides in the landslide-prone shale.

There are almost no outcrops for the remaining 35 or so miles to Alzada, with the exception of horizontal layers of gray Cretaceous sediments where MT 323 crosses the Little Missouri River. The tiny community of Alzada, population twenty-eight in 2016 (probably including nearby ranches), in this remote, southeastern corner of Montana, is just hanging on.

MT 200
Bowmans Corners (US 287 Junction)—Great Falls
50 miles

Almost the entire stretch of MT 200 between Bowmans Corners and Great Falls crosses rather subdued terrain on soft Cretaceous sediments deposited in and along the Western Interior Seaway. Most of the layers are horizontal, but one small thrust fault breaks the layers about 6 miles northeast of Bowmans Corners.

The dark hills visible along the southeastern horizon from high vantage points between Bowmans Corners and Simms are the eroded remnants of the old Adel Mountains volcano. This volcano fed the laccoliths—the big steep-sided buttes—that punctuate the skyline just south of the road. Those buttes consist almost entirely of dark volcanic rocks, mostly shonkinite with blocky crystals of shiny black augite. The laccoliths were intruded about 75 million years ago. Although you might assume the magma came up from below each butte, mapping shows that the magma squirted out laterally from the volcanic neck under the main Adel Mountains volcano.

The magma spread outward radially as much as 22 miles along thin, vertical dikes that remain as segments of resistant rock "walls" cutting through the flat sediments of the plains. At their outer ends, the dikes curve over to one side into flat sills, or laccoliths, blisters of magma that bulged the overlying sediments; each was fed by a single dike. Eagle Sandstone once lay directly above the intruding laccolith magma and prevented it from rising to the surface. The laccoliths probably formed late in

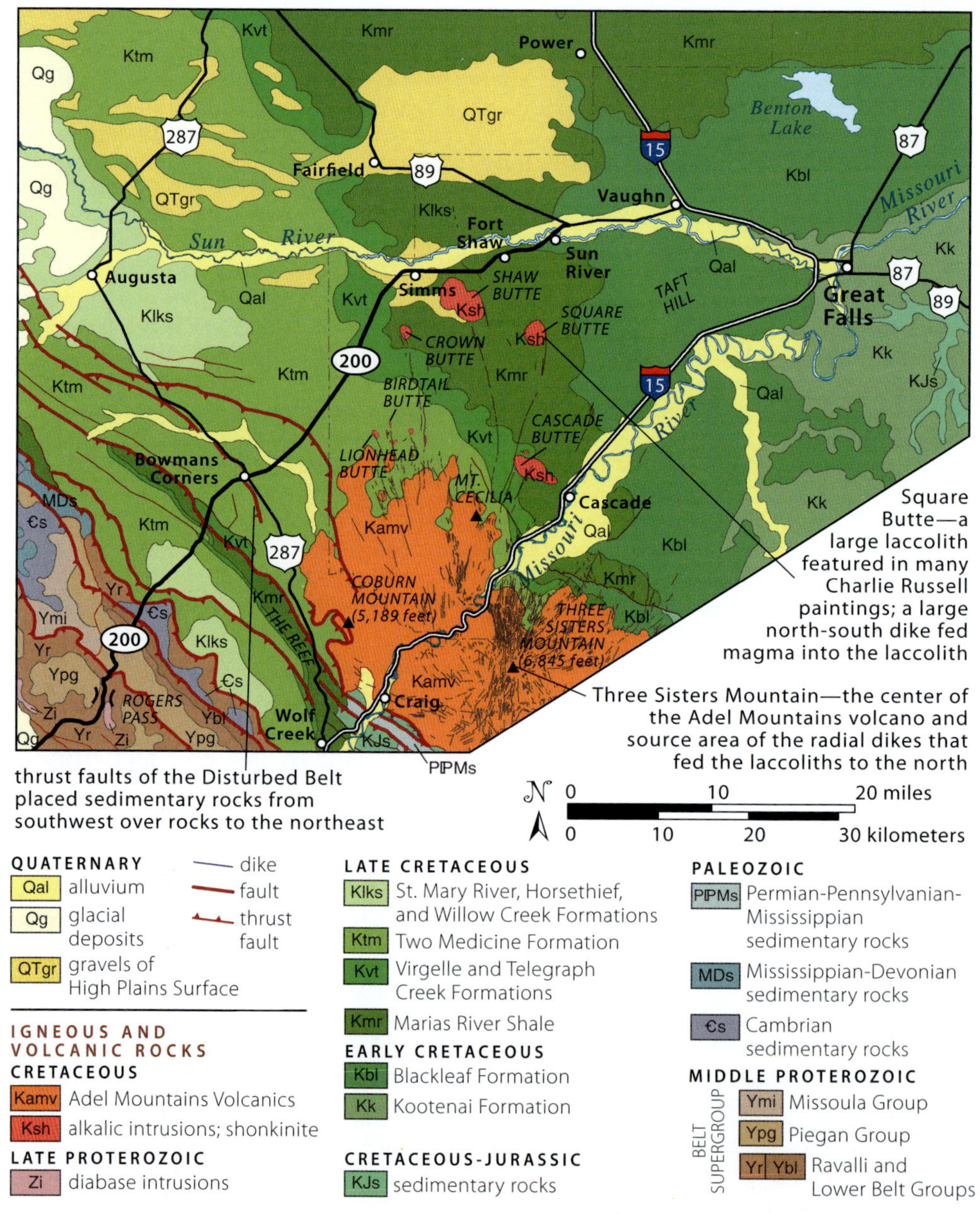

Geology along MT 200 between Bowmans Corners and Great Falls.

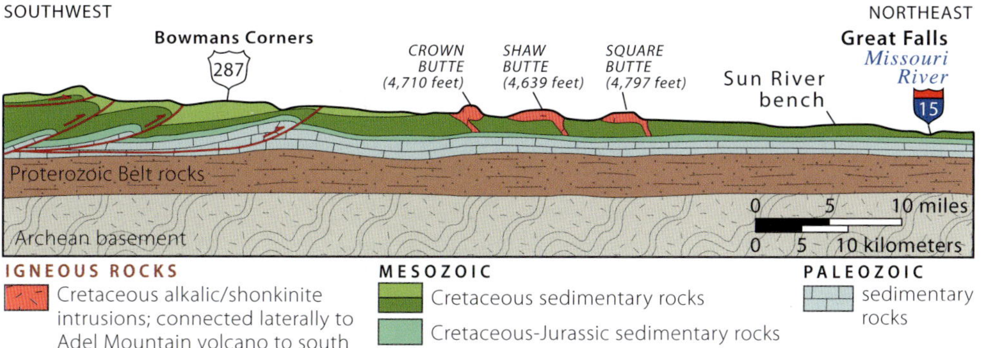

Southwest-northeast cross section near the line of MT 200 from west of Bowmans Corners (junction of MT 200 and US 287) to Great Falls, showing the complexly thrust-faulted front of the Rocky Mountains fading into the undisturbed rocks under the High Plains. The buttes northeast of Bowmans Corners are laccoliths fed laterally by magma from the Adel Mountains volcano to the south.

the growth of the Adel Mountains volcano. Modeling experiments have shown that the load of the volcano above its magma chamber dictated both the formation of the radial dikes and the laccoliths at their ends. From the air the dikes look like the broken spokes of an enormous wagon wheel. Erosion has since stripped off the overlying soft Cretaceous rocks, leaving most of the hard shonkinite of the buttes standing high and exposed at the surface.

Crown Butte, an example of one of these laccoliths, lies east of MT 200 a few miles directly south of Simms. Standing 1,000 feet above the surrounding prairie, the butte looks like a gigantic sawn-off tree stump. The black upper part is the laccolith; the pale grass-covered slopes below are the Cretaceous sedimentary rocks under it. The

The Crown Butte laccolith southwest of Simms. The inset shows columnar jointing in the shonkinite at the summit. —Inset photo courtesy of Rod Benson, Bigskywalker.com

gaping gulley in these rocks exposes faint layering. From the top of the laccolith to the present surrounding countryside, more than 1,100 feet of surrounding sedimentary rock was eroded from this landscape in the 75 million years since the laccolith intruded. A trail from the south side of the butte leads to the "tall-grass prairie" at the top of the butte (about a two-hour hike). The Nature Conservancy owns the butte.

Shaw Butte, an oversize version of Crown Butte, is very close to the highway about 3 miles west of Simms. Square Butte, the best known of the three laccoliths because cowboy artist Charlie Russell made it famous in his paintings, is about 5 miles southeast of Fort Shaw. Its steep sides and internal layering are similar to that of the other buttes. Each butte also has a single feeder dike along one side of it.

The highest Glacial Lake Great Falls shoreline reached 4 miles southwest of Simms along MT 200. Although we're near the lake's former thin edge, patches of lake silts are preserved near stream level near the town of Fort Shaw. See the road guides for I-15 and US 87 for discussions of the Great Falls area.

MT 200
Fairview—Sidney—Jordan
153 miles

Fairview is virtually at the North Dakota border, where the Yellowstone River leaves Montana. The river meanders through its half-mile-wide floodplain, which in turn meanders through a much broader 4-mile-wide ice-age valley. Fairview is built on ice-age river gravels. The Paleocene-age Tongue River Member of the Fort Union Formation rises a couple of hundred feet at the west edge of town and is exposed near the road between Fairview and Sidney.

In Sidney, a farming and ranching community, farmers grow sugar beets in alluvial gravels of the broad Yellowstone River valley, as well as wheat and other crops on the gravels that cover the higher terrace. Settlers showed up in the 1870s, and Sidney's post office was established in 1888. A Montana-Dakota Utilities electric power plant on the Yellowstone River, at the edge of Sidney, has provided electric power to eastern Montana and North and South Dakota since 1958, first using low-grade coal and more recently natural gas.

Sidney lies in the western part of the broad, gently sagging bowl known as the Williston Basin, which contains about 16,000 feet of sedimentary rocks and spans an area of 300 by 470 miles. It contains rocks from every geologic period down to the Archean basement. The sedimentary layers slope gently toward the center of the basin in northwestern North Dakota. The Late Devonian–Early Mississippian Bakken Formation, 8,500 to 10,500 feet beneath the plains, is the prolific, major source of oil. Production from the Bakken was very limited until 2000, when a combination of horizontal drilling and hydraulic fracking permitted drillers to force the oil through tiny fractures in the rock to pumping wells. The producing formation is dolomite, a magnesium-rich variant of limestone. The Bakken is only about 45 feet thick, so precise horizontal drilling is required. The actual production zone is often only 8 to14 feet thick, making it a small target.

The Elm Coulee oil field in Richland County, west-northwest of Sidney, is well within the Williston Basin and had the largest total production of any onshore field in the lower forty-eight states. You'll see numerous oil pump jacks between Sidney

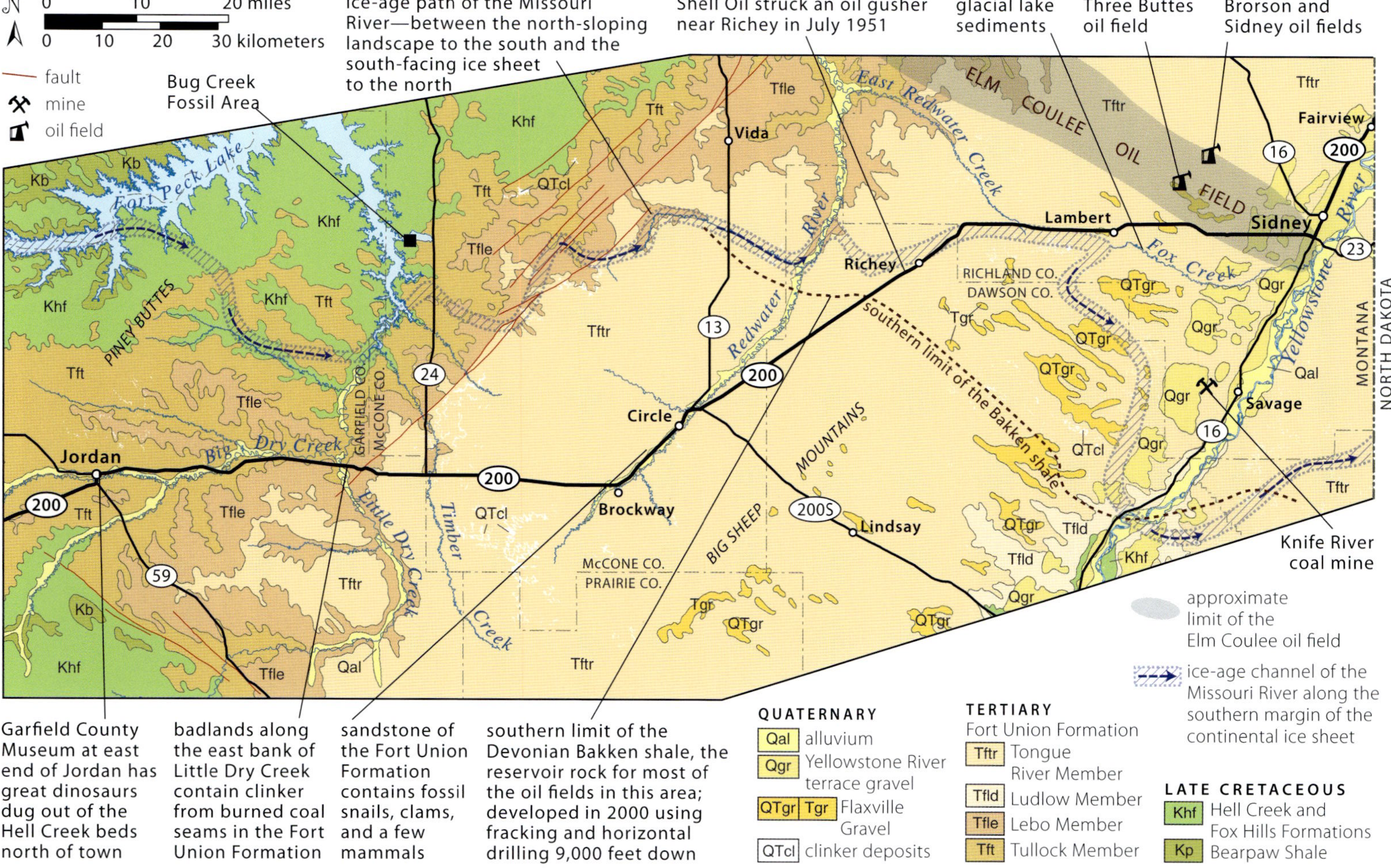

Geology along MT 200 between Fairview and Jordan.

and Lambert, some on the broad, flat, old Yellowstone River terrace, though some aren't pumping.

From the Yellowstone River at Sidney, MT 200 crosses the soft, beige Tongue River Member, which has been eroded into gullied hills. The nearly flat upland is covered by glacial till. The last advance of the Pleistocene continental ice sheet (Late Wisconsin glaciation) covered the area between Sidney and Lambert. A tongue of ice extended southwest up the Yellowstone River valley to as far as Glendive.

With the ice blocking drainages, the Yellowstone River backed up south of Glendive to around Fallon, forming Glacial Lake Glendive. About 15 miles west of Sidney, MT 200 crosses the North Fork of Fox Creek. Glacial lake sediments fill this valley for 3 miles upstream to the northwest. A small lake ponded here against the continental ice sheet to its north. During the earlier Illinoian glaciation, between 190,000 and 130,000 years ago, the continental ice sheet pushed the Missouri River out of its channel. It ran along the southern edge of the ice, carving a large valley there. MT 200 follows that river valley, now dry, between Lambert and Richey. Oil is also produced from the Mississippian Charles Formation limestone and Devonian beds in the Richey area and to the west.

Soft mudstones and sandstones of the Fort Union Formation lie beneath the flat plains here. Very little of this rock resists erosion well enough to form bedrock outcrops, but about 6 miles west of Lambert, just south of the highway, red clinker caps low, ragged hills composed of the Tongue River Member of the Fort Union. A few miles east of Richey, old valleys that were cut in the Fort Union Formation contain gravel left by northeast-flowing streams that occupied them as the Late Wisconsin-age glaciers melted from the region; the valleys are dry now. Southeast of Richey, big cuts in beige Tongue River sandstone are exposed on most hillcrests, with some exposed in gullied arroyos.

Isolated monuments of white sandstone of the Fort Union Formation, exceptions to the lack of exposed rock, stand in a depression at the crest of a ridge near Richey. The wind is snatching the sandstone away grain by grain to excavate the depression and carve the monuments of rock.

A prominent sandstone tower 3 miles east of Richey is Tongue River sandstone.

Well-layered beds of sandstone and shale of the Tongue River Member just east of Richey, near milepost 35.2. The dark layer is an impure coal bed.

Black ironstone (chert) concretions in Fort Union rocks were precipitated from iron-rich groundwater, often around a small nucleus, such as a shell. Concretions are found in many Late Cretaceous and early Tertiary formations in eastern Montana.

Circle is named for the Mabry Cattle Company, which began here in 1884 with a circle-shaped cattle brand. In 1951 Shell Oil struck oil, and the struggling community boomed. Trucks carried oil to railcars, and by 1955 the oil was being shipped to refineries to the south. The Northern Pacific Railway finally pulled out in the 1990s.

At Circle, near milepost 80, MT 200 crosses the Redwater River, which it follows for about 11 miles before turning west at Brockway, where the terrain becomes hilly. The highway continues in the yellowish to light-brown shale and sandstone of the Tongue River Member for about 25 miles.

The rest area at the junction of MT 200 and MT 24 offers more than just facilities. Boulders display some interesting features of the Tongue River Member. Look for collapse breccia, angular fragments of rock, now cemented together; the fragments collapsed into the space left by burned-out coal seams in the Tongue River rocks.

MT 24 heads north to Fort Peck and Glasgow and provides access to the Bug Creek Fossil Area, a National Natural Landmark at the edge of Fort Peck Lake. Dinosaur fossils have been found here, but the site is best known for one of the most well-studied exposures of the Cretaceous-Tertiary extinction boundary anywhere. The boundary layer contains iridium-rich clay and shocked quartz, both strong evidence that a large asteroid struck the Earth about 66 million years ago. The extinction boundary occurs at the contact between the Cretaceous Hell Creek Formation and the overlying Tullock Member of the Paleocene Fort Union Formation. Both formations were deposited in streams, floodplains, and sometimes swamps and estuaries near the shoreline of the Western Interior Seaway between 67 and 64 million years ago.

The site has contributed greatly to our understanding of the environmental and biological changes leading up to the demise of dinosaurs and the initial evolution and spread of mammals after the extinction event. Prior to the extinction, dinosaurs, including tyrannosaurs and triceratops, lived among crocodiles, turtles, lizards, and small mammals in the Bug Creek area. Immediately after the extinction event, the same environments were occupied by many of the same animals, but the dinosaurs were gone. In their place were mammals—more than twenty-four species of them. Garfield and McCone Counties are the source of half of the tyrannosaur skeletons ever found. Nice displays of the dinosaurs are in the Fort Peck Interpretive Center at Fort Peck Dam and in the Garfield County Museum in Jordan, along the highway at the east edge of town.

Where MT 200 begins to descend into the valley of Little Dry Creek, it crosses into Lebo Member shale of the Fort Union Formation, deposited in a broad lake environment. Badlands in the shale become especially spectacular where the road crosses the east bank of Little Dry Creek and passes through a miniature desert. Splashing raindrops and running surface water have carved the fantastic shapes of this desert as they selectively strip off softer material, leaving the more resistant parts of the rock

Thin beige-and-gray layers of the Tullock Member near the valley of Little Dry Creek and Road 462 (about 27 miles east of Jordan and 26 miles west of Brockway).

standing in relief. The hard, colorful exposures of the Fort Union Formation were baked by burning coal seams. Little Dry Creek flows north from here into Fort Peck Lake, which is only about 14 miles from MT 200.

West of the Little Dry Creek crossing, MT 200 follows Big Dry Creek to Jordan, passing through a rough landscape with numerous isolated buttes and patches of badlands topography along the valley walls. About 12 miles east of Jordan, the ragged small buttes north of the road are capped by hard clinker left by burned-out coal seams.

Scattered roadcuts and gullies expose soft beige to red sediments of the Tullock Member 14 miles west of the junction of MT 200 and MT 24 (5 miles west of Little Dry Creek). The Lebo Member shale is somewhat upslope of MT 200 both north and south of the road. This is Black Angus country, with herds everywhere.

MT 200
Jordan—US 87 at Grass Range
99 miles

Jordan, on Big Dry Creek, is in the Tullock Member of the Paleocene Fort Union Formation, which was deposited after the demise of the dinosaurs. The first *Tyrannosaurus rex* skeleton ever found was in the Hell Creek Formation north of Jordan, in 1902. The Garfield County Museum in Jordan exhibits a T. rex skull and a full-size triceratops.

West of Jordan, MT 200 crosses a vast expanse of the Tullock Member for about 35 miles, to a few miles west of Sand Springs. Exceptions are where the Lebo Member shale caps low ridges, such as 7 or 8 miles west of Jordan and especially 1 to 3 miles east of Sand Springs. The soft beige shale is exposed in big roadcuts at the ridge crests. These formations are everywhere horizontal, and the Lebo Member, being younger, rests on top of the Tullock Member.

The Tullock Member about 2 miles west of Jordan. Water runs off the water-resistant clay and removes the weakly cemented sediment it crosses, forming rivulets. The log-like feature at the top is a not a fossil but a solid concretion that formed as hard water precipitated a carbonate cement along porous bedding planes.

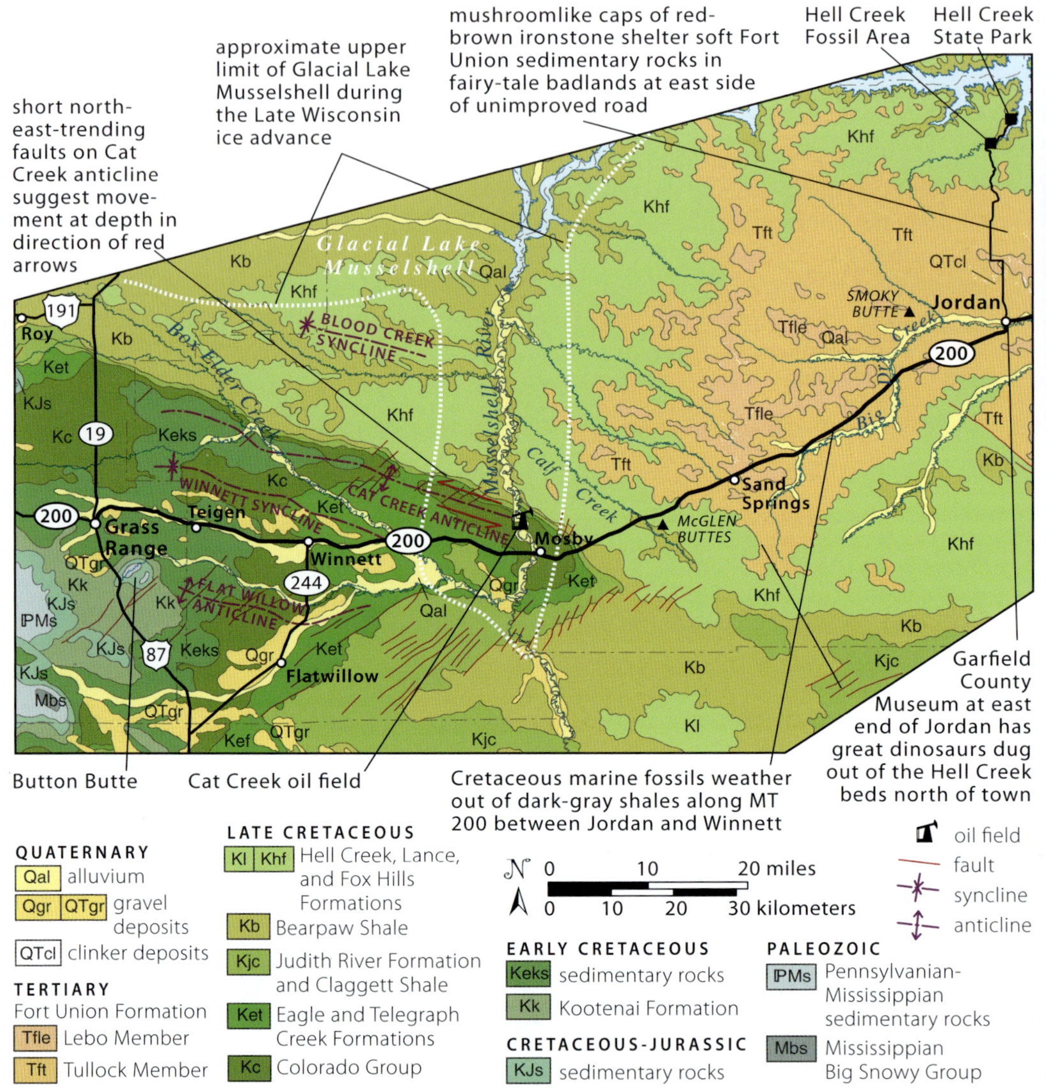

Geology along MT 200 between Jordan and US 87.

Smoky Butte, northwest of the highway and 8 miles west of Jordan, is a diatreme, a small, potassium-rich igneous intrusion that punched its way up into the crust from 60 to 120 miles down in the Earth's mantle. As it got close to the surface, the intrusion apparently spread out along a fracture, forming a dike oriented southwest-northeast. The igneous rock of the diatreme and the baked Fort Union Formation around it resist erosion well enough to enable Smoky Butte to stand slightly above the surrounding plains. Some diatremes contain diamonds, though none have been found here. Most of Montana's diatremes are in the Central Montana Alkalic Province; this is the easternmost known outpost of such activity. Based on the age of those intrusions, Smoky Butte was likely emplaced between 51 and 46 million years ago.

HELL CREEK FOSSIL AREA

If the weather is dry, the short side trip north from Jordan on MT 543 to a fossil site and the namesake area of the dino-bearing Hell Creek Formation is worth the time. Areas off the gravel road are not recommended for passenger cars; if even a little rain is expected, dirt roads can quickly become impassible in this region, even for four-wheel-drive vehicles.

About 4 or 5 miles north of town, the ragged tops of small buttes are held up by clinker, heated by burned-out coal beds in the Tullock Member of the Fort Union Formation. About 11 miles north of town, right next to the east side of the road, are wild-looking erosional forms in the Tullock. Harder cemented-sandstone caprock concretions protect the pale "stems" of these mushroomlike formations from raindrop erosion. The caps have thin cross beds, inclined internal layering within a larger, horizontally layered bed. These were once sandy stream channels; downstream flow was in the direction of the sloping surface of the cross beds. Black, glossy rubble layers are debris from ironstone concretions, formed when iron oxide minerals precipitated around a nucleus, such as a shell or a bit of organic matter, in the sediment. The iron-cemented mass is harder than the surrounding calcite-cemented rock and thus weathers out of its host.

Mushroomlike concretions in the Tullock Member of the Fort Union Formation 11 miles north of Jordan on MT 543.

Several miles farther north, the road drops through a badlands landscape in the latest Cretaceous Hell Creek Formation, the burial environment for the last of Earth's dinosaurs. The road leads another 10 miles to Hell Creek State Park, at Fort Peck Lake. The very lumpy lower slopes around Fort Peck Lake are landslide deposits in the Hell Creek beds; the cliffs above are the landslide slip surfaces, where sliding is promoted by slippery smectite (bentonite) in the sediments.

About 4 miles south of the park is the Hell Creek Fossil Area. Digging or collecting fossils is not permitted in any state park, but closely supervised digging is permitted in conjunction with a few research groups outside of the parks. If you're interested in observing paleontologists in action or helping with a dig, inquire at Makoshika State Park in Glendive or at the Two Medicine Dinosaur Center in Bynum.

Smoky Butte provided Montana with two of its most interesting geologic discoveries. A French mineralogist examining samples of the fine-grained igneous rock from Smoky Butte found armalcolite, a mineral otherwise known only in rocks collected on the Moon. The mineral is named after the three astronauts who made the first landing on the moon—Armstrong, Aldrin, and Collins.

The other discovery is in the sedimentary rocks exposed at Smoky Butte; the thin layer of dark sediment that marks the Cretaceous-Tertiary boundary here contains some of the strongest evidence anywhere to support the theory that dinosaurs did indeed die from a major asteroid impact. In addition to iridium, an element that is abundant in some kinds of asteroids but extremely rare on Earth's surface, the fatal layer at Smoky Butte also contains shocked quartz, mineral grains full of microscopic fractures known to occur only in rocks shattered in extremely violent explosions—far more violent than any volcanic explosion. It also contains an extremely rare mineral called stishovite, a distinctive variety of quartz that likewise forms only in explosion-shocked rocks.

Most of the countryside for miles around Sand Springs, near milepost 179, is nearly flat, open range with grass, some sagebrush, and huge fields of grain. White alkali salts, or saline seeps, left by the evaporation of alkali-rich groundwater, mark gullies. Four or five miles west of Sand Springs, MT 200 drops into the Cretaceous Hell Creek Formation, the older sedimentary rocks below the Fort Union Formation. In so doing it crosses the Cretaceous-Tertiary boundary into the rocks marking the sedimentary environment of the dinosaurs. The road remains in this formation for about 3 miles. Then, in the next mile or so, the road drops farther—into the Cretaceous Fox Hills Sandstone, barrier island sands deposited during a retreat of the Western Interior Seaway. This formation forms the McGlen Buttes at the south edge of the highway. The road continues down into the marine Bearpaw Shale for a mile or so in the bottom of intermittent Calf Creek. About 3 miles west of Calf Creek, the highway climbs through the Fox Hills Sandstone to the Hell Creek Formation across the top of a ridge, then descends through the same sequence on the southwest side.

MT 200 crosses the axis of the Cat Creek anticline 2.5 miles east of Mosby. This steplike fold trends east-southeast and is the eastward extension of a very broad structural arch that likely formed during Eocene time, about 50 million years ago as fault blocks in the basement were raised during the Laramide orogeny. The overlying sedimentary rocks were draped over the blocks. The structure forms the Big Snowy, Judith, and Little Snowy Mountains farther west. Cretaceous rocks, mostly shales, are exposed in the core of the fold, along with some sandstones.

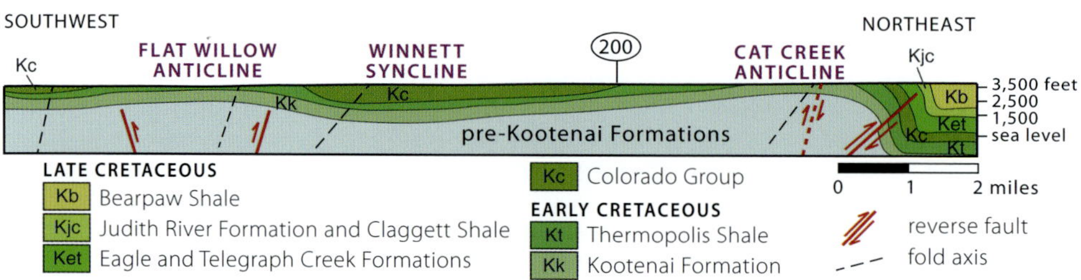

Southwest-northeast cross section across the Cat Creek anticline.
—Porter and others, 1999

Northeast-dipping Late Cretaceous shale and calcareous shale on the north flank of the Cat Creek anticline about 3 miles east of Mosby. The inset is a 4-inch-thick black, almost flintlike chert layer (pen on right for scale). The silica for the chert likely came from groundwater migrating through the sediment after it was deposited.

These tilted layers of sedimentary rock attracted the attention of early oil geologists, who quickly located the crests of the folds. A wildcat well discovered oil in the crest of the Cat Creek anticline in 1920, at a depth of 998 feet. That was the second oil discovery in the region but Montana's first major discovery. The oil was rich enough to use directly in Model-T Fords and farm tractors. Numerous other wells quickly followed in a frenzied drilling boom; production peaked in 1921. Although daily production is modest by some standards, the field has produced an immense amount of oil during its long history. It undoubtedly includes the two most profitable square miles in this part of Montana.

East of Mosby, over a distance of 7 or 8 miles, MT 200 descends through pine forests with some sagebrush and grass to the Musselshell River. In doing so, it goes down the Cretaceous stratigraphic section, through older and older rocks in the center of the anticline, from the Hell Creek beds through the Fox Hills Sandstone, Bearpaw Shale, Judith River Formation sandstone, brownish-gray Claggett Shale, white Eagle Sandstone, Telegraph Creek Formation shale and sandstone, dark-brownish-gray shale of the Niobrara Formation, dark-gray to brownish Carlile Shale, and a little Greenhorn Formation shale at the community of Mosby on the Musselshell River. Soft, gullied, black Carlile Shale is the marine deposit exposed in a big roadcut at Mosby. Plesiosaurs, long-necked swimming reptiles, have been recovered from this formation in Kansas.

The marine Carlile Shale along the Musselshell River about 1.5 miles west of Mosby and 1 mile south of MT 200. The black color is due to small bits of organic material buried in the marine sediment.

The Missouri River and the north-flowing Musselshell River were blocked by the continental ice sheet that covered most of Canada and the northern part of Montana during Pleistocene time. At the ice's maximum during the Late Wisconsin glaciation, the Missouri River backed up to form Glacial Lake Jordan and its downstream extension, Glacial Lake Musselshell, to an elevation of 2,600 feet. It was centered on today's Fort Peck Lake, which now has an elevation of 2,250 feet. The extra 350 feet of water during the Late Wisconsin glaciation spread out over a much broader area, including up the Musselshell River valley to just south of MT 200. Mosby would have been under about 85 feet of water.

West of the Musselshell River, MT 200 goes back up through most of the same Cretaceous sequence, but it tops out on the Eagle Sandstone and remains in it to about 2 miles east of Winnett. Between Winnett and Grass Range, Late Cretaceous shales and some sandstones, all laid down in and along the Western Interior Seaway, were gently warped into west-trending folds. An anticline lies south of the road, and a syncline to the north. Dips on the beds are less than about 10 degrees both north and south of the road.

GLOSSARY

alkalic. Igneous rocks that contain an uncommonly large amount of sodium or potassium, or both.

alluvial fan. A fan-shaped deposit of sediment produced where a stream flows from a mountain out onto a valley floor. Mudflows can also deposit them.

ammonite. A coiled-shell marine animal related to the octopus. Lived from Devonian to Cretaceous time.

amphibolite. A metamorphic rock containing a high percentage of the mineral amphibole (e.g., hornblende); it is often a metamorphosed basalt.

andesite. A volcanic rock intermediate between rhyolite and basalt in composition, color, and eruptive behavior. Andesite is the common rock of large stratovolcanoes.

angular unconformity. An unconformity in which flat layers are deposited on tilted, eroded layers of older rock.

anticline. An arch folded in layered rock. The two sides of an anticline, called its limbs, dip outward from the crest.

Archean. The unit of time from about 4.6 to 2.5 billion (2,500 million) years ago.

arête. A knifelike ridge formed by erosion by glaciers on either side of the ridge.

asteroid. A large mass of rock or ice and dust (comet) in space orbiting the Sun.

asthenosphere. The mechanically weak (ductile or "plastic") part of Earth's upper mantle that lies below the more rigid lithosphere and above the deeper mantle.

augite. A common calcium-magnesium-iron silicate mineral variety of the pyroxene group; it is generally black or dark green and forms rather stumpy crystals. It typically makes up about half of basalt and gabbro.

badlands. A region of dry, soft, sedimentary rock severely eroded by wind and rain; characterized by ragged, deeply gullied landscapes, often marked by hoodoos and mushroom-shaped rocks.

basalt. A common volcanic igneous rock. Basalt is black and consists mostly of the minerals plagioclase and augite.

basement rocks. A general term referring to ancient, strongly metamorphosed rocks that underlie at deeper levels most rocks exposed at the surface. In Montana they are older than the Belt Supergroup and commonly occur as schist, gneiss, and ancient granite.

Basin and Range Province. A broad, generally north-south area of broad basins and ranges west of the high Rockies and east of the Idaho batholith. It formed by stretching of the continental crust.

batholith. Any large mass of granite or similar rock, technically one that covers an area greater than about 40 square miles. A **stock** is smaller than 40 square miles.

bedding. The layered structure of sedimentary rocks formed during sediment deposition and caused by variations in sediment size and composition over time.

bedrock. The solid rock of the Earth's crust, sometimes underlying loose sediment or soil.

Belt rocks. A very thick section of hard sedimentary rocks deposited in the Belt Sea between 1.47 and 1.40 billion years ago and as much as 12 miles thick in parts of northwestern Montana. Formal name is Belt Supergroup.

bentonite. A clay mineral, also called smectite, that expands when it gets wet, leading to very slippery conditions. It generally forms by weathering of silica-rich volcanic ash.

Big Sky orogeny. A mountain-building event produced by the closure of an ocean basin followed by a continental collision between the Medicine Hat block and the Wyoming craton from about 1.9 to 1.7 billion years ago.

biotite. The black to dark brown potassium-magnesium-iron variety of mica common in granites and light-colored gneisses and schists.

brachiopod. A clam-like marine animal that appeared in Cambrian time and continues to exist today. They can be distinguished from clams by their shell symmetry, which is perpendicular to the hinge line.

braided river. A river with a floodplain composed of many interconnected, shifting channels separated by coarse sand or gravel bars.

breccia. A rock composed of fragments, often angular.

Bull Lake glaciation. The earlier of the two main ice advances in the Rocky Mountains; occurred between 160,000 and 130,000 years ago and reached its maximum extent around 136,000 years ago. Approximately coeval with the Illinoian glaciation on the Great Plains, which reached its maximum extent around 140,000 years ago.

butte. An isolated hill with steep sides and a flat top.

calcite. A mineral composed of calcium carbonate. It is a major component of limestone.

caldera. A large collapsed depression that sank into the magma chamber of a volcano. This is in contrast to a crater, which is produced by explosive eruption.

carbonate. Rocks such as limestone or dolostone that are formed with calcium carbonate or magnesium carbonate.

Cenozoic time. The geological era from 66 million years ago to the present day.

chert. A sedimentary rock composed of quartz crystals too small to see with the naked eye. It forms as a chemical deposit, found primarily as nodules precipitated from water percolating through limy sediments after deposition, but it can also be precipitated directly from water on the seafloor.

cirque. A very steep-walled basin eroded by a glacier near the top of a mountain. Several cirques may coalesce to form an arête or a horn.

climate change. Ongoing changes in world's climate resulting from increases in carbon dioxide and methane, for example, and mostly caused by human actions.

clinker. Rock baked hard above a burning coal seam. Most clinker is either red or yellow and has a bubbly texture.

coarse-grained. Said of sedimentary rocks that have clasts, or particles, that are relatively large, usually averaging 2 millimeters in diameter or larger. Also said of igneous rocks with relatively large crystals.

columnar joints. Columns typically five or six sided in cross section and formed in a lava flow during contraction while the lava crystallizes.

compressional fault. *See* reverse fault.

conglomerate. A sedimentary rock consisting of rounded pebbles surrounded by finer grains.

contact metamorphism. Temperature-dominated metamorphism caused by the nearby intrusion of hot magma.

continental plate. One of Earth's lithospheric plates consisting of continental crust, which is composed of silica- and alkali-rich rocks.

convection current. Currents in a liquid (for example water or magma) that are driven by rise of the hotter, lighter part of the liquid.

craton. The ancient core of a continent; usually composed of Archean to Proterozoic-age rocks.

crinoid. A shallow marine animal growing from the seafloor on a thin stem.

crystalline rock. Grainy igneous or metamorphic rock, such as granite, schist, or gneiss.

cross beds. Internal sedimentary layers more steeply oriented than and at an angle to the main layers. Produced by movement (bouncing) of grains carried in moving water or air.

crust. The Earth's outer layer; about 4 miles thick under the oceans and tens of miles thick under the continents.

cyanobacteria. Bacteria that obtain their energy through photosynthesis and produce oxygen as a result. They first appeared in Proterozoic time and built layered sedimentary structures called stromatolites.

debris flow. A large, downslope flow of water, mud, and rocks.

décollement. A low-angle thrust fault with a large displacement; it is usually the lowest fault in a stack of thrust faults.

detachment fault. A nearly horizontal or gently dipping fault associated with large-scale extensional tectonics and often with large displacement (tens of miles) that juxtapose lightly metamorphosed rocks above from higher-grade metamorphic rocks below.

diabase. An igneous rock that resembles basalt except that individual crystals of black augite and pale plagioclase feldspar are visible without using a microscope. Grain size intermediate between gabbro and basalt.

diatreme. A small igneous intrusion generally shaped like a dike or vertical cylinder, composed of peridotite or similar dark rock derived directly from the earth's mantle.

dike. A body of igneous rock that formed as magma filled an open fracture either across layers or in massive rock. Compare to **sill**.

diorite. A plutonic igneous rock with a silica composition intermediate between granite and gabbro.

dip. The slope of a geologic surface, such as a fault or bedding plane, measured down from the horizontal.

dip-slip fault. A fault on which the upper side of the fault moved downslope on the fault surface.

Disturbed Belt. The subdued but still deformed area immediately east of the Overthrust Belt; similar to the Overthrust Belt, but less deformed and less mountainous.

dolomite. A sedimentary carbonate rock composed of the calcium-magnesium carbonate mineral dolomite.

drumlin. A streamlined hill composed of till shaped by a moving glacier.

dunes. Fine sand grains accumulated in piles by bouncing in the wind.

erratic. A boulder left by a melting glacier; a rock type that differs from that of the underlying bedrock indicates it was eroded from elsewhere.

esker. A ridge of gravel deposited by a stream flowing beneath a glacier.

extension. Stretching or pulling apart of Earth's crust.

extensional basin. A large topographic depression created by the stretching or pulling apart of Earth's crust.

extensional fault. A dip-slip fault, generally with a steep angle, that results from the extension of the crust. Also called a normal fault.

faceted spurs. Sharp-faced mountainside ridges truncated by a normal fault so they show triangular shapes. Typically indicates recent or active faulting

Farallon plate. A remnant tectonic crustal plate spreading eastward from the East Pacific Rise (just off the Washington coast), subducting under western North America, and causing the Laramide and Sevier orogenies from Jurassic to early Tertiary time.

fault. A fracture in Earth's crust along which the opposite sides slipped past each other.

fault scarp. The steep slope produced by slip on a fault surface.

fault zone. A relatively narrow region in which multiple faults, more or less parallel but often interconnected, have developed.

feldspar. A group of abundant aluminum silicate minerals. Potassium feldspar, called orthoclase, is typically white, beige, or pink. Sodium-calcium feldspar, called plagioclase, is typically white, greenish white, or gray. Most feldspars occur in rectangular crystals.

fine-grained. Said of sedimentary rocks that have clasts, or particles, that are relatively small, usually averaging less than 2 millimeters in diameter. Also said of igneous rocks with relatively small, hard-to-see crystals.

fold-and-thrust belt. A long zone of deformation of layered sedimentary rocks crumpled into folds and broken by thrust faults.

foliation. Sheetlike or flaky texture caused by oriented micas and other minerals in metamorphic rocks.

foreland basin. A big elongate depression formed by crustal loading at the margin of a newly formed mountain range.

formation. A formal stratigraphic unit of sedimentary rocks having similar characteristics, such as rock composition and similar age.

fossil. Any trace of a plant or animal preserved in rock.

fracking. Artificial fracturing of a petroleum-bearing layer to facilitate release and movement of the petroleum to the well. Sand is injected to keep the fractures open. Chemicals are injected to help petroleum movement.

gabbro. A coarse-grained igneous rock composed principally of plagioclase feldspar and augite; a coarse-grained equivalent of basalt.

gneiss. Coarse-grained metamorphic rocks having a banded or streaky appearance. Gneisses form through recrystallization at high temperature of various kinds of sedimentary or igneous rocks.

graben. A large elongate block of rock dropped to form a valley between two normal faults.

grade. The degree of metamorphism of a rock (generally due to temperature or sometimes pressure).

granite. A coarse-grained igneous rock composed of more than 35 percent alkali feldspar and more than 20 percent quartz. An igneous rock is said to be **granitic** if it has coarse-grained, light-colored crystals.

granodiorite. A rock similar to granite but with less alkali feldspar; between granite and diorite in composition.

groundwater. Water in the ground, mostly derived from rainfall and snowmelt.

group. A formal rock unit containing two or more formations.

gumbo. Wet sticky clay, often containing smectite (bentonite), a swelling, slippery clay.

half graben. A large dropped block in an extensional fault zone, marked by only one normal fault.

High Line. The colloquial name for the northern fringe of Montana, along US 2 and the Burlington Northern Railroad line.

High Plains. A subregion of the Great Plains, immediately east of the Overthrust Belt, characterized by relatively flat topography, high elevation, and a cold semiarid climate.

High Plains Surface. Remnants of a broad plain that formed during a dry period about 15 million years ago.

Holocene. The geological epoch after the end of the Pleistocene ice ages; the last 11,700 years.

hoodoos. Tall erosional pinnacles, most commonly in loose, poorly cemented sediments.

horn. A sharp mountain peak sculpted by cirque glaciers on three sides.

hornfels. A very fine-grained hard metamorphic rock that looks like broken porcelain when struck with a hammer.

hornblende. A common silicate mineral that crystallizes into glossy black needles in light-colored igneous and metamorphic rocks such as granite, schist, gneiss, and amphibolite.

hot spot. The site of a giant volcano erupted above a deep mantle plume, or a deep zone of extension that permits the rise of mantle magma. Yellowstone is an example.

hydraulic mining. Using water under pressure to wash down gravel banks and flush their contents through sluices to recover gold.

hydrothermal. Pertaining to hot water circulated by the heat of a nearby magma or by deep-Earth heat.

ice cap. The large expanse of ice over the north or south pole or over a high mountain range.

igneous. Rocks that crystallized from molten magma.

Illinoian glaciation. An advance of the continental ice sheet that occurred between 190,000 and 130,000 years ago, reaching its maximum extent around 140,000 years ago. Approximately coeval with the Bull Lake glaciation in the Rocky Mountains.

intrusion. Igneous rocks that have intruded older rocks.

ironstone. Iron-rich cementation of sandstone, or other porous rock, that leads to a dark red, brown, or black, very hard, impervious rock that resists erosion.

isotope. An atom of an element with a different number of neutrons. Different radioactive isotopes, for example carbon 14 (^{14}C), can be used to determine the age of young carbon-bearing materials less than about 40,000 years old.

jasper. An extremely fine-grained red or brown variety of chert (silica).

karst. Weathering of limestone to form cavities and sharply ragged surfaces.

kettle lake. A small pond in a depression formed when a patch of ice at the toe of a receding glacier is buried in outwash sediments before melting.

K-T boundary. The Cretaceous-Tertiary (or Cretaceous-Paleogene) boundary in sediments, formed 66 million years ago when the dinosaurs were exterminated.

laccolith. An igneous intrusion that formed as a blister of magma injected between layers of sedimentary rock near Earth's surface.

laminated. Said of a fine-grained sedimentary rock with thin layers.

Laramide deformation. Deformation, ranging from Late Cretaceous to earliest Tertiary time, that involved the rise of thick basement-cored crustal blocks on steep reverse faults. Perhaps initiated by crustal drag on a shallow subducting slab.

laterite. Rock deeply weathered under moist, tropical conditions that leached out most of its silica, calcium, sodium, and potassium, leaving mostly aluminum and iron oxides.

Late Wisconsin glaciation. The most recent ice age in North America. The ice advance between about 35,000 and 11,700 years ago, with a maximum extent around 23,000 years ago. Approximately coeval with the Pinedale glaciation in the Rocky Mountains.

Laurentide ice sheet. The continent-scale ice sheet of Pleistocene time that covered most of central and eastern Canada and extended south into the northern United States.

lava. Molten rock, or magma, that has erupted onto the Earth's surface.

left-lateral fault. A strike-slip fault on which the side across the fault moved to the left.

Lewis overthrust. A large thrust fault in Northwest Montana that carried a 300-mile-long, probably well over 2-mile-thick slab of Proterozoic-age Belt rock as much as 80 miles east over Cretaceous sediments on a virtually horizontal surface.

lignite. The softest, lowest-grade coal.

limestone. A carbonate sedimentary rock composed of the calcium carbonate mineral calcite.

lineation. An internal texture within certain rocks that gives them directionality in the same way that a handful of uncooked spaghetti noodles has directionality.

listric normal fault. A normal fault on which the fault surface gradually flattens deeper in the Earth.

lithosphere. The rigid outer rind of the Earth, approximately 60 miles thick, including the crust and uppermost mantle.

mafic. Said of an igneous rock characterized by a high percentage of dark minerals (gabbro or basalt).

magma. Molten rock beneath Earth's surface; magma becomes lava if it pours out on the surface.

magma chamber. A large mass of molten magma that may feed a volcano.

mantle. A deep internal layer below Earth's crust, making up most of the volume of the Earth.

marble. A grainy metamorphic rock formed by recrystallization of a carbonate sedimentary rock, such as limestone or dolomite.

marine. Related to the sea or ocean.

massive. Said of a rock layer (such as a thick sandstone bed) or rock type (such as granite) without evident layering.

meander. The sinuous back and forth sweep of a river or stream.

member. A formal stratigraphic unit representing some specially developed part of a formation. It is ranked lower than a formation and higher than a bed.

Mesozoic. The geological era between 252 and 66 million years ago; between the Paleozoic and Cenozoic Eras.

metamorphic core complex. High-grade regional metamorphic rocks exposed at the surface after the overlying low-grade rocks moved off them along a deep mylonite zone and a shallow detachment fault.

metamorphism. Recrystallization of a rock at high temperature or pressure to produce a new rock that commonly differs considerably in texture from the original, but not so much in composition. **Meta**, as a prefix, means a particular rock type is metamorphic.

mica. A group of silicate minerals that tend to split into thin flakes. The common varieties are black biotite and colorless muscovite. Both typically occur in generally light-colored igneous and metamorphic rocks such as granite, gneiss, or schist.

mineralized vein. A fracture filled with quartz or calcite that contains minerals of economic interest.

moraine. A deposit of glacial till left at the fringe of a melting glacier.

muscovite. The colorless potassium variety of mica common in granites and light-colored gneisses and schists.

mylonite. A generally fine-grained, intensely sheared rock in a fault zone or at the top of a metamorphic core complex.

normal fault. A fault along which one fault block slides down a sloping fault surface relative to the fault block on the other side of the fault. A normal fault forms as the result of extensional forces that are pulling an area apart.

North American plate. The major lithospheric tectonic plate that includes North America and the western half of the Atlantic Ocean.

offset. Used to describe the distance of separation along a fault.

oil shale. A fine-grained sedimentary rock that will yield oil if it is roasted. Oil shale is typically dark brown or black, has a waxy appearance and feel, and burns with a yellow, smoky flame.

ore. Any kind of rock that can be profitably mined.

orogeny. A major mountain-building episode caused by the collision between two tectonic plates.

orthoclase. A common feldspar mineral that contains potassium. Orthoclase is generally white, beige, or pink and typically occurs in light-colored igneous and metamorphic rocks.

outwash. Sediment deposited from glacial meltwater. Like other stream sediments, outwash typically consists of distinct layers of well-sorted sand and gravel.

Overthrust Belt. A tectonic belt of the Rocky Mountain front in which regional compression has pushed thick slabs of sedimentary rock over other rocks. It is more deformed and more mountainous than the Disturbed Belt, on its east toward the High Plains.

overthrust fault. A nearly horizontal (0 to 15 degrees) fracture surface along which the rocks above slid over those below. The Sevier deformation of the Rocky Mountain west produced overthrust faults.

Pacific plate. The major lithospheric tectonic plate that includes most of the Pacific Ocean. Its northeastern edge, along the coast of North America, is bounded by active subduction zones in some places and strike-slip faults in others.

Paleozoic. The geological era between 541 and 252 million years ago.

peat. An accumulation of partially decayed plant material. Buried peat can eventually turn into coal.

pediment. A broad, very gently dipping erosional surface, often at the mouth of one or more mountain canyons, and sometimes covered by a layer of gravel.

pegmatite. An extremely coarse-grained igneous rock, most commonly of granite composition (feldspars and quartz) and formed in the later stages of crystallization of granite magma, with a high-water content, which minimizes nucleation of individual mineral grains.

peridotite. An igneous rock composed essentially of pyroxene (augite or hypersthene) and olivine, a green silicate mineral.

piedmont glacier. A broad low-lying glacier formed near a mountain front by the coalescence of valley glaciers.

pillow lava. Basalt lava that chilled into rounded blobs or tongues a few feet across as it entered water or wet mud.

Pinedale glaciation. The most recent ice age in the Rocky Mountains. The ice advance between about 30,000 and 12,000 years ago, with a maximum extent between 19,000 and 15,000 years ago. Approximately coeval with the Late Wisconsin glaciation on the Great Plains.

placer. A concentration of heavy minerals near the base of a layer of sand or gravel in a stream or beach deposit.

plagioclase. An extremely common feldspar mineral that contains calcium and sodium in varying proportions. It is typically white, greenish white, or gray and occurs in a wide variety of igneous and metamorphic rocks.

plate. One of several large segments of the lithosphere, the rigid upper part of Earth's crust and upper mantle.

Pleistocene. The geological time period that includes the ice ages, between about 2.6 million and 11,700 years ago.

pluton. A solidified body of intrusive igneous rock, such as granite, that crystallizes beneath the surface. The adjective **plutonic** contrasts with volcanic, which refers to igneous rocks that form at the surface.

Powder River Basin. A very large continental sedimentary basin in Wyoming and southeastern Montana that produces oil from early Cretaceous sandstones and coal from the Paleocene Fort Union Formation.

Proterozoic. The eon 2,500 million to 541 million years ago; immediately before Cambrian time.

pumice. A light-colored, frothy, sponge-like volcanic rock, blown out of a silica-rich magma full of expanding gases.

pyroclastic flow. A dense mixture of volcanic ash, fragments, and gas that rushes, 100 or more miles per hour, down the flank of a volcano. These flows collapse from tall columns of erupted ash.

pyroxene. A black or dark green silicate mineral that typically makes up about half of basalt, gabbro, or shonkinite. It is rich in calcium and magnesium, with some iron and aluminum. Augite is a common variety.

quartz. A very common mineral composed of silicon dioxide. Quartz comes in many varieties, of which the most common is clear and colorless, like glass.

quartzite. A sandstone metamorphosed by heat or pressure so that the rock breaks across individual quartz sand grains rather than around them.

radiocarbon age. The age of a carbon-bearing material less than 40,000 years old, as determined by measuring its carbon isotopes.

radiometric age. The absolute age of a rock material as determined by measuring its radiometric isotopes, such as those of uranium-lead (U-Pb), potassium-argon (K-Ar), and carbon 14 (^{14}C).

reverse fault. A steeply dipping fault in which the overlying block appears to have moved up relative to the block underlying the fault. Reverse faults commonly form as a result of crustal compression.

rhyolite. A fine-grained or glassy, pale volcanic rock, either ash or lava, that contains quartz. Rhyolite has about the same composition as granite.

Richter magnitude scale. A logarithmic scale of earthquake-wave magnitude or shaking height. Thus, magnitude 7 is 10 times the size of magnitude 6 and magnitude 8 is 100 times the size of magnitude 6.

rift. A long, narrow, down-dropped zone where Earth's crust is separating. The process of a tectonic plate being pulled apart or splitting is called **rifting**.

right-lateral fault. A strike-slip fault on which the side across the fault moved to the right.

rimrocks. The sheer cliffs at the edge of a plateau or the top of a narrow canyon.

saline seep. White salty crusts caused by salty groundwater rising to the surface and evaporating.

sandstone. A deposit of sand hardened into rock, either by cementation or metamorphism.

scarp. A linear cliff produced by faulting or erosion. The term is an abbreviation of escarpment.

schist. A metamorphic rock that contains enough parallel mica flakes to make it split into sheets, or enough needle-shaped minerals to make it splintery. Schists form through recrystallization at high temperature of various kinds of sedimentary rocks.

sedimentary rocks. Deposited material such as sand, clay, mud, gravel, or skeletal fragments hardened into rock.

Sevier deformation. Thin-skinned deformation of sedimentary rocks in the Northern Rockies, primarily during Jurassic, Cretaceous, and earliest Tertiary time, involving compression to form thrust faults and folds, typically located above the basement.

shonkinite. A dark igneous rock composed mostly of augite and potassium feldspar, and essentially similar in composition to basalt but having a much larger potassium content.

silica. Silicon dioxide, the compound that makes up quartz in all its varieties, including chert.

sill. A layer of igneous rock imjected between layers of sedimentary rocks.

shale. A soft, flaky rock composed largely of clay.

skarn. A kind of metamorphic rock that forms where granitic magma reacts with limestone. Skarns typically consist of large crystals of garnet, epidote, and other calcium silicate minerals.

slate. A fine-grained metamorphic rock formed from the weak metamorphism of shale. It is tougher than shale because it contains microscopic mica in place of clay minerals.

sluice box. A long, sloping box with horizontal wood slats or riffles across the bottom. Heavier grains, such as gold or garnet, will settle to be caught at the upper edges of the riffles when sand or gravel is flushed with water down the length of the box.

smectite. *See* bentonite.

smelter. A large mill in which a metallic mineral concentrate is roasted to separate molten metal from sulfur as sulfur-oxide gases.

sorting. A term used to describe the uniformity of grain size in a sedimentary rock. A well-sorted rock has a uniform grain size.

stalactite. The limestone dripstone that grows down from the roof of a limestone cavern.

stalagmite. A column that grows upward from a limestone cavern floor where drips from the roof land and evaporate.

stock. A granite intrusion smaller than 40 square miles in area. A cluster of stocks may be large enough to form a granite batholith.

stratovolcano. A large volcano, usually andesite to basalt in composition, and consisting of layers of volcanic ash and lavas.

striation. A scratch on a rock that results from friction of rocks moving on a fault plane or by abrasion of rock beneath a moving glacier.

strike-slip fault. A near-vertical fault on which the two sides slipped laterally with respect to one another.

strip mine. A mine in which waste rock is thin enough to strip off with heavy equipment to reach the generally near-horizontal and near-surface deposit.

stromatolite. A cyanobacterial (algae-like) limestone deposit, most abundant in Proterozoic time. Calcium carbonate builds up in shallow water by cementation or precipitation facilitated by cyanobacteria to form irregularly laminated deposits, including those that in cross section resemble big cabbage heads.

subduction zone. A major tectonic collision zone, most commonly between a descending oceanic plate and an overriding continental plate. It can cause major earthquakes and produce explosive volcanoes in the upper plate.

sulfide. A mineral compound of a metallic element and sulfur, for example pyrite (iron sulfide) or chalcocite (copper sulfide).

supercontinent. An amalgamation of continents; for example, Pangaea.

swelling clay. Smectite (bentonite), a clay mineral that swells and becomes a very slippery gumbo when it gets wet.

syenite. A coarse-grained plutonic igneous rock consisting dominantly of orthoclase (potassium-sodium feldspar).

syncline. A downfold or trough folded in layered rocks. The two sides of a syncline are called its limbs.

tailings. Waste rock produced during mining.

talc. An extremely soft magnesium silicate mineral that occurs in metamorphic rocks recrystallized at relatively low temperature.

tarn. A lake in a glacial cirque.

tectonic. Mountain-building deformation processes involving folds or faults, including compressional, extensional, and strike-slip movements.

tephra. Volcanic ash and other fragmental material blown from a volcano.

terrace. Older stream deposits standing at a somewhat higher level than the active stream deposits.

terrane. A fault-bounded crustal fragment with a geologic history that differs from adjacent fragments.

Tertiary. The geological period from 66 million years ago to 2.6 million years ago. It has been formally replaced with two periods: the Paleogene and Neogene.

thick-skin deformation. Laramide deformation between 75 million years ago and 55 to 35 million years ago, driven by shallow-slab subduction and crustal block uplifts including basement rocks.

thin-skin deformation. Sevier deformation during Jurassic through earliest Tertiary time, dominated by thrust faults moving sedimentary cover above the crystalline basement rocks.

thrust fault. A gently dipping (less than 45 degrees) fracture along which the rocks above the fault surface moved up over those beneath. The hanging wall above a thrust fault is termed a **thrust sheet**, where it is relatively thin relative to its areal extent. Also called a thrust plate and thrust slab.

till. Sediment dumped directly from glacial ice. Till typically consists of partly rounded rocks of all sizes and shapes and does not have internal layering.

trace fossil. A fossil, such as a burrow, a fecal pellet, or animal tracks, that preserves evidence of past life but is not the actual skeletal remains.

trilobite. An extinct group of arthropods with external shells that appeared about 520 million years ago in Cambrian time and was exterminated in the global extinction at the end of Permian time, 252 million years ago.

tuff. A soft volcanic rock composed of volcanic ash, most commonly rhyolite.

unconformity. A gap in the rock record, either because sediments were not deposited or because they eroded before the deposition of overlying rocks.

varves. Alternating annual light-colored layers (summer silt) and dark-colored layers (winter clay and organic material) in a glacial lake.

vein. A fracture that has been filled, usually with quartz or calcite. Many veins contain valuable minerals.

vermiculite. A black or brown mica that expands greatly when heated. It is used as fire-resistant insulation and as a soil conditioner to hold nutrients. Some vermiculite contains asbestos and can cause lung cancer.

volcanic arc. A chain of volcanoes that formed above an ocean-floor subduction zone.

vug. A small to medium-size cavity in a rock. Quartz or calcite crystals sometimes grow inward from the walls.

Western Interior Seaway. A broad seaway extending from the Arctic through central and eastern Montana and the Dakotas to the Gulf of Mexico, during Cretaceous and earliest Tertiary time.

wildcat hole. A drill hole in an area of undiscovered oil or gas.

Williston Basin. A very large continental sedimentary basin in western North Dakota, northeastern Montana, and southern Saskatchewan.

Wyoming craton. The Archean basement rocks of south-central Montana and Wyoming.

xenolith. A fragment (often angular) of older rock enclosed in an igneous rock.

REFERENCES

Nontechnical References

Alt, D., 2001, *Glacial Lake Missoula and its humongous floods*, Mountain Press Publishing Co., Missoula, MT.

Alt, D. D., and D. W. Hyndman, 1986, *Roadside geology of Idaho*, Mountain Press Publishing Co., Missoula, MT.

Alt, D. D., and D. W. Hyndman, 1989, *Roadside geology of Montana*, 1st ed., Mountain Press Publishing Co., Missoula, MT.

Alt, D. D., and D. W. Hyndman, 1995, *Northwest exposures: A geologic story of the Northwest*, Mountain Press Publishing Co., Missoula, MT.

Baker, V. R., and R. C. Barker, 1985, Cataclysmic late Pleistocene flooding from Glacial Lake Missoula: A review, *Quaternary Science Reviews*, v. 4, no. 1, 1–41.

Benson, R., no date, Montana's earth science pictures, http://formontana.net.

Bretz, J. H., 1969, The Lake Missoula floods and the Channeled Scabland, *Journal of Geology*, v. 77, no. 5, 505–543.

Fritz, W. J., and J. W. Sears, 1993, Tectonics of the Yellowstone hot spot wake in southwestern Montana, *Geology*, v. 21, no. 5, 427–430.

Fritz, W. J., and R. C. Thomas, 2011, *Roadside geology of Yellowstone Country*, 2nd ed., Mountain Press Publishing Co., Missoula, MT.

Hendrix, M. S., 2011, *Geology underfoot in Yellowstone Country*, Mountain Press Publishing Co., Missoula, MT.

Hodges, M., 2016, *Rockhounding Montana: A guide to 100 of Montana's best rockhounding sites*, Falcon guides, Guilford, CT.

Hyndman, D. W., and D. Alt, 1987, Radial dikes, laccoliths, and gelatin models, *Journal of Geology*, v. 95, no. 6, 763–774.

Kiver, E. P., 1989, Lake Missoula floods and the Channeled Scabland, in *Geologic guidebook for Washington and adjacent areas*, information circular, ed. N. L. Joseph, State of Washington, Dept. of Natural Resources, Division of Geology and Earth Resources, 307–308.

Lund, H. L., Weight, W., and D. R. Lopach, 2015, Fracking in Montana: Asking questions, finding answers, Montana Farmers Union, Great Falls, MT.

Montana Dept. of Transportation, no date, Geologic road signs, www.mdt.mt.gov/travinfo/geomarkers.shtml.

Silkwood, J. T., 1998, Glacial Lake Missoula and the Channeled Scabland: A digital portrait of landforms of the last ice age, Washington, Oregon, northern Idaho, and western Montana, US Forest Service, Northern Region, Minerals and Geology, Geology Program Digital Mapping, scale 1:1,000,000.

Smyers, N. B., and R. M. Breckenridge, 2003, Glacial Lake Missoula, Clark Fork ice dam, and the floods outburst area: Northern Idaho and western Montana, in *Western Cordillera and adjacent areas*, field guide 4, ed. T. W. Swanson, Geological Society of America, 1–15.

Spritzer, D., 1999, *Roadside history of Montana*, Mountain Press Publishing Co., Missoula, MT.

Tracy, A. W., 1921, *Guide through Montana*, Montana Division of the Theodore Roosevelt International Highway.

Trexler, D., 2012, *Becoming dinosaurs: A prehistoric perspective on climate change today*, Sweetgrass Books, Helena, MT.

US Geological Survey, no date, The national map, https://viewer.nationalmap.gov/basic.

Walker, J. D., Geissman, J. W., Bowring, S. A., and L.E. Babcock, compilers, 2018, Geologic Time Scale v. 5.0: Geological Society of America.

Technical References

Amidon, L., 1997, Paleoclimate study of Eocene fossil woods and associated paleosols from the Gallatin Petrified Forest, Gallatin National Forest, SW Montana, unpublished master's thesis, University of Montana, Missoula.

Anastasio, D. J., Majerowicz, C. N., Pazzaglia, F. J, and C. A. Regalla, 2010, Late Pleistocene–Holocene ruptures on the Lima Reservoir fault, SW Montana, *Journal of Structural Geology*, v. 32, no. 12, 1996–2008.

Anders, M. H., and N. H. Sleep, 1992, Magmatism and extension: The thermal and mechanical effects of the Yellowstone hotspot, *Journal of Geophysical Research*, v. 97, no. B11, 15379–15393.

Baker, V. R., 1973, Paleohydrology and sedimentology of Lake Missoula flooding in eastern Washington, special paper 144, Geological Society of America.

Berg, R., 1979, Talc and chlorite deposits in Montana, memoir 45, Montana Bureau of Mines and Geology.

Berg, R. B., 2008, Geologic map of the Choteau 30' x 60' quadrangle, north-central Montana, open-file report 571, Montana Bureau of Mines and Geology.

Berg, R. B., Lonn, J. D., and W. W. Locke, 1999, Geologic map of the Gardiner 30' x 60' quadrangle, south-central Montana, open-file report 387, Montana Bureau of Mines and Geology.

Berg, R. B., Lopez, D. A., and J. D. Lonn, 2000, Geologic map of the Livingston 30' x 60' quadrangle, south-central Montana, open-file report 406, Montana Bureau of Mines and Geology.

Bonnichsen, B., Leeman, W. P., Honjo, N., McIntosh, W. C., and M. M. Godchaux, 2007, Miocene silicic volcanism in southwestern Idaho: Geochronology, geochemistry, and evolution of the central Snake River Plain, *Bulletin of Volcanology*, v. 70, no. 3, 315–342.

Bonnichsen, B., and others, 1989, Silicic volcanic rocks in the Snake River Plain—Yellowstone Plateau Province, in *Field excursions to volcanic terranes in the western United States*, v. 2, Cascades and Intermountain West, memoir 47, eds. C. E. Chapin and J. Zidek, New Mexico Bureau of Mines and Mineral Resources, 135–153.

Boudreau, A. E., and I. S. McCallum, 1986, Investigations of the Stillwater Complex: III, the Picket Pin Pt/Pd deposit, *Economic Geology*, v. 81, no. 8, 1953–1975.

Boyer, S. E., and D. Elliott, 1982, Thrust systems, *American Association of Petroleum Geologists Bulletin*, v. 66, no. 9, 1196–1230.

Brady, J. B., Burger, H. R., Cheney, J. T., and T. A. Harms, eds., 2004, Precambrian geology of the Tobacco Root Mountains, Montana, special paper 377, Geological Society of America.

Brown, P. L., and others, 1983, Saline-seep diagnosis, control, and reclamation, conservation research report 30, US Dept. of Agriculture.

Buffet, B., and D. Archer, 2004, Global inventory of methane clathrate: Sensitivity to changes in the deep ocean, *Earth and Planetary Science Letters*, v. 227, no. 3–4, 185–199.

Byrne, J. M., Fagre, D. B., MacDonald, R., and C. C. Muhlfeld, 2015, Climate change and the Rocky Mountains, in *Impact of global changes on mountains: Responses and adaptation*, eds. V. I. Grover, A. Borsdorf, J. H. Breuste, P. C. Tiwari, and F. W. Frangetto, CRC Press, Boca Raton, FL, 432–463.

Calbeck, J. M., 1975, Geology of the central Wise River valley Pioneer Mountains Beaverhead County Montana, unpublished master's thesis, University of Montana, Missoula.

Carrara, P. E., Short, S. K., and R. E. Wilcox, 1986, Deglaciation of the mountainous region of northwestern Montana, USA, as indicated by late Pleistocene ashes, *Arctic and Alpine Research*, v. 18, no. 3, 317–325.

Chadwick, R. A., 1970, Belts of eruptive centers in the Absaroka-Gallatin volcanic province, Wyoming-Montana, *Geological Society of America Bulletin*, v. 81, no. 1, 267–274.

Chambers, R. L., 1971, Sedimentation in Glacial Lake Missoula, master's thesis, University of Montana, Missoula.

Cheney, J. T., Webb, A. A. G., Coath, C. D., and K. D. McKeegan, 2004, In situ ion microprobe $^{207}Pb/^{206}Pb$ dating of monazite from Precambrian metamorphic suites, Tobacco Root Mountains, Montana, in *Precambrian geology of the Tobacco Root Mountains, Montana*, special paper 377, eds. J. B. Brady, H. R. Burger, J. T. Cheney, and T. A. Harms, Geological Society of America, 151–179.

Cheney, J. T., and others, 2004, Proterozoic metamorphism of the Tobacco Root Mountains, Montana, in *Precambrian geology of the Tobacco Root Mountains, Montana*, special paper 377, eds. J. B. Brady, H. R. Burger, J. T. Cheney, and T. A. Harms, Geological Society of America, 105–129.

Christiansen, R. L., 2001, The Quaternary and Pliocene Yellowstone Plateau volcanic field of Wyoming, Idaho, and Montana, professional paper 729-G, US Geological Survey.

Christiansen, R. L., Foulger, G. R., and J. R. Evans, 2002, Upper mantle origin of the Yellowstone hot spot, *Geological Society of America Bulletin*, v. 114, no. 10, 1245–1256.

Christiansen, R. L., and R. R. Wahl, 1999, Digital geologic map for Yellowstone National Park, Idaho, Montana, and Wyoming and vicinity, open-file report 99-0174, US Geological Survey.

Christiansen, R. L., and others, 2007, Preliminary assessment of volcanic and hydrothermal hazards in Yellowstone National Park and vicinity, professional paper 729-B, US Geological Survey.

Christopherson, E., 1962, *The night the mountain fell: The story of the Montana-Yellowstone earthquake*, Yellowstone Publications, West Yellowstone, MT.

Chu, R., and others, 2010, Mushy magma beneath Yellowstone, *Geophysical Research Letters*, v. 37, L0136.

Clark, A. M., Fagre, D. B., Peitzsch, E. H., Reardon, B. A., and J. T. Harper, 2017, Glaciological measurements and mass balances from Sperry Glacier, Montana, USA, years 2005–2015, *Earth System Science Data*, v. 9, 47–61.

Constenius, K. N., 1996, Late Paleogene extensional collapse of the Cordilleran foreland fold and thrust belt, *Geological Society of America Bulletin*, v. 108, no. 1, 20–39.

Czamanske, G. K., and M. L. Zientek, 1985, *The Stillwater Complex, Montana: Geology and guide*, special publication 92, Montana Bureau of Mines and Geology.

DeCelles, P. G., 2004, Late Jurassic to Eocene evolution of the Cordilleran thrust belt and foreland basin system, western USA, *American Journal of Science*, v. 304, no. 2, 105–168.

DeCelles, P. G., and K. N. Constenius, 2012, Regional structure and kinematic history of the Cordilleran fold-thrust belt in northwestern Montana, USA, *Geosphere*, v. 8, no. 5, 1104–1128.

DeCelles, P. G., and others, 1991, Kinematic history of a foreland uplift from Paleocene synorogenic conglomerate, Beartooth Range, Wyoming and Montana, *Geological Society of America Bulletin*, v. 103, no. 11, 1458–1475.

Diemer, J. A., and E. S. Belt, 1991, Sedimentology and paleohydraulics of the meandering river system of the Fort Union Formation, southeastern Montana, *Sedimentary Geology*, v. 75, no. 1–2, 85–108.

Dorf, E. 1980, *Petrified forests of Yellowstone*, US Government Printing Office, Washington, DC.

Downey, J. S., and G. A. Dinwiddie, 1988, The regional aquifer system underlying the northern Great Plains in parts of Montana, North Dakota, South Dakota, and Wyoming—summary, professional paper 1402-A, US Geological Survey.

Du Bray, E. A., and S. S. Harlan, 1998, Geology, petrology, and age of the Cretaceous Sliderock Mountain stratovolcano, Montana, professional paper 1602, US Geological Survey.

English, J. M., and S. T. Johnston, 2004, The Laramide orogeny: What were the driving forces? *International Geology Review*, v. 46, no. 9, 833–838.

Fagre, D. B., no date, History of glaciers in Glacier National Park, US Geological Survey, available online from the Northern Rocky Mountain Science Center at www.usgs.gov/centers/norock.

Feeley, T. C., 2003, Origin and tectonic implications of across-strike geochemical variations in the Eocene Absaroka Volcanic Province, *Journal of Geology*, v. 111, no. 3, 329–346.

Feinstein, S., Kohn, B., Osadetz, K., and R. A. Price, 2007, Thermochronometric reconstruction of the prethrust paleogeothermal gradient and initial thickness of the Lewis thrust sheet, southeastern Canadian Cordillera foreland belt, in *Whence the mountains? Inquiries into the evolution of orogenic systems: A volume in honor of Raymond A. Price*, special paper 433, eds. J. W. Sears, T. A. Harms, and C. A. Evenchick, Geological Society of America, 167–182.

Ferreira, R. F., Cannon, M. R., and R. E. Davis, 1986, Montana ground-water quality, open-file report 87-0736, US Geological Survey, https://pubs.usgs.gov/of/1987/0736/report.pdf.

Finn, T. M., and others, 2010, Cretaceous-Tertiary composite total petroleum system (503402), Bighorn Basin, Wyoming and Montana, in *Petroleum systems and geologic assessment of oil and gas in the Bighorn Basin Province, Wyoming and Montana*, digital data series DDS–69–V, US Geological Survey, Bighorn Basin Province assessment team, https://pubs.usgs.gov/dds/dds-069/dds-069-v/REPORTS/69_V_CH_3.pdf.

Flores, R. C., and F. G. Ethridge, 1985, Evolution of intermontane fluvial systems of Tertiary Powder River Basin, Montana and Wyoming, in *Cenozoic paleogeography of the west-central United States: Rocky Mountain paleogeography symposium*, eds. R. M. Flores and S. S. Kaplan, v. 3, Rocky Mountain Section, Society of Economic Paleontologists and Mineralogists, Denver, CO, 107–127.

Flores, R. M., 1986, Styles of coal deposition in Tertiary alluvial deposits, Powder River Basin, Montana and Wyoming, in *Paleoenvironmental and tectonic controls in coal-forming basins of the United States*, special paper 210, eds. P. C. Lyons and C. L. Rice, Geological Society of America, 79–104.

Flores, R. M., and T. A. Cross, 1991, Cretaceous and Tertiary coals of the Rocky Mountains and Great Plains regions, in *Economic geology*, Decade of North American geology series, v. P-2, eds. H. J. Gluskoter, D. D. Rice, and R. B. Taylor, Geological Society of America, 547–571.

Flores, R. M., Ochs, A. M., Bader, L. R., Johnson, R. C., and D. Volver, 1999, Framework geology of the Fort Union coal in the Powder River Basin, professional paper 1625-A, US Geological Survey.

Foose, R. M., Wise, D. U., and G. S. Garbarini, 1961, Structural geology of the Beartooth Mountains, Montana and Wyoming, *Geological Society of America Bulletin*, v. 72, no. 8, 1143–1172.

Foster, D. A., and C. M. Fanning, 1997, Geochronology of the northern Idaho batholith and the Bitterroot metamorphic core complex: Magmatism preceding and contemporaneous with extension, *Geological Society of America Bulletin*, v. 109, no. 4, 379–394.

Foster, D. A., Grice Jr., W. C., and T. J. Kalakay, 2010, Extension of the Anaconda metamorphic core complex: $^{40}Ar/^{39}Ar$ thermochronology and implications for Eocene tectonics of the northern Rocky Mountains and the Boulder batholith, *Lithosphere*, v. 2, no. 4, 232–246.

Foster, D. A., Hyndman, D. W., and J. L. Anderson, 1990, Magma mixing and mingling between synplutonic mafic dikes and granite in the Idaho–Bitterroot batholith, in *The nature and origin of cordilleran magmatism*, memoir 174, ed. J. L. Anderson, Geological Society of America, 347–358.

Foster, D. A., Mueller, P. A., Heatherington, A., Gifford, J. N., and T. J. Kalakay, 2012, Lu-Hf systematics of magmatic zircons reveal a Proterozoic crustal boundary under the Cretaceous Pioneer batholith, Montana, *Lithos* 142–143: 216–225.

Foster, D. A., Mueller, P. A., Mogk, D. W., Wooden, J. L., and J. J. Vogl, 2006, Proterozoic evolution of the western margin of the Wyoming craton: Implications for the tectonic and magmatic evolution of the northern Rocky Mountains, *Canadian Journal of Earth Sciences*, v. 43, no. 10, 1601–1619.

Foster, D. A., Mueller, P. A., Vogl, J. J., Gifford, J., and D. W. Mogk, 2013, The Little Belt Arc of the Great Falls tectonic zone: A key component of the Belt Basin basement, *Northwest Geology* 42: 107–110.

Foster, D. A., Schafer, C., Fanning, C. M., and D.W. Hyndman, 2001, Relationships between crustal partial melting, plutonism, orogeny, and exhumation: Idaho-Bitterroot batholith, *Tectonophysics* 342: 313–350.

Foulger, G. R., 2010, *Plates vs. plumes: A geological controversy*, Wiley-Blackwell, Hoboken, NJ.

Fournier, R. O., Christiansen, R. L., Hutchinson, R. A., and K. L. Pierce, 1994, A field-trip guide to Yellowstone National Park, Wyoming, Montana, and Idaho: Volcanic, hydrothermal, and glacial activity in the region, bulletin 2099, US Geological Survey.

Frank, T. D., Lyons, T. W., and K. C. Lohmann, 1997, Isotopic evidence for the paleoenvironmental evolution of the Mesoproterozoic Helena Formation, Belt Supergroup, Montana, USA. *Geochimica et Cosmochimica Acta*, v. 61, no. 23, 5023–5041.

Friends of the Beartooth All-American Road, no date, Welcome to the Beartooth Highway, wayfinding map, www.beartoothhighway.com.

Fritz, W. J., and others, 2007, Cenozoic volcanic rocks of southwestern Montana, *Northwest Geology*, v. 36, 91–110.

Fuentes, F., DeCelles, P. G., Constenius, K. N., and G. E. Gehrels, 2011, Evolution of the cordilleran foreland basin system in northwestern Montana, USA, *Geological Society of America Bulletin*, v. 123, no. 3–4, 507–533.

Fullerton, D. S., Colton, R. B., and C. A. Bush, 2012, Continental glaciation of northern Montana, open-file report 1217-508, US Geological Survey.

Fullerton, D. S., Colton, R. B., and C. A. Bush, 2013, Quaternary geologic map of the Shelby 1° x 2° quadrangle, Montana, open-file report 2012–1170, US Geological Survey, scale 1:250,000.

Fullerton, D. S., Colton, R. B., Bush, C. A., and A. W. Straub, 2004, Map showing spatial and temporal relations of mountain and continental glaciations on the northern plains, primarily in northern Montana and northwestern North Dakota, scientific investigations map 2843, US Geological Survey, scale 1:1,000,000.

Gammons, C. H., Metesh J. J., and T. E. Duaime, 2006, An overview of the mining history and geology of Butte, Montana, *Mine Water and the Environment*, v. 25, no. 2, 70–75.

Gifford, J. N., Mueller, P. A., Foster, D. A., and D. W. Mogk, 2014, Precambrian crustal evolution in the Great Falls tectonic zone: Insights from xenoliths from the Montana Alkali Province, *Journal of Geology* 122: 531–548.

Gifford, J. N., Mueller, P. A., Foster, D. A., and D. W. Mogk, 2018, Extending the realm of Archean crust in the Great Falls tectonic zone: Evidence from the Little Rocky Mountains, Montana, *Precambrian Research* 315: 264–281.

Gordon, L., and R. I. Tilling, 2008, This dynamic planet: A teaching companion, US Geological Survey, available online at https://volcanoes.usgs.gov.

Greenwalt, D., Goreva, Y., Siljeström, S., Rose, T., and R. E. Harbach, 2013, Hemoglobin-derived porphyrins preserved in a middle Eocene blood-engorged mosquito, *Proceedings of the National Academy of Sciences*, v. 110, no. 46, 18496–18500.

Greenwalt, D., and C. Labandeira, 2013, The amazing fossil insects of the Eocene Kishenehn Formation in northwestern Montana, *Rocks and Minerals*, v. 88, no. 5, 434–441.

Gridley, D. J., and M. S. Smith, 2004, The Gardiner basalts: Petrological and geochemical comparison with the Hepburn Mesa and Yellowstone National Park basalts, *Geological Society of America Abstracts with Programs*, v. 36, no. 2, 103.

Guilbert, J., and C. Park, Jr., 1986, *The geology of ore deposits*, Waveland Press, Long Grove, IL.

Hamilton, W., and W. Myers, 1967, The nature of batholiths, professional paper 554-C, US Geological Survey.

Hamilton, W., and W. Myers, 1974, Nature of the Boulder Batholith of Montana. *Geological Society of America Bulletin* 85: 365–378.

Harlan, S. S., 2006, $^{40}Ar/^{39}Ar$ dates from alkaline intrusions in the northern Crazy Mountains, Montana: Implications for the timing and duration of alkaline magmatism in the central Montana alkalic province, *Rocky Mountain Geology*, v. 41, no. 1, 45–55.

Harms, T. A., Brady, J. B., Burger, H. R., and J. T. Cheney, 2004, Advances in the geology of the Tobacco Root Mountains, Montana, and their implications for the history of the northern Wyoming province, in *Precambrian geology of the Tobacco Root Mountains*, Montana, special paper 377, eds. J. B. Brady, H. R. Burger, J. T. Cheney, and T. A. Harms, Geological Society of America, 227–243.

Harrison, J. E., Griggs, A. B., and J. D. Wells, 1974, Tectonic features of the Precambrian Belt Basin and their influence on post-Belt structures, professional paper 866, US Geological Survey.

Hauge, T. A., 1993, The Heart Mountain detachment, northwestern Wyoming: 100 years of controversy, in *Geology of Wyoming*, memoir 5, eds. A. W. Snoke, J. R. Steidtmann, and S. M. Roberts, Geological Survey of Wyoming, 530–571.

Hearn, B. C., Jr., 1976, Geologic and tectonic maps of the Bearpaw Mountains area, north-central Montana, US Geological Survey Miscellaneous Geologic Investigations Map I-919, scale 1:125,000.

Hearn, B. C., Jr., and others, eds., 1989, *Montana high-potassium igneous province: Crazy Mountains to Jordan, Montana*, field trip guidebook 346, American Geophysical Union, Washington, DC.

Heise, B., and others, 2010, Chief Mountain and Glacier National Park, Glacier National Park 100th anniversary, Montana geology 2010, Montana Bureau of Mines and Geology calendar, 1 sheet.

Hoffman, P. F., 1988, United plates of America, the birth of a craton: Early Proterozoic assembly and growth of Laurentia, *Annual Review of Earth and Planetary Science*, v. 16, 543–603.

Horner, J. R., and J. Gorman, 1988, *Digging dinosaurs*, Workman Publishing, New York.

Horodyski, R. J., 1989, Stromatolites of the Belt Supergroup, Glacier National Park, Montana, in *Middle Proterozoic Belt Supergroup, western Montana, Great Falls, Montana, to Spokane, Washington, July 20–28, 1989*, field trip guidebook 334, eds. D. Winston, R. J. Horodyski, and J. W. Whipple, American Geophysical Union, Washington, DC, 27–45.

Hyndman, D. W., 1983, The Idaho batholith and associated plutons, Idaho and western Montana, in *Circum-Pacific plutonic terranes*, memoir 159, ed. J. A. Roddick, Geologic Society of America, 213–240.

Hyndman, D. W., D. Alt, and J. W. Sears, 1988, Post-Archean metamorphism and tectonic evolution of western Montana and northern Idaho, in *Metamorphism and crustal evolution of the western United States*, Rubey v. 7, ed. W. G. Ernst, Prentice-Hall, Englewood Cliffs, NJ, 332–361.

Hyndman, D. W., and D. A. Foster, 1988, The role of tonalites and mafic dikes in the generation of the Idaho batholith, *Journal of Geology*, v. 96, no. 1, 31–46.

Hyndman, D. W., and S. A. Myers, 1988, The transition from amphibolite facies mylonite to chloritic breccia and the role of the mylonite in formation of Eocene epizonal plutons, Bitterroot dome, Montana, *Geologische Rundschau*, v. 77, no. 1, p. 201–226.

James, H. L., 1995, *Geologic and historic guide to the Beartooth Highway, Montana and Wyoming*, special publication 110, Montana Bureau of Mines and Geology.

Janecke, S. U., 1994, Sedimentation and paleogeography of an Eocene to Oligocene rift zone, Idaho and Montana, *Geological Society of America Bulletin*, v. 106, no. 8, 1083–1095.

Kalakay, T. J., Foster, D. A., and J. D. Lonn, 2014, Polyphase collapse of the cordilleran hinterland: The Anaconda metamorphic core complex of western Montana, in *Exploring the northern Rocky Mountains*, field guide 37, eds. C. A. Shaw and B. Tikoff, Geological Society of America, 145–159.

Kalakay, T. J., Foster, D. A., and R. C. Thomas, 2003, Geometry and timing of deformation in the Anaconda extensional terrane, west-central Montana, *Northwest Geology*, v. 32, 124–133.

Kalakay, T. J., John, B. E., and D. R. Lageson, 2001, Fault-controlled pluton emplacement in the Sevier fold-and-thrust belt of southwest Montana, USA, *Journal of Structural Geology*, v. 23, no. 6–7, 1151–1165.

Katz, M., Cramer, B. S., Mountain, G. S., Katz, S., and K. G. Miller, 2001, Uncorking the bottle: What triggered the Paleocene/Eocene thermal maximum methane release? *Paleoceanography*, v. 16, no. 6, 549–562.

KellerLynn, K., 2011, Bighorn Canyon National Recreation Area: Geologic resources inventory report, natural resource report NPS/NRSS/GRD/NRR—2011/447, National Park Service, Fort Collins, CO.

Kellogg, K. S., Ruleman, C. A., and S. M. Vuke, 2007, Geologic map of the central Madison Valley (Ennis area), southwestern Montana, open-file report 543, Montana Bureau of Mines and Geology, scale 1:50,000.

Kellogg, K. S., Schmidt, C. J., and S. W. Young, 1995, Basement and cover-rock deformation during Laramide contraction in the northern Madison Range (Montana) and its influence on Cenozoic basin formation, *American Association of Petroleum Geologists Bulletin*, v. 79, no. 8, 1117–1137.

Kellogg, K. S., and V. S. Williams, 2005, Geologic map of the Ennis 30' x 60' quadrangle, Madison and Gallatin Counties, Montana, and Park County, Wyoming, open-file report 529, Montana Bureau of Mines and Geology, 27 p., scale 1:100,000.

Kuenzi, W. D., and R. W. Fields, 1971, Tertiary stratigraphy, structure, and geologic history, Jefferson basin, Montana, *Geological Society of America Bulletin*, v. 82, no. 12, 3374–3394.

Lageson, D. R., 1987, Structural geology of the Sawtooth Range at Sun River canyon, Montana Disturbed Belt, Montana, in *Rocky Mountain Section of the Geological Society of America*, centennial field guide, v. 2, ed. S. S. Beus, Geological Society of America, 37–39.

Lageson, D. R., and J. G. Schmitt, 1994, The Sevier orogenic belt of the western United States: Recent advances in understanding its structural and sedimentologic framework, in *Mesozoic systems of the Rocky Mountain region, USA*, eds. M. V. Caputo, J. A. Peterson, and K. J. Franczyk, Society for Sedimentary Geology, Rocky Mountain Section, Denver, CO, 27–64.

Lewis, R. S., Kiilsgaard, T. H., Bennett, E. H., and W. E. Hall, 1987, Lithologic and chemical characteristics of the central and southeastern part of the southern lobe of the Idaho batholith, in *Geology of the Blue Mountains region of Oregon, Idaho, and Washington: The Idaho batholith and its border zone*, professional paper 1436, eds. T. L. Vallier and H. C. Brooks, US Geological Survey, 171–196.

Lewis, R. S., and others, 2014, Hells Canyon to the Bitterroot front: A transect from the accretionary margin eastward across the Idaho batholith, in *Exploring the northern Rocky Mountains*, field guide 37, eds. C. A. Shaw and B. Tikoff, Geological Society of America, 1–50.

Livacarri, R. F., 1991, Role of crustal thickening and extensional collapse in the tectonic evolution of the Sevier-Laramide orogeny, western United States, *Geology*, v. 19, no. 11, 1104–1107.

Locke, W. W., 1990, Late Pleistocene glaciers and the climate of western Montana, USA, *Arctic and Alpine Research*, v. 22, no. 1, 1–13.

Locke, W. W., and others, 1995, The middle Yellowstone valley from Livingston to Gardiner, Montana: A microcosm of northern Rocky Mountain geology, *Northwest Geology*, v. 24, 1–65.

Locke, W. W., and L. N. Smith, 2004, Pleistocene mountain glaciation in Montana, USA, *Developments in Quaternary Science* 2:125–129.

Lofgren, D. L., and others, 2017, Coprolites and mammalian carnivores from Pipestone Springs, Montana, and their paleoecological significance, *Annals of Carnegie Museum*, v. 84, no. 4, 265–285.

Longrich, N. R., Bhullar, B.-A. S., and J. A. Gauthier, 2012, Mass extinction of lizards and snakes at the Cretaceous–Paleogene boundary, *Proceedings of the National Academy of Sciences of the United States of America*, v. 109, no. 52, 21396–21401.

Longrich, N. R., Tokaryk, T., and D. J. Field, 2011, Mass extinction of birds at the Cretaceous–Paleogene (K–Pg) boundary, *Proceedings of the National Academy of Sciences of the United States of America*, v. 108, no. 37, 15253–15257.

Lonn, J. D., and A. R. English, 2002, Preliminary geologic map of the eastern part of the Gallatin Valley, Montana, open-file report 457, Montana Bureau of Mines and Geology, scale 1:50,000.

Lonn, J. D., McDonald, C., Sears, J. W., and L. N. Smith, 2010, Geologic map of the Missoula east 30' x 60' quadrangle, western Montana, open-file report MBMG 593, Montana Bureau of Mines and Geology.

Lonn, J. D., and others, 2000, Geologic map of the Lima 30' x 60' quadrangle, SW Montana, open-file report 408, Montana Bureau of Mines and Geology.

Lopez, D. A., 2001, Preliminary geologic map of the Red Lodge 30' x 60' quadrangle, south-central Montana, open-file report 423, Montana Bureau of Mines and Geology.

Lopez, D. A., and J. C. Reiten, 2003, Preliminary geologic map of Paradise Valley, south-central Montana, open-file report 480, Montana Bureau of Mines and Geology.

Lowenstern, J. B., and S. Hurwitz, 2008, Monitoring a supervolcano in repose: Heat and volatile flux and the Yellowstone Caldera, *Elements*, v. 4, no. 1, 35–40.

Lyons, J. B., 1944, Igneous rocks of the northern Big Belt Range, Montana, *Geological Society of America Bulletin*, v. 55, no. 4, 445–472.

MacLean, J. S., and J. W. Sears, 2016, Belt Basin: Window to Mesoproterozoic Earth, special paper 522, Geological Society of America.

Malone, D. H., and J. P. Craddock, 2008, Recent contributions to the understanding of the Heart Mountain detachment, Wyoming, *Northwest Geology*, v. 37, 21–40.

Mastin, L. G., Van Eaton, A. R., and J. B. Lowenstern, 2014, Modelling ash fall distribution from a Yellowstone supereruption, *Geochemistry, Geophysics, Geosystems*, v. 15, no. 8, 3459–3475.

McCallum, I. S., and others, 1980, Investigations of the Stillwater Complex: Part I. Stratigraphy and structure of the banded zone, *American Journal of Science*, v. 280-A, 59–87.

McDonald, C., Elliott C. G., Vuke, S. M., Lonn, L. D., and R. B. Berg, 2012, Geologic map of the Butte south 30' x 60' quadrangle, southwestern Montana, open-file report 622, Montana Bureau of Mines and Geology.

Methner K., and others, 2016, Rapid middle Eocene temperature change in western North America, *Earth and Planetary Science Letters*, v. 450, 132–139.

Mitrovica, J. X., Beaumont, C., and G. T. Jarvis, 1989, Tilting of continental interiors by the dynamical effects of subduction, *Tectonics*, v. 8, no. 5, 1079.

Mogk, D. W., and D. J. Henry, 1988, Metamorphic petrology of the northern Archean Wyoming province, southwestern Montana, in *Metamorphism and crustal evolution of the western United States*, Rubey v. 7, ed. W. G. Ernst, Prentice-Hall, Englewood Cliffs, NJ, 362–382.

Mogk, D. W., Mueller, P., and J. L. Wooden, 1992, The nature of Archean terrane boundaries: An example from the northern Wyoming province, *Precambrian Research*, v. 55, no. 1–4, 155–168.

Montagne, J., and D. R. Lageson, 1995, History of geologic exploration of the Yellowstone River valley (Paradise Valley), southwest Montana, *Northwest Geology*, v. 24, 315–330.

Mudge, M. R., 1972, Structural geology of the Sun River canyon and adjacent areas, northwestern Montana, professional paper 663-B, US Geological Survey.

Mudge, M. R., and R. L. Earhart, 1978, The Lewis thrust fault and related structures in the Disturbed Belt, northwestern Montana, professional paper 1174, US Geological Survey.

Mueller, P. A., Mogk, D. W., Henry, D. J., Wooden, J. L., and D. A. Foster, 2008, Geologic evolution of the Beartooth Mountains: Insights from petrology and geochemistry, *Northwest Geology*, v. 37, 5–20.

National Earthquake Hazards Reduction Program, 2019, Montana earthquake scenarios, www.nehrpscenario.org/completed/montana-earthquake-scenarios.

Nichols, D. J., 2003, Palynostratigraphic framework for age determination and correlation of the nonmarine lower Cenozoic of the Rocky Mountains and Great Plains region, in *Cenozoic systems of the Rocky Mountain region*, eds. R. G. Raynolds and R. M. Flores, Society for Sedimentary Geology, Rocky Mountain Section, 107–134.

O'Brien, H. E., Irving, A. J., McCallum, I. S., and M. F. Thirlwall, 1995, Strontium, neodymium, and lead isotopic evidence for the interaction of post-subduction asthenospheric potassic mafic magmas of the Highwood Mountains, Montana, USA, with ancient Wyoming craton lithospheric mantle, *Geochimica et Cosmochimica Acta*, v. 59, no. 21, 4539–4556.

O'Connor, J. E., and V. R. Baker, 1992, Magnitudes and implications of peak discharges from Glacial Lake Missoula, *Geological Society of America Bulletin*, v. 104, no. 3, 267–279.

O'Neill, J. M., 1995, Early Proterozoic geology of the Highland Mountains, southwestern Montana, and field guide to the basement rocks that compose the Highland Mountain gneiss dome, *Northwest Geology*, v. 24, 85–97.

O'Neill, J. M., and Christiansen, R.L., 2004, Geologic map of the Hebgen Lake quadrangle, Beaverhead, Madison, and Gallatin Counties, Montana, Park and Teton Counties, Wyoming, and Clark and Fremont Counties, Idaho, scientific investigations map 2816, US Geological Survey, scale 1:100,000.

O'Neill, J. M., Lonn, J. D., Lageson, D. R., and M. J. Junk, 2004, Early Tertiary Anaconda metamorphic core complex, southwestern Montana, *Canadian Journal of Earth Sciences*, v. 41, no. 1, 63–72.

O'Neill, J. M., and D. A. Lopez, 1985, Character and regional significance of Great Falls tectonic zone, east-central Idaho and west-central Montana, *American Association of Petroleum Geologists Bulletin*, v. 69, no. 3, 437–447.

Pană, D. I., and B. A. van der Pluijm, 2014, Orogenic pulses in the Alberta Rocky Mountains: Radiometric dating of major faults and comparison with the regional tectono-stratigraphic record, *Geological Society of America Bulletin*, v. 127, no. 3–4, 480–502.

Pardee, J. T., 1942, Unusual currents in Glacial Lake Missoula, Montana, *Geological Society of America Bulletin*, v. 53, no. 11, 1569–1599.

Pardee, J. T., 1950, Late Cenozoic block faulting in western Montana, *Geological Society of America Bulletin*, v. 61, no. 4, 359–406.

Peale, A. C., 1896, Three Forks, Montana sheet, geological atlas, US Geological Survey, folio 24, 4 maps.

Pearson, D. A., Schaefer, T., Johnson, K. R., Nichols, D. J., and J. P. Hunter, 2002, Vertebrate biostratigraphy of the Hell Creek Formation in southwestern North Dakota and northwestern South Dakota, in *The Hell Creek Formation and the Cretaceous-Tertiary boundary in the northern Great Plains: An integrated continental record of the end of the Cretaceous*, special paper 361, eds. J. H. Hartman, K. R. Johnson, and D. J. Nichols, Geological Society of America, 145–167.

Pederson, G. T., and others, 2011, The unusual nature of recent snowpack declines in the North American cordillera, *Science*, v. 333, no. 6040, 332–335.

Peterson, J. A., 1988, Phanerozoic stratigraphy of the northern Rocky Mountain region, in *Sedimentary cover—North American craton: US, the geology of North America*, v. D-2, ed. L. L. Sloss, Geological Society of America, 83–107.

Petrik, F. E., 2008, Scarp analysis of the Centennial normal fault, Beaverhead County, Montana, and Fremont County, Idaho, unpublished master's thesis, Montana State University, Bozeman.

Pierce, K. L., 2003, Pleistocene glaciations of the Rocky Mountains, *Developments in Quaternary Sciences*, v. 1, 63–76.

Pierce, K. L., Chesley-Preston, T. L., and R. L. Sojda, 2014, Surficial geologic map of the Red Rock Lakes area, southwest Montana, open-file report 2014-1157, US Geological Survey.

Pierce, K. L., Licciardi, J. M., Krause, T. R., and C. Whitlock, 2014, Glacial and Quaternary geology of the northern Yellowstone area, Montana and Wyoming, in *Exploring the northern Rocky Mountains*, field guide 37, eds. C. A. Shaw and B. Tikoff, Geological Society of America, 189–203.

Pierce, K. L., and L. A. Morgan, 1992, The track of the Yellowstone hotspot: Volcanism, faulting, and uplift, in *Regional geology of eastern Idaho and western Wyoming*, memoir 179, eds. P. K. Link, M. A. Kuntz, and L. B. Platt, Geological Society of America, 1–53.

Pierce, K. L., and L. A. Morgan, 2009, Is the track of the Yellowstone hotspot driven by a deep mantle plume? Review of volcanism, faulting, and uplift in light of new data, *Journal of Volcanology and Geothermal Research*, v. 188, no. 1–3, 1–25.

Pollastro, R. M., Roberts, L. N. R., and T. A. Cook, 2010, Geologic assessment of technically recoverable oil in the Devonian and Mississippian Bakken Formation, in *Petroleum systems and geologic assessment of oil and gas in the Bighorn Basin Province, Wyoming and Montana*, digital data series DDS–69–V, US Geological Survey, Bighorn Basin Province assessment team, https://pubs.usgs.gov/dds/dds-069/dds-069-w/contents/REPORTS/69_W_CH_5.pdf.

Porter, K. W., Wilde, E. M., and S. M. Vuke, 1996, Preliminary geologic map of the Big Snowy Mountains 30' x 60' quadrangle, central Montana, open-file report 341, Montana Bureau of Mines and Geology.

Porter, K. W., Wilde, E. M., and S. M. Vuke, 1999, Geologic map of the Winnett 30' x 60' quadrangle, central Montana, open-file report 307, Montana Bureau of Mines and Geology.

Puskas, C. M., and R. B. Smith, 2009, Intraplate deformation and microplate tectonics of the Yellowstone hot spot and surrounding western US interior, *Journal of Geophysical Research: Solid Earth*, v. 114, B04410.

Raedeke, L. D., and I. S. McCallum, 1984, Investigations in the Stillwater Complex: Part II. Petrology and petrogenesis of the ultramafic series, *Journal of Petrology*, v. 25, no. 2, 395–420.

Rasmussen, D. L., 2003, Tertiary history of western Montana and east-central Idaho: A synopsis, in *Cenozoic systems of the Rocky Mountain region*, eds. R. G. Raynolds and R. M. Flores, Society for Sedimentary Geology, Rocky Mountain Section, 459–477.

Raup, O. B., Earhart, R. L., Whipple, J. W., and P. E. Carrara, 1983, Geology along Going-to-the-Sun Road, Glacier National Park, Montana, Glacier Natural History Association, West Glacier, MT.

Reheis, M., 1985, Evidence for Quaternary tectonism in the northern Bighorn Basin, Wyoming and Montana, *Geology*, v. 13, no. 5, 364–367.

Reiten, R., and J. Thamke, 2012 energy development and water needs in the Williston Basin, American Water Resources Association, Montana Section meeting, Anaconda, MT, October 12, available online at www.montanaawra.org.

Retallack, G. J., Dunn, K., and J. Saxby, 2013, Problematic Mesoproterozoic fossil *Horodyskia* from Glacier National Park, Montana, USA, *Precambrian Research*, v. 226, 125–142.

Reynolds, M. W., and T. R. Brandt, 2005, Geologic map of the Canyon Ferry Dam 30' x 60' quadrangle, west-central Montana, scientific investigations map 2860, US Geological Survey, scale 1:100,000.

Reynolds, M. W., and T. R. Brandt, 2007, Preliminary geologic map of the White Sulphur Springs 30' x 60' quadrangle, Montana, open-file report 2006-1329, US Geological Survey.

Richmond, T., 2012, Montana's new energy frontier, Economic Outlook, http://www.bber.umt.edu/pubs/energy/12energy.pdf.

Roberts, L. N. R., and M. A. Kirschbaum, 1995, Paleogeography of the Late Cretaceous of the western interior of middle North America: Coal distribution and sediment accumulation, professional paper 1561, US Geological Survey, https://pubs.usgs.gov/pp/1561/report.pdf.

Roe, W. P., 2010, Tertiary sediment of the Big Hole Valley and Pioneer Mountains, southwestern Montana: Age, provenance, and tectonic implications, unpublished master's thesis, University of Montana, Missoula.

Ross, C. P., 1959, Geology of Glacier National Park and the Flathead region, northwestern Montana, professional paper 296, US Geological Survey.

Ross, G. M., and M. Villeneuve, 2003, Provenance of the Mesoproterozoic (1.45 Ga) Belt Basin (western North America): Another piece in the pre-Rodinia paleogeographic puzzle, *Geological Society of America Bulletin*, v. 115, no. 10, 1191–1217.

Ruleman, C. A., Larsen, M., and M. C. Stickney, 2014, Neotectonics and geomorphic evolution of the northwestern arm of the Yellowstone tectonic parabola: Controls on intracratonic extensional regimes, southwest Montana, in *Exploring the northern Rocky Mountains*, field guide 37, eds. C. A. Shaw and B. Tikoff, Geological Society of America, 65–87.

Ruppel, E. T., Lopez, D. A., and M. J. O'Neill, 1993, Geologic map of the Dillon 1 degree by 2 degree quadrangle, Idaho and Montana, IMap 1803-H, US Geological Survey, scale 1:250,000.

Scannella, J. B., Fowler, D. W., Goodwin, M. B., and J. R. Horner, 2014, Evolutionary trends in *Triceratops* from the Hell Creek Formation, Montana, *Proceedings of the National Academy of Sciences of the United States of America*, v. 111, no. 28, 10245–10250.

Schellart, W. P., Stegman, D. R., Farrington, R. J., Freeman, J., and L. Moresi, 2010, Cenozoic tectonics of western North America controlled by evolving width of Farallon slab, *Science*, v. 329, no. 5989, 316–319.

Schmidt, C. J., O'Neill, J. M., and W. C. Brandon, 1988, Influence of Rocky Mountain foreland uplifts on the development of the frontal fold and thrust belt, southwestern Montana, in *Interaction of the Rocky Mountain foreland and the Cordilleran thrust belt*, memoir 171, eds. C. J. Schmidt and W. J. Perry Jr., Geological Society of America, 171–201.

Sears, J. W., 2006, Montana transform: A tectonic cam surface linking thin- and thick-skinned Laramide shortening across the Rocky Mountain foreland, *Rocky Mountain Geology*, v. 41, no. 2, p. 65–76

Sears, J. W., 2009, Influence of Proterozoic Belt-Purcell rift system on Phanerozoic facies, isopachs, structures, and hydrocarbon prospectivity in Montana, *AAPG Search and Discovery* article #40386.

Sears, J. W., 2013, Late Oligocene–early Miocene Grand Canyon: A Canadian connection? *GSA Today*, v. 23, no. 11, 4–10.

Sears, J. W., 2014, Tracking a big Miocene river across the Continental Divide at Monida Pass, Montana/Idaho, in *Exploring the northern Rocky Mountains*, field guide 37, eds. C. A. Shaw and B. Tikoff, Geological Society of America, 89–99.

Sears, J. W., 2019, Mesoproterozoic Belt Basin: Crustal-scale anomaly in the Northern Rockies: oral presentation. EarthScope in the Northern Rockies Workshop, powerpoint available online at https://serc.carleton.edu/earthscoperockies/2019_wksp.html.

Sears, J. W., and W. J. Fritz, 1998. Cenozoic tilt-domains in southwest Montana: Interference among three generations of extensional fault systems, in *Accommodation zones and transfer zones: The regional segmentation of the Basin and Range Province*, special paper 323, eds. J. E. Faulds and J. H. Stewart, Geological Society of America, 241–247.

Sears, J. W., Hendrix, M. S., Thomas, R. C., and W. J. Fritz, 2009, Stratigraphic record of the Yellowstone hot spot track, Neogene Sixmile Creek Formation grabens, southwest Montana, *Journal of Volcanology and Geothermal Research*, v. 188, no. 1–3, 250–259.

Sears, J. W., Khudoley, A., and R. A. Price, 2004, Linking the Mesoproterozoic Belt-Purcell and Udzha basins across the west Laurentia-Siberia connection, *Precambrian Research*, v. 129, no. 3, 291–308.

Sears, J. W., and P. C. Ryan, 2003, Cenozoic evolution of the Montana cordillera: Evidence from paleovalleys, in *Cenozoic systems of the Rocky Mountain region*, eds. R. G. Raynolds and R. M. Flores, Rocky Mountain Section, Society of Economic Paleontologists and Mineralogists, Denver, CO, 289–301.

Sears, J. W., and R. C. Thomas, 2007, Extraordinary middle Miocene crustal disturbance in southwest Montana: Birth record of the Yellowstone hot spot? *Northwest Geology*, v. 36, 133–142.

Sims, P. K., and others, 2004, Precambrian basement geologic map of Montana: An interpretation of aeromagnetic anomalies, scientific investigations map 2829, US Geological Survey.

Sivakumar, M. V. K., 2007, Interactions between climate and desertification, *Agricultural and Forest Meteorology*, v. 142, no. 2–4, 143–155.

Skipp, B., Lageson, D. R., and W. J. McMannis, 1999, Geologic map of the Sedan quadrangle, Gallatin and Park Counties, Montana, geologic investigations series I-2634, US Geological Survey.

Sloan, R. E., 1987, Upper Cretaceous–Paleocene sequence, Bug Creek area, northeastern Montana, in *Rocky Mountain Section of the Geological Society of America*, centennial field guide, v. 2, ed. S. S. Beus, Geological Society of America, 45–48.

Smedes, H. W., and H. J. Prostka, 1972, Stratigraphic framework of the Absaroka Volcanic Supergroup in the Yellowstone National Park region, professional paper 729C, US Geological Survey.

Smith, L. N., 2003, Late Pleistocene stratigraphy and implications for deglaciation and subglacial processes of the Flathead lobe of the Cordilleran ice sheet, Flathead Valley, Montana, USA, *Sedimentary Geology*, v. 165, no. 3–4, 295–332.

Smith, L. N., 2006, Stratigraphic evidence for multiple drainings of Glacial Lake Missoula along the Clark Fork River, Montana, USA, *Quaternary Research*, v. 66, no. 2, 311–322.

Smith, L. N., and M. A. Hanson, 2014, Sedimentary record of Glacial Lake Missoula along the Clark Fork River from deep to shallow positions in the former lakes: St. Regis to near Drummond, Montana, in *Exploring the northern Rocky Mountains*, field guide 37, eds. C. A. Shaw and B. Tikoff, Geological Society of America, 51–63.

Smith, R. B., and L. Braile, 1994, The Yellowstone hot spot, *Journal of Volcanology and Geothermal Research*, v. 61, no. 3, 121–187.

Smith, R. B., and others, 2009, Geodynamics of the Yellowstone hot spot and mantle plume: Seismic and GPS imaging, kinematics, and mantle flow, *Journal of Volcanology and Geothermal Research*, v. 188, no. 1–3, 26–56.

Sonderegger, J. L., Schofield, J. D., Berg, R. B., and M. L. Mannick, 1982, The upper Centennial Valley, Beaverhead and Madison Counties, Montana: An investigation of resources utilizing geological, geophysical, hydrochemical, and geothermal methods, memoir 50, Montana Bureau of Mines and Geology.

St. Clair, J., and others, 2016, Geologic conceptional model, Snake River Geothermal Consortium, FORGE, US Dept. of Energy, available online at www.energy.gov.

Stickney, M. C., and M. J. Bartholomew, 1987, Seismicity and late Quaternary faulting of the northern Basin and Range Province, Montana and Idaho, *Bulletin of the Seismological Society of America*, v. 77, no. 5, 1602–1625.

Stickney, M. C., Haller, K. M., and M. N. Machette, 2000, Quaternary faults and seismicity in western Montana, special publication 114, Montana Bureau of Mines and Geology, scale 1:750,000.

Stroup, C. N., 2008, Provenance of Cenozoic continental sandstones in southwest Montana: Evidence from detrital zircon, unpublished master's thesis, Idaho State University, Pocatello.

Taylor, W. J., and others, 2000, Relations between hinterland and foreland shortening: Sevier orogeny, central North America cordillera, *Tectonophysics*, v. 19, no. 6, 1124–1143.

Thomas, R. C., 1995, Tectonic significance of Paleogene sandstone deposits in southwestern Montana, *Northwest Geology*, v. 24, 237–244.

Thomas, R. C., 2007, A field guide to the Cambrian section at Camp Creek, southwest Montana, *Northwest Geology*, v. 36, 231–244.

Thomas, R. C., and S. Roberts, 2000, Geology of the Lewis and Clark Trail: The Three Forks of the Missouri River to Camp Fortunate, Montana, in *Geologic field trips, western Montana and adjacent areas*, Rocky Mountain Section of the Geological Society of America meeting, April 15–20, eds. S. Roberts and D. Winston, University of Montana, Missoula, 209–322.

Thomas, R. C., and S. Roberts, 2007, A summary of the stratigraphy and depositional setting of Paleozoic rocks in the Dillon area, *Northwest Geology*, v. 36, 35–56.

Thomas, R. C., Smith, L. N., and S. Roberts, 2007, A field guide to the glacial geology of the Birch Creek and Rattlesnake Creek areas, Beaverhead County, Montana, *Northwest Geology*, v. 36, 261–268.

Thompson, G., 2011, The Pryor Mountains: Geology tour, http://www.pryormountains.org/natural-history/geology-2/geology.

Thompson, G., and D. Walton, 2014, The Pryor Mountains: Vermillion valley geology, http://www.pryormountains.org/natural-history/geology-2/vermillion-valley-geology.

Thornberry-Ehrlich, T., 2004, Glacier National Park, Geologic resource evaluation report, natural resource report NPS/NRPC/GRD/NRR—2004/001, National Park Service, Denver, CO, https://irma.nps.gov/DataStore/DownloadFile/426435.

Tilling, R. I., 1974, Composition and time relations of plutonic and associated volcanic rocks, Boulder batholith region, Montana, *Geological Society of America Bulletin*, v. 85, no. 12, 1925–1930.

Trexler, D., 2001, Two Medicine Formation, Montana: Geology and fauna, in *Mesozoic vertebrate life*, eds. D. H. Tanke and K. Carpenter, Indiana University Press, Bloomington, 298–309.

Tysdal, R. G., 1990, Geologic map of the Sphinx Mountain quadrangle and adjacent parts of the Cameron, Cliff Lake, and Hebgen Dam quadrangles, Montana, miscellaneous investigations series map I-1815, US Geological Survey, scale 1:62,500.

Underwood, S. J., and others, 2014, The Yellowstone and Regal talc mines and their geologic setting in southwestern Montana, in *Exploring the northern Rocky Mountains*, field guide 37, eds. C. A. Shaw and B. Tikoff, Geological Society of America, 161–187.

US Dept. of Energy, National Renewable Energy Laboratory, 2002, Montana wind power resource estimates, http://windeis.anl.gov/guide/maps/images/mt50mwind.pdf.

US Energy Information Administration, accessed 2017, Montana Field production of crude oil, available online at www.eia.gov.

US Geological Survey, Earthquake Hazards Program, no date, Quaternary fault and fold database of the United States, https://earthquake.usgs.gov/hazards/qfaults.

Vuke, S. M., 1986, Geology of Cedar Creek anticline in Baker, Wibaux, and Glendive, 30' x 60' quadrangles, eastern Montana and adjacent North Dakota, *AAPG Bulletin*, v. 70, 8.

Vuke, S. M., 2004, Geologic map of the Divide area, southwest Montana, open-file report 502, Montana Bureau of Mines and Geology, scale 1:50,000.

Vuke, S. M., 2006, Geologic map of Cenozoic deposits of the lower Jefferson valley, southwestern Montana, open-file report 537, Montana Bureau of Mines and Geology, scale 1:50,000.

Vuke, S. M., 2009, Geologic map of the southern Townsend basin, Broadwater and Gallatin Counties, Montana, open-file report 586, Montana Bureau of Mines and Geology, scale 1:24,000.

Vuke, S. M., 2013, Landslide map of the Big Sky area, Madison and Gallatin Counties, Montana, open-file report 632, Montana Bureau of Mines and Geology, scale 1:24,000.

Vuke, S. M., and R. B. Colton, 2003, Geologic map of the Terry 30' x 60' quadrangle, eastern Montana, open-file report 477, Montana Bureau of Mines and Geology, scale 1:100,000.

Vuke, S. M., Lonn, J. D., Berg, R. B., and C. J. Schmidt, 2014, Geologic map of the Bozeman 30' x 60' quadrangle, southwestern Montana, open-file report 648, Montana Bureau of Mines and Geology, scale 1:100,000.

Vuke, S. M., Porter, K. W., Lonn, J. D., and D. A. Lopez, 2007, Geologic map of Montana, geologic map 62, Montana Bureau of Mines and Geology, scale 1:500,000, http://www.mbmg.mtech.edu/pdf/GM62_2007Booklet.pdf.

Waite, G. P., Smith, R. B., and R. L. Allen, 2006, VP and VS structure of the Yellowstone hot spot from teleseismic tomography: Evidence for an upper mantle plume, *Journal of Geophysical Research: Solid Earth*, v. 111, BO4303.

Waitt, R. B., Jr., 1985, Case for periodic, colossal jökulhlaups from Pleistocene Glacial Lake Missoula, *Geological Society of America Bulletin* 96, no. 10, 1271–1286.

Walker, B., Powell, A., Rollins, D., and R. Shaffer, 2006, Elm Coulee field Middle Bakken Member (lower Mississippian/upper Devonian), Richland County, Montana, Search and Discovery Article #20041, adapted from presentation at AAPG Rocky Mountain Section, Billings, MT, June 11–13, http://www.searchanddiscovery.com/documents/2006/06131walker.

Ward, E. M. G., and J. W. Sears, 2007, Reinterpretation of fractures at Swift Reservoir, Rocky Mountain thrust front, Montana: Passage of a Jurassic forebulge? in *Whence the mountains? Inquiries into the evolution of orogenic systems: A volume in honor of Raymond A. Price*, special papers 433, eds. J. W. Sears, T. A. Harms, and C. A. Evenchick, Geological Society of America, 197–210.

Whipkey, C. E., Cavaroc, V. V., and R. M. Flores, 1991, Uplift of the Bighorn Mountains, Wyoming and Montana: A sandstone provenance study, bulletin 1917-D, US Geological Survey.

Whipple, J. W., 1992, Geologic map of Glacier National Park, miscellaneous investigations series map I-1508-F, US Geological Survey, scale 1:100,000.

Winston, D. W., 1986, Belt Supergroup stratigraphic correlation sections showing general lithologies and history of formal usage, geologic map 40, Montana Bureau of Mines and Geology.

Winston, D. W. 1989, Introduction to the Belt, in *Volcanism and plutonism of western North America*, v. 2, field trip T334, Middle Proterozoic Belt Supergroup, western Montana, trip leaders D. Winston, R. J. Horodyski, and J. W. Whipple, in the collection of field trips for the 28th International Geological Congress, American Geophysical Union, 1–6.

Winston, D. W., 2007, Revised stratigraphy and depositional history of the Helena and Wallace Formations, mid-Proterozoic Piegan Group, Belt Supergroup, Montana and Idaho, USA, in *Proterozoic geology of western North America and Siberia*, special publication 86, eds. P. K. Link and R. S. Lewis, Society for Sedimentary Geology, 65–100.

Wise, D. U., 2000, Laramide structures in basement and cover of the Beartooth uplift near Red Lodge, Montana, *American Association of Petroleum Geologists Bulletin*, v. 84, no. 3, 360–375.

Witkind, I. J., 1960, The Hebgen Lake, Montana, earthquake of August 17, 1959, in *West Yellowstone: Earthquake area*, Billings Geological Society 11th annual field conference guidebook, 31–44.

Witkind, I. J., 1975, Geology of a strip along the Centennial fault, southwestern Montana and adjacent Idaho, miscellaneous investigations series map I-890, US Geological Survey, scale 1:62,500.

Witkind, I. J., and others, 1962, Geologic features of the earthquake at Hebgen Lake, Montana, August 17, 1959, *Bulletin of the Seismological Society of America*, v. 52, no. 2, 163–180.

INDEX

Page numbers in bold include photographs.

Donald W. Hyndman is a Professor Emeritus of the University of Montana (UM) in Missoula. He earned a BA in Geological Engineering at the University of British Columbia and a PhD in Geology at the University of California, Berkeley. Recognized as a Distinguished Teacher and Distinguished Scholar by UM, he is also a Fellow of the Geological Society of America and the American Geophysical Union. He mentored innumerable graduate students at UM and wrote more than one hundred papers in professional journals. His research focused on the origin and intrusion of granite magmas and their relationships to deep crustal metamorphic rocks, mountain-building processes, and the formation of radial dike swarms and laccoliths. Don also wrote widely adopted university textbooks on igneous and metamorphic rocks and more recently, with his son, five editions of *Natural Hazards and Disasters*.

Robert C. Thomas is a Professor of Geology in the Environmental Sciences Department at the University of Montana Western in Dillon. He earned a BA from Humboldt State University, an MS from the University of Montana, and a PhD from the University of Washington. A Montana Regents Professor and Teaching Scholar, Montana Educator of the Year, and Carnegie US Professor, Rob is also a Fellow of the Geological Society of America and recipient of their Distinguished Service Award. He has authored or coauthored over 75 publications, including *Roadside Geology of Yellowstone Country*. As a field geoscientist, he engages students in real projects designed to benefit society and the environment. His personal passions are family, mountain recreation, and playing guitar around a campfire.